Lecture Notes in Computer Sci

Edited by G. Goos and J. Hartmanis

Advisory Board: W. Brauer D. Gries J. Stoer

Lecture Notes in Computer Science

Edited by G. Goos and J. Hartmanis

Advisory Board: W. Brauer D. Gries J. Stoer

Jeffrey J. Joyce   Carl-Johan H. Seger (Eds.)

Higher Order Logic Theorem Proving and Its Applications

6th International Workshop, HUG '93
Vancouver, B.C., Canada, August 11-13, 1993
Proceedings

Springer-Verlag

Berlin Heidelberg New York
London Paris Tokyo
Hong Kong Barcelona
Budapest

Series Editors

Gerhard Goos
Universität Karlsruhe
Postfach 69 80
Vincenz-Priessnitz-Straße 1
D-76131 Karlsruhe, Germany

Juris Hartmanis
Cornell University
Department of Computer Science
4130 Upson Hall
Ithaca, NY 14853, USA

Volume Editors

Jeffrey J. Joyce
Carl-Johan H. Seger
Department of Computer Science, The University of British Columbia
2366 Main Mall, Vancouver, B.C., Canada V6T 1Z4

CR Subject Classification (1991): F.3.1, F.4.1, D.2.2, I.2.2-3, B.2.2, B.6.3, B.7.2,
C.2.2

ISBN 3-540-57826-9 Springer-Verlag Berlin Heidelberg New York
ISBN 0-387-57826-9 Springer-Verlag New York Berlin Heidelberg

CIP data applied for

Typesetting: Camera-ready by author
SPIN: 10131976   45/3140-543210 - Printed on acid-free paper

Preface

In an era of increasing specialization, the reader may be surprised to discover areas as diverse as domain theory and VLSI CAD represented under a single cover. Yet, this may not be so surprising when one considers the generality and demonstrated utility of the "glue" that binds together the diverse contents of this proceedings – namely, higher order logic and the software tools that have been developed to support the use of higher order logic as a method of formalization and formal reasoning. The contents of this proceedings are also pleasantly diverse with respect to their sources. There is a healthy mixture of international contributions from academia and industry. Some of the contributors have an extensive background in higher order logic. Others have expertise in areas other than higher-order logic – their contribution here typically describes how higher order logic has been applied in their own area of expertise. As the editors of this proceedings, we are also pleased to have had contributions from students whose work demonstrates how higher order logic can serve as a springboard into areas such as programming language theory, formal hardware verification, algebra, CASE, protocol specification, and machine-assisted theorem-proving.

The papers in this proceedings were presented at the *1993 HOL User's Group Workshop* held August 11-13 at the University of British Columbia. The workshop was sponsored by the Centre for Integrated Computer System Research and by travel grants from the British Council. The workshop was the sixth in the series of annual workshops dedicated to the topic of Higher-Order-Logic theorem proving, its usage in the HOL system and its applications. Previous workshops have taken place in Cambridge UK (1988 and 1989), Aarhus Denmark (1990), Davis California USA (1991), and Leuven Belgium (1992). The workshop was attended by 70 researchers and practitioners, and the presentations included 39 refereed papers plus two invited talks by David Parnas, McMaster, Hamilton, Canada, and Paul Gilmore, University of British Columbia, Vancouver, Canada.

Much of the work represented in this proceedings is based upon a public domain software system called "HOL" which was originally developed in the Computer Laboratory at Cambridge University. Indeed, this series of workshop began in 1988 as a very small scale users group meeting for HOL. The workshop has since broadened to welcome a wider range of contributions – including those that involve the use of software tools other than HOL.

The proceedings begins with a group of papers which consider the higher order logic embedding of programming languages or similar kinds of notations. The second group of papers focus on the use of logic in formal approaches to the specification and verification of hardware. The development of infrastructure for the HOL system is discussed in a third group of papers. A fourth group consists of a challenge to the machine-assisted theorem-proving community in the form of examples of simple-but-practical theorems that occur in context of validating formal specifications software systems. The fifth group of papers describes the integration of verification tools based on higher order logic with other kinds of verification methods and tools such as computer algebra and model-checking.

The sixth group of papers shifts attention back to applications of logic in formal approaches to the specification and verification of hardware. The seventh group of papers returns to the subject of semantic embedding – in particular, the semantic embedding of theories, special-purpose logics, and specification notations in higher order logic. This is followed by a single paper that describes a theorem-proving tool for type theory. The proceedings then returns once again to the subject of formal hardware verification – this time focusing more on the development of infrastructure for formal hardware verification. The ninth group of papers describes the development of general-purpose infrastructure for the HOL system and, more generally, mechanised logics. The final group of papers describes more applications of higher order logic and machine-assisted theorem-proving techniques.

The papers in these proceedings were revised by the lecturers after the workshop took place.

February 1994

Jeffrey Joyce and Carl Seger

Workshop Co-Chairs

Program Committee

Carl-Johan Seger, (University of British Columbia, Canada), Chair
Jim Alves-Foss, (University of Idaho, USA)
Luc Claesen, (IMEC & Catholic University Leuven, Belgium)
John Herbert, (SRI International, UK)
Jeffrey Joyce, (University of British Columbia, Canada)
Tom Melham, (Computer Laboratory, University of Cambridge, UK)
David Shepherd, (INMOS Ltd, UK)
Phillip Windley, (Brigham Young University, USA)

Contents

Program Verification using HOL-UNITY

Flemming Andersen and Kim Dam Petersen and Jimmi S. Pettersson

Tele Danmark Research, Lyngsø Allé 2, DK-2970 Hørsholm

Abstract. HOL-UNITY is an implementation of Chandy and Misra's UNITY theory in the HOL88 and HOL90 theorem provers. This paper shows how to verify safety and progress properties of concurrent programs using HOL-UNITY. As an example it is proved that a lift-control program satisfies a given progress property. The proof is compositional and partly automated. The progress property is decomposed into basic safety and progress properties, which are proved automatically by a developed tactic based on a combination of Gentzen-like proof methods and Pressburger decision procedures. The proof of the decomposition which includes induction is done mechanically using the inference rules of the UNITY logic implemented as theorems in HOL. The paper also contains some empirical results of running the developed tactic in HOL88 and HOL90, respectively. It turns out that HOL90 in average is about 9 times faster than HOL88. Finally, we discuss various ways of improving the tactic.

1 Introduction

This paper presents an approach to verify safety and progress properties of concurrent programs using HOL-UNITY [2, 3], an implementation of Chandy and Misra's UNITY theory [6] in the HOL theorem provers [12, 20]. By combining a representation of UNITY programs in HOL and the HOL-UNITY theory, it is possible in a structured way to prove that UNITY programs satisfy required properties.

Various aspects of this approach is demonstrated by proving that a lift control program satisfies a given (**leadsto**) progress property. The example distinguishes itself by containing both boolean, arithmetic and quantified program constructs. The progress property formalises that every request for the lift is eventually serviced. To prove this kind of property, it is necessary to prove a number of basic properties satisfied by the transitions of the program [6, 16]. The property is then proved from the basic properties using the inference rules of UNITY logic supported as theorems in HOL-UNITY. Proving that a program satisfies the basic properties is tedious. This is due to the high degree of details in the proofs and the need for interaction with the verifier caused by the lack of automated support in HOL-UNITY. In order to relieve the verifier, automated verification methods needs to be applied.

Based on a combination of Gentzen and Pressburger methods [5, 9, 18, 21] a straight forward implementation of a proof tactic is used to automatically prove the basic properties of the lift control program. Some of the design decisions when making the tactic and consequences thereof are discussed and proposals are given that may be used for improvement.

The paper is organized as follows. In Section 2, UNITY and its implementation in HOL are briefly described. In Section 3, the example is described and the UNITY program for controlling the lift is presented together with its representation in HOL. Section 4 presents the decomposition of the required progress property into basic properties which are provable by the implemented tactic. Section 5 presents an overview of the developed tactic. Finally, section 6 contains some empirical results of verifying the lift example in both HOL88 and HOL90. It turns out that HOL90 in average is 9 times faster than HOL88 for this example. Section 7 concludes the paper discussing what remains to be done in order to achieve a tool applicable to a larger class of examples.

2 HOL-UNITY

UNITY [6] is a theory for specifying and verifying programs. The theory consists of a simple language for specifying programs, a logic to express safety and progress properties which a program must satisfy, and a proof system for proving that programs satisfy such properties. This section presents the formalization of UNITY in the higher order logic [7] supported by HOL [12].

Programs

UNITY programs basically consists of a set of variables, a set of *initialization* equations on the program variables, and a set of conditionally enabled multiple assignments, *actions*. Programs are modeled by (unlabeled) transition systems, which consists of a state space Σ, defined by the variables, the initial condition given by the state predicate init, and a set of actions $\mathcal{A}$.

A state $\sigma \in \Sigma$ is a function from variable names to their respective values. Variable names are represented by the values of an enumeration type corresponding to their unique naming. Values are represented by a type which is defined as the disjoint union of the types given by the variables.

For a program with a boolean variable b and an array variable a of booleans, the type of variable names are defined using Tom Melham's [15] type package: $var = b \mid a$ and the type of values by: $val = Bool$ bool $\mid Array$ (num$\rightarrow$bool). For each value a destructor is introduced, e.g., $destBool\,(Bool\,x) = x$ such that extracting the value of variables in a state σ can be done by $destBool\,(\sigma\,b)$. Hence, a state dependent variable v_* of type T is defined as: $v_*\,\sigma = destT\,(\sigma\,v)$, where the suffix priming $*$ represents the state lifting of the variable name v. Evidently all program constructs become state lifted. For readability HOL-UNITY defines a collection of state lifting operators. A constant c may be state lifted by prefixing it with the predefined HOL combinator K, which satisfies: $K\,c\,\sigma = c$. The logic connectors, e.g. $\wedge, \vee, \Rightarrow$ are similarly lifted to state abstracted connectors: $\wedge_*, \vee_*, \Rightarrow_*$ e.g.: $(p \wedge_* q)\ \sigma = p\ \sigma \wedge q\ \sigma$ and, for array variables the indexing operator at is defined as: $(a\ \mathbf{at}\ e)\ \sigma = (a\ \sigma)\ (e\ \sigma)$.

Assignments

A (multiple) assignment in UNITY consists of updating a state, by changing the values of one or more of its variables or array elements. If all variables are simple (i.e. non-arrays) the assignment may be performed by first building a list of variable names and computed values and next building the new state by stepping through the list. It is defined by the combinator **asg**, using the notation: $(b \rightarrow c \mid e)$ denoting the conditional expression: **if** b **then** c **else** e:

$\mathbf{asg}\ ((v', e') :: l)\ \sigma\ v = (v = v') \rightarrow e'\ \sigma \mid \mathbf{asg}\ l\ \sigma\ v$

If several array elements are to be assigned to the same array variable in one assignment, they must be collected such that the updated array reflects all changes assigned to the variable. The operator **upd** supports multiple update of an array at multiple element positions:

$(a\ \mathbf{upd}\ ((i', e') :: l))\ \sigma\ i = (i\ \sigma = i'\ \sigma) \rightarrow e'\ \sigma \mid (a\ \mathbf{upd}\ l)\ \sigma\ i$

Notice, **asg** and **upd** returns the old state for empty lists. As an example the UNITY assignment: b, $a[0]$, $a[1] := true$, $a[1]$, $a[0]$ is represented in HOL-UNITY by:

$\mathbf{asg}\ [(b,\ Bool\ \circ (K\ T));$
$\qquad (a,\ Array \circ (a_* \mathbf{upd}\ [((K\ 0),\ a_* \mathbf{at}\ (K\ 1));\ ((K\ 1),\ a_* \mathbf{at}\ (K\ 0))]))]$

Actions

A UNITY action is a conditionally enabled multiple assignment. The representation of a multiple assignment without enabling conditions has been shown above. The conditional enabling of a multiple assignment is defined in terms of the combinator **when**:

$(a\ \mathbf{when}\ g)\ \sigma = g\ \sigma \rightarrow a\ \sigma \mid \sigma$

The presented definitions are sufficient to model all language constructs possible in the UNITY programming language of which the variable types are naturally extended to the types in HOL.

Properties

The UNITY logic contains two safety properties: **unless** and **invariant** and two progress properties: **ensures** and **leadsto**. The logic properties in UNITY can be formally defined in terms of modal relations between sets of states expressed by state dependent predicates, where a state dependent predicate p is a function from states to booleans. The three basic properties **unless**, **invariant** and **ensures** are defined for a given program $F = (\Sigma,\ \text{init},\ \mathcal{A})$ as follows.

p **unless** q $\equiv \forall \alpha: \{p \wedge \neg q\} \ \alpha \ \{p \vee q\}$
$\qquad\qquad \equiv \forall \alpha: \ \forall \sigma: \ p \, \sigma \, \wedge \, \neg q \, \sigma \Rightarrow p \, (\alpha \, \sigma) \ \vee \ q \, (\alpha \, \sigma)$

p **invariant** $\equiv (\text{init} \Rightarrow p) \ \wedge \ \forall \alpha: \{p\} \ \alpha \ \{p\}$
$\qquad\qquad \equiv \forall \sigma: \ (\text{init} \, \sigma \Rightarrow p \, \sigma) \ \wedge \ (\forall \alpha: \ \forall \sigma: \ p \, \sigma \Rightarrow p \, (\alpha \, \sigma))$

p **ensures** $q \equiv p$ **unless** $q \ \wedge \ \exists \alpha: \{p \wedge \neg q\} \ \alpha \ \{q\}$
$\qquad\qquad \equiv p$ **unless** $q \ \wedge \ (\exists \alpha: \ \forall \sigma: \ p \, \sigma \, \wedge \, \neg q \, \sigma \Rightarrow q \, (\alpha \, \sigma))$

where α ranges over the set of actions in the program ($\alpha \in \mathcal{A}$), init is the predicate which characterizes the initial states, and σ ranges over the complete state space of the program, i.e., $\sigma \in \Sigma$. The progress property **leadsto** is defined inductively as the transitive, disjunctive closure of **ensures** properties, i.e., it is defined as the least fixed point solution to the following three implications [3, 4]:

p **ensures** q $\qquad\qquad\qquad\qquad\qquad \Rightarrow p$ **leadsto** q
$(\exists r: \ p$ **leadsto** $r \wedge r$ **leadsto** $q)$ $\qquad\quad \Rightarrow p$ **leadsto** q
$(\exists P: \ (p = \bigvee P) \wedge (\forall p' \in P: \ p'$ **leadsto** $q)) \Rightarrow p$ **leadsto** q

where $\bigvee P$ is the disjunction of all the elements in the possibly infinite set of predicates P.

The UNITY logic also includes a number of inference rules proved as theorems in HOL-UNITY. These rules allow us to derive new properties from other properties. E.g., to prove that a program satisfies a specific **leadsto** property we must find some basic properties (unless, invariant and ensures properties) which can be proved from the individual actions of the program. From these basic properties the **leadsto** property can then be proved using the inference rules mentioned above. Hence, we have two kinds of proofs in UNITY: (1) proofs of the basic properties which amounts to proving assertions over individual actions, and (2) proofs of properties derived from other properties. Doing proof by hand, proofs of the first kind are usually left out, the basic properties are just assumed to be valid. However, using HOL requires proving both.

3   The lift-control program

Here the lift-control program is presented (Figure 1). The system consists of a lift that moves between a number of floors to serve requests on these. A request on a floor is served by eventually moving to and stopping the lift at the floor and opening the doors. The bottom and top floors are indicated by two parameters min and max, satisfying $min \leq max$.

The necessary elements of a lift to reflect its behavior are: floor position, door- and movement conditions, and request indicators. The actions of the lift can be represented by assigning values to variables, e.g., opening the door by setting a variable $open$ to true. The state space of the lift can be represented by 6 variables: $floor$ (the current position of the lift), $open$ (whether the door is open at $floor$), $stop$ (whether the lift is stopped at $floor$), req (for each floor whether the lift is requested at that floor), up (the current direction of movement for the lift), and $move$ (whether moving the lift takes precedence over opening of the doors). The first four variables defines the *observable* parts of the lift. The two remaining variables have been introduced to enable a *control* of the lift that ensures fair progress of the lift as discussed below.

The progress requirement of the lift is that any request is eventually satisfied. The strategy for achieving this is to let the lift continue in one direction as long as there are requests to be served in that direction. During this movement, the lift will serve the floors with requests. When there are no more requests in the current direction, the lift starts to move in the other direction, provided there are any in either direction outstanding requests in that direction. When no requests are outstanding the lift is stopped until new requests are made. To avoid a continued request on a floor from blocking the lift at that floor, the moving of the lift to another floor with a request takes precedence over re-opening the door. The variable *move* is used to indicate this.

```
program Lift(min,max)
   declare
      floor                    : min..max
      up,move,stop,open        : bool
      req                      : array[min..max] of bool
   initially
      floor                    = min
[t]   up,move,stop,open        = false,true,true,false
      (∀i ∈ min..max: req[i] = false)
   always
      above                    = ∃i: floor < i ≤ max ∧ req[i]
      below                    = ∃i: min ≤ i < floor ∧ req[i]
      queueing                 = above ∨ below
      goingup                  = above ∧ (up ∨ ¬below)
      goingdown                = below ∧ (¬up ∨ ¬above)
      ready                    = stop ∧ ¬open ∧ move
   assign
      stop,move              := true,false           when ¬stop ∧ req[floor]                    (1)
   [] open,req[floor],move   := true,false,true      when  stop ∧ ¬open ∧ req[floor] ∧
                                                           ¬(move ∧ queueing)                    (2)
   [] open                   := false                when  open                                 (3)
   [] stop,floor,up          := false,floor + 1,true when ready ∧ goingup                       (4)
   [] stop,floor,up          := false,floor − 1,false when ready ∧ goingdown                    (5)
   [] floor                  := floor + 1            when ¬stop ∧  up ∧ ¬req[floor]              (6)
   [] floor                  := floor − 1            when ¬stop ∧ ¬up ∧ ¬req[floor]             (7)
end
```

Fig. 1. UNITY program for lift control

Initialization

Basically the *observable* variables should reflect "physically" valid values, and the *control* variables should be set properly. Hence, we assume the lift is initially in a state where it is stopped at the bottom floor without any requests. This is reflected by the **initially** clause of the program in Figure 1.

Actions

The lift is controlled by a collection of *actions* prescribing the updating of its state. Each action is conditionally *enabled* by a predicate, which indicates when the action can change the state. The lift consists of 7 actions (Figure 1): (1) stops the lift when the present floor has a request, (2) opens the door, when the lift is stopped at a floor with request, and opening request has precedence over moving, (3) closes the door, (4) and (5) starts the lift in the presence of a request in up- or downwards direction, when the doors are closed and moving takes precedence over opening them, and (6) and (7) continues moving the lift when it is already in motion and there is no request at the present floor.

Representation of the program in HOL-UNITY

The types representing the lift variable names, and their values are:

$var = floor \mid up \mid move \mid stop \mid open \mid req$
$val = Bool\ \textbf{bool} \mid Num\ \textbf{num} \mid Bits\ (\textbf{num} \rightarrow \textbf{bool})$
from which the state dependent variables $floor_*$, up_*, $move_*$, $stop_*$, $open_*$ and req_* are defined as described above, e.g., the $floor_*$ variable:

$floor_*\ \sigma = destNum\,(\sigma\ floor)$
The initial condition is defined by the predicate:

$init = (floor_* =_* \mathbf{K}\,min) \wedge_* (up_* \quad =_* \mathbf{K}\,\mathbf{F}) \wedge_* (move_* =_* \mathbf{K}\,\mathbf{T}) \wedge_*$
$\qquad (stop_* \quad =_* \mathbf{K}\,\mathbf{T}) \quad \wedge_* (open_* =_* \mathbf{K}\,\mathbf{F}) \wedge_*$
$\qquad (\forall_* i\colon\ req_*\,\text{at}\,(\mathbf{K}\,i) \quad =_* \quad \mathbf{K}\,\mathbf{F})$
Each individual program action is represented in HOL-UNITY as a conditionally enabled multiple assignment. Action (1), for example, is represented by:

$\textbf{asg}\,[(stop,\quad Bool \circ (\mathbf{K}\,\mathbf{T}));$
$\qquad (move,\ Bool \circ (\mathbf{K}\,\mathbf{F}))]\ \textbf{when}\ (\neg_* stop_* \wedge_* req_*\,\text{at}\,floor_*)$
In this way, the complete program is defined as a list of **asg** actions.

4 Mechanized proof of progress property

The progress property which we will prove is the requirement that any request for service will eventually be served. In UNITY logic this property can be formalized as:

$\forall n: \ min \leq n \leq max \Rightarrow req[n]$ **leadsto** $open \land floor = n$

It turns out that the lift-control program *Lift* does not have this property in HOL-UNITY. The problem is that in HOL-UNITY, the **leadsto** property must be satisfied for any state in the *total* state space spanned by the program variables. The progress property, however, is only required to be satisfied for *reachable* states of the program. This means, that the lift program may only guarantee the HOL-UNITY progress for a subset of the total state space. This subset which we will call the *valid* states includes the reachable states of the program.

The valid states may be represented by a predicate. Using this predicate, a strengthened **leadsto** property can be formalized and proved in HOL-UNITY:

$\forall n: \ min \leq n \leq max \Rightarrow req[n] \land valid$ **leadsto** $open \land floor = n$

In the following a characterization of the *valid* predicate adequate for proving the given **leadsto** property of the *Lift* program is described.

Valid state space

The *valid* states of a program represented by a predicate may be described as a conjunction of predicates, each characterizing a requirement to the acceptable states of the lift program. This also implies that a *valid* predicate must be a program invariant. In UNITY, this requirement is expressed as:

valid **invariant**

In general, the conjuncts in the *valid* predicate may be classified as the different properties: (a) Directly given safety requirements such as lower and upper bound of variables (b) Implicit given safety dependency requirements describing value dependencies between variables and (c) Builtin fairness properties which guarantee progress.

The valid *Lift* states is the conjunction of the following predicates:

(1) $min \leq floor \leq max$, the lift must be within bottom and top floors.

(2) $open \Rightarrow stop$, the lift must be stopped when the doors are open.

(3) $open \Rightarrow move$, a floor has been served when the doors are open.

(4) $stop \land \neg move \Rightarrow req[floor]$, when the lift is stopped and *floor* has not been served, there must be a request on *floor*.

(5) $\neg stop \land up \Rightarrow \exists f: \ floor \leq f \leq max \land req[f]$, when the lift is moving upwards, there must be a request for service in this direction.

(6) $\neg stop \land \neg up \Rightarrow \exists f: \ min \leq f \leq floor \land req[f]$ when the lift is moving downwards, there must be a request for service in that direction.

In this conjunct (1) belongs to class (a), conjuncts (2), (4), (5) and (6) belong to (b) and (3) belongs to class (c). We are now ready to present a method for proving the strengthened leadsto property.

Progress Decomposition

The first step in proving the leadsto property is to decompose it into simpler
UNITY properties from which the required leadsto property can be deduced
using the transitivity, disjunctivity and derived inference rules for leadsto, e.g.
the following variant of the disjunction rule called case split:

$$\frac{p \wedge b \textbf{ leadsto } q \, , \, p \wedge \neg b \textbf{ leadsto } q}{p \textbf{ leadsto } q}$$

For displaying the decomposition we use a generalized version of Owicki-Lamport
proof-lattices [16]. A UNITY proof lattice for the wanted progress property is
shown in Figure 2.

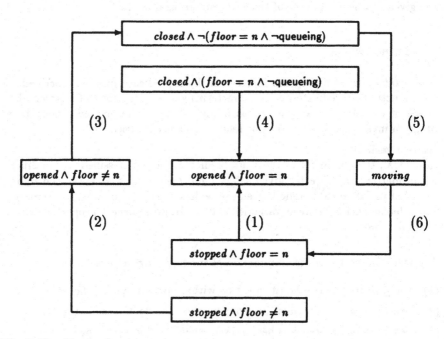

Fig. 2. Proof lattice for the progress property

The arrows in the figure represent leadsto properties to be proved, and the
nodes are state predicates just as in the Owicki-Lamport method. However, the
directed acyclic graph making up a UNITY proof lattice does not necessarily
have a single entry node. In fact, for the proof lattice in Figure 2 every node
is a possible entry node. The proof lattice has only one exit node though, just
like in Owicki and Lamports proof lattices. Owicki and Lamport describe how
invariants propagate through leads-to properties in proof lattices for temporal
logic [16]. This result also applies to UNITY logic.

The property $valid \land req[n]$, is assumed to be conjuncted to all predicates in the proof lattice. Hence, an arrow: $p \land valid \land req[n]$ **leadsto** $q \land req[n]$ is written as: p **leadsto** q.

We have partitioned the valid states into four classes in the proof lattice: *moving* (the lift is moving), *stopped* (the lift is stopped, but the door has not been opened yet), *opened* (the door has been opened after the lift is stopped), and *closed* (the door has been closed again, but the lift has not started moving yet). The predicates describing these classes are given below.

$$
\begin{aligned}
moving &= \neg stop \land \neg open \\
stopped &= stop \land \neg open \land \neg move \\
opened &= stop \land open \land move \\
closed &= stop \land \neg open \land move
\end{aligned}
$$

The proof lattice actually describes the proof of the **leadsto** property:

$$(moving \lor stopped \lor opened \lor closed) \land valid \land req[n] \text{ leadsto } opened \land floor = n$$

from which the wanted property can be easily derived. The arrows (1) to (4) in Figure 2 all follow from corresponding **ensures** properties. The arrow (5) is proved by a state split into the cases *goingup* and its negation; each of these cases follows from corresponding **ensures** properties. The arrow (6) is proved by induction as discussed below.

Metric Induction

The metric induction principle of UNITY is based on the traditional decreasing metric technique, which is formally expressed by the inference rule:

$$\frac{p \land (Metric = N) \text{ leadsto } (p \land (Metric < N)) \lor q}{p \text{ leadsto } q}$$

The leadsto property represented by the arrow (6) in the proof lattice of Figure 2, viz.,

$moving$ **leadsto** $stopped \land floor = n$

can using this induction principle be deduced from the following induction property:

$$moving \land (metric_n = N) \text{ leadsto } (moving \land metric_n < N) \lor (stopped \land floor = n)$$

The predicate $valid \land req[n]$ is just as in Figure 2, implicitly conjuncted to the predicates of the property. The metric function $metric_n$ denotes the longest distance to move from the *floor* where the lift is currently located to the n'th floor counted in numbers of floors. $metric_n$ is defined as the worst case number of floors the lift has to move before it arrives at the wanted floor:

$$
\begin{aligned}
metric_n = \ &(up \land floor < n) \rightarrow (n - floor) \mid (\neg up \land n < floor) \rightarrow (floor - n) \\
\mid \ &(up \land n < floor) \rightarrow (max - floor) + (max - n) \\
\mid \ &(\neg up \land floor < n) \rightarrow (floor - min) + (n - min) \mid 0
\end{aligned}
$$

A proof-lattice for the induction property is given in Figure 3. For this proof

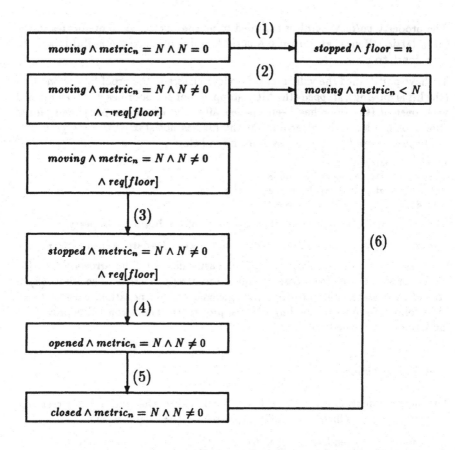

Fig. 3. Proof lattice for induction property

lattice the three nodes in the top left corner of the figure are the possible entry nodes as the disjunction of the three predicates is equivalent to the predicate

$moving \land metric_n = N$

and the two nodes in the upper right corner are the exit nodes as the disjunction of the two predicates is equal to the consequence of the induction property.

The arrows (1), (3), (4) and (5) in Figure 3 follow directly from corresponding ensures properties. The arrow (2) is proved using state split considering the cases: *up* and its negation; the two cases follow from corresponding ensures properties. The arrow (6) is proved using state split considering the cases: goingup and its negation; again the cases follow from corresponding ensures properties.

It can be checked by HOL-UNITY that these proof lattices actually form a proof of the wanted leadsto property, i.e., that the leadsto property can be derived from the ensures properties present as arrows in the proof lattices using the derived inference rules in HOL-UNITY, including the rule induction. Proving that the lift-control program actually satisfies the ensures properties can be automated to a large degree, as described in the next section.

5   A Program Property Tactic

HOL-UNITY is a higher order logic theory, since UNITY is defined as a theory within a typed higher order logic. However, by restricting the expressions of HOL-UNITY programs to consist of state lifted boolean, number or array values, a HOL-UNITY property that must be satisfied by programs is in fact restricted to first order logic extended with arithmetic relations.

As a consequence, proving the program properties **unless**, **invariant** and **ensures** in HOL-UNITY can be automated by first eliminating the UNITY operators, then simplifying the state lifted operators, and finally applying standard Gentzen-like techniques attempting to prove the logic parts of the proof goal, and if this fails applying the Pressburger decision procedure on the arithmetic relations.

Based on this idea a tactic is developed for automating the proof of such properties. An overview of some simple strategies applied by the proof tactic to speed up the proving is presented.

Reduction of HOL-UNITY properties

Every **unless**, **invariant** and **ensures** property that must hold for a program, i.e. is applied to the program which is represented as a list of (asg) state transition actions can be rewritten into a conjunction of properties each applied to only one action. Utilizing this knowledge we can reduce the size of the goals to be proved and obtain a proof algorithm which is linear in the number of actions in the program. In addition to eliminating the operators by inserting their definitions, we may further reduce the terms when applying a conditionally enabled action a **when** g to the properties:

- The property p **unless** q is reduced to:
  $$(p \wedge_* \neg_* q \wedge_* g) \Rightarrow_* ((p \vee_* q) \circ a)$$
  where it is utilized that a is a function from states to states.
- The property **invariant** p is reduced to:
  $$(\text{init} \Rightarrow_* p) \wedge ((p \wedge g) \Rightarrow_* (p \circ a))$$
- The existential part of p **ensures** q is reduced to:
  $$(p \wedge_* \neg_* q \wedge_* g) \Rightarrow_* (q \circ a), \text{ and } (p \wedge_* \neg_* q \wedge_* \neg_* g) \Rightarrow_* q$$

Actions without enabling conditions are reduced similarly with g equal to true.

Example:

- The **unless** property
  $$(v = n \land 0 \leq v) \text{ unless } (v = n + 1) [v := v + 1 \text{ when } 0 \leq v]$$
  is reduced to the predicate:
  $$(v_* =_* (K\,n) \land_* (K\,0) \leq_* v_*) \land_* \neg_* (v_* =_* (K\,n) +_* (K\,1)) \land_* ((K\,0) \leq_* v_*) \Rightarrow_*$$
  $$(v_* =_* (K\,n) \land_* (K\,0) \leq_* v_*) \lor_* (v_* =_* (K\,n) +_* (K\,1))) \circ (\text{asg}[v, v_* +_* (K\,1)]) \sigma$$

Notice that all operators in the above predicates are state lifted as indicated by the suffix priming $*$.

Reduction of Program Constructs

It is possible to apply reductions to the defined language constructs like for the HOL-UNITY properties. It is assumed that a state is either a variable σ or state derived from a state variable σ by an application $\text{asg}[v_1, e_1; \ldots] \sigma$ where v_i are variable names and e_i are expressions in which **asg** does not occur. The following rewrite rules are applied in the tactic:

- For each variable v, in the above $\text{asg}[v_1, e_1; \ldots] \sigma$:
  $$(v_* (\text{asg} ((v, T \circ e) :: l)) \sigma = e\,\sigma)$$
- For each variable u different from v:
  $$(v_* (\text{asg} ((u, T \circ e) :: l)) \sigma = v_* (\text{asg}\, l) \sigma)$$

Example (continued):

- The result predicate presented above is now reduced to:
  $$((v_* \sigma = n \land 0 \leq v_* \sigma) \land \neg (v_* \sigma = n + 1) \land (0 \leq v_* \sigma)) \Rightarrow$$
  $$((v_* \sigma + 1 = n \land 0 \leq v_* \sigma + 1) \lor (v_* \sigma + 1 = n + 1))$$

Other Reductions

It may be convenient to get rid of subtraction, conditional expressions and arithmetic relations different from $<$. Boulton has automated this process for subtraction and conditional expressions as part of his arithmetic library [5]. Hence, only the inequality operators: $>$, $\leq$ and $\geq$ need to be replaced by terms of $<$, e.g., $a \leq b = a < b + 1$. This is also done by the tactic.

More about the tactic

As a pre-processing, the automation tactic utilizes reduction rules as described above before any attempt is done to use a Gentzen like way of proving a property. If the tactic fails to prove the goal in this way it will apply a strategy, including quantifiers, for proving lower and upper bound cases implied by arithmetic inequalities. Finally if everything else fails the Pressburger decision procedure is applied.

6 Empirical results

The example was originally initiated in HOL88 but completed in HOL90 because it turned out that HOL90 was considerably faster for this kind of proving. Table 6 shows the results in CPU-seconds of executing the proof tactic in HOL88 and HOL90, respectively, on a DECstation 5000/200 running Ultrix V4.2A.

Proof obligation	hol90	hol88	speedup
invariance of valid	97.0	407.4	4.20
(1)	53.7	137.6	2.56
(2)	54.7	159.3	2.91
(3)	76.4	248.4	3.25
(4)	80.9	318.7	3.93
(5)	160.5	472.1	2.94
(1)	93.0	256.7	2.76
(2)	430.2	2751.0	6.39
(3)	102.6	185.0	1.80
(4)	114.4	306.4	2.67
(5)	103.3	237.2	2.29
(6)	1444.7	19433.0	13.45
Total	2811.4	24912.8	8.86

The first line in the table shows the time for proving that the predicate *valid* is invariant. The next 5 lines show the times for proving the ensures properties corresponding to the arrows marked (1) to (5) in the proof lattice of Figure 2. Note that the arrow marked (5) corresponds to two ensures properties. The last six lines in the table shows the times for proving the ensures properties corresponding to the arrows marked (1) to (6) in the proof lattice of Figure 3. Here (2) and (6) corresponds to two ensures properties. These two properties are also induction steps which accounts for their relatively high execution times. The last column of the table shows the number of times that the execution of the tactic in HOL90 is faster than HOL88.

It follows that for invariant and non-induction steps HOL90 is 3-4 times faster than HOL88. However, for the two induction steps HOL90 is 6 and 13 times faster than HOL88. The two induction steps includes the largest amount of arithmetic proofs. In average HOL90 is 9 times faster than HOL88.

7 Discussion

In this paper we have shown how to prove that a lift-control program satisfies a given progress property using HOL-UNITY [3]. We have presented a tactic for automatically proving the basic **unless, invariant** and **ensures** properties. From these properties the required **leadsto** property is derived as shown in the proof lattices in Figure 2 and 3 using the inference rules of the UNITY [6] logic built into HOL-UNITY as proved theorems.

Combining a Pressburger proof procedure, recently introduced as a library [5, 13] in the HOL system, with a Gentzen-like way of proving, it has been possible to implement the tactic in HOL enabling automatic proofs of the basic properties of the lift-control example.

The example proof of a single leadsto progress property contains 120 applications of the tactic to prove the necessary ensures and invariance properties satisfied by the transitions of the lift program. Thus, the tactic does a great deal of tedious work for us, enabling us to concentrate on the difficult aspects of the proof, viz. finding the invariant and decomposing the leadsto property. How to find the invariants and decompose the leadsto properties has not been discussed here.

In proving a basic property satisfied by a program we utilized that it is possible to first separately prove the property satisfied for each action of the program and then from these properties derive a proof of the property for the entire program. This proof method is possible, since the basic properties satisfy the theorem:

$$\frac{Pr = [st_1; \; ...; \; st_n], \; R \; p \; q \; [st_1], \; ..., \; R \; p \; q \; [st_n]}{R \; p \; q \; Pr}$$

where R is one of the properties **unless**, **invariant** or **ensures**. Using this method has the advantage that every single proof is of minimum size, but the number of proofs are linear in the number of program actions.

In view of the timing results, this may put limitations on proving properties of larger examples. To speed up the currently implemented tactic, it contains strategies looking for a certain class of contradictions, specialized optimizations for lower and upper bound analysis of arithmetic terms and term normalization. These design decisions speed up the search for a proof in many of the occasions in the example. Another perhaps better solution is adapting the $\mathcal{F}aust$ prover [17] to combine it with the Pressburger tool.

Our next goal is to finish the ongoing development of a compiler for UNITY. The compiler will support generating the HOL-UNITY representation and the proof obligations. Until it is available, the generation of HOL-UNITY information has to be done manually.

References

1. S. Agerholm. Mechanizing Program Verification in HOL. Master's thesis, Computer Science Department, University of Århus, Denmark, September 1991.
2. F. Andersen. A Definitional Theory of UNITY in HOL. In *Summary of talks at the Third Annual HOL User Meeting*, PB 340, pages 151–162. DAIMI, Århus University, october 1990.
3. F. Andersen. *A Theorem Prover for UNITY in Higher Order Logic*. PhD thesis, Technical University of Denmark, 1992. Also published as TFL RT 1992-3.

4. F. Andersen and K. D. Petersen. Recursive Boolean Functions in HOL. In *1991 International Tutorial and Workshop on the HOL Theorem Proving System and its Applications*, pages 367–377. IEEE Computer Society, August 1991.

5. R. Boulton. The HOL arith Library. Technical report, Computer Laboratory University of Cambridge, July 1992.

6. K. M. Chandy and J. Misra. *Parallel Program Design: A Foundation*. Addison-Wesley, 1988.

7. A. Church. A Formulation of the Simple Theory of Types. *Journal of Symbolic Logic*, 5, 1940.

8. U. Engberg, P. Grønning, and L. Lamport. Mechanical Verification of Concurrent Systems with TLA. In *Fourth International Workshop on Computer Aided Verification*, 1992.

9. J. H. Gallier. *Logic for Computer Science*. Foundations of Automatic Theorem Proving. Harper & Row, Publishers, 1986.

10. S. Garland, J. Guttag, and J. Staunstrup. Verification of VLSI circuits using LP. Technical report, DAIMI PB-258, University of Århus, Denmark, July 1988.

11. D. M. Goldschlag. Mechanically Verifying Concurrent Programs with the Boyer-Moore Prover. *IEEE Transactions on Software Engineering*, 16(9):1004–1023, September 1990.

12. M. J. C. Gordon. *HOL - A Proof Generating System for Higher-Order Logic*. Cambridge University, Computer Laboratory, 1987.

13. J. Harrison. The HOL reduce Library. Technical report, Computer Laboratory University of Cambridge, June 1991.

14. C.S. Jutla, E. Knapp, and J.R. Rao. A Predicate Transformer Approach to Semantics of Parallel Programs. *ACM Symposium on Principles of Distributed Computing*, 1989.

15. T. Melham. Automating Recursive Type Definitions in Higher Order Logic. Technical Report No. 146, Computer Laboratory, University of Cambridge, Sept. 1988.

16. S. Owicki and L. Lamport. Proving Liveness Properties of Concurrent Programs. *ACM Transactions on Programming Languages and Systems*, 4(3), July 1982.

17. K. Schneider, R. Kumar, and T. Kropf. New Concepts in Faust. In *1992 International Workshop on Higher Order Logic Theorem Proving and its Applications*, pages 471–493. imec Interuniversity Micro-Electronics Center, September 1992.

18. R. E. Shostak. Deciding Combinations of Theories. *JACM*, 31:1–12, 1984.

19. Beverly A. Sanders. Eliminating the Substitution Axiom from UNITY Logic. *Formal Aspects of Computing*, 3(2):189–205, April-June 1991.

20. K. Slind. HOL90 Users Manual. Technical report, 1992.

21. R. Smullyan. *First Order Logic*, volume 43 of *Ergebnisse der Mathematik und ihrer Grenzgebiete*. Springer-Verlag, second printing 1971 edition, 1968.

Graph model of LAMBDA in
Higher Order Logic
(Progress Report)

Kim Dam Petersen[*]

Tele Danmark Research, Lyngsø Allé 2, DK-2970 Hørsholm

Abstract. This paper documents some preliminary steps made to achieve a *safe* theorem prover for reasoning about functions that occur in denotational semantics. A theorem prover is safe when: (1) its underlying logic is consistent and (2) it enforces extensions of the logic to preserve consistency.

The main idea is to extend a general purpose theorem prover with a theory of the Graph (or $P\omega$) model of Dana Scott's LAMBDA language. This theory is extended to a theory of a typed version of LAMBDA with general recursive types.

A short summary of the Graph model is given along with details of how it has been formalised in a theorem prover.

The model has been constructed using the HOL theorem prover which supports a polymorphic, strongly typed higher order logic based on Church's simple type theory.

1 Introduction

The need for recursive types appears almost everywhere in denotational semantics of programming languages. As an example, the denotation of a type V with truth values and number values, arbitrary pairing of values, and endogenuous functions must satisfy:

$$V = \mathsf{T} + \mathsf{N} + (V \times V) + (V \to V)$$

Many systems exists for doing formal proofs about functions, e.g. [3, 8, 9, 10].

In HOL the type system is not powerful enough to express functions over the above type [6]. What we may do is to define a theory within HOL which allows such types to be expressed. A subset of these types may be embedded directly in HOL.

We may achieve a safe extension of a logic with recursively specified types by formally constructing a minimal set of values within the logic (corresponding to a least fixed point) that satisfies the recursion property of the type, and introduce this set of values as a new type.

---

[*] email: kimdam@tfl.dk

There exist several methods for constructing denotations of recursive types. They are all based on the assumption that each recursive type is specified by a type equation. Given a type equation, they construct a minimal set of values that satisfies the type equation.

One method, based on "inverse limits", is described in e.g. [2, 11]. With this method the minimal set is constructed as the limit of the type sequence $\langle \bot, \mathcal{F}(\bot), \ldots, \mathcal{F}^n(\bot), \ldots \rangle$, where $\bot$ denotes a special "one-element" set.

Another method is based on "retracts" as described in [12]. Using this method a "universe" of values which is closed w.r.t. abstraction and application of continuous functions is constructed. This universe is defined from the powerset of numbers, and continuous functions are represented using an encoding of their graphs. The universe is also called $P\omega$ or the "Graph model". Given a type equation it is then possible to find a minimal subset of the universe that satisfies the type equation.

To achieve security we want the construction to be done in a mechanised typed higher order logic which supports consistency preserving constant- and type extensions. The mechanised logic HOL [8] satisfies these requirements. We have chosen to perform the computer proved constructions in this mechanised logic.

In order to apply the inverse limit construction within a logic it must be possible to represent the sequence $\langle \bot, \mathcal{F}(\bot), \ldots, \mathcal{F}^n(\bot), \ldots \rangle$ as an object in the logic. As $\mathcal{F}$ transforms one type into another type, the elements of the sequence above will in general have different types. Hence sequences of differently typed elements must be representable in the logic. The retract method on the other hand, does not impose such demands. The universe of the retract theory is constructed as the powerset of numbers. The "closing" of the universe and the recursive types may be constructed using just non-recursive functions on powerset of numbers.

The type discipline of HOL demands all elements of a sequence to be of the same type. This has the consequence that the inverse limit method cannot be mechanised in HOL. Fortunately, the retract method imposes no demands which cannot be met by HOL. Hence, we have chosen to construct recursive types based on the retract method.

We have succeeded in constructing a model of $P\omega$ and a model of untyped lambda calculus in the HOL system, and have been able to prove some of the important properties about these models. To our knowledge, it is the first time, that a model of untyped lambda calculus has been constructed in a mechanised higher order logic and has been proved consistent on a computer.

Based on this model it is the plan within HOL to introduce a notion of types based on the retract method found in [12], and to introduce the basic concepts for specifying and reasoning about programs using denotational semantics.

A work similar to ours is currently carried out by Sten Agerholm [1]. However, his work is based on information systems [13], whereas the work presented here is based on $P\omega$.

In this paper we outline how the models of $P\omega$ and untyped lambda calculus have been constructed using the HOL logic and we sketch the structure of some of the basic proofs.

The paper is organised as follows. In §2 the background of the semantics of the Graph model is presented. The section specifies how the powerset of numbers $P\omega$ can be organised as a reflexive cpo.

In §3 we introduce Scott's language LAMBDA. The syntax and semantics of LAMBDA are defined. The results from §2 entails that LAMBDA satisfies the β-reduction rule. The section concludes with a description of a notion of type, including how general recursive types, can be introduced in LAMBDA.

The HOL-theory used in formalising the Graph model is explained in §4. We show some interesting details of the formalisation in some of the involved HOL-theories. The section also contains a discussion on how Scott's LAMBDA language can be represented in HOL based on the theory of the Graph model.

The LAMBDA language constitutes an object theory for reasoning about denotational semantics within HOL. In §5 we discuss how a class of LAMBDA-functions can be embedded in HOL as HOL-functions.

In §6 we summarise the work done and outline possible future directions.

2 The Graph model

This section provides some basic background on the semantic model of the language LAMBDA. This model is also called the Graph model. The powerset of numbers $P\omega$ may be organised as a reflexive cpo. In [2] this is done in detail. Below the basic steps are repeated:

1. Prove that $P\omega$ ordered by inclusion ($\subseteq$) is a cpo. The bottom element of the cpo is the empty set {} and the least upper bound of a set $X \subseteq P\omega$ is the union $\bigcup X$ of the sets in X.
2. Define the functions $\mathcal{F}$ and $\mathcal{G}$ that decodes respectively encodes continuous functions from $P\omega$ to $P\omega$:

$$\mathcal{F}(s)(x) = \{m | \exists n \cdot (\mathcal{S}^{-1}(n)) \subseteq x \land (\mathcal{P}(n,m)) \in s\}$$
$$\mathcal{G}(f) = \{\mathcal{P}(n,m) | m \in f(\mathcal{S}^{-1}(n))\}$$

where $\mathcal{P}$ denotes the bijective function from pairs of numbers to numbers:

$$\mathcal{P}(n,m) = \frac{(m+n)*(m+n+1)}{2} + m$$

and $\mathcal{S}$ denotes the bijective function from finite sets of numbers to numbers:

$$\mathcal{S}(\{n_1, \ldots, n_k\}) = \sum_{i=1 \ldots k} 2^{n_i}$$

where $n_1 < \ldots < n_k$. Prove that $\mathcal{P}$ and $\mathcal{S}$ are bijective.

3. Prove that $P\omega$ is a *reflexive* cpo by $\mathcal{F}$ and $\mathcal{G}$, i.e. that:

$$\mathcal{F} : P\omega \to [P\omega \to P\omega]$$
$$\mathcal{G} : [P\omega \to P\omega] \to P\omega$$
$$\mathcal{F} \circ \mathcal{G} = \mathrm{id}_{[P\omega \to P\omega]}$$

The notation $[D \to D']$ denotes the continuous functions from D to D'.

In [2] it is shown how a reflexive cpo can be used as a model for untyped lambda calculus. In [12] the name Graph model is introduced for the particular reflexive cpo based on $P\omega$.

We have used the encoding scheme suggested by Scott. However, arbitrary encoding schemes satisfying the equation in step 3 can be used. It is possible to invent a encoding scheme which makes the model extensional [5]. In an extensional model η-reduction becomes valid.

3 The language LAMBDA

The language LAMBDA is described by Scott in [12]. LAMBDA is a variant of lambda calculus with its semantics defined in terms of $P\omega$. The core LAMBDA language is untyped, however, Scott describes how it may be extended with a notion of type by using *retracts*.

3.1 Syntax and semantics

The LAMBDA terms includes the variables, application and abstraction constructs known from the lambda calculus. In addition to this, LAMBDA includes a notation for the empty and total set of elements in $P\omega$, a notation for natural numbers, and a notation for successor, predecessor and test of natural numbers. In Table 1 we list the notation of LAMBDA terms and their interpretation.

The symbols f, g, x, y, and z denotes LAMBDA terms, the symbols n and k denote natural numbers, and the symbols u and v denote variables.

The LAMBDA terms above the line denotes proper terms; the terms below the line are convenient shorthands for some specific LAMBDA terms. An abstraction term is defined only if the function $(\lambda v.x) : P\omega \to P\omega$ is continuous. Fortunately, this is always the case if x is a LAMBDA term.

3.2 Reduction

For application and abstraction to be of any use they have to satisfy some reduction rule similar to the β-reduction of λ calculus. In §2 we saw that the functions $\mathcal{F}$ and $\mathcal{G}$ organise $P\omega$ as a reflexive cpo [2]. A consequence of this is the graph theorem [12]:

$$\forall f : [P\omega \to P\omega].\mathcal{F}(\mathcal{G}f) = f$$

Table 1. Semantics of LAMBDA

Notation	Semantics	Description
$\bot$	$\{\}$	The empty set
$\top$	ω	The total set
v	v	Variable
$\underline{n}$	$\{n\}$	Number constant
succ x	$\{k+1 \mid k \in x\}$	Successor
pred x	$\{k \mid k+1 \in x\}$	Predecessor
$z \Rightarrow x \mid y$	$\{m \in x \mid 0 \in z\} \cup \{m \in y \mid \exists k.k+1 \in z\}$	Test for 0
$f \cdot x$	$\mathcal{F}fx$	Application
$\underline{\lambda}v.x$	$\mathcal{G}(\lambda v.x)$	Abstraction
$\langle\rangle$	$\{\}$	The empty list
$x :: l$	$\underline{\lambda}z.z \Rightarrow x \mid l \cdot (\text{pred } z)$	List prepending
$\langle x_1, \ldots, x_n \rangle$	$x1 :: (\ldots (x_n :: \langle\rangle))$	A finite list
$z \rightarrow x \mid y$	$z \Rightarrow (z \Rightarrow x \mid \top) \mid (z \Rightarrow \top \mid y)$	Double strict test 0
$f \circ g$	$\underline{\lambda}x.f \cdot (g \cdot x)$	Function composition
Y	$\underline{\lambda}u.(\underline{\lambda}x.u \cdot (x \cdot x)) \cdot (\underline{\lambda}x.u \cdot (x \cdot x))$	Fixed point

From the graph theorem the following β-reduction rule can be derived:

$$\forall(\lambda v.x) : [P\omega \rightarrow P\omega].(\underline{\lambda}v.x) \cdot y = x[y/v]$$

Because we use the variable concept of the underlying logic (HOL) α-reduction is also valid. The model, however, is not extensional, hence, η-reduction is not valid for the model. As mentioned in §2 it is possible to construct $\mathcal{F}$ and $\mathcal{G}$ such that an extensional model is achieved.

3.3 Types

A notion of types may be introduced in LAMBDA. To characterise the elements that make up a given type we use special elements of $P\omega$, the *retracts*, i.e. those elements $a \in P\omega$, which, satisfy $a = a \circ a$. The impact of the equation $a = a \circ a$ is that a, considered as a function, maps $P\omega$ onto some subset of $P\omega$ and that a on this subset is the identity function.

We let the range of each retract a be the notion of a type in LAMBDA. Type-membership is then easy to test for since u belongs to the range of a if and only if $u = a \cdot u$. We will use the notation $u : a$ to indicate that u belongs to type a.

By using a retract a as a filter in front of a function f, one gets the functions $f \circ a$; which have the property that f will only be applied to elements belonging to type a, no matter what element $f \circ a$ is applied to. This is a way of specifying that f takes arguments of type a. We will use the notation $\underline{\lambda}v : a.x$ to indicate that $\underline{\lambda}v.x$ takes arguments of type a.

We have summarised the notation for types in Table 2.

Table 2. Syntax of typed LAMBDA

Notation	Semantics	Description
$u : a$	$u = a \cdot u$	u has type a
$\underline{\lambda}v : a.x$	$(\underline{\lambda}v.x) \circ a$	restrict arguments to type a

3.4 Standard types

The standard types of numbers and booleans can be defined as retracts. The elements of the number type consists of all singleton sets of $P\omega$ representing the genuine numbers and the two elements $\perp$ and $\top$ which may be considered representing "undefined" and "overdefined" numbers respectively. The boolean values consists of the singleton sets $\{0\}$ and $\{1\}$ representing true and false respectively and the two elements $\perp$ and $\top$. A singleton element type, unit, can be defined as a retract. The standard type constructors: product, sum and function can be defined as functions in LAMBDA. Table 3 provides notation and corresponding retract definitions for the standard types and type constructors.

Table 3. LAMBDA retracts

Notation	Semantics	Description
bool	$\underline{\lambda}u.u \rightarrow \underline{0}\|\underline{1}$	The type of booleans
int	$Y(\underline{\lambda}i.\underline{\lambda}u.u \rightarrow \underline{0}\|((i \cdot (\text{pred } u)) \rightarrow u\|u))$	The type of numbers
unit	$\underline{\lambda}u.\perp$	The singleton element type
$a \circ\!\!\rightarrow b$	$\underline{\lambda}u.b \circ u \circ a$	The functions from a to b
$a \otimes b$	$\underline{\lambda}u.\langle a \cdot (u \cdot \underline{0}), b \cdot (u \cdot \underline{1})\rangle$	The product of a and b
$a \oplus b$	$\underline{\lambda}u.u \cdot \underline{0} \rightarrow \langle \underline{0}, a \cdot (u \cdot \underline{1})\rangle\|\langle \underline{1}, b \cdot (u \cdot \underline{1})\rangle$	The sum of a and b

The traditional operators on the standard types can be defined in LAMBDA. As an example the operators on sums can be defined as:

$$\text{inleft} = \underline{\lambda}x.\langle \underline{0}, x\rangle$$
$$\text{inright} = \underline{\lambda}x.\langle \underline{1}, x\rangle$$
$$\text{outleft} = \underline{\lambda}u.u \cdot \underline{0} \rightarrow u \cdot \underline{1}\|\perp$$
$$\text{outright} = \underline{\lambda}u.u \cdot \underline{0} \rightarrow \perp\|u \cdot \underline{1}$$
$$\text{which} = \underline{\lambda}u.u \cdot \underline{0}$$

Some type properties of the sum operators are:

$$(a \oplus b) \circ \text{inleft} \circ a : a \circ\!\!\rightarrow a \oplus b$$

$$a \circ \text{outleft} \circ (a \oplus b) : a \oplus b \rightarrowtail a$$
$$\text{which} \circ (a \oplus b) : (a \oplus b) \rightarrowtail \text{bool}$$

Some properties satisfied by the sum operators are:

$$\text{outleft} \cdot (\text{inleft} \cdot x) = x$$
$$\text{which} \cdot (\text{inleft} \cdot x) = \underline{0}$$

3.5 Recursive types

Recursive types can be defined using the Y combinator. The graph F of a continuous function mapping retracts to retracts: $c = \text{Y} \cdot F$ is a retract. This allows us to use the Y for defining the meaning of recursive type.

As an example: $F = \underline{\lambda}t.\text{unit} \oplus (t \otimes t)$, considered a function, maps retracts to retracts, hence $\text{tree} = \text{Y} \cdot F$ is a retract. The recursion property of Y has the consequence that $\text{tree} = F \cdot c$, i.e. $\text{tree} = \text{unit} \oplus (\text{tree} \otimes \text{tree})$.

Constructors and destructors can be defined on tree values:

$$\text{mkunit} = \text{inleft} \cdot \bot$$
$$\text{mkbranch} = \text{inright}$$
$$\text{isunit} = \text{which} \rightarrow \underline{1}|\underline{0}$$
$$\text{isbranch} = \text{which} \rightarrow \underline{0}|\underline{1}$$
$$\text{outbranch} = \text{outright}$$

As expected these functions satisfy the equations:

$$\text{isunit} \cdot \text{mkunit} = \underline{1}$$
$$\text{outbranch} \cdot (\text{mkbranch} \cdot \langle l, r \rangle) = \langle l, r \rangle$$

4 Constructing the Graph model in HOL

In this section we will show how the Graph model construction can be formalised in HOL. First a brief summary of HOL is given, then an overview of the formalisation phase is given, and finally we present some details of the formalisation.

4.1 HOL

The HOL system [7, 8] is a theorem prover for a polymorphic and strongly typed higher order logic. The HOL system offers a collection of predefined concepts, which include: polymorphic total functions, natural numbers and polymorphic lists.

The HOL logic includes a variant of the axiom of choice as follows: a polymorphic *choice* function $\Delta : (\alpha \rightarrow \text{bool}) \rightarrow \alpha$ is provided that satisfy $\forall P x.P(x) \rightarrow$

$P(\Delta(P))$. In other words when applied to a predicate P which is satisfied for some element the choice function always yields a fixed element that satisfies P. Because HOL functions are total the choice function will yield some arbitrary proper typed element when applied to a predicate that always yields false.

The HOL logic allows new constants and types to be introduced. To ensure that consistency is preserved a set of conditions must hold before a new constant or type can be accepted. The conditions for introducing a new constant is basically that the constant can be specified as a closed term in the existing logic. The condition for introducing a new type is basically that the values of the new type can be specified as a predicate in the existing logic and that this predicate is satisfied by at least one element.

The HOL system has a mechanism for organising complex theories in a hierarchy. In each HOL-theory types and constants can be introduced and properties of these can be proven.

4.2   Organisation of HOL-theories

The properties of the Graph model is based on a number of mathematical concepts. To handle the complexity of these concepts we have built up a hierarchy of HOL-theories describing: partial functions, partial orders, complete- and algebraic partial orders, topology (includes the concept of continuity), Scott topology, structured complete partial orders of power set, product and function space, $P\omega$, and LAMBDA.

4.3   Constructing the HOL-theories

In the following we give a summary of the HOL-theories. For some of the more interesting theories we show the details of the construction and discuss alternative methods of construction.

Partial orders Based on the theory of sets the theory of partial orders is constructed. A binary relation on a set is traditionally represented by the set of element pairs in the relation. A partial order is a special kind of binary relation. In order to specify its properties it is necessary to refer to the underlying set, hence a partial order is represented by a pair, the first component being the underlying set, and the second being the set that represents the binary relation.

The theory also defines the concept of bottom (least element) and lub (least upper bound). To represent these concepts care must be taken because not every partial order has a bottom, and not every partially ordered set has a lub. We have chosen to represent the properties of the two concepts by predicates. The bottom and lub are then represented, using the axiom of choice, as chosen elements that satisfy the bottom and lub properties respectively. Uniqueness of bottom and lub properties can then be proven. The theories on bottom and lub are all proven under the assumption that these elements exist.

Complete Partial Orders The theory of complete partial orders is based on the theory of partial orders. The theory defines the concepts of directed set and complete partial order. A directed set is represented by a partial order over a set and a subset of that set. A complete partial order is represented like a partial order. The required properties of a directed set respectively a complete partial order are represented by predicates.

Functions Based on the theory of sets the theory of functions is constructed. All functions in the HOL logic are total functions from one type to another. There is no representation of a function from one set to another. To avoid confusion, a function in the HOL logic will be called a HOL-function and a function from one set to another a set-function. The theory of functions introduces a representation for set-functions.

An obvious approach is to allow a set-function to be represented by any HOL-function that yields the proper result on elements in the domain of the set-function. A drawback of this approach is that there are several HOL-functions corresponding to each set-function; this has the fatal consequence that set-functions cannot be organised as a cpo – it is not possible to achieve reflexivity.

To pick out a single of these HOL-functions we require that it must yield a fixed value outside the domain of the set-function. Using the axiom of choice we arbitrarily choose the fixed value of a HOL-function to be the one returned by the choice function on the set of all elements in the type of the range of the set-function. To define properties of set-functions it is furthermore necessary to indicate the domain and range.

The theory defines the concept of a set-function and its concepts of injectivity, surjectivity and bijectivity.

The theory also defines the concept of an inverse set-function. As this requires the set-function to be bijective the inverse image of each element is unique, so we can apply the same strategy as used to define bottom and lub.

In the following we use function as an abbreviation for set-function.

Topology The theory of topology is based on the theory of functions. The theory defines the concepts of a topology and continuous functions. A topology over a set X is a collection $T \subseteq X$ of subsets, which has the empty set and X as members, and which is closed w.r.t. union and finite intersection. In the theory it is proved that composition of continuous functions yields a continuous function.

Scott topology Based on the theories of topology and complete partial order the theory of Scott topology is constructed. A set O belongs to the Scott topology if and only if (1) O is closed upwards, i.e. contains all elements greater than elements belonging to O and (2) any directed set with lub in O contains an element in O. This theory defines the concepts Scott topology and Scott continuous functions, and proves that these concepts are topologies and complete partial orders. In the theory it is proved that the Scott continuous functions can be characterised as usual, i.e. they preserve lubs of directed sets.

Algebraic Complete Partial Orders The theory of algebraic complete partial orders is based on the theory of Scott topology. This theory defines the concepts of compact element and algebraic complete partial order. An element x is compact if and only if every directed set with lub greater than x contains an element greater than x. A complete partial order is algebraic if and only if for any element x the compact elements smaller than x make up a directed set. It is proved that Scott continuous functions are characterised by their value of argument being the lub of the function applied to the compact elements less than the argument.

Powersets The theory of powersets is based on the theory of algebraic complete partial orders. In this theory it is proved that the power set of any type ordered by inclusion is an algebraic complete partial order. The compact elements of this partial order are the finite sets. The following corollary is proved: The value of a Scott continuous function on a set is the union of the values of the finite sets included in this set.

$P\omega$ The theory $P\omega$ is based on the theory of powersets. In this theory the functions F and G that make $P\omega$ a reflexive complete partial order are constructed. In the theory it is proved that $P\omega$ is a reflexive complete partial order.

LAMBDA The theory LAMBDA is based on the theory $P\omega$. In the LAMBDA theory the inductively defined LAMBDA terms are constructed.

The language LAMBDA could be represented by introducing a syntax for the language constructs and then defining the semantics of each construct by a recursive function denoting a $P\omega$ value. Another approach is to represent each LAMBDA term explicitly by a HOL function, which operates directly on $P\omega$ values.

The first approach has the advantage that all language constructs are explicitly listed, which simplifies the validation of recursively defined language properties. The disadvantage of this approach is that the notion of variables has to be represented syntactically, i.e. environments have to be used when the semantics of LAMBDA values is defined. The second approach has the advantage that reasoning can be performed directly on the semantic level, i.e. the syntactic level disappears.

We have chosen the second approach to avoid the use of a variable environment. In Table 4 we show the HOL representation of the LAMBDA terms.

The HOL constants **Fun** and **Graph** represents the functions $\mathcal{F}$ and $\mathcal{G}$ on $P\omega$. For readability we have chosen to make << an infix, and L a binding operator. Hence, application and abstraction of LAMBDA is expressed in HOL as follows:

```
f << x = Fun f x
$L f   = Graph f
```

The remaining HOL constants can be defined directly from their definitions in Table 1.

Table 4. HOL representation of LAMBDA

LAMBDAterm	Representation
$\perp$	{}
$\top$	All
v	v
$\underline{n}$	Num n
succ x	Succ x
pred x	Pred x
$z \Rightarrow x\|y$	Cond $z\,x\,y$
$f \cdot x$	$f \ll x$
$\lambda v.x$	L($\backslash v . x$)
$\langle\rangle$	<>
$x :: l$	$x :: l$
$z \to x\|y$	DCond $z\,x\,y$
$u \circ v$	$u\,O\,v$

Continuity of a given LAMBDA function composed from the constructs in Table 4 can be proved by iterated application of the theorems:

```
!k.       Cont (\x. k)
          Cont (\x. x)
!f.       Cont f ==> Cont (\x. Succ (f x))
!f.       Cont f ==> Cont (\x. Pred (f x))
!f g h.   Cont f g h ==> Cont (\x. Cond (f x) (g x) (h x))
!f g.     Cont g ==> Cont (\x. (f x) << (g x))
!f.       Cont f ==> Cont (\x. $L (f x))
```

5 Embedding LAMBDA in HOL

This section presents some ideas on how LAMBDA can be embedded in HOL. We stress that none of the ideas have actually been carried out in HOL.

The language LAMBDA as described until now constitutes a theory of its own. HOL is only used as a meta-language. To be able to use the reasoning tools in HOL on LAMBDA functions, we would like a method for transforming LAMBDA functions to HOL functions and back again.

The idea of achieving such a transformation is to exploit the type information and strictness. Consider a LAMBDA function $f : a \circ\!\!\to b$. The type information of f has the consequence that whenever f is applied to an argument of type a it will yield a result of type b. Now if we can find two HOL types, say $\hat{a}$ and $\hat{b}$, that are isomorphic to the LAMBDA types a and b, i.e. that have the same number of elements, we can construct a HOL-function $\hat{f}$ that is equivalent to f by using the type isomorphism to *convert* values of the two functions.

Almost every LAMBDA type includes the extreme values $\perp$ and $\top$, corresponding to under- and overdefined function results. These values have no counterparts in HOL. If a LAMBDA function f is known only to yield an extreme value if applied to a non-extreme value we can construct $\hat{f}$ as sketched considering only the non-extreme values of LAMBDA types a and b.

As an example consider the LAMBDA type int $= \{\underline{n}|n \in \omega\} \cup \bot \cup \top$ (recall: $\underline{n} = \{n\}$). Excluding the extreme values we get the set $\{\underline{n}|n \in \omega\}$ of LAMBDA values which evidently is isomorphic to the set of numbers in HOL, i.e. isomorphic to the HOL type of numbers. Let us denote the isomorphisms by HOL-functions: NumCode : **num** → **Pomega** and NumDecode : **Pomega** → **num**. The HOL type **Pomega** is defined as **num** -> **bool**. The isomorphisms may be defined, using the choice function Δ, as:

$$\text{NumCode}(n) \equiv \{n\}$$
$$\text{NumDecode}(N) \equiv \Delta n.n \in N$$

Using these functions we can now express transformations between a LAMBDA function $f : a \circ\!\!\rightarrow b$ and its equivalent HOL-function $\hat{f}$:

$$\hat{f} \equiv \text{NumDecode} \circ f \circ \text{NumCode}$$
$$f \equiv \text{NumCode} \circ \hat{f} \circ \text{NumDeCode}$$

The two functions are related by the following theorems:

```
(f Of (NumType FunType NumType)) /\
(Continuous f)
(NumStrict f) /\ % !n. ?m. f {n} = {m} %
(f x = y)
==> ((LAMBDA_to_HOL f) (NumDecode x) = (NumDecode y))

(Continuous (HOL_to_LAMBDA f))
==> (HOL_to_LAMBDA f) Of (NumType FunType NumType)

(f x = y) /\
(Continuous (HOL_to_LAMBDA f))
==> ((HOL_to_LAMBDA f) (NumCode x) = (NumCode y))
```

The method described here can be used for any HOL type isomorphic to a LAMBDA type. By considering the cardinality of $P\omega$ we immediately observe that HOL types with countable elements can be isomorphic to LAMBDA types. Hence, a LAMBDA function on union or pairs constructed from boolean and number types are potentially equivalent to some HOL function.

6 Conclusion

We have constructed the Graph model for LAMBDA using the HOL theorem prover. All properties of the model have been proved using HOL.

The perspective of this work has been to construct a theorem prover in which recursive types, as found in denotational semantics, can be defined automatically; and reasoning about functions on these can be mechanised.

Ideas on how to embed the LAMBDA model within the theorem prover HOL have been sketched. Verifying these ideas remains to be done.

References

1. Sten Agerholm. *Formalising Domain Theory in HOL*. Computer Science Department, Aarhus University. DRAFT, March, 1993.
2. Hendrik P. Barendregt. *The Lambda Calculus: Its Syntax and Semantics*. North Holland, 1984.
3. R. S. Boyer and J. S. Moore. *A Computational Logic*. Academic Press, 1979.
4. Alonzo Church. *A Formulation of the Simple Theory of Types*. Journal of Symbolic Logic, Vol. 5 (1940), pp. 56–68.
5. Anders Gammelgaard and Kim D. Petersen. *Unpublished note*.
6. Elsa L. Gunter. *Why we can't have SML Style datatype Declarations in HOL*. Proceedings from International Workshop on Higher Order Logic Theorem Proving and its Applications, 1992.
7. Mike J. Gordon. *HOL – A Proof Generating System for Higher-Order Logic*. VLSI Specification, Verification and Synthesis, Kluwer, 1987.
8. *The HOL System: DESCRIPTION*, Version HOL88.2.01 (1992). HOL88 documentation.
9. Larry C. Paulson and Tobias Nipkow. *Isabelle Tutorial and User's Manual*. Unpublished, Isabelle documentation, 1988.
10. Mike J. Gordon, Robin Milner and Christopher P. Wadsworth. *Edinburgh LCF: A Mechanised Logic of Computation*. LNCS 78, Springer Verlag, 1979.
11. David Schmidt. *Denotational Semantics: A Methodology for Language Development*. Allyn and Bacon, 1986.
12. Dana Scott. *Data Types as Lattices*. SIAM J. Comput. Vol. 5, No. 3, September 1976.
13. Glynn Winskel. *The Formal Semantics of Programming Languages*. MIT Press, Cambridge, 1993.

Mechanizing a Programming Logic for the Concurrent Programming Language microSR in HOL *

Cui Zhang, Rob Shaw, Ronald A. Olsson, Karl Levitt,
Myla Archer, Mark R. Heckman and Gregory D. Benson

Department of Computer Science
University of California, Davis, CA 95616
email: zhang@cs.ucdavis.edu, verification-lab@cs.ucdavis.edu

Abstract. This paper presents our current effort to formally derive, using HOL, a sound Hoare logic for the concurrent programming language microSR, a derivative of SR. Our methodology is built on Gordon's work on mechanizing programming logics for a small sequential programming language. The constructs of microSR include those basic to common sequential programming languages, in addition to an asynchronous *send* statement, a synchronous *receive* statement, a guarded communication *input* statement, and a *co* statement for specifying concurrent execution. This language has the appearance of a high-level system programming language that supports distributed applications. The Hoare logic for microSR with concurrency features presented in this paper has been formally proven to be sound within HOL. The logic we derived allows one to reason and state formal assertions about concurrently executing processes that do not share any data objects, but communicate through shared channels.

1 Introduction

The work described in this paper is part of a long term project called the Silo project that is currently being undertaken in the UC Davis Verification Lab. The Silo project is aimed at verifying, by mechanized layered formal proof, a complete distributed computer system from hardware to programming language. Each layer will be formally modeled as an interpreter that interacts with the other layers. This layered interpreter approach will allow us to verify distributed applications with respect to the entire system. The CLI stack [4] has shown the feasibility of full system verification using a layered proof technique, but their model does not allow for concurrency and distributed programming, nor have they fully integrated the operating system into their "stack".

This paper concentrates on the mechanization of a Hoare logic in HOL for the concurrent programming language microSR, which is a derivative of SR [1, 3]. SR has several aspects that make it a good language to be used in our research.

---

* This work was sponsored in part by the US Department of Defense.

Foremost is SR's expressiveness within the realm of concurrent programming paradigms. Secondly, SR's constructs are clean enough to be described with simple, intuitive formal semantics. We selected a subset of SR called microSR for our research, with primary emphasis on preserving the concurrency and message-passing features. Figure 1 shows the overview of our work involving microSR; the dotted box indicates this paper's focus.

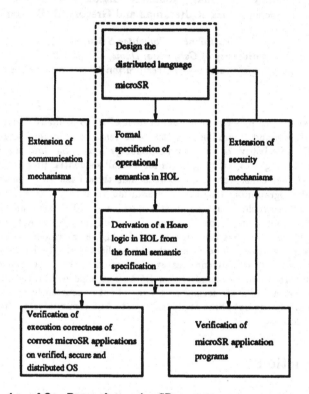

Fig. 1. Overview of Our Research on microSR

Our effort to formally derive, using HOL, a sound Hoare logic from microSR semantics is a generalization of similar work by Gordon for a small sequential language [6]. Our goal is to develop a feasible methodology for mechanizing Hoare-style programming logics for concurrent programming languages; our current results represent steps toward that goal, although they may not prove to be the best formulation. We are concerned with just partial correctness and therefore use semantic relations, rather than functions, in our formal specification for microSR constructs; doing so obviates the possible need for powerdomains in the state abstraction for microSR programs due to the inherent non-determinism that accompanies interleaved executing processes and the explicit non-determinism of some sequential constructs. Our logic system employs global invariants to handle the interference problem arising from concurrent execution. Our axiomatic logic for microSR allows one to reason and state formal assertions about concurrently executing processes that do not share any data objects, but communicate through shared channels. Our current work differs significantly from the early work in our lab by Harrison [7]. Not only is microSR a more practical high-level language than Harrison's, but also the Hoare logic has been formally proven to be sound within HOL, i.e., axioms and inference rules are all mechanically derived in HOL as the logical implication of the formal specification of microSR semantics, without introducing any possible logical inconsistency by postulating new axioms.

Section 2 describes the microSR language and its formal semantic specification in HOL. Section 3 gives our interpretation of a Hoare-style partial correctness specification, and outlines the Hoare logic derived in HOL for microSR. Section 4 concludes our work to date and discusses our future work.

2   Semantics of microSR in HOL

2.1   microSR

As mentioned earlier, microSR supports distributed programming with explicit mechanisms for concurrency. The processes, however, do not execute in shared memory. They communicate only via message passing. Because in microSR terminology, a shared communication channel is called an operation, we will use "operation" and "channel" interchangeably in this paper. The constructs of microSR include those basic to common programming languages in addition to an asynchronous send statement, a synchronous receive statement, a guarded communication input statement similar to Hoare's CSP [8], and a co statement for specifying concurrent execution. The statement level syntax of microSR used in this paper is specified by the BNF below. In this grammar, the non-terminals E, E1, and E2 range over integer expressions; B, B1, and B2 range over boolean expressions; S, S1, S2, ... , Sn range over statements; V ranges over program variables with values as numerals, and the terminals op, op1, and op2 range over operation names. The alternative (if) statement, iteration (do) statement, and guarded communication input (in) statement have explicitly non-deterministic semantics.

> S ::= skip | V := E | send op (E) | receive op (V)
> | if B1 → S1 [] B2 → S2 fi | do B1 → S1 [] B2 → S2 od
> | in op1 (V) → S1 [] op2 (V) → S2 ni | co S1 // ... // Sn oc

For the syntactic and semantic details of expressions, we use an abstract representation in HOL. Integer expressions are abstracted as functions of type $(num)list \rightarrow num$. Similarly, boolean expressions are abstracted as functions of type $(num)list \rightarrow bool$. Any integer expression instance is represented as "E, vl", where E is of type $(num)list \rightarrow num$ and vl is the variable list of type $(string)list$, and any boolean expression instance is represented as "B, vl", where

B is of type *(num)list → bool* and vl of type *(string)list*. As shown below, the actual values of expression's variables come from the current set of bindings that is the local state of the process. This abstraction provides us flexibility in future language extensions and the generalization of our method to other concurrent programming languages with similar semantics but different syntax.

The following HOL definitions give the meaning of expressions. "local_state" is a mapping from variables to their values. "MAP" is a HOL constant function and "MAP ls vl" returns the list obtained by applying ls to the element of vl in turn. Integer expressions and boolean expressions are both expressions in microSR language and formulas in predicate logic, and preconditions and post-conditions of program variables are formulas in predicate logic. Their meanings are defined in the same way. A HOL-like syntax is used in this paper, except that ∀, ∃, and λ are used instead of !, ?, and \.

$$m\_i\_expression = \vdash def\, \forall(e:(num)list \to num)\, (vl:(string)list)\, (ls:local\_state).$$
$$m\_i\_expression\ (e, vl)\ ls = e\ (MAP\ ls\ vl)$$
$$m\_b\_expression = \vdash def\, \forall(b:(num)list \to bool)(vl:(string)list)\, (ls:local\_state).$$
$$m\_b\_expression\ (b, vl)\ ls = b\ (MAP\ ls\ vl)$$
$$m\_pre\_post\_cond = \vdash def\, \forall(t:pre\_post\_cond)\, (vl:(string)list)\, (ls:local\_state).$$
$$m\_pre\_post\_cond\ (t, vl)\ ls = t\ (MAP\ ls\ vl)$$

2.2 Semantic Model

Figure 2 shows schematically the system state abstraction for our microSR operational semantics. Just as the microSR programmer views the executing code as a collection of processes (each with a private local state) that communicate over shared channels, so is our global system state a collection of processes that share a global "message pool". This pool is divided into sets of FIFO queues which consist a set of separate FIFO sub-queues for each process' messages sent to this microSR operation. These queues allow us to formalize the main property about communication channels that microSR guarantees: over each point-to-point link, messages arrive at the destination in the same order as they were sent from the source. The dotted arcs indicate the effects of send/receive-transitions. Actually, performing a receive from a microSR operation entails a non-deterministic choice of exactly one subqueue, followed by the receipt of the earliest message from the chosen subqueue. This model also contains auxiliary information, such as message counts and an abstract notion of global time which characterize certain properties that one would like to reason about, such as the ordering of events, the absence of message loss and duplication. Below is the type definition of the global system state for microSR operational semantics. Every message data is tagged with a process id, proc_id, which is the sending process' id. It is also tagged with a message id, m_id, which is the abstract sending time at which the message is sent. m_id is unique with respect to the sub-queue into which the message was sent.

global_state:	((process_state)list)#msg_pool#global_time
msg_pool:	(op_name#(proc_id#(message)list)list)list
message:	data#proc_id#m_id
process_state:	name#proc_id#local_state#op_list#rcv_count
local_state:	string → num

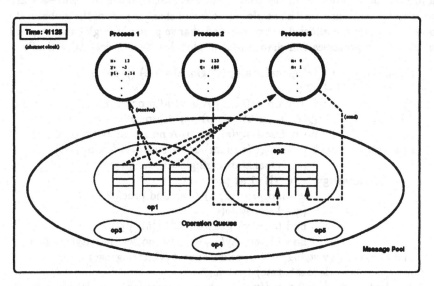

Fig. 2. System State Representation for microSR Semantics

In programming logics, the interference problem is simply that one process' alteration of the shared state (in our case, the communication channel contents) may falsify another process' assertions. Programmers must prove that, regardless of the run-time execution interleaving, no interference is present at any assertions. In order to handle interference, we have introduced into our logic the well-known strategy of global invariants and modeled the only potential interference hazard of communication channels as "sent-sets" and "received-sets" as in [2, 7]. Messages existing in the FIFO queue associated with a channel are the messages that have already arrived in this channel's "sent-set" but not yet in its "received-set". The benefit is that any sub-queue for a process' messages sent to a given operation is allowed to be accessed at any time by one process for one purpose, either to send a message into it or receive a message from it. Consequently, when used in a point-to-point manner, the two communicating processes for a given operation are interference-free. Programmers can state global invariant properties of each operation in terms of its "sent-set" and "received-set" to characterize their programs' communication behaviors, such as the ordering of messages in the queues, the absence of message loss and duplication.

2.3 Semantics of Statements

For all statements except co, the meaning of the statement S is a predicate m_S of type *global_state* → *global_state* → *proc_id* → *bool*. "m_S s1 s2 p" indicates whether global state s2 can be reached from global state s1 after executing statement S in a process with proc_id p. As shown below, the "m_S" relation provides a precise specification to model concurrent executions. When one process starts to execute a statement with one global state and changes part of the global state, other processes are allowed to execute and change part of the global state, and, moreover, all processes can change part of global state non-deterministically.

$$m\_skip = \vdash def \forall \text{ (s1:global\_state) (s2:global\_state) (p:proc\_id)}.$$
$$m\_skip \text{ s1 s2 p} =$$
$$\text{is\_valid\_proc(s1, p)} \wedge \text{is\_valid\_proc(s2, p)} \wedge$$
$$\text{(get\_proc\_state(s1, p)} = \text{get\_proc\_state(s2, p))} \wedge$$
$$\text{no\_snd\_activity(s1, s2, p)} \wedge \text{no\_rcv\_activity(s1, s2, p)}$$
$$m\_assign = \vdash def \forall \text{ (v:string) (e:(num)list} \rightarrow \text{num) (vl:(string)list)}$$
$$\text{(s1:global\_state) (s2:global\_state) (p:proc\_id)}.$$
$$m\_assign \text{ (v, e, vl) s1 s2 p} =$$
$$\text{is\_valid\_proc(s1, p)} \wedge \text{is\_valid\_proc(s2, p)} \wedge$$
$$\text{(get\_local\_state(s2, p)} =$$
$$\text{bnd (v, e, vl, get\_local\_state(s1, p)))} \wedge$$
$$\text{no\_snd\_activity(s1, s2, p)} \wedge \text{no\_rcv\_activity(s1, s2, p)}$$
$$bnd = \vdash def \forall \text{ (v:string) (e:(num)list} \rightarrow \text{num) (vl:(string)list)}$$
$$\text{(ls:local\_state) (x:string)}.$$
$$bnd \text{ (v, e, vl, ls) x} = ((x = v) \Rightarrow \text{m\_i\_expression(e, vl, ls)} \mid \text{ls x })$$

Predicate "is_valid_proc" checks the existence and uniqueness of the process in the global state. "get_proc_state" returns process state out of the global state. "get_local_state" returns the function mapping from process' variables to their values. Predicates "no_snd_activity" and "no_rcv_activity" are defined as "sanity checks". They assert that for each construct of non-communication that the content of message pool pertaining to the executing process p is not changed unexpectedly, i.e., there is no message loss, duplication, modification, or miraculous creation. "no_snd_activity" asserts that there is no "new" message in the message pool tagged as sent out by process of proc_id p ("new" means they exist in s2, but not in s1). "no_rcv_activity" likewise asserts that this process does not receive any message and its own receive count is unchanged. These last two predicates also guarantee the uniqueness of messages by asserting that for any message msg1 in some sub-queue of some operation's queue in message pool of s1, if there is a message msg2 with the same abstract time msg_id in the same sub-queue of the same operation's queue in s2, this msg2 must be unique and the whole messages of msg1 and msg2 are the same as well.

As shown in the BNF in section 2.1, the alternative (if) and iteration (do) statements contain two guarded statements. The two guards are evaluated in some arbitrary order in a certain amount of time during which other processes may execute and change the global_state. So the following specification asserts the existence of some intermediate global state s3 after the guard evaluation. "global_time_later_than" asserts that the abstract global time of s3 is later than the time of s1. "get_local_state", "no_snd_activity" and "no_rcv_activity" assert that this process has no effect in this time period either on its own local_state or on the shared message_pool. If only one guard Bi is found to be true, the process starts to execute the corresponding statement Si in global state s3, rather than in s1. If both guards happen to be true, the choice of which statement is executed is non-deterministic. If no guard is true, execution of the alternative statement has no effect on the local state of the process which executes it, i.e., the execution of this statement terminates with s3. Similarly, the execution of the iteration statement terminates with global state s3 when neither guard is true. The disjunction clause is used in the specification to represent the non-deterministic effect on the global state by the execution of one or none of the two guarded statements. Notice that intermediate global states could be introduced into the specification for assignment statement as well, however, the execution will not terminate in any intermediate state s3.

$$
\begin{aligned}
&\text{m\_if} = \vdash \textit{def}\, \forall\ (\text{b1:(num)list} \rightarrow \text{bool})\ (\text{vl1:(string)list}) \\
&\qquad\qquad (\text{b2:(num)list} \rightarrow \text{bool})\ (\text{vl2:(string)list}) \\
&\qquad\qquad (\text{sl1:(statement)list})\ (\text{sl2:(statement)list}) \\
&\qquad\qquad (\text{s1:global\_state})\ (\text{s2:global\_state})\ (\text{p:proc\_id})\ . \\
&\quad \text{m\_if}\ (\text{b1, vl1, b2, vl2, sl1, sl2})\ \text{s1 s2 p} = \\
&\qquad \exists\ (\text{s3:global\_state})\ . \\
&\qquad \text{is\_valid\_proc(s1, p)}\ \wedge\ \text{is\_valid\_proc(s2, p)}\ \wedge\ \text{is\_valid\_proc(s3, p)}\ \wedge \\
&\qquad (\text{get\_local\_state(s1,p)} = \text{get\_local\_state(s3,p)})\ \wedge \\
&\qquad \text{global\_time\_later\_than s3 s1}\ \wedge \\
&\qquad \text{no\_snd\_activity(s1, s3, p)}\ \wedge \text{no\_rcv\_activity(s1, s3, p)} \\
&\qquad ((\neg\ \text{m\_b\_expression (b1, vl1) (get\_local\_state(s1,p))}\ \wedge \\
&\qquad \neg\ \text{m\_b\_expression (b2, vl2) (get\_local\_state(s1,p))}\ \wedge\ (\text{s2} = \text{s3}))\ \vee \\
&\qquad (\text{m\_b\_expression (b1, vl1) (get\_local\_state(s1,p))}\ \wedge \\
&\qquad \text{m\_stmt\_list sl1 s3 s2 p})\ \vee \\
&\qquad (\text{m\_b\_expression (b2, vl2) (get\_local\_state(s1,p))}\ \wedge \\
&\qquad \text{m\_stmt\_list sl2 s3 s2 p})) \\[4pt]
&\text{m\_do} = \vdash \textit{def}\, \forall\ (\text{b1:(num)list} \rightarrow \text{bool})\ (\text{vl1:(string)list}) \\
&\qquad\qquad (\text{b2:(num)list} \rightarrow \text{bool})\ (\text{vl2:(string)list}) \\
&\qquad\qquad (\text{sl1:(statement)list})\ (\text{sl2:(statement)list}) \\
&\qquad\qquad (\text{s1:global\_state})\ (\text{s2:global\_state})\ (\text{p:proc\_id})\ . \\
&\quad \text{m\_do}\ (\text{b1, vl1, b2, vl2, sl1, sl2})\ \text{s1 s2 p} = \\
&\qquad \exists\ (\text{n:num})\ . \ \text{iter n b1 vl1 b2 vl2 sl1 sl2 s1 s2 p} \\[4pt]
&\text{iter} = \vdash \textit{def}\, (\forall\ (\text{b1:(num)list} \rightarrow \text{bool})\ (\text{vl1:(string)list}) \\
&\qquad\qquad (\text{b2:(num)list} \rightarrow \text{bool})\ (\text{vl2:(string)list}) \\
&\qquad\qquad (\text{sl1:(statement)list})\ (\text{sl2:(statement)list}) \\
&\qquad\qquad (\text{s1:global\_state})\ (\text{s2:global\_state})\ (\text{p:proc\_id})\ . \\
&\quad \text{iter 0 b1 vl1 b2 vl2 sl1 sl2 s1 s2 p} = \text{F}\ )\ \wedge \\
&\quad (\forall\ (\text{n:num})\ (\text{b1:(num)list} \rightarrow \text{bool})\ (\text{vl1:(string)list}) \\
&\qquad (\text{b2:(num)list} \rightarrow \text{bool})\ (\text{vl2:(string)list})
\end{aligned}
$$

(sl1:(statement)list) (sl2:(statement)list)
(s1:global_state) (s2:global_state) (p:proc_id) .
iter (SUC n) b1 vl1 b2 vl2 sl1 sl2 s1 s2 p =
∃ (s3:global_state) (s4:global_state) .
is_valid_proc(s1,p) ∧ is_valid_proc(s2,p) ∧
is_valid_proc(s3,p) ∧ is_valid_proc(s4,p) ∧
(get_local_state(s1,p) = get_local_state(s3,p)) ∧
global_time_later_than s3 s1 ∧
no_snd_activity(s1, s3, p) ∧ no_rcv_activity(s1, s3, p)
((¬ m_b_expression (b1, vl1) (get_local_state(s1,p)) ∧
¬ m_b_expression (b2, vl2) (get_local_state(s1,p)) ∧ (s2 = s3)) ∨
(m_b_expression (b1, vl1) (get_local_state(s1,p)) ∧
m_stmt_list sl1 s3 s4 p ∧ iter n b1 vl1 b2 vl2 sl1 sl2 s4 s2 p) ∨
(m_b_expression (b2, vl2) (get_local_state(s1,p)) ∧
m_stmt_list s12 s3 s4 p ∧ iter n b1 vl1 b2 vl2 sl1 sl2 s4 s2 p))

In microSR, a process sends a message to an operation by executing the statement "send op (E)". Executing this statement evaluates the expression and puts a message containing the resulting data into this process' queue in the operation in the message pool. In the following specification, the existence of a new message with a m_id of abstract global time is asserted. This new message, which exists in the message pool in state s2 but not in the message pool in state s1, contains the resulting data after evaluating the expression, the proc_id of the process executing this send statement, and the m_id representing the global time. "(SND(SND msg) ≥ global_time s2)" asserts that the m_id in this new message is later than the global time of s2 because of the possible system time delay for the send transition. "new_m_id_of_opQ (s1, op, p, SND(SND msg))" asserts that this message sent by process of proc_id p is the newest one in this process' queue in the operation. This guarantees the total order of messages existing in any process' queue in any operation for the point-to-point message passing.

A process of proc_id p receives a message from an operation by executing "receive op (V)". The execution of this statement is delayed until there exists at least one message in the operation op sent by some process with proc_id p'. In our semantic specification, we do not have to include actual assertions for this delay. We get it "for free" by virtue of how semantic relations work. "earliest_m_id_opQ" asserts that the message received by the process is the first one in the front of one of the queues associated with this operation. After executing this receive statement, this received message is removed from the queue. This process' own receive count is increased by 1. The local state of this process in the global state s2 maps variable V to its new value, the data of the received message. "non_interferring_send_or_receive" asserts that any process' queue of any operation in the message pool is allowed at any time to be accessed by only one process, although different operations are allowed to be accessed by different processes simultaneously. So send and receive statements can be viewed as calls to a monitor that protects communication channels. Putting a message into or picking up a message from a process' queue in an operation is the monitor's atomic action taken with respect to the accessed process' queue in the given operation as in [2].

m_send = ⊢ *def* ∀ (op:op_name) (e:(num)list → num) (vl:(string)list)
(s1:global_state) (s2:global_state) (p:proc_id) .
    m_send (op, e, vl) s1 s2 p =
        ∃ (msg:message) (m_id:num) .
        is_valid_proc(s1,p)∧is_valid_proc(s2,p)∧ is_valid_op(s1,p,op)∧
        (get_local_state(s1,p) = get_local_state(s2,p))) ∧
        non_interferring_send_or_receive(s1,s2,op,p) ∧
        (msg=(m_i_expression (e,vl,get_local_state(s1,p)),p,m_id)) ∧
        (SND(SND msg) ≥ global_time s2) ∧
        new_m_id_of_opQ (s1, op, p, SND(SND msg)) ∧
        ¬ msg_of_p_of_op_in_pool(s1, op, p, msg) ∧
        msg_of_p_of_op_in_pool(s2, op, p, msg) ∧
        no_rcv_activity(s1, s2, p)
m_receive = ⊢ *def* ∀ (op:op_name) (v:string)
(s1:global_state) (s2:global_state) (p:proc_id) .
    m_receive (op, v) s1 s2 p =
    is_valid_proc(s1,p) ∧ is_valid_proc(s2,p) ∧ is_valid_op(s1,p,op) ∧
    non_interferring_send_or_receive(s1, s2, op, p) ∧
        ∃ (msg:message) (p':proc_id) .
        earliest_m_id_opQ(s1, op, p',SND(SND msg)) ∧
        msg_of_p_of_op_in_pool(s1, op, p', msg) ∧
        ¬ msg_of_p_of_op_in_pool(s2, op, p', msg) ∧
        (get_local_state(s2,p)=bnd(v,FST msg,get_local_state(s1,p))∧
        no_snd_activity(s1, s2, p) ∧ rcv_count_plus1(s1, s2, p)

The input statement of "in op1(v) → s1 [] op1(v) → s2 ni" provides processes with guarded communication service. Similar to Ada's select/accept construct, an input statement delays the executing process until some invocation is selectable. Then the corresponding statement list is executed. In our initial work, "selectable" means simply that an invocation of the operation is pending, i.e., sent but not yet serviced. If invocations of both operations are pending, then the choice as to which invocation is serviced is non-deterministic. So a process can communicate with more than one processes through different channels with a non-deterministic order.

As shown in section 2.1, we specify concurrent execution by the co statement. The co statement "co S1//...//Sn oc" executes all Si of sequential programs concurrently. Execution of co statement terminates when all of the Si have terminated. As co statement is the top level statement in microSR, its meaning predicate m_co is of type *global_state* → *global_state* → *(proc_id)list* → *bool*.

m_in = ⊢ *def* ∀ (op1:op_name) (m1:string)
(op2:op_name) (m2:string)
(sl1:(statement)list) (sl2:(statement)list)

$$(\text{s1:global\_state}) \ (\text{s2:global\_state}) \ (\text{p:proc\_id}) \ .$$

m_in (op1, op2, v, SL1, SL2) s1 s2 p =
is_valid_proc(s1, p) $\wedge$ is_valid_proc(s2, p) $\wedge$
    $\exists$ (s3:global_state) . is_valid_proc(s3, p) $\wedge$
    ((m_receive(op1, v) s1 s3 p $\wedge$ m_stmt_list sl1 s3 s2 p) $\vee$
    (m_receive(op2, v) s1 s3 p $\wedge$ m_stmt_list sl2 s3 s2 p))

m_co = $\vdash$ def $\forall$ (Si_list:((statement)list)list)
    (s1:global_state) (s2:global_state)
    (proc_list:(proc_id)list) .

m_co Si_list s1 s2 proc_list =
$\forall$(i:num). is_valid_proc(s1,EL i proc_id)$\wedge$is_valid_proc(s2,EL i proc_id)
$\Rightarrow$ m_stmt_list (EL i Si_list) s1 s2 (EL i proc_list)

3 Hoare Logic for microSR

3.1 Partial Correctness Specification

The partial correctness specification in our system has two levels. Firstly, we give our interpretation of {P_and/or_GI} S {Q_and/or_GI}, the inter-process partial correctness specification, where S is the microSR statement, P and Q are assertions on program variables, GI is the assertion on operations, associated with executing S and taken with respect to a particular process. In our Hoare-like logic for concurrent microSR, the interpretation is given by the definition of predicate SPEC shown below. Secondly, we give our interpretation of the global partial correctness specification {{(P_list) $\wedge$ GI}} S {{(Q_list) $\wedge$ GI}}, where S is the top level statement for specifying concurrent executions, GI is the assertion on operations, P_list and Q_list are assertion lists on program variables and the ith elements of the two lists is taken with a particular process for executing the ith sequential program within the S statement. The following is the definition of predicate G_SPEC. All arguments of SPEC and G_SPEC are abbreviated forms of their meaning functions.

SPEC (P_and/or_GI, S, Q_and/or_GI) = $\vdash$ def
    $\forall$ (s1:global_state) (s2:global_state) (p:proc_id) .
    is_valid_proc(s1, p) $\wedge$ is_valid_proc(s2, p) $\wedge$
    P_and/or_GI s1 p $\wedge$ S s1 s2 p
    $\Rightarrow$ Q_and/or_GI s2 p
G_SPEC ((P_list) $\wedge$ GI, S, (Q_list) $\wedge$ GI) = $\vdash$ def
    $\forall$ (s1:global_state) (s2:global_state) (proc_list:(proc_id)list) .
    ($\forall$ (i:num) .
    is_valid_proc(s1,EL i proc_list)$\wedge$is_valid_proc(s2,EL i proc_list)$\wedge$
    (El i P_list) s1 (El i proc_list) $\wedge$ GI s1 (EL i proc_list)) $\wedge$
    S s1 s2 proc_list
    $\Rightarrow$ ($\forall$ (i:num) .
    (El i Q_list) s2 (El i proc_list)$\wedge$GI s2 (EL i proc_list))

3.2 Axioms and Inference Rules

We now present a Hoare-like logic derived for the concurrent language microSR. Again, this logic has been formally proven to be sound within HOL as the logic implication of the formal specification of microSR semantics. As in Gordon's work on sequential languages, no additional axioms are introduced into the logic precluding any possible logic inconsistency. The term GI in the logic is the global invariant property of channels in the message pool. The global invariant for each op is a predicate involving the "sent-set" σ and "received-set" ρ denoting all messages ever sent and received on that channel. μ is simply a message constructor function for converting an entity of type integer into one of type message.

- Global Part
  - Co Rule

$$\frac{\{GI \wedge Pi\}\; \text{SLi}\; \{GI \wedge Qi\}}{\{\{GI \wedge P\_list\}\}\; \text{co SL1} \; // \; \ldots \; // \; \text{SLn oc}\; \{\{GI \wedge Q\_list\}\}}$$

  - GI Introducing Rule

$$\frac{\{P\}\, \text{S}\, \{Q\}\, ,\, \{GI\}\text{S}\, \{GI\}}{\{GI \wedge P\}\, \text{S}\, \{GI \wedge Q\}}$$

- Intra-process Part
  - Skip Axiom

$$\{P\}\, \text{skip}\, \{P\}$$

  - Assignment Axiom

$$\{P_E^v\}\, \text{v} \; := \; \text{E}\, \{P\}$$

  - Send Axiom

$$\{P \wedge GI \wedge GI_{\sigma_{op} \cup \mu(E)}^{\sigma_{op}}\}\, \text{send op (E)}\, \{P \wedge GI\}$$

  - Receive Rule

$$\frac{P \wedge GI \wedge \mu(E) \in \sigma_{op} \Rightarrow Q_E^v \wedge GI_{\rho_{op} \cup \mu(E)}^{\rho_{op}}}{\{P \wedge GI\}\, \text{receive op(v)}\, \{Q \wedge GI\}}$$

  - In Rule

$$\frac{\{P\}\text{receive}\, \neg\text{p1(v)}\{R1\}\text{S1}\{Q\}, \{P\}\text{receive op2(v)}\{R2\}\text{S2}\{Q\}}{\{P\}\, \text{in op1(v) -> S1} \; \square \; \text{op2(v) -> S2 ni}\, \{Q\}}$$

  - If Rule

$$\frac{P \wedge \neg(B_1 \vee B_2) \Rightarrow Q\, ,\, \{P \wedge B_1\}\, \text{S1}\, \{Q\}\, ,\, \{P \wedge B_2\}\, \text{S2}\, \{Q\}}{\{P\}\, \text{if B1 -> S1} \; \square \; \text{B2 -> S2 fi}\, \{Q\}}$$

- Do Rule

$$\frac{\{LI \wedge B_1\} \text{ S1 } \{LI\} \,, \{LI \wedge B_2\} \text{ S2 } \{LI\}}{\{LI\} \text{ do B1 -> S1 } \square \text{ B2 -> S2 od } \{LI \wedge \neg(B_1 \vee B_2)\}}$$

- Sequencing Rule

$$\frac{\{P\} \text{ S1 } \{Q\} \,, \{Q\} \text{ S2 } \{R\}}{\{P\} \text{ S1;S2 } \{R\}}$$

- Precondition Strengthening Rule

$$\frac{P' \Rightarrow P \,, \{P\} \text{ S } \{Q\}}{\{P'\} \text{ S } \{Q\}}$$

- Postcondition Weakening Rule

$$\frac{\{P\} \text{ S } \{Q\} \,, Q \Rightarrow Q'}{\{P\} \text{ S } \{Q'\}}$$

One can see that this logic allows microSR programmers to reason and state formal assertions about both the sequential behaviors of each process, as well as the communication activity among them. Our preliminary attempts at manual proof of microSR programs indicate that common situations such as critical sections with mutual exclusion, resource management processes, and producer-consumer synchronization are all provable using this logic. Figure 3 is a sample microSR program proof outline that gives the flavor of how to prove microSR programs in the logic we derived.

4 Conclusions

We have presented our research on formally deriving, using HOL, a sound Hoare logic from the concurrent microSR semantics. This research will now serve as a basis of our further research on semantic formalization and programming logic derivation for the complete SR language. Following our incremental and iterative approach (see Figure 1), we expect that our final language will be close to its parent language in its expressive power for distributed computing. For instance, our later extended version will also support rendezvous in addition to message passing. Our research so far indicates that SR concurrency features (dynamic process creation, message-passing, remote procedure calls, and rendezvous) are all amenable to a Hoare-like programming logic because the components of our semantic model for microSR have already formalized most of entities and behaviors that SR programmers as well as other modern concurrent programmers must consider during their design process [2, 5].

Figure 4 gives an overview of our work on microSR program verification in our logic system. We will continue to examine the expressive power of our logic with more program proofs in a more rigorous manner. Our early results with manual proofs has motivated us to develop a verification methodology on how to use this logic to prove and deduce concurrent microSR programs. Our goal

$\{GI \wedge START\}$
`receive in1(m1)` $\{GI \wedge START' \wedge m_1 = \bar{\mu}_{in1}^{(p_1,1)}\}$
`i:=1` $\{GI \wedge START' \wedge m_1 = \bar{\mu}_{in1}^{(p_1,i)}\}$
`receive in2(m2)` $\{GI \wedge START'' \wedge m_1 = \bar{\mu}_{in1}^{(p_1,i)} \wedge m_2 = \bar{\mu}_{in2}^{(p_2,1)}\}$
`j:=1` $\{GI \wedge START'' \wedge m_1 = \bar{\mu}_{in1}^{(p_1,i)} \wedge m_2 = \bar{\mu}_{in2}^{(p_2,j)}\}$ $\{GI \wedge LI\}$
`do m1 <= m2 ->` $\{GI \wedge LI \wedge m_1 \leq m_2\}$ $\{GI \wedge GI_{\sigma_{out} \cup \mu(m_1)}^{\sigma_{out}} \wedge LI\}$
$\quad\quad\quad$ `send out(m1)` $\{GI \wedge LI\}$
$\quad\quad\quad$ `receive in1(m1)` $\{GI \wedge LI'\}$
$\quad\quad\quad$ `i++` $\{GI \wedge LI\}$
`[] m1 >= m2 ->` $\{GI \wedge LI \wedge m_2 \leq m_1\}$ $\{GI \wedge GI_{\sigma_{out} \cup \mu(m_2)}^{\sigma_{out}} \wedge LI\}$
$\quad\quad\quad$ `send out(m2)` $\{GI \wedge LI\}$
$\quad\quad\quad$ `receive in2(m2)` $\{GI \wedge LI''\}$
$\quad\quad\quad$ `j++` $\{GI \wedge LI\}$
`od` $\{GI\}$

$$START \equiv \sigma_{out} = \emptyset \wedge \rho_{in2} = \emptyset \wedge \rho_{in1} = \emptyset$$
$$START' \equiv \sigma_{out} = \emptyset \wedge \rho_{in2} = \emptyset$$
$$START'' \equiv \sigma_{out} = \emptyset$$
$$GI \equiv GI_{in1} \wedge GI_{in2} \wedge GI_{out}$$
$$GI_{in1} \equiv \forall i. (\exists \mu, \mu'. \mu = \mu_{in1}^{(p_1,i)} \wedge \mu' = \mu_{in1}^{(p_1,i+1)}) \Rightarrow \bar{\mu}_{in1}^{(p_1,i)} \leq \bar{\mu}_{in1}^{(p_1,i+1)}$$
$$GI_{in2} \equiv \forall i. (\exists \mu, \mu'. \mu = \mu_{in2}^{(p_2,i)} \wedge \mu' = \mu_{in2}^{(p_2,i+1)}) \Rightarrow \bar{\mu}_{in2}^{(p_2,i)} \leq \bar{\mu}_{in2}^{(p_2,i+1)}$$
$$GI_{out} \equiv \forall i. (\exists \mu, \mu'. \mu = \mu_{out}^{(p_3,i)} \wedge \mu' = \mu_{out}^{(p_3,i+1)}) \Rightarrow \bar{\mu}_{out}^{(p_3,i)} \leq \bar{\mu}_{out}^{(p_3,i+1)}$$
$$LI \equiv m_1 = \bar{\mu}_{in1}^{(p_1,i)} \wedge m_2 = \bar{\mu}_{in2}^{(p_2,j)} \wedge \forall \mu \in \sigma_{out}. m_1 \geq \bar{\mu} \wedge m_2 \geq \bar{\mu}$$
$$LI' \equiv m_1 = \bar{\mu}_{in1}^{(p_1,i+1)} \wedge m_2 = \bar{\mu}_{in2}^{(p_2,j)} \wedge \forall \mu \in \sigma_{out}. m_1 \geq \bar{\mu} \wedge m_2 \geq \bar{\mu}$$
$$LI'' \equiv m_1 = \bar{\mu}_{in1}^{(p_1,i)} \wedge m_2 = \bar{\mu}_{in2}^{(p_2,j+1)} \wedge \forall \mu \in \sigma_{out}. m_1 \geq \bar{\mu} \wedge m_2 \geq \bar{\mu}$$

Fig. 3. Sample Proof Outline: Stream Merge Maintains Ordering Property

is to establish a systematic method for creating annotated microSR programs including assertions of global invariants on the shared message pool. Another challenging task is to develop in HOL an interactive backward microSR prover of LCF [10] style proof tactics based on our logic. We also intend to apply verification to the microSR applications and the microSR implementation in a logically consistent approach; i.e., the formal specification of microSR operational semantics used for deriving Hoare logic will also serve as the specification against which the microSR code generator/runtime support is verified, so as to guarantee that microSR concurrent applications that are correct in our Hoare logic will run correctly in our Silo system with semantics preserved. We expect our verification of concurrent microSR implementation will extend the related works by Joyce[9] and Young[11].

References

1. G.R. Andrews, R.A. Olsson, M. Coffin, I.J.P. Elshoff, K. Nilsen, T. Purdin, and G. Townsend.: An Overview of the SR Language and Implementation. ACM Transactions on Programming Languages and Systems 10, 1 (January 1988), 51-86.

42

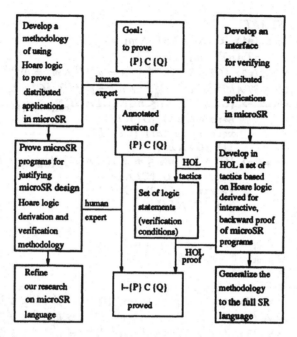

Fig. 4. Verification of microSR Programs

2. G.R. Andrews.: Concurrent Programming: Principles and Practice. The Benjamin/Cummings Publishing Company, Inc. Redwood City, CA, 1991.
3. G.R. Andrews and R.A. Olsson.: The SR Programming Language: Concurrency in Practice. Benjamin/Cummings Publishing Company, Inc. Redwood City, CA, 1993.
4. W.R. Bevier, W.A. Hunt, J.S. Moore, and W.D. Young.: An approach to systems verification. Journal of Automated Reasoning, 5:411–428, 1989.
5. M. Chandy and J. Misra.: Parallel Program Design: A Foundation of Programming Logic. Addison-Wesley Publishing Company, Inc. 1988.
6. M. J. C. Gordon.: Mechanizing Programming Logics in Higher Order Logic. In: Current Trends in Hardware Verification and Automated Theorem Proving. Springer-Verlag, New York, 1989.
7. W. Harrison, K. Levitt, and M. Archer.: A HOL Mechanization of the Axiomatic Semantics of a Simple Distributed Programming Language. In: Higher Order Logic Theorem Proving and Its Applications. North-Holland, Netherlands, 1993.
8. C.A.R. Hoare.: Communicating Sequential Processes. Prentice-Hall, Englewood Cliffs, N.J., 1985.
9. J.J. Joyce.: A Verified Compiler for a Verified Microprocessor. Technical Report No.167, Computer Laboratory, University of Cambridge, March 1989.
10. L. C. Paulson.: Logic and Computation: Interactive Proof with Cambridge LCF. Cambridge ; New York : Cambridge University Press, 1987.
11. W. D. Young.: A Mechanically Verified Code Generator. Journal of Automated Reasoning, Vol. 5: 493–518, 1989.

Reasoning with the Formal Definition of Standard ML in HOL

Donald Syme[12]

[1] Department of Computer Science, The Australian National University, GPO Box 4 Canberra 2601, Australia
[2] Division of Information Technology, CSIRO, GPO Box 664 Canberra 2601, Australia

Abstract

This paper describes the results of a project to embed the Formal Definition of the Standard ML Core language into the HOL mechanized reasoning system. HOL theories of the Core syntax and dynamic semantics are presented, including a purely definitional representation of the semantic inference rules. The correctness of the representation is demonstrated by the derivation of several important language properties, including determinacy. A symbolic evaluator to assist in proving properties of Standard ML program fragments is also described. Some sample applications of the theory in the area of program verification are given.

1 Introduction

The Formal Definition of Standard ML (called "The Definition" hereafter) was completed by Milner, Tofte and Harper in 1990 [11]. Its aim was to give a rigorous and complete mathematical description of all aspects of the Standard ML language, from its syntax to the semantics of interactive operation. The Definition describes the language on three levels - the Core language, the Module system and Programs. In turn each of these sections is described on three levels: *syntax*, *static semantics* and *dynamic semantics*. The main concern of this paper is with the dynamic semantics of the Core language, that is to say, with the runtime semantics of the basic constructs of any Standard ML program. Little substantial reasoning can be performed on the dynamic aspects of a Standard ML program without a useful representation of these semantics.

The semantics of the Core language are described in the Formal Definition using Natural Semantics [11, 9, 8]. The basic constructs in the semantics are sentences of the form $A \vdash phrase \Rightarrow R$. A sentence $A \vdash phrase \Rightarrow R$ denotes that in some environment A the execution of *phrase* will give the result R. A set of *inference rules* define which sentences are inferable. The inference rules are the heart of the semantics, and take the form

$$\frac{A_1 \vdash phrase_1 \Rightarrow R_1 \cdots A_n \vdash phrase_n \Rightarrow R_n}{A \vdash phrase \Rightarrow R}$$

There are 57 inference rules in the Core dynamic semantics, each of which deals with a particular case of evaluation.

In the case of the Core language, the background and result objects (A and R) are composed of *semantic types* such as environments E, states s, values *val*, closures C, exception packets p and records *rec*. Syntactic fragments of Standard ML code are represented by *syntactic types* such as expressions *exp* and patterns *pat*. The set of phrases is the union of all the syntactic types.

The other major aspect of the dynamic semantics to be modeled in this work is the semantics of the fundamental values provided by Standard ML. These include arithmetic and comparison operators such as +, − and =. The semantics of the standard I/O facilities and of real numbers are omitted.

Section 2 of this paper deals with the construction of HOL representations for the syntactic and semantic types used in the Definition. Section 3 deals with the construction of a representation for the inference rules and other semantic notions. In Section 4 a series of useful properties are proven in HOL based on the theories developed in Sections 2 and 3. Section 5 describes the construction of a symbolic evaluator for the Core language. Section 6 describes a mechanism for reasoning cleanly about concretely-typed Standard ML values within HOL based theorem proving. Lastly, in Section 7 the theory and tools constructed are applied to several areas. Correctness theorems for two pieces of Standard ML code are stated and verified in HOL, some computation properties such as termination and no-state-change are defined and used, and the equivalence of two small program fragments is demonstrated.

This paper is a summary of subthesis completed by Donald Syme in partial fulfillment of the requirements of an Honours degree at the Australian National University in 1992. Some details have been omitted, and a full description of the work can be found in [13]. That subthesis was based in part on work performed by Matthew Hutchins for his Honours project in 1990, also at the Australian National University [5].

2 Embedding the Types in HOL

The first task to be faced in embedding the semantics of the Core into HOL is to construct a representation of the syntactic and semantic types that are used in the Core semantics.

2.1 Representing the Syntactic Types

A small subset of the Standard ML syntax is shown below.

$AtExp$	= AtExp_Const *Const*
		AtExp_Paren *Exp*
Exp	= Exp_AtExp *AtExp*
		Exp_Appl *AtExp Exp*

Constructing the HOL representations of these types is conceptually quite easy, as concrete syntax trees are well understood and to some extent supported in

HOL. The Standard ML syntax, however, poses one substantial problem - it is a mutually recursive syntax, and no automated tools are built into HOL for the definition of recursive grammars. A package written in HOL90 by Claudio Russo of Edinburgh [12] was ported to HOL88 and used to define both the types themselves and various mutually recursive functions over the types. Making the definitions introduces 14 new HOL types, including *Exp*, *At_Exp*, *Pat*, *Const* and *Phrase*.

2.2 Representing the Semantic Types

The semantic types pose more substantial problems for the type definition mechanisms of the HOL system. Not only are the semantic types mutually recursive, but they are also non-concrete, as they contain *finite maps*. In the Core semantics, finite maps are used to represent records, memory and environments. Finite maps are not directly supported in HOL, and so a *finite_map* theory was created. This theory is in the spirit of the *finite_set* theory already present in the HOL system. Finite maps are also known as finite partial functions or lookup tables. The theory constructed for finite maps contains operators such as DOMAIN, RANGE, APPLY, EXT (for extension of the map) and TRANSFORM. A substantial number of useful theorems were proven about these operators, including useful results allowing induction over the domain of the map. For example, the following theorems demonstrate basic results about applying the map:

$$\vdash \text{APPLY } z \text{ ZIP} = \text{FAILURE}$$
$$\vdash \text{APPLY } z \text{ (EXT } (z, v) \text{ } f) = \text{RESULT } v$$
$$\vdash \text{APPLY } z \text{ (EXT } (x, v) \text{ } f) = ((x = z) \Rightarrow \text{RESULT } v \mid \text{APPLY } x \text{ } f)$$

Finite maps are essentially a HOL construct used to avoid the syntactic and semantic obfuscation involved with using functions or lists to represent these objects. They do not solve the fundamental problem involved with the non-concrete mutual recursion present within the semantic types. A solution to this problem involves type theory beyond the scope of this project, and in the end a simple axiomatization of the properties of the types was used. This axiomatization was based on examples derived by other HOL users [7]. Significant steps were made toward justifying this axiomatization via a fully definitional approach.

For both the syntactic and semantic types a full pretty printer and parser were constructed, both of which were invaluable in later work. The pretty-printer was implemented by modifying the appropriate HOL88 Lisp code to gain maximum output efficiency.

The following illustrates the use of some semantic type constructors to make a *Val* which is a *Closure* consisting of a match and two environments.

```
# let tm = "Val_Closure (Closure m e v)";;
tm = "[m,e,v]" : term
    %< pretty-printed version of a closure >%

# let tm2 = dest_Val_Closure tm;;
tm2 = "Closure m e v" : term
```

```
# let tm3 = dest_Closure tm2;;
tm3 = ("m", "e", "v") : (term # term # term)
```

3   Embedding the Semantics in HOL

3.1   Incorrect Approaches

The second stage in embedding the Core language definition into HOL involves constructing appropriate representations of the semantic inference rules and some associated constructs.

A first attempt at creating a theory capturing these constructs was made by Matthew Hutchins at the Australian National University in 1990 [5]. He used an axiomatic approach to model the logical properties of the inference rules of the semantics. Initially each inference rule was asserted as an implicative axiom. Thus a simple rule, such as that for parenthesized expressions:

$$\frac{s, E \vdash exp \Rightarrow v, s'}{s, E \vdash (exp) \Rightarrow v, s'}$$

was asserted as the axiom

$\vdash$ INFERABLE$(s, E, atexp, val, s') \supset$ INFERABLE$(s, E, (atexp), val, s')$

Here INFERABLE stands for the inferability of a particular sentence. A similar approach has been used in other verification environments based on the Formal Definition of Standard ML [2]. However, it soon becomes obvious within such schemes that important properties cannot be proven. The axioms do not fully capture the fact that a sentence is inferable *if and only if* it can be inferred via one of the inference rules. Within such an axiomatization scheme there is no way to prove that a sentence is *not* inferable, and thus there is no guarantee of determinacy or any other such reasonable language property.

After these problems were identified, Hutchins used another set of axioms which he thought corrected this problem. However, in this new scheme the same essential problem became manifest with infinite computations - there was no way of proving whether a sentence involving an infinite computation was or was not inferable. This effectively equates to the lack of a principle of computational induction. The incompleteness of the axiomatization was by no means easy to spot, and points to the inherent dangers in using axiom based approaches.

3.2   A Correct Representation

Full tools are now becoming widely used for inductively defining relationships in HOL, and it has been demonstrated that it is possible to use this method to define relationships derived from a Natural Semantics definition [6, 10]. However, in this project a different, definitional approach was used to construct a correct representation of the semantics, based on the notion of *inference trees*. As noted in the Commentary on the Formal Definition of Standard ML, successive applications of the inference rules may be thought of as constructing an *inference tree* for a sentence [9]. The children of each node in an inference tree are the sub-evaluations that contribute to the evaluation of the phrase at the node. Thus an inference tree is a tree labeled with sentences where each sentence is directly deducible via inference rules from its children. A small inference tree is shown in Figure 1 for the declaration "**val x = y**" where **y** has value 1. The conclusion of the inference tree states that **x** gets mapped to the value 1 as a result of evaluating the declaration.

Inference trees can be used to define inferability in the following way:

Definition 1. A sentence is inferable if and only if there exists a valid finite inference tree whose root node is that sentence.

Validity of an inference tree is defined as follows:

Definition 2. Given an inference tree T with root node

$$s, obj \vdash phrase \Rightarrow res, s'$$

and children $T_1, \ldots, T_n$, then T is valid if and only if its node can be deduced via an inference rule from its children, and each if its children are also valid.

Thus, an inference tree is valid if and only if it represents a valid computation according to the inference rules.

Inference trees are modeled in HOL using the *ltree* theory and a set of definitions which capture the logical content of each individual inference rule. As an example, the predicate which tests for validity at an node containing a single variable is:

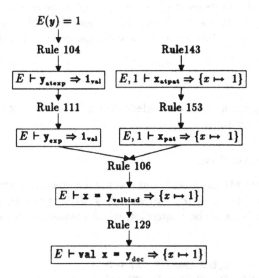

Fig. 1. Inference Tree for "**val x = y**" where $E(\mathbf{y}) = 1$

$\vdash_{def}$ VALID_INFR_Var *children* (s, E, var, val, s') =
   $(children = [])$ ∧ $(s' = s)$ ∧ $(E(var) = $ RESULT $val)$

A similar predicate is defined for each type of phrase in the syntax, and these are collected into the predicate VALID_NODE. The definitions of validity and inferability can then be made:

$\vdash_{def}$ VALID_TREE (Node sent *ch*) =
   VALID_NODE (MAP ROOT *ch*) sent ∧ EVERY VALID_TREE *ch*

$\vdash_{def}$ INFERABLE *sent* =
   ∃*tree*. VALID_TREE *tree* ∧ (ROOT *tree* = *sent*)

There is an obvious one-to-one correspondence between sequences of inferences based on the inference rules and inference trees. Thus the definition of inferability given above precisely captures the entire set of inferable sentences specified by the Formal Definition. All the reasoning about the Core language in the following sections is based on the definition of inferability given above. In those sections a large number of more convenient results about inferability will be derived. Ultimately these eliminate the need to refer directly to inference trees at all.

3.3 Eliminating Free Choice

The dynamic semantics of the Core given in the Formal Definition have the unfortunate property that they are inherently non-deterministic. This is due to the fact that at certain times a free choice is allowed of names in a name-space. The semantics make no demand on the particular name chosen, except that it may not currently be in use. To model this notion, the Hilbert-choice operator was used to ensure determinism and yet still capture the idea that no specific choice is enforced.

  $\vdash$ CHOOSE_Name *used'names* = $\varepsilon n.$ ¬(n IN *used'names*)

The essential property is that no knowledge is available of the exact choice made, and this is still maintained.

3.4 The Dynamic Basis

The basic Standard ML arithmetic and comparison operators are defined in the HOL theory by reference to the corresponding HOL operators and types. For example, the semantics of the + operator, which operates on a paired value, are defined by:

$\vdash_{def}$ plus val =
    let $n = $ MLNUM_TO_NUM (MLPAIR_FST *val*)
    and $m = $ MLNUM_TO_NUM (MLPAIR_SND *val*) in
     OUT_OF_RANGE($n + m$) => Packet 'Sum' | NUM_TO_MLNUM ($n + m$)

With the definitions of the basic Standard ML operations in place, the definition of the **APPLY** function specified in Appendix B of the Formal Definition can be made:

$\vdash_{def}$ APPLY BasVal_plus val = plus val $\land$
APPLY BasVal_minus val = minus val $\land$

$\cdots$

APPLY BasVal_geq val = geq val

4 Proving Language Properties

In one sense a complete HOL theory of Standard ML has now been constructed, for *any* fact which may be logically deduced from the Core dynamic semantics can now be proven in HOL. The semantics are essentially captured in terms of the INFERABLE predicate.

It is useful at this time to draw a parallel with reasoning about the natural numbers. The HOL type *num* is defined in terms of 0 and SUC. However, it is essential that this primitive theory be enriched with theorems and a whole range of derived results, some of which will be interesting in themselves and others of which are merely building blocks for other results. This is effectively the aim of the following section - to expand the present theory of Standard ML with theorems, rules, tactics and definitions.

4.1 Backward Rewrite Rules

The first set of results to be proven are those which describe exactly under what conditions a sentence may be inferred, in terms of its sub-inferences. These results are crucial in eliminating the need to constantly refer back to inference trees in our reasoning.

The following HOL goal captures the notion that the sentence $s, E \vdash var \Rightarrow val, s'$, which represents the evaluation of a variable, may be inferred if and only if the variable itself evaluates to the value *val* in the environment E, and if $s = s'$:

"INFERABLE($s, E,$ AtExp_Var var, val, s') = ($s' = s$) $\land$ ($E(var)$ = RESULT val)"

This result is proved quite easily by expanding the definition of INFERABLE. Note what the result is saying - that a sentence with a variable as its phrase can be inferred if and only if the conditions on the right hold.

There are 43 such rules, one for each type of phrase in the Standard ML syntax, with most of them being far more complex than that above. The proofs of all of these are streamlined through one tactic PROVE_BWDRULE_TAC. Once the results are proven they form the basis for a range of tactics and rules such as BWDRULE_REWRITE_TAC, which automatically determines the correct rule to apply to rewrite individual subterms in a goal.

4.2 Forward Inference Rules

The natural way to think about an inference rule is as an implicative theorem, where the premises of the rule imply the conclusion. This is how the inference rules were originally modeled as axioms in the previous section. There are 86 forward inference rules in total, corresponding to the inference rules in the Formal Definition. For example, one of the rules for parenthesized expressions is:

$$\vdash \text{INFERABLE}(s, E, atexp, val, s') \supset \text{INFERABLE}(s, E, (\ atexp\ ), val, s')$$

The proofs of these results follow fairly easily from the backward rewrite rules. The forward inference rules are crucial in allowing the inferability of sentences to be proven, and they form the basis of the symbolic evaluator described in the following section.

4.3 The Pattern Matching Theorem

We now proceed to outline two properties of the Standard ML language, both of which are described in the Commentary on the Formal Definition [9]. These are the Pattern Matching theorem and the Determinacy theorem. Both of these properties have been proven in HOL using the theories constructed in Section 2 and 3.

As with many functional languages, pattern matching lies at the heart of function application in Standard ML. Pattern matching should have no side-effects, as no expressions are evaluated. The Commentary on Standard ML states the property in the following way:

Theorem 3 Pattern Matching. *Let E, v and pat be any environment, value and pattern. Suppose that*

$$s, E, v \vdash pat \Rightarrow r, s'$$

can be inferred for states s, s' and result r. Then s = s'.

That is to say, no state change occurs due to the evaluation. Stating this as a HOL goal:

$$\text{"INFERABLE}(s, E, pat, res, s') \supset (s = s')\text{"}$$

This goal was proven by structural induction over the syntax of patterns. This induction employs the mutual recursion theorems from the Standard ML syntax. Each step of the induction holds because of the nature of the kind of inferences that can be made on that particular type of pattern.

4.4 The Determinacy Theorem

The second of the two theorems is perhaps the single most useful result proved in this project. It is the Determinacy Theorem, which states that any two computations over the same phrase starting in the same background will produce the same result. After due consideration, it becomes obvious from the Formal Definition that the evaluation process is in fact deterministic, apart from the free choice mentioned in Section 3. The theorem can be stated as follows:

Theorem 4 Determinacy. *Let the two sentences*

$$cs, A \vdash phrase \Rightarrow R, s'$$
$$s, A \vdash phrase \Rightarrow R', s''$$

both be inferred. Then $R = R'$ and $s' = s''$.

The proof of this theorem uses computational induction on the length of the inference process. The goal for the Determinacy Theorem is quite simple to state in HOL:

```
"INFERABLE(s,E,ph,res,s') ∧ INFERABLE(s,E,ph,res',s'') ⊃
   (res = res') ∧ (s' = s'')"
```

The goal asserts the equality of the results and final states of the two inferable sentences.

Any proof of this theorem must use some form of *computational* induction. In the theory the finiteness of computation is captured by the finite size of the inference trees on which the definition of inferability is based. Computational induction is thus equivalent to an induction on the depth of the inference trees involved.

The proof of the theorem in HOL is very lengthy and can only be performed by automating each of the many sub-cases of the induction. It is an excellent example of a theorem that can only be formally proven by machine.

4.5 The Significance of Determinacy

The Determinacy Theorem ensures that each evaluation has *at most* one result. This result may not be strengthened to *exactly* one result, for two reasons. Firstly, there are infinite computations for which no result can be inferred. Secondly, many computations do not make sense under the typing rules of the language, and these in general have no valid inferences.

Determinism thus justifies the following definition of evaluation as a partial function:

$$\vdash_{def} \text{EVAL } (s,E) \ ph =$$
$$(\exists res \ s'. \ \text{INFERABLE}(s,E,ph,res,s') \Rightarrow$$
$$\text{RESULT } (\varepsilon(res,s'). \ \text{INFERABLE}(s,E,ph,res,s')) \ | \ \text{FAILURE}$$

The following theorem, which captures the intuitive relationship between evaluation and inferability, can then be proven:

$$\vdash \texttt{INFERABLE}(s, E, phrase, res, s') = (\texttt{EVAL}\ (s, E)\ phrase = \texttt{RESULT}\ (res, s'))$$

5 The Symbolic Evaluator

5.1 Basic Symbolic Evaluation

Evaluation of programming languages is, of course, a well understood process. As well demonstrated by Boyer and Moore [1], evaluation is a natural technique for reasoning about the properties of functional programs. The usefulness of evaluation is increased when it is conducted *symbolically*. In this way an evaluator can be used to prove properties of infinite families of computations. As an example, the expression "let val x = n in 1 end" will always evaluates to the value 1 regardless of n. By symbolic evaluation the inferability of the sentence $s, E \vdash$ "let val x = n in 1 end" $\Rightarrow 1, s$ can be proven.

A fully symbolic evaluator based on the Core semantics developed in the previous sections was implemented in HOL as part of this project. As the evaluator proceeds through the execution tree, theorems about the evaluation are proven via the forward inference rules derived in the previous section. An example of the invocation of the symbolic evaluator to prove the inferability of the sentence shown above is demonstrated below. Here P is the program "let val x = n in 1 end".

```
# SML_EVAL (s, E) P;
```

$$\vdash \texttt{INFERABLE}(s,\ E,\ P,\ 1,\ s)$$
$$\vdash \texttt{EVAL}(s,\ E)\ P = 1,\ s$$

5.2 Multiple Evaluation Paths

Often the symbolic evaluator will come to an evaluation which it cannot compute. One technique for handling this is to split the computation into two or more paths, each of which makes a different assumption about the symbols involved. For instance, when evaluating the match $M = $ true => false | false => true, against an arbitrary value v, the evaluator produces three different theorems:

$$(v = \text{true}) \qquad\qquad \vdash \texttt{EVAL}\ (s,\ E,\ v)\ M = \text{false},\ s$$
$$\neg(v = \text{true}) \wedge (v = \text{false}) \vdash \texttt{EVAL}\ (s,\ E,\ v)\ M = \text{true},\ s$$
$$\neg(v = \text{true}) \wedge \neg(v = \text{false}) \vdash \texttt{EVAL}\ (s,\ E,\ v)\ M = [\text{Match}],\ s$$

Because Standard ML is strongly typed the last theorem can, eventually, be ignored. However, as we shall see in the next section, this involves specifying some information about the value v.

5.3 Other Features of the Evaluator

The symbolic evaluator is capable of using extra information to aid it in the evaluation process. In particular, it accepts a list of theorems of the form

$$\vdash \forall v1..vn. \text{ EVAL } (s, E) \text{ phrase } = res, s'$$

which provide the evaluator with particular evaluation results. Before the evaluator attempts the evaluation of any phrase it tries to instantiate one of these theorems to give an answer for the evaluation being considered. One main use of this facility in section 7 is to provide induction assumptions to the evaluator.

The evaluator is designed to allow extensions to be made whenever new and more complex evaluation patterns need to be added. For instance, a module was added to allow variables to be evaluated in partially undefined environments . This module was based on results proven about *finite_maps*. Another module was added to allow the evaluation of phrases involving the Core base values such as + and =.

The evaluator is also capable of reasoning about exception-producing expressions, and if the result of a computation is unknown it will split into two paths according to whether it assumes an exception has occurred or not.

5.4 An Evaluation Tactic

It is very useful to define an HOL tactic EVAL_TAC which invokes the evaluator in an attempt to solve a particular goal. The goal is expected to be of the form

```
"EVAL (s, E) phrase = res, s'"
```

EVAL_TAC supplies the arguments s, E and *phrase* to the evaluator, and if enough information is available the evaluator will return the evaluation theorem which can then be used to help prove the goal.

6 Reasoning with Standard ML Types

It is now almost possible to begin applying the theories and tools developed in the previous chapter to more substantial problems. However, any attempts to do so run into immediate difficulties due to an inability to talk about typed values in any systematic manner. For the Core language, it is not not necessary to know any type information to evaluate a Standard ML program. However, the information gained from knowing that Standard ML is strongly typed is crucial in proving general theorems about Standard ML programs.

6.1 Concrete Types as Subsets of *Val*

The type *Val* which is defined in the semantics encompasses all possible values
that may be assigned to variables within Standard ML programs, as well as
other, non-realizeable values. Each Standard ML type is associated with a subset
of *Val* which realizes the type. In general, Standard ML *function* types such as
int -> int are difficult to characterize in the context of the dynamic semantics.
It would be helpful to be able to relate such a type to its subset of *Val*. However,
any particular function type has an infinite number of corresponding possible
closure-values to which it is semantically related. Furthermore, it is not possible
to easily characterize this set, as this would involve embedding the entire typing
system into HOL.

However, Standard ML concrete types such as bool and 'a list do correspond to easily identified subsets of *Val*. It is with these types that we shall be
concerned. A table of some simple types and their corresponding *Val*-subsets is
shown in Table 1. [3] It is necessary to deduce the properties of these subsets of
Val to begin to reason about program-fragments using these types.

SML Type	Corresponding *Val*s
int	"Val_Con n"
'a * 'b	"Val_Record $\{1 \mapsto L, 2 \mapsto R\}$"
bool	"Val_Con 'true'"
	"Val_Con 'false'"
'a list	"Val_Con 'nil'"
	"Val_Con_Val 'cons' (Val_Record $\{1 \mapsto L, 2 \mapsto R\}$)"

Table 1. Some Standard ML types and their *Val*-subsets

6.2 Using an Isomorphic HOL Type

The main technique used in this work to define the properties of a *Val*-subset
is via reference to an isomorphic HOL type. For instance, the *Val*-subset for
Standard ML lists is isomorphic to the HOL type :list(Val). Similarly, the
values for booleans are isomorphic to *bool*.

The *Val*-subset is defined by a predicate over the type *Val*. To relate this
subset and its HOL type, bijections need to be defined between the two. An
example of this for numbers is shown in Figure 2. The properties of the HOL
type may then be "lifted" via the bijections to equivalent results over the *Val*-
subset.

---

[3] For polymorphic types such as 'a list it is generally sufficient to use the *Val* subset
containing all well-formed lists, as well as badly typed lists such as 1, 'a'. An
application of this can be seen in the next section, where the correctness of the
append function is proven for all list values, even some that cannot arise in actual
programs.

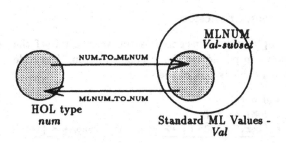

Fig. 2. Relating a Standard ML Type to an Isomorphic HOL type

6.3 Automatic Definition and Theorem Lifting

By using the bijections it is quite easy to define operations over the *Val*-subset by reference to the corresponding HOL definition. For instance, addition over *Val* is defined by:

$$\vdash_{def} m \text{ MLNUM\_ADD } m' = \text{NUM\_TO\_MLNUM(MLNUM\_TO\_NUM } m + \text{MLNUM\_TO\_NUM } m')$$

It is also possible to implement a proof technique which will automatically "lift" the properties of the HOL type to the *Val*-subset. This can automate the lifting of both definitions and theorems as shown below.

```
# lift_definition "n < m";;
```
$$\vdash_{def} n \text{ MLNUM\_LT } m = (\text{MLNUM\_TO\_NUM } n) < (\text{MLNUM\_TO\_NUM } m)$$

```
# LESS_TRANS;;
```
$$\text{LESS\_TRANS} = \vdash n < n' \wedge n' < n'' \supset n < n''$$

```
# lift_theorem LESS_TRANS;;
```
$$\vdash \text{IS\_MLNUM } p \wedge \text{IS\_MLNUM } q \wedge \text{IS\_MLNUM } r \supset$$
$$(p \text{ MLNUM\_LT } q \wedge q \text{ MLNUM\_LT } r \supset p \text{ MLNUM\_LT } r)$$

This automated-lifting technique was a late innovation which could prove quite useful in other areas.

7 Verifying Standard ML Programs and Beyond

We are now in a position to apply the HOL theory of the Core language that have been developed to prove some illustrative results about Standard ML.

7.1 The Append Function

The first example to be considered is the verification of the correctness of a simple function - in particular the **append** function. The code for this function is shown below in the syntax of the Core.

```
val rec append = fn
    {1=nil, 2=M} => M |
    {1=cons {1=x,2=L},2= M} => cons {1=x,2=append {1=L,2=M}};
```

It is most convenient to state the correctness of a function in terms of its match, M_{append}, shown underlined above. The **append** function is correct if and only if:

- The application of M_{append} in **append**'s local environment E_{append} to (v_1, v_2) evaluates to v_1 appended to v_2 for all list-values v_1 and v_2.
- There is no change of state due to the evaluation.

In HOL this goal is stated as:

```
"∀s v₁ v₂.
    IS_MLLIST v₁ ∧ IS_MLLIST v₂ ⊃
    EVAL (s, E_append, MLPAIR v₁ v₂) M_append = MLLIST_APPEND v₁ v₂, s"
```

Here MLLIST_APPEND is the HOL logical definition of appending two values, lifted by the techniques described in the previous section from the definition for APPEND.

7.2 The Proof in HOL

The proof in HOL follows by induction over v_1. Because we have assumed that v_1 is an **MLLIST**, this is very similar to an ordinary list induction.

The base case is the following goal:

```
"EVAL (s, E_append, MLPAIR MLLIST_NIL v₂) M_append
    = MLLIST_APPEND MLLIST_NIL v₂, s"
```

By calling **SML_EVAL_TAC** the goal can now be solved by symbolic evaluation. The evaluator proves that the call to **append** returns v_2 without a change in state, and the result follows easily:

```
# SML_EVAL_TAC THEN REWRITE_TAC [MLLIST_APPEND_NIL];;
goal proved...
```

The step case also follows by using symbolic evaluation and a simple theorem about MLLIST_APPEND. The induction hypothesis is employed to determine the result of the recursive call to **append** within the **append** function itself. With the proof of correctness in the step case the entire proof is completed.

Perhaps the most remarkable thing about the proof is that it makes almost no reference to the details of the semantics of the language. The symbolic evaluator performs all the reasoning steps about the language constructs that are needed.

The evaluator has applied approximately 100 inference rules and many more minor results. The final form of the automated proof took about an hour to compose and a few minutes to execute, a proof by hand would take much longer.

7.3 Another Verification Result

One more verification proof that has been performed using the HOL theory of Standard ML will now be outlined. The function to be verified is a small interpreter over the language given by the **datatype** declarations below.

```
datatype IExp = If   of (BExp * IExp * IExp)
              | Num  of int
      and BExp = Eq   of (IExp * IExp)
              | Bool of bool
```

The interpreter itself is shown in Figure 4. The statement of correctness can be given in a manner similar to that for the **append** function, using the small denotational semantics shown in Figure 3 to supply the required meaning functions. The interpreter was deliberately written in a slightly convoluted style to demonstrate that the code need not look identical to the denotational semantics for the proof to work.

$$\mathcal{M}_I(\underline{\text{If}}\ b\ e_1\ e_2) = \mathcal{M}_I(e_1) \text{ if } \mathcal{M}_B(b)$$
$$\mathcal{M}_I(e_2) \text{ if } \neg\mathcal{M}_B(b)$$
$$\mathcal{M}_I(\underline{\text{Num}}\ n) = n$$

$$\mathcal{M}_B(\underline{\text{Eq}}\ e_1\ e_2) = (\mathcal{M}_I(e_1) = \mathcal{M}_I(e_2))$$
$$\mathcal{M}_B(\underline{\text{Bool}}\ b) = b$$

Fig. 3. The Denotational Semantics for the Small Interpreter

The full details of the proof shall not be shown here. Essentially it proceeds by structural induction over the different possible programs that may be input to the interpretation function. For each step and base case of the induction, symbolic evaluation is used over the appropriate constructs to prove that the correct result is produced. The induction hypotheses are employed to deduce the value of the sub-evaluations in each call.

7.4 Computation Properties

We now move on from the topic of verification to briefly discuss how the semantics can be used to reason about the language in other ways. Termination can be defined by:

```
val choose = fn (true,  (i1,_)) => i1 |
                (false, (_,i2)) => i2;

val rec interp_iexp =  fn
  If (bexp, (iexp1,iexp2)) =>
    let path = choose (interp_bexp bexp,(iexp1,iexp2))
    in interp_iexp path  |
  Num n => n

and interp_bexp = fn
  Eq (iexp1,iexp2) => interp_iexp iexp1 = interp_iexp iexp2 |
  Bool b => b;
```

Fig. 4. The Interpreter

Definition 5 Termination. The evaluation of a phrase *ph* in a background consisting of *s* and *E* terminates if and only if there exists a result *res* and state *s′* such that $s, E \vdash ph \Rightarrow res, s'$ is inferable.

In HOL this is stated as:

$$\vdash_{def} \text{TERMINATES } (s,E) \ phrase = (\exists res \ s'. \ \text{INFERABLE}(s,E,phrase,res,s'))$$

It is similarly easy to state what it means for a computation to cause no state changes and no exceptions:

$$\vdash_{def} \text{NO\_SIDE\_EFFECTS } (s,E) \ phrase =$$
$$\forall res \ s'. \ \text{INFERABLE}(s,E,phrase,res,s') \supset \ (s = s'))$$

$$\vdash_{def} \text{NO\_EXCEPTIONS } (s,E) \ phrase =$$
$$\forall res \ s' \ p. \ \text{INFERABLE}(s,E,phrase,res,s') \supset \ \text{NOT\_PACK } res$$

Via symbolic evaluation and the correctness theorem it is quite easy to prove that the append function satisfies all these conditions. These results also hold for the interpreter that has been verified. Similarly, it is possible to prove that the code

```
let val tmp = !store in
  (store := 6; store := tmp)
end;
```

causes no side effects, despite its use of imperative features.

7.5 The Associativity of append

The last result to be presented in this paper is a proof of the associativity property of the **append** function. This result is not a verification proof, for it asserts the semantic equivalence of two program fragments:

append $(X,$ append$(Y, Z)) \equiv$ append (append$(X, Y), Z)$

X, Y and Z are arbitrary expressions, which may be assumed to evaluate to some values $xval$, $yval$ and $zval$. The evaluation of these expressions may also have side effects. Calling the two fragments P_1 and P_2, equivalence is stated by requiring that if P_1 evaluates to some value, then P_2 must evaluate to the same value, and vice-versa. The goal is stated in HOL as:

```
"IS_MLLIST xval ∧ IS_MLLIST yval ∧ IS_MLLIST zval ∧
 EVAL(s₀, E₀) append = Closureₐₚₚₑₙd, s₀ ∧
 EVAL(s₀, E₀) X = xval, s₁ ∧
 EVAL(s₁, E₀) Y = yval, s₂ ∧
 EVAL(s₂, E₀) Z = zval, s₃ ⊃
    ((EVAL (s₀, E₀) P₁ = res, sfinal) = (EVAL (s₀, E₀) P₂ = res, sfinal))"
```

The proof begins by applying the correctness result for the **append** function and an associativity result for HOL lists. This reduces the goal to the following subgoal:

```
"EVAL (s₀, E₀) P₁ = res, sfinal) = EVAL (s₀, E₀) P₂ = res, sfinal)"
  [ EVAL (s₀, E₀) P₁ = MLLIST_APPEND xval (MLLIST_APPEND yval zval), s₂ ]
  [ EVAL (s₀, E₀) P₂ = MLLIST_APPEND xval (MLLIST_APPEND yval zval), s₂ ]
```

The two assumptions give the results of the evaluations for the left and right hand sides of the goal. The result now follows by applying the Determinacy Theorem. Note that without the Determinacy Theorem the result could not be concluded.

8 Summary

The aim of this work was to investigate aspects of machine assisted reasoning about the programming language Standard ML. The method used was to construct a HOL theory of the Core dynamic semantics of the language, based on its Formal Definition. A series of important results have been presented: the construction of an embedding of the semantics into HOL; the development of a useful set of primitive rules from this basis; the proof of the Pattern Matching and Determinacy Theorems; the construction of a symbolic evaluator; the development of methods by which to reason with Standard ML concrete types in the context of the HOL theory; the proof of a substantial set of results including correctness proofs, termination proofs and the equivalence of program fragments. The work has demonstrated that the application of a tool such as HOL to such reasoning problems is quite feasible.

References

1. Robert S. Boyer and J Strother Moore. *A Computational Logic.* ACM Monograph Series. Academic Press, 1979.
2. A. Cant and M.A. Ozols. A verification environment for ML. In *Proceedings of the ACM SIGPLAN Workshop on ML and its Applications,* pages 151–155, June 1992.
3. Michael J. C. Gordon. Mechanizing programming languages in Higher Order Logic. *HOL System Documentation,* 1988.
4. Michael J. C. Gordon et al. *The HOL System Description,* December 1989.
5. Matthew Hutchins. *Machine Assisted Reasoning About Standard ML Using HOL.* Australian National University, 1990. Honours Thesis.
6. Tom Melham. A package for inductive relation definitions in HOL. *Proceedings of the 1991 HOL Workshop,* 1991.
7. Tom Melham. HOL-INFO mailing list archives, 1992.
8. Robin Milner. Lecture notes on language semantics. Computer Science 3, University of Edinburgh.
9. Robin Milner and Mads Tofte. *Commentary on Standard ML.* MIT Press, 1990.
10. Rachel Onate-Ruxos. Mutually recursive inductive relation definitions in HOL. *HOL 1993 Workshop,* 1993. Preprint of submitted paper.
11. Robin Milner Robert Harper and Mads Tofte. *The Definition Of Standard ML, Version 2.* MIT Press, 1990.
12. Claudio V. Russo. *Automating Mutually Recursive Type Definitions in HOL.* University Of Edinburgh, June 1992. Honours Thesis.
13. Donald Syme. *Reasoning with the Formal Definition of Standard ML in HOL.* Australian National University, 1992. Honours Thesis.

HOL-ML

Myra VanInwegen[*1] and Elsa Gunter[2]

[1] University of Pennsylvania Computer and Info Science Dept
Philadelphia PA 19104, USA
myra@saul.cis.upenn.edu
[2] AT&T Bell Laboratories, Rm.#2A-432
600 Mountain Ave., Murray Hill, N.J. 07974, USA
elsa@research.att.com

Abstract. We describe here HOL-ML, an encoding of a subset of SML and its dynamic semantics (as described by *The Definition of Standard ML*) in HOL. This encoding, which is the first stage in a project that will include typechecking and SML Modules, allows the formal study of the evaluation of a real programming language including state and control constructs (exceptions). In this paper we describe the subset of SML that we encoded and the semantic objects needed for its evaluation. We explain how we defined the evaluation rules and how we proved that evaluation is deterministic. We describe briefly the next step, which is to define a larger language that includes type declarations and to define the typechecking rules on it. Finally, we give a short description of the mutually recursive type definition package that we wrote to enable us to define the types we needed to create the HOL-ML grammar.

1 Introduction

This paper describes HOL-ML, our encoding of the dynamic Core of SML in the HOL interactive theorem prover. The purpose of this work is to act as a foundation for a system for formally specifying and reasoning about SML programs. The work described here deals only with the dynamic Core language and the evaluation of programs expressed in it. As such it is one piece of a larger project that will eventually include a system for reasoning about elaboration (*e.g. type-checking*) and evaluation of programs from both the Core and the module system. Even though it is intended that this work be a piece of a larger whole, it is already a system allowing us to carry out significant reasoning about SML. This work enables us to rigorously prove routinely assumed facts, such as that evaluation as described in [3] is deterministic. It also gives us a framework for rigorously proving that two expressions evaluate to the same value. Such an ability would be necessary, for example, to use the system for safely optimizing HOL described in [4]. Perhaps most significantly, this work will allow us to state and prove facts about programs that alter state, including local state. While we have only begun to carry out exercises such as those mentioned above, the

---

* Partially supported by AT&T

encoding of the formal semantics of the dynamic Core of SML and its evaluation relations is a necessary major first step.

In the remainder of this section, we describe the subset of SML we encoded. In Sect. 2 we describe in detail the HOL-ML grammar and semantic objects (the states, environments, and possible values for HOL-ML phrases). In Sect. 3 we describe how we defined the evaluation rules. In Sect. 4 we explain how we proved that evaluation is deterministic. In Sect. 5 we briefly describe how encoding the static Core language and defining typechecking will be handled in a manner analogous to the work described here, and how the static Core relates to the dynamic Core. In the appendix there is a brief description of the mutually recursive type definition package we created to enable us to define the types needed for the HOL-ML grammar.

We encoded a subset of Standard ML, as described in *The Definition of Standard ML* ([3]). We often refer to this book as the *Definition* and enclose references to it in brackets ([·]). Our starting point is the syntax of the Core [Chap. 2]. This work only deals with the dynamic Core of SML and the evaluation relations. As such we restrict our language to the reduced syntax described in [Sect. 6.1]. This reduction removes all type information, including declarations of new types, type tags in expressions and patterns, and indications of argument types for exception constructors. The dynamic Core also lacks information concerning fixity of operators: it is assumed that infixes have been resolved during the parsing phase. Thus we also eliminate rules pertaining to the declaration and use of infixes. One further reduction we made, which is not in keeping with the *Definition*, is that we pared down the variety of basic datatypes, allowing only integers and strings (not reals).

We do not enforce the syntactic restrictions of [Sect. 2.9], such as the restriction that a pattern may not contain the same variable more than once; however we assume in our definition of the evaluation rules that program phrases satisfy these restrictions. Similarly, we assume that the programs are type-correct, in the sense that they are the reduced-syntax versions of type-correct programs. In Sect. 5 of this paper we describe more clearly the relationship between the full syntax and the reduced one we use here when we examine the typechecking application.

The syntax of the Core is primitive, not including derived syntax such as **fun** declarations, boolean operators, and the **if** construct. As HOL-ML language phrases are created with constructors rather than parsed from some concise concrete syntax as is the case with SML compilers, HOL-ML programs are large and quite complicated to read. However, it is possible to write a parser and pretty-printer for HOL-ML so that programs could be entered using SML syntax.

In addition to encoding a description of the dynamic Core grammar, we also need an encoding of the semantic objects involved in the evaluation of the dynamic Core. These are discussed in the next section.

2 HOL-ML Grammar and Semantic Objects

The HOL-ML grammar and semantic object definitions are presented in Backus-Naur Form (BNF) in Figs. 1 and 2. Each of the language phrases and semantic objects represents an HOL type (the appendix explains how to transform these descriptions into HOL type definitions).

```
var    ::= VAR string
con    ::= CON string
scon   ::= SCINT integer | SCSTR string
excon  ::= EXCON string
label  ::= LABEL string
strid  ::= STRID string

'a nonemptylist  ::= ONE 'a | MORE 'a ('a nonemptylist)
'a long          ::= BASE 'a | QUALIFIED strid ('a long)

atexp   ::= SCONatexp scon | VARatexp (var long) | CONatexp (con long) |
            EXCONatexp (excon long) | LETatexp dec exp | PARatexp exp |
            RECORD1atexp | RECORD2atexp exprow
exprow  ::= EXPROW1 label exp | EXPROW2 label exp exprow
exp     ::= ATEXPexp atexp | APPexp exp atexp | HANDLEexp exp match |
            RAISEexp exp | FNexp match
match   ::= MATCH1 mrule | MATCH2 mrule match
mrule   ::= MRULE pat exp
dec     ::= VALdec valbind | EXCEPTdec exbind | LOCALdec dec dec |
            OPENdec ((strid long) nonemptylist) | SEQdec dec dec | EMPTYdec
valbind ::= PLAIN1valbind pat exp | PLAIN2valbind pat exp valbind |
            RECvalbind valbind
exbind  ::= EXBIND1 excon | EXBIND2 excon exbind |
            EXBIND3 excon (excon long) | EXBIND4 excon (excon long) exbind
atpat   ::= WILDCARDatpat | SCONatpat scon | VARatpat var |
            CONatpat (con long) | EXCONatpat (excon long) | RECORD1atpat |
            RECORD2atpat patrow | PARatpat pat
patrow  ::= DOTDOTDOT | PATROW1 label pat | PATROW2 label pat patrow
pat     ::= ATPATpat atpat | CONpat (con long) atpat |
            EXCONpat (excon long) atpat | LAYEREDpat var pat
```

Fig. 1. HOL-ML grammar

The grammar in Fig. 1 refers to the HOL types **integer** and **string**, both of which are defined in libraries that must be loaded prior to defining the grammar.

Definitions for the identifiers **var** (variables), **con** (value constructors), **scon** (special constants), **excon** (exception constructors), **label** (labels) and **strid** (structure identifiers) are presented before the actual grammar. The polymorphic types of **nonemptylist** and **long** are also given before the grammar. The type of **nonemptylist** is used for those constructs in the grammar that take a collection of one or more arguments of predefined type. The type **long** is used to construct various kinds of long identifiers. The HOL-ML grammar itself consists of the language phrases **atexp** (atomic expressions), **exprow** (expression rows, which form record expressions), **exp** (expressions), **match** (lists of alternatives that form function definitions), **mrule** (match rules), **dec** (declarations), **valbind** (value bindings), **exbind** (exception bindings), **atpat** (atomic patterns), **patrow** (pattern rows, which form patterns that match records), and **pat** (patterns).

The syntax of the Core language, as defined in [3], is also given by a BNF, but of a different form than is found in Fig. 1. The most immediate difference is the use of constructors for each case of each clause of the HOL-ML grammar, versus the use of either concrete syntax fragments or nothing in the cases of the clauses in the grammar of the *Definition*. The representation in Fig. 1 closely mirrors the mutually recursive type specification that defines the types corresponding to the language phrases. Another difference is that our grammar has been reduced to the dynamic Core language. Thus, for example, where the *Definition* [3] has the clause

$$exp ::= atexp$$
$$exp\ atexp$$
$$exp_1\ id\ exp_2$$
$$exp\ :\ ty$$
$$exp\ \textbf{handle}\ match$$
$$\textbf{raise}\ exp$$
$$\textbf{fn}\ match$$

for expressions, the HOL-ML grammar has

```
exp ::= ATEXPexp atexp | APPexp exp atexp | HANDLEexp exp match |
        RAISEexp exp | FNexp match
```

(The cases $exp_1\ id\ exp_2$ and $exp\ :\ ty$ are removed by the reduction to the dynamic Core language.)

Below we sometimes give the SML concrete syntax version of a language construct. This is for sake of exposition only; HOL-ML uses only the abstract syntax in Fig. 1.

There are more well-formed terms in HOL-ML than are possible results of parsing and elaboration of concrete SML syntax. An example is that the parsing rules of the *Definition* [Sect. 2.4] prevent a string from being used as a variable when it is in the scope of a constructor using the same string. In the declaration **val C = (5,C)**, parsing prevents the first C from being a variable while the second is a constructor. In HOL-ML this is not a problem, since the constructors (**VAR** and **CON**) distinguish them. In addition, it is obvious that we can construct terms in this grammar that cannot come from typechecked terms. As an eventual consequence, most theorems we will want to prove in later uses of HOL-ML will need to carry with them the hypothesis that the appropriate HOL-ML expressions are the translation of type-checked expressions from the full Core.

```
sval          ::= SVINT integer | SVSTR string
addr          ::= ADDR num
basval        ::= Size | Chr | Ord | Explode | Implode |
                  Abs | Div | Mod | Neg | Times | Plus | Minus |
                  Eql | Noteql | Less | Greater | Lesseql | Greatereql
exname        ::= EXNAME num
val           ::= ASSGval | SVALval sval | BASval basval | CONval con |
                  APPCONval con val | EXVALval exval | RECORDval record |
                  ADDRval addr | CLOSUREval closure
record        ::= NONErec | SOMErec label val record
exval         ::= NAMEexval exname | NAMEVALexval exname val
pack          ::= PACK exval
closure       ::= CLOSURE match env varenv
mem           ::= NONEmem | SOMEmem addr val mem
exnameset     ::= EXNAMESET (exname set)
state         ::= STATE mem exnameset
env           ::= ENV strenv varenv exconenv
strenv        ::= NONEstrenv | SOMEstrenv strid env strenv
varenv        ::= NONEvarenv | SOMEvarenv var val varenv
exconenv      ::= NONEexconenv | SOMEexconenv excon exname exconenv
val_pack      ::= VALvp val | PACKvp pack
record_pack   ::= RECORDrp record | PACKrp pack
val_pack_fail ::= VALvpf val | PACKvpf pack | FAILvpf
env_pack      ::= ENVep env | PACKep pack
varenv_pack   ::= VARENVvep varenv | PACKvep pack
exconenv_pack ::= EXCONENVeep exconenv | PACKeep pack
varenv_fail   ::= VARENVvef varenv | FAILvef
```

Fig. 2. HOL-ML evaluation semantic objects

The definition of the semantic objects pose a few problems not present for the grammar. The semantic objects form a mutual recursion, as does the grammar. However, the structure of this recursion is more complex than that of the grammar. Several of the semantic types are given as finite functions between others which are mutually recursive with those types. For example, in [Fig. 13], a record is described as a finite map from labels to values, while a value is described as a disjoint sum of records with other things. This kind of mutual recursion is well beyond the scope of any package for defining mutually recursive types that currently exists, and it seemed that trying to support this would take us too far afield. Therefore, wherever the *Definition* specifies the type of semantic objects as finite functions, we have used lists of pairs instead. In each case, it is possible to give a linear ordering on the elements in the domain of the desired finite functions, and we have used this ordering to define functions for inserting pairs into the lists so that only canonical lists will be generated during evaluation. If, at a later time, support is added to HOL for such complex mutual recursions as ones involving recursive occurrences within finite function type constructors, then our encoding can easily be re-encoded with finite functions replacing lists and function update replacing insertion.

The other problem with defining the types of semantic objects arises from the fact that several of them are left under-specified in the definition. Both addresses and exception names are only specified as infinite types. We could parameterize by these, but we chose to use a concrete instance instead; we chose to represent represent addresses (**addr**) and exception names (**exname**) as natural numbers wrapped with appropriate constructors.

We conclude this section with a quick summary of the rest of the semantic objects. A special value (**sval**) is an integer or a string. Basic values (**basval**) are the basic functions bound to predefined identifiers. Values (**val**) are evaluations of expressions. Records (**record**) are evaluations of expression rows. Packets (**pack**) are raised exceptions. Closures (**closure**) are evaluations of functions. Memory (**mem**) is a finite mapping from addresses to values, represented as a list. A state (**state**) consists of memory and the exception names used. An environment (**env**) is the evaluation of a declaration. A structure environment (**strenv**) is a finite mapping from structure identifiers to environments. A variable environment (**varenv**) is the evaluation of a pattern phrase type. An exception constructor environment (**exconenv**) is the evaluation of an exception binding.

The last several semantic objects are the results of evaluating language phrases. The name of the semantic object is composed of two or three other semantic objects (or **fail**, indicating that the value doesn't match the pattern) separated by underscores, such as **val_pack**. Each of these is just the disjoint sum of the components in its name.

3 Evaluation

To define evaluation, we must define the relations, one for each language phrase in the grammar, that determine the evaluation of an HOL-ML program. These relations are named **eval_phrase** for each language phrase *phrase*. Each relation is an HOL function taking a collection of terms to booleans. The arguments of the functions are the language phrase being evaluated and various semantic objects, including evaluation results, states, environments, and, for patterns, the value against which the pattern is being matched. The relation for a phrase type holds if, in the context given by the state, environment, and (for patterns) value, the phrase evaluates to the result, in the process yielding the new state.

In order to define our evaluation relations we take a somewhat roundabout route. These relations are mutually recursive functions, since the evaluation of a language phrase depends on the result of the evaluation of the subparts of the phrase. Unfortunately, we could not use **define_mutual_functions** (a part of our mutually recursive types definition package that is used to define mutually recursive functions over these types) to define the evaluation relation because evaluation is not primitive recursive. To see why, look at rules (117) and (118) for function evaluation in [3]: there we evaluate a subpart (the match) of the result of a previous evaluation (the closure), not a subphrase of the phrase we're evaluating.

Instead, we define predicates on potential evaluation relations (which are any collection of functions of the appropriate HOL types) that hold of those collections of relations that satisfy the evaluation rules in [Chap. 6].

3.1 Defining eval_pred

Examining the rules for evaluation in [Sect. 6.7] of the *Definition*, we note that there are actually three separate classes of inductively defined relations given there. There are the rules for evaluating exception bindings; the rules for evaluating atomic patterns, pattern rows, and patterns; and the rules for evaluating everything else, such as expressions and declarations. The latter class depends upon the previous two, but not the other way around. Therefore, we can break up our definitions of the rules into three separate smaller definitions of mutually inductive relations. The cases are all fairly similar and they could be handled in one large definition. However, there are several pragmatic reasons for breaking them up. If we decompose the rules into three separate classes, we will get three smaller principles of induction to be used later for proving properties about the evaluation relations. In particular, if we wish to prove a property of the evaluation of a pattern, we will not have to consider a collection of irrelevant cases involving expressions, declarations and so on. Another pragmatic reason for breaking up the rules into these classes is to shorten the computation time required to prove that the relations we are defining actually satisfy the rules.

Therefore, to define the evaluation relations, we define three separate predicates, eval_exbind_pred, eval_pat_pred and eval_pred, each of which takes as arguments potential evaluation relations and returns T (true) if the relations satisfy the evaluation rules. The biggest difference in how the definitions of these predicates are derived from the rules in [3] comes from the implicit rules concerning the generation of packets (raised exceptions) rather than values as results of evaluations of program phrases. The type information given in the *Definition* for the rules concerning the evaluation of patterns indicates that it is not possible for those evaluations to result in packets. Therefore, there are no implicit rules concerning the treatment of packets in those cases. To be clearer about all this, let us examine the most complicated predicate, eval_pred, in greater detail.

In order to give a general flavor of what eval_pred looks like, we will explain how one of the rules, Rule (107), is reflected in its conjuncts. Fig. 3 shows the general form of the term defining eval_pred and the three conjuncts corresponding to Rule (107).

Rule (107) is as follows:

$$\frac{\langle E \vdash exprow \Rightarrow r \rangle}{E \vdash \{\langle exprow \rangle\} \Rightarrow \{\}\langle + r \rangle \text{ in Val}} \tag{107}$$

This has two explicit cases, one for evaluating an empty record and one for evaluating a record with fields. To each of these two rules we are then required

```
eval_pred
    (eval_atexp:atexp->state->env->state->val_pack->bool)
    (eval_exprow:exprow->state->env->state->record_pack->bool)
    (eval_exp:exp->state->env->state->val_pack->bool)
    (eval_match:match->state->env->val->state->val_pack_fail->bool)
    (eval_mrule:mrule->state->env->val->state->val_pack_fail->bool)
    (eval_dec:dec->state->env->state->env_pack->bool)
    (eval_valbind:valbind->state->env->state->varenv_pack->bool) =
. . .
(* Rule 107a *)
    (!s E. eval_atexp RECORD1atexp s E s (VALvp (RECORDval NONErec))) /\
(* Rule 107b *)
    (!s1 E s2 exprow r.
    eval_exprow exprow s1 E s2 (RECORDrp r) ==>
        eval_atexp (RECORD2atexp exprow) s1 E s2
                    (VALvp (RECORDval (add_record NONErec r)))) /\
    (!s1 E s2 exprow p.
    eval_exprow exprow s1 E s2 (PACKrp p) ==>
        eval_atexp (RECORD2atexp exprow) s1 E s2 (PACKvp p)) /\
. . .
```

Fig. 3. Part of eval_pred

to apply what the *Definition* refers to as the *state convention* and the *exception convention*. Once we have done so, the preceding two rules become the following three rules:

$$s, E \vdash \{\} \Rightarrow \{\} \text{ in Val}, s$$

$$\frac{s_1, E \vdash exprow \Rightarrow r, s_2}{s_1, E \vdash \{exprow\} \Rightarrow \{\} + r \text{ in Val}, s_2}$$

$$\frac{s_1, E \vdash exprow \Rightarrow p, s_2}{s_1, E \vdash \{exprow\} \Rightarrow p, s_2}$$

where s, s_1, and s_2 are states and p is a packet. These three rules now correspond to the three conjuncts shown in Fig. 3.

During the course of encoding the rules of evaluation into HOL, we observed some errors and some peculiarities. The two significant problems, mentioned below, were known (see [1]) at the time, but not to us. A trivial typographical error, but one that was caught by the typechecker of HOL is a missing v on the lefthand side of the turnstile in the conclusion of Rule (126). Another problem arises with Rule (116), the rule for evaluating the application of a base constant to an expression. This rule as stated only deals with the case when the application of the corresponding base value to the expression's value results in a value; it does not handle the case when the application results in an exception being raised. Such a case is not handled by the exception convention and needs a rule of its own. We took the liberty of correcting these problems when encoding the rules.

A somewhat stickier problem arises in the case of evaluations involving the constructor **ref**. Note that the constructor **ref** is given special treatment in the pattern rules (154), (155), and (158), as it is in our translation of those rules. This also occurs in rules (112) and (114) for evaluating expressions. Because of this, one cannot rebind **ref** to be a constructor for another type, although the static semantics (the typechecking rules) allow it. Stephen Kahrs, in [1], suggests a way to resolve this conflict between the static and dynamic (evaluation) semantics. It requires retaining datatype definitions in the reduced syntax used for evaluation and keeping information on constructors in the variable environment. Although we became aware of this problem during our endeavors, on this point we chose to encode the rules as stated, despite their deficiency. We felt that this problem merited further study before determining what the best solution is.

We would like to take the opportunity to express our gratitude for the very considerable care and rigor that the authors put into the *Definition*. It is a truly remarkable piece of work, and without their attention to detail, our task would have been impossible.

3.2   The Evaluation Relations

Note that **eval_pred** only specifies when the potential evaluation relations must return T, so functions satisfying **eval_pred** may return T even when the rules do not justify it. Because of this we must define the evaluation relations to be the smallest relations satisfying **eval_pred**, that is, the intersection of all relations satisfying **eval_pred**. Thus, in the definition of each evaluation relation, we specify that a tuple is in the relation if and only if it is in every possible evaluation relation satisfying **eval_pred**. For.example, the term used to define **eval_exp** is given as follows:

```
eval_exp ex s1 e s2 vp =
!poss_eval_atexp poss_eval_exprow poss_eval_exp poss_eval_match
 poss_eval_mrule poss_eval_dec poss_eval_valbind.
eval_pred poss_eval_atexp poss_eval_exprow poss_eval_exp
 poss_eval_match poss_eval_mrule poss_eval_dec poss_eval_valbind
   ==> poss_eval_exp ex s1 e s2 vp
```

After defining the evaluation relations, we prove that the resulting relations do indeed satisfy **eval_pred**. That is, the evaluation relation satisfy all the rules of the *Definition*.

In addition to the evaluation relations, to actually have the whole story for evaluation, we also give the definitions of the functions defined in the initial dynamic basis (the dynamic basis provides environments needed for evaluation) described in [Appendix D]. One of these functions (**not**) is shown below.

3.3 Writing programs with HOL-ML

The declarations of even the simplest HOL-ML programs are long and hard to read. Take, for example, the definition of not. Its definition, using the subset of the SML grammar we've encoded is:

```
val not = fn true => false | false => true
```

This is quite concise compared to its equivalent in HOL-ML, shown in Fig. 4.

```
VALdec
(PLAINlvalbind
 (ATPATpat (VARatpat (VAR "not")))
 (FNexp (MATCH2 (MRULE (ATPATpat (CONatpat (CON "true")))
                       (ATEXPexp (CONatexp (CON "false"))))
               (MATCH1 (MRULE (ATPATpat (CONatpat (CON "false")))
                             (ATEXPexp (CONatexp (CON "true")))))))))
```

Fig. 4. HOL-ML definition of not

As part of this project we intend to write an SML program that, given an expression and an environment and state in which to evaluate it, will figure out the result of the evaluation and final state (if the evaluation halts) and prove, using the evaluation relations, that the phrase evaluates to this.

4 Accomplished Proofs

Our largest proof to date has been the proof that evaluation is deterministic. As a major lemma, we proved that in our translation of the evaluation rules, the conclusion holds if and only if the hypothesis holds. Since eval_pred (or eval_pat_pred or eval_exbind_pred) holds of the rules, we know that the hypothesis implies the conclusions. The other direction is necessary for proofs by induction: in order to use facts about the evaluations of subparts in the hypothesis of the rule to prove facts about the evaluation in the conclusion of the rule, we need to know that the hypothesis holds if the conclusion does.

In order to do proofs by induction, we used a series of theorems that are based on the definitions of the evaluation rules. The "induction theorem" for expressions is as follows:

```
|- !atexp_prop exprow_prop exp_prop match_prop
    mrule_prop dec_prop valbind_prop.
    eval_pred atexp_prop exprow_prop exp_prop match_prop
     mrule_prop dec_prop valbind_prop ==>
    (!ex s1 e s2 vp. eval_exp ex s1 e s2 vp ==>
                     exp_prop ex s1 e s2 vp)
```

This theorem says that if **eval_pred** holds of some properties that have the same types as the evaluation relations, and **eval_exp** holds of some tuple, then the property on expressions, **exp_prop**, also holds of this tuple. The theorem is true because **eval_exp** is the smallest relation satisfying **eval_pred**.

In order to prove that evaluation is deterministic, we needed to prove a collection of properties, one for each phrase type, that says that given the context of an environment and a state, a phrase evaluates to at most one final state and result. The property for expressions (called **exp_det**) is as follows:

```
!ex s1 e s2 vp s2' vp'.
    (eval_exp ex s1 e s2 vp /\ eval_exp ex s1 e s2' vp') ==>
    (s2 = s2') /\ (vp = vp')
```

In order to use our induction theorem, however, we had to rephrase our goal as the following equivalent statement, called **exp_det2**:

```
!ex s1 e s2 vp. eval_exp ex s1 e s2 vp ==>
            !s2' vp'. eval_exp ex s1 e s2' vp' ==>
                    ((s2 = s2') /\ (vp = vp'))
```

Comparing this goal with the induction theorem, we see that **exp_prop** must be:

```
!s2' vp'. eval_exp ex s1 e s2' vp' ==> ((s2 = s2') /\ (vp = vp'))
```

We defined similar properties for all phrase types and then proved that **eval_pred** holds of those properties (this is the hard part). After this, it only required simple manipulations to get the determinacy properties (of which **exp_det** is one) we wanted.

5 The Next Step: Typechecking

In this section we discuss a planned extension to HOL-ML. We have encoded the dynamic Core of SML and defined its evaluation rules. However, as we have noted, the evaluation relations will work correctly only on *type-correct* programs, for some meaning of type-correct. One of two things will happen when we try to evaluate programs that are not type-correct. The first is that the program can evaluate to garbage. For example, we can prove that (op ::)5 (the application of the *cons* operator to 5) evaluates to the value (::, 5). This is because the evaluation rule

$$\frac{s, E \vdash exp \Rightarrow con, s' \quad con \neq \mathbf{ref} \quad s', E \vdash atexp \Rightarrow v, s''}{s, E \vdash exp\, atexp \Rightarrow (con, v), s''} \tag{112}$$

is applicable despite the fact that the constructor :: ought to be applied only to records $\{1 = x, 2 = L\}$ where L is a list and x is an item of the same type as the items in the list.

The second thing that can happen is that the program may fail to evaluate at all because no rule applies to it. This is the case with the program

$$@\{1 = (\mathbf{op}\,::)5, 2 = \mathbf{nil}\}$$

which is an attempt to append (op ::)5 to the empty list. Recall that the
definition of append, without using any derived forms, is

```
val rec @ = fn {1 = nil, 2 = M} => M
           | {1 = :: {1 = x, 2 = L}, 2 = M} =>
                    ::{1 = x, 2 = @ {1 = L, 2 = M}}
```

According to the rules for evaluating function application, the first match rule
is tried first, so an attempt is made to match the value to which (op ::)5
evaluates, which is (::, 5), against the pattern nil. This results in a FAIL,
so the second match rule is tried. Here, an attempt is made to match (::, 5)
against the pattern {1 = x, 2 = L}. The rule for evaluating record patterns
has as a side condition the requirement that the value is a record. This is not
the case here. Thus no rule applies, so the program does not evaluate. One of
the most important theorems one can prove about typed languages is that if
the program typechecks, then its evaluation will not "get stuck"; that is, in the
search for a proof that the program evaluates to a value, there will always be a
rule that applies. In the *Commentary on Standard ML*, the authors state that
they believe this result to be true for Standard ML, but have not proved it.
We hope to eventually prove it for HOL-ML. Doing so will require additional
machinery such as the definition of the length of an evaluation, which we do not
currently have defined.

Despite the need for programs to be in some way type-correct in order to
evaluate correctly, one cannot define what it means to be type-correct using
the grammar of HOL-ML because it includes no type information. Thus type-
correctness must be defined in terms of an expanded language that includes type
declarations and type tags on language phrases. We could define the typechecking
rules for this language in a manner similar to that for the evaluation rules for
HOL-ML. Given this, we could define a type-correct program in this language
to be one that can be proven to have a type using the typechecking rules. A
translation to HOL-ML could be defined that keeps only the information relevant
to evaluation. A type-correct HOL-ML program would then be defined to be the
translation of a type-correct program in the expanded language into the reduced
syntax of HOL-ML. It would be possible to write a function in that would
automatically prove to what term in HOL-ML a term in the full typed language
translates to, thereby easing the burden of the users of this system.

A   Mutually Recursive Types

The easiest way to explain how to use our mutually recursive types definition
package is to show an example. The goal will be to produce a set of types with
the BNF description

```
Aty = A1 num | A2 num Bty
Bty = B1 Aty
```

The input to our mutually recursive types definition package is shown below. The structure `TypeInfo` defines `type_info`, which has the definition `datatype type_info = existing of hol_type | being_defined of string`.

```
val num = ==':num'==;
structure SampleInput  : MutRecTyInputSig =
struct
structure TypeInfo = TypeInfo
open TypeInfo
val mut_rec_ty_spec =
[{type_name = "Aty",
  constructors =
  [{name = "A1", arg_info = [existing num]},
   {name = "A2", arg_info = [existing num, being_defined "Bty"]}]},
 {type_name = "Bty",
  constructors =
  [{name = "B1", arg_info = [being_defined "Aty"]}]}]
end (* struct *);
```

We invoke our mutually recursive type definition package with the following statement:

```
structure SampleDef =
  MutRecDefFunc
    (structure ExtraGeneralFunctions = ExtraGeneralFunctions
     structure MutRecTyInput = SampleInput
     structure SimpleDefineType = SimpleDefineType)
```

The result is that the HOL types `Aty` and `Bty` and the constructors `A1`, `A2`, and `B1` are defined. The contents of `SampleDef` are three theorems. The first is `New_Ty_Induct_Thm`:

```
|- !Aty_Prop Bty_Prop.
     (!x1. Aty_Prop (A1 x1)) /\
     (!x1 x2. Bty_Prop x2 ==> Aty_Prop (A2 x1 x2)) /\
     (!x1. Aty_Prop x1 ==> Bty_Prop (B1 x1)) ==>
     (!x1. Aty_Prop x1) /\ (!x2. Bty_Prop x2)
```

This theorem allows the user to prove inductive theorems about the members of the types by proving that the properties are true of the base cases (such as that for `A1`) and the inductive steps (such as that for `B1`).

The second is `New_Ty_Existence_Thm`:

```
|- !A1_case A2_case B1_case.
     ?fn1 fn2.
       (!x1. fn1 (A1 x1) = A1_case x1) /\
       (!x1 x2. fn1 (A2 x1 x2) = A2_case (fn2 x2) x1 x2) /\
       (!x1. fn2 (B1 x1) = B1_case (fn1 x1) x1)
```

This allows the definition of mutually recursive functions. It is this theorem that is passed as an argument to **define_mutual_functions**.

The third is **New_Ty_Uniqueness_Thm**:

```
|- (((!x1. fn1 (A1 x1) = A1_case x1) /\
     (!x1 x2. fn1 (A2 x1 x2) = A2_case (fn2 x2) x1 x2) /\
     (!x1. fn2 (B1 x1) = B1_case (fn1 x1) x1)) /\
     (!x1. fn1' (A1 x1) = A1_case x1) /\
     (!x1 x2. fn1' (A2 x1 x2) = A2_case (fn2' x2) x1 x2) /\
     (!x1. fn2' (B1 x1) = B1_case (fn1' x1) x1) ==>
     (fn1 = fn1') /\ (fn2 = fn2')
```

This theorem states that if two sets of mutually recursive functions are defined using **New_Ty_Existence_Thm**, then they are equal.

The SML function **define_mutual_functions** simplifies the definition of mutually recursive functions by allowing the user to simply describe the properties desired of the functions. As an example, we define the mutually recursive functions count_A and count_B:

```
val count_DEF =
  define_mutual_functions SampleDef.New_Ty_Existence_Thm
  (--'(count_A (A1 n) = n) /\
      (count_A (A2 n b) = n + (count_B b)) /\
      (count_B (B1 a) = (count_A a))'--);
```

The result is that the functions are defined, and the return value is a theorem that looks very much like the definition:

```
val count_DEF =
  |- (!x1. count_A (A1 x1) = x1) /\
     (!x1 x2. count_A (A2 x1 x2) = x1 + count_B x2) /\
     (!x1. count_B (B1 x1) = count_A x1) : thm
```

References

1. Stefan Kahrs. *Mistakes and Ambiguities in the Definition of Standard ML*, unpublished manuscript.
2. Robin Milner, Mads Tofte. *Commentary on Standard ML*, The MIT Press, Cambridge, Mass, 1991.
3. Robin Milner, Mads Tofte, Robert Harper. *The Definition of Standard ML*, The MIT Press, Cambridge, Mass, 1990.
4. Konrad Slind, Adding new rules to an LCF-style logic implementation. In L. J. M. Claesen and M. J. C. Gordon, editors, *Higher Order Logic Theorem Proving and its Applications*, pages 549 – 559. North-Holland, 1993.

Structure and Behaviour in Hardware Verification

K. G. W. Goossens

Laboratory for Foundations of Computer Science, Department of Computer Science,
University of Edinburgh, The King's Buildings, Edinburgh EH9 3JZ, U.K.

Abstract. In this paper we review how hardware has been described
in the formal hardware verification community. Recent developments in
hardware description are evaluated against the background of the use
of hardware description languages, and also in relation to programming
languages. The notions of structure and behaviour are crucial to this
discussion.

1 Introduction

Hardware has long been described using hardware description languages (HDLs).
More recently, in the field of hardware verification logic-based notations have
been used. In this paper we explore how the relationship between the structure
and behaviour of circuits has been perceived over time in the formal verification
field. The structure of this paper is as follows: we give our view of HDLs and
simulation prior to the advent of formal methods, then we comment on formal
logic methods used to describe and reason about hardware. Connections with
conventional programming languages are also explored.

Hardware Description Languages and Simulation

The first rôle of HDLs was to document hardware designs and facilitate communication between designers [11, 36]. It was soon realised, however, that these
descriptions could be used to *simulate* the realisations of the designs they described [34]. The shift from the use of HDLs as documentation to their use as
behavioural descriptions is important. A structural description of the physical
realisation of the system has been replaced by the behavioural description of the
design of the system.

In the former situation there is an explicit understanding that every construct
in the language stands for, or represents, a real hardware component. (In fact,
in [54] an HDL was given a semantics in these terms. See also PMS [36].) In
the presence of simulation, however, an HDL description requires a model that
defines the behaviour of the basic components of the language. The gap between
an HDL description and the behaviour of one of its implementations is filled by
the simulation model, or model of hardware. Features that the model abstracts
away from cannot be reasoned about, and if the model is unrealistic or incorrect,
the behaviours associated with an HDL program are also invalid. While this
has always been clearly understood in areas such as device modelling where
simulation programs have been used extensively [53] and system-level modelling,
this was not always so obvious in formal hardware verification [17].

A separate development addressed the need to document and design systems
at higher levels of abstraction. Behavioural notations such as ISP [60], closer to

conventional programming languages, were defined for this purpose. By definition, this type of description does not relate to any particular implementation. Simulation of higher-level descriptions is less contentious than that of structural descriptions because the former do not relate to an underlying physical implementation via a model like the latter. Note that a design written in a behavioural HDL can only be interpreted indirectly, using a simulator.

The two distinct developments of simulation and emphasis on behaviour, together with the ability to generate structural descriptions from low-level behavioural hardware descriptions using synthesis tools, diffused the original intention of hardware description languages: to document circuit implementations. Formal hardware verification started from these premises, and it is therefore not surprising that structure and behaviour were not cleanly separated until recently.[1] In the remainder of this section we review how hardware has been described in the formal hardware verification community until recently. Some research explicitly addressing these issues is then discussed.

Structure and Behaviour in Formal Hardware Verification

Where proof assistants have been used in the hardware verification community, the following schema has generally been employed:

$$\vdash \textit{implementation} \text{ IMPLEMENTS } \textit{specification}$$

The relation IMPLEMENTS expresses that the implementation satisfies the specification. IMPLEMENTS has been interpreted as equivalence ($\leftrightarrow$ or $=$), and implication ($\rightarrow$). Although more sophisticated notions have been investigated [7], logical implication is used predominantly. Nearly always *implementation* is a relation between input and output signals, describing the behaviour of the design under consideration. This behavioural description of the implementation is commonly regarded as a structural description. However, in purely structural descriptions there is no behavioural information: $and(x,y,z)$ means only that in the corresponding place in the implementation there is 'a piece of hardware commonly called an AND gate.'

In the approach taken by researchers using the Boyer-Moore theorem prover the circuit description *b-and x y* already denotes a particular behaviour — that normally associated with an AND gate. Consider the following representative example from [12] below. The description has been broken down into small components that we associate immediately with their usual gate-level implementations but the description remains behavioural.

(defn b-not a) (if (equal a F) T F)
(defn b-and a b) (if (and (boolp a) (boolp b))
*   (and (equal a T) (equal b T)) F)*
(defn b-nand a b) (b-not (b-and a b))

---

[1] Although, of course, a major reason for formal hardware verification was the early realisation that simulation alone would not be feasible for the verification of hardware [11]. Note that mathematical logic may be considered as a sufficiently expressive behavioural HDL not to require animation.

The example consists of a composition of constants that already have an interpretation. We insist on commencing with the uninterpreted syntax of a structural language; behaviour is a secondary concept, and is provided in an explicit manner [23, 24]. This highlights the fundamental difference between the structure of hardware and the behaviour of the hardware, when it is abstracted using a particular model. In the Boyer-Moore system only Brock and Hunt have used this approach [12]. Their work is discussed in Section 4. Other Boyer-Moore work provides interpretations such as the one given above; the hardware description is a recursive function which is intended to model the behaviour of the design. The use of tail recursion to represent the advance of time was introduced by Hunt [35], and has generally been used by hardware verification research based on the Boyer-Moore theorem prover.

In higher-order logic proof assistants such as LAMBDA [22] and HOL [29], nearly all work has been in terms of similar direct interpretations [32, *e.g.* Section 4]. Exceptions are discussed later. Consider the usual HOL definition of an AND gate: $\vdash$ and$(x, y, z) = (z = x \wedge y)$. It defines a three-place relation between booleans. It may be composed with a similarly defined NOT gate as follows:

$$\vdash \text{ and}(x, y, a) \wedge \text{not}(a, z)$$

Although this looks conspicuously like a structural description it is a behavioural description, composed of the two very simple relational descriptions and and not. Consider another implementation of a NAND gate:

$$\vdash \text{ not}(x, a) \wedge \text{not}(y, b) \wedge \text{or}(a, b, z)$$

These two descriptions are logically equivalent, but are intended to denote structurally different circuits. The identical behaviour (at this level of abstraction) is captured, but the structural distinction is lost. For this reason we introduce a description that is truly structural:

$$\vdash \text{ P}(\text{strand}(x, y, a) \text{ \&\& } \text{strnot}(a, z))$$

There is a considerable difference between the first relational behavioural description, and this purely structural description. **strand** is an object denoting a purely structural AND gate. **&&** is an operator combining structural descriptions, with result type *structural*. The purely structural description is not a truth valued expression, like the relational descriptions: we have to say something about the structural expression, which is what the context P indicates. For example, we could give a meaning to the structural description using a semantics, synthesise circuits, *etc.* See Section 4 for more details.

In our opinion a proper separation between the structural and behavioural aspects of a circuit description is crucial. In the remainder of this section we review research that has explicitly addressed this issue.

Research Addressing These Issues

In [33] Hanna and Daeche present the VERITAS hardware verification approach. Theories are used to define new notions such as a theory of *gate behaviours* containing basic gates. It is important to note that only behaviours are defined; there is no mention of structure. For example, if *wf* is the type of waveforms,

$$\vdash \text{ANDBEHAV} : \quad characteristics \rightarrow (wf \times wf \times wf) \rightarrow bool \; = \; definition$$

is a parametrised relational definition of the behaviour of an AND gate. The association of structure with behaviour can only be completed after a theory of *simple structures* has been given. This theory defines the structural aspects of a circuit. Elements of a type correspond to implementations; subtypes are used to axiomatise input and output ports, components, and interconnections. Projection functions are used to extract characteristics from structural entities. For example, we use the function in_i : *andgate* $\rightarrow$ *inport* to obtain the ith input port of an AND gate. We associate an AND gate behaviour ANDBEHAV, as defined in the gate behaviour theory, with a particular simple structure g of type *andgate* as follows:

$$\vdash \forall g : andgate. \; \text{ANDBEHAV} \; (characteristics \; g) \; (in_1 \; g) \; (in_2 \; g) \; (out \; g)$$

This axiom states that every purely structural AND gate g with its particular properties, in this case *characteristics* g, input and output ports, satisfies the behaviour of an AND gate as axiomatised by ANDBEHAV. Finally, a theory of *compound structures* defines composite structures, properties of which, such as subgates and their interconnections, are again obtained by applying projection functions. The behaviour of composite structures may be derived from the behaviours of subcomponents. This work is a good example of the separation of structure and behaviour. It is distinctive in its use of projection functions to extract the composition of non-simple structures. Usually subcircuits are combined explicitly using composition and hiding operators (*e.g.* CIRCAL [45] and LCF_LSM [30]).

Wang [61] describes a Hardware Synthesis Logic which also maintains a clear distinction between structure and behaviour. Circuit structures are composed in a simple structural algebra, called the implementation language, containing a structural connective $\&$, which is comparable to $\&\&$ introduced previously. A logic called the specification language is used to reason about properties of implementations and about specifications. The calculus is independent of a particular specification logic, although a higher-order logic is used in the example below. The implementation and specification languages are related through a so-called construction logic, which contains some inference rules and axiom schemas. The latter define, using the specification language, the behaviour S of basic terms I in the implementation language. This is denoted by the use of the connective in $I \models S$. For example:

$$Register(i, c, o) \models \forall t. \; o(S \; t) = if \; c \; t \; then \; i \; t \; else \; o \; t$$

The structural conjunction $\&$ is preserved by $\models$, so that the following inference rule is part of the calculus:

$$\frac{\vdash I_1 \models S_1 \qquad \vdash I_2 \models S_2}{\vdash I_1 \& I_2 \models S_1 \wedge S_2}$$

Wang proves a number of meta-results relating the implementation, specification, and construction logics.

2 Programming Language Semantics

The structure versus behaviour issues discussed above have been investigated for conventional programming languages using formal semantics. Three types of semantics have been proposed to give meaning to programs; axiomatic [20], denotational [51], and operational [49]. The three types of semantics may be viewed as progressively more concrete, and therefore suited to different applications [51]. Axiomatic semantics map programs directly onto properties characterising their behaviours. Denotational semantics map programs onto functions, from which input-output behaviours may be derived. Operational semantics allow a behaviour to be derived through the sequence of transitions a program may perform.

In Section 3 axiomatic and denotational approaches to hardware description are presented, whereas Section 4 contains operational methods. In both sections informal, partially formal, and formal methods are distinguished.

3 Extracting Behaviour From Circuit Descriptions

The intuitive solution to the structure-behaviour division is to *extract a behaviour* from a circuit description directly. We have a function **behaviour** : *structural* → *bool*. In other words, **behaviour** maps a hardware description to a logical formula characterising its behaviour. For example:

$$\text{behaviour } (\text{delay}(c, in, out)) \ = \ (out\ 0 = c \ \wedge \ \forall t.\ out(\mathsf{S}\ t) = in\ t) \qquad (1)$$

Here **delay(c,in,out)** is an HDL description for a unit transport delay. Let us first assume that this equation is entirely outside a proof system. This definition raises the following question: what is the relation between **in** and *in*? The former is a syntactic structural object, whereas the latter is part of the formal system in which the behaviour is expressed. The situation is clarified by giving explicit types to the various components:

$$\text{behaviour} : structural \rightarrow bool$$
$$\text{delay} : (value \times name \times name) \rightarrow structural$$
$$\text{in} : name$$
$$in : signal = time \rightarrow value$$

We would like behaviour functions to always produce formulae that are consistent, *i.e.* do not contain contradictions. If this were not the case, a particular circuit for which an inconsistent behaviour description was produced would satisfy any specification. We note that in principle the range of the behaviour function may be anything, as long as it allows us to express our intuitions about the behaviour of circuits. If the result is truth-valued then the behaviour function could be called axiomatic, or denotational otherwise.

The definition of **behaviour** could be an entirely informal exercise, but rather than using an *ad hoc* implementation of the manipulation of behaviours later work advocated mapping the extracted behaviour into a proof system. This lead to a clean separation of conceptually different processes, namely the extraction of the behaviour and the formal reasoning about this behaviour. We may view Equation 1 in this light; the right hand side could be inside the proof system. One fundamental problem remains: the behaviour function itself resides outside the proof system. This means that we cannot reason about it within the proof system. In particular, we will have to accept the correctness of the implementation of the behaviour function in good faith. The HDL description is also informal, which means we cannot reason about structural terms. We can only use the behaviour of the design, and no structural aspects, inside the proof system. This becomes a problem where we want to reason about general properties possessed by all, or a set of circuits. The solution is to move the behaviour function into the proof system also. For example, some hardware models may satisfy the property that for every input an output exists. It is preferable to prove a general theorem of the form

$$\vdash \forall e : structural. \ \forall i : const. \ \exists o : const. \ simulation \ e \ i = o$$

rather than a number of instantiations. It is important to note that the type *structural*, representing terms of type circuit, resides inside the proof system. Thus the structural circuit description may be manipulated independently from its behaviour; we discuss this in more detail in Section 4. Whether formal or informal, there is a real separation between the description of the circuit and its behaviour.

The remainder of this section refers to research that has some aspect of explicitly relating structural descriptions to behaviour.

Informal Behaviour Extraction Functions

Early research into hardware verification was informal and rather *ad hoc*. Most efforts took the form of a software system that given a hardware description and a specification would try to show their equivalence. From a historical perspective we may consider these efforts as primitive behaviour extraction functions. In the late 1970s and early 1980s a number of efforts were directed at *functional abstraction*; this is to the process of extracting a behaviour from a circuit description [38, 4, 39].

Pitchumani and Stabler [48] used a Floyd-Hoare style semantics to give a definition for a register transfer-level HDL. The language which is described in [48] does not have an explicit notion of time. Rather, time is introduced in the semantics through the use of a distinguished variable t that represents time. It may be used in pre- and post-conditions, but not in programs. This precludes assignments to the time variable, but does allow temporal information to be given in the specification. Consider the NULL statement with its conventional semantics $\{P\}$ NULL $\{P\}$. When time is involved this becomes $\{P[t+1/t]\}$ NULL $\{P\}$. Thus NULL has no effect other than to pass time.

Partially Formal Behaviour Extraction Functions

In [8] Borrione and Paillet recognise the need for a formal system to unambiguously express the semantics of an HDL. They outline the design of a system to translate VHDL descriptions to a representation of their behaviour in a proof system. The behaviour is represented by a set of simultaneous functional equation, in the Boyer-Moore and REVE proof systems.

Boulton [10] describes a behaviour extraction function from a subset of ELLA[2] to the HOL proof assistant. The behaviour function and its abstract syntax tree input are outside the formal part of the HOL proof system. ELLA constructs are mapped to high-level behaviours in HOL. For example, consider the **case** statement in ELLA:

```
[case in of lo: hi, hi: lo] =
CASE [in] [OF [[lo: hi]; [hi: lo]]] (UNLIFT UU) =
CASE in [OF [CONST lo, SIGNAL LIFT_hi;
            CONST hi, SIGNAL LIFT_lo]] (UNLIFT UU)
```

The behaviour function [·] gives a semantics to the structural description **case in of lo: hi, hi: lo**. CASE and OF are HOL functions that, given the subcomponents' behaviours [lo: hi] and [hi: lo], represent the behaviour of the whole **case** statement. Because this behaviour function is itself not part of HOL, the **case** statement is informal and the variable **in** has no explicit relation to in (cf. Equation 1).

Other related work includes [58, 19] which describe mapping VHDL into HOL and SDVS respectively. SILAGE has also been given a HOL semantics as above [27]. In [47] behaviours of CASCADE descriptions are mapped into the Boyer-Moore and TACHE theorem provers. Recently Umbreit has used LAMBDA to map VHDL programs onto formally defined ML descriptions [57].

Formal Behaviour Extraction Functions

In [41] Melham describes a formal behaviour function in HOL. He defined an abstract data type representation of CMOS circuit descriptions inside the HOL proof assistant. Part of this data type is given below.

$$circ ::= \textbf{pwr } str \mid \textbf{ntran } str \; str \; str \mid \textbf{join } circ \; circ \mid \ldots \qquad (2)$$

join $c \; c'$ is structural composition, comparable to **&&** introduced earlier. Switch-level model and threshold model semantics were defined using primitive recursion functions. A fragment of the former is:

$$\vdash \textbf{Sm } (\textbf{pwr } p) \; e \quad = (e \; p = T)$$
$$\vdash \textbf{Sm } (\textbf{ntran } g \; s \; d) \; e = (e \; g \supset (e \; d = e \; s)) \qquad (3)$$
$$\vdash \textbf{Sm } (\textbf{join } c_1 \; c_2) \; e \quad = \textbf{Sm } c_1 \; e \wedge \textbf{Sm } c_2 \; e$$

Sm $: circ \rightarrow (str \rightarrow bool) \rightarrow bool$ is the function mapping circuits with environments to a formula describing their switch-level behaviour. The term $e : str \rightarrow$

---

[2] ELLA is a trademark of the Secretary of State for Defence, United Kingdom.

bool is the environment, mapping strings *str*, denoting wire names, to their values. As we briefly indicated in Section 3 because the data type expressions are ordinary proof system terms we may quantify over structural descriptions. This feature was used to relate the switch-level and threshold models of hardware formally, *i.e.* as a theorem in HOL.

In [5] Basin uses the NUPRL proof assistant [18], which implements a constructive type theory. He uses the *proofs-as-circuits* paradigm, which is an adaptation of the *propositions-as-types* idea. A constructive proof contains computable evidence, *e.g.* a circuit, of the truth of the proposition it proves. Different proofs correspond to different implementations. Proving

$$>> \forall i, o. \exists c. S(i, o, c)$$

entails exhibiting a witness *c* that satisfies the specification $S(i, o, c)$. ($>>$ is NUPRL's judgement.) There is no guarantee, however, that realisation *c* has a particular form, or *intention*; we only know that it has behaviour, or *extension*, *S*. We would like *c* to be a circuit description, not just any old proof term. To force the realisation to have a particular form, or to be at a particular level of abstraction, a type of circuit terms is introduced. This type *trans* is a recursively defined data type. An interpreter $Interp_{trans} : trans \rightarrow env \rightarrow bool$ is defined to give a meaning to these terms. *trans* and $Interp_{trans}$ correspond to Melham's *circ* (Equation 2) and S⧫ (Equation 3) respectively.

The VERITAS approach, discussed in the introduction, corresponds to an axiomatic approach fully within a proof system.

4 Deriving Behaviour via a Semantics

In the previous section we showed how behaviour could be extracted directly from circuit descriptions. This is a high-level approach with no indication of an underlying model of how the behaviour is arrived at. Industrial HDLs usually have a simulator to animate hardware descriptions. It makes sense not to state properties directly about circuit descriptions, but to derive properties using the simulator. That is, we take a more operational stance. Taken at face value, this would seem to imply that we can only derive properties using simulation; exactly what we are trying to get away from. This is not the case, however: if we provide an operational semantics for the HDL, we may prove *general properties* about the simulator model. For example, we can characterise the domain on which the simulator is a total function. An operational semantics gives us a firm mathematical grip on the simulator model.[3] Often we can prove more detailed properties using operational semantics than with other types of semantics because we can refer to the simulation method.

As with behaviour functions earlier, we can define an operational semantics on paper or use a proof system. In this case, however, there is no half-way stage: either everything is on paper or everything is in a proof system. The reason for this is clear when we consider a fragment of an operational semantics.

---

[3] Particular implementations of this algorithm may still be incorrect. [25] shows how formal simulation overcomes this problem.

opsem env (wire n) $= env\ n$

opsem env (parcomp $(c_1,\ c_2)$) $=$ (opsem $env\ c_1$, opsem $env\ c_2$)

opsem env (mux $(c_1,\ c_2,\ c_3)$) $=$ *if* opsem $env\ c_1$ *then* opsem $env\ c_2$ *else* opsem $env\ c_3$

wire n returns the value on the wire n, parcomp is parallel composition, and mux a multiplexor. Although it is conceivable to map from outside into a proof system this does not really make sense because the same objects and types occur in both the domain and range of the semantics. This was not necessarily the case for the axiomatic behaviour function of Equation 1.

The discussion that follows applies equally to 'paper' and embedded operational semantics. The difference is of a more pragmatic nature; it is possible to use an operational semantics on paper but it quickly becomes tedious and error-prone.

To embed an operational semantics in a proof system circuits, input and output values, and the semantic rules must be encoded. Auxiliary objects such as environments and wire names are also needed. The structure and behaviour of hardware are kept separate by providing a structural description language, which is given a meaning through the use of a semantics (*cf.* Section 3). Operational semantics relate a circuit and its inputs to an output according to some simulation model. A type of the semantics could be the following (for simplicity we allow only one input):

$$\text{opsem} : (structural \times value) \times value$$

Here the concept of state is missing; most circuits contain latches, which retain a value between clock cycles. Adding an explicit state yields the following (*cf.* [59]):

$$\text{opsem} : (structural \times state \times value) \times (value \times state)$$

An alternative view is to dispense with the state, and evolve the circuit itself so that the state is part of the circuit description [23]:

$$\text{opsem} : (structural \times value) \times (value \times structural)$$

Introduced by Milner [43], this type was used by Gordon as the basis for LCF_LSM [30]. State transition functions of state machines have a similar type. This view is also common in process algebras such as CCS [44], CIRCAL [45], and HOP [26], which use labelled transition systems.

After the structural aspects of the HDL have been defined they can be manipulated using proof system facilities. This leads to a number of possible applications: we may quantify over circuits, expressing properties that hold for all or particular classes of circuits. Circuits expressions can contain free variables, corresponding to plug-in components [23]. Circuits may be operated on by transformation functions, which may be proven correct [24], or be the result of formal hardware synthesis functions [12]. Interactive synthesis, perhaps based on the operational semantics rules [24], or refinement-based strategies [21] is also possible. Operational semantics based formal simulation is another powerful application [25]. It may be very useful to have multiple semantic functions emphasising different aspects of the structural description [12, 46].

Embeddings of hardware notations are more powerful but also harder to use than extracting the behaviour directly because one has to resort to the semantics to obtain any behavioural information (see *e.g.* [9, Section 7.8]).

Recently a number of HDLs have been given formal semantics. With a few exceptions, these have all been paper exercises [55, 2, 62], although [3, 56] provide computer support. Below we discuss work in conjunction with proof systems.

Compiler Correctness in Proof Systems

Correctness proofs of compiler (algorithms) in proof systems use the same techniques as those for embedding HDLs in proof systems. To reason about programs their syntax and semantics have to be encoded in the proof system.

Milner and Weyhrauch used the Stanford LCF proof checker to check the correctness of a simple compiler algorithm [42]. The source and target languages were axiomatised in the system through the use of constructors and destructors. Aiello *et al.* encoded a denotational semantics for Pascal in the Stanford LCF in a similar manner [1]. Using the Edinburgh LCF [28] Cohn proved a compiler correct with respect to the denotational semantics of imperative source and target languages [16]. Other research involving compiler correctness proofs using proof systems includes Sokolowski's LCF work [52]. Joyce verified a compiler using HOL with as target machine a non-idealised formally verified computer Tamarack [31] taking into account finite storage [37]. A group at Computational Logic [6] has used the Boyer-Moore theorem prover to verify a code generator, assembler and linker to a verified microprocessor FM8502.

Embedded Hardware Description Notations

Brock and Hunt [12] describe a simple combinatorial logic hardware description language in the Boyer-Moore theorem prover. This is the earliest research known to us that defines an operational semantics for an HDL in a proof system. Circuits are encoded as list constants, which are interpreted by a semantic function. For example, a full adder is described as follows.

```
'(half-adder (a b) (sum carry) (((sum) (b-xor a b))
                                ((carry) (b-and a b))))
'(full-adder (a b c ) (sum carry)
        (((sum1 carry1) (half-adder a b))
         ((sum carry2)  (half-adder sum1 c))
         ((carry)       (b-or carry1 carry2))))
```

The circuit *half-adder* is defined as having two inputs *a* and *b*, and two outputs *sum* and *carry*. *b-xor* and *b-and* represent primitive XOR and AND gates respectively. A well-formedness predicate is defined to check that these definitions are purely combinatorial. The output value of the circuit description is computed by an operational semantics, which is encoded as a total recursive function. The conceptual type of the semantic function is as follows:

$$heval : \quad name \rightarrow signalenv \rightarrow circuitenv \rightarrow value\ list$$

name consists of the name of the top-level component and its inputs. The environment *circuitenv* contains the definitions of non-primitive functions, such as *half-adder*, and *signalenv* is used to store the values of input, output, and internal variables such as *sum1*. To evaluate the half adder with inputs x and y with values F and T respectively, we use:

(heval '(half-adder x y) (list (cons 'x F) (cons 'y T)) (list '(half-adder (a b)...

A recent extension to this work allows sequential circuits with delayed feedback loops and explicit state holding components [13].

Goossens [23] describes the embedding of a formal static and dynamic operational semantics for a subset of the industrial HDL ELLA [50]. The HDL contains unit delays, generalised multiplexors, and allows both delayed and delayless feedback loops. In common with other work he defines a data type to define the abstract syntax of the HDL. Due to restrictions of LAMBDA version 3.2 the operational semantics is given as a function that is defined structurally on abstract syntax terms. This limits proofs to structural induction on program terms. A number of meta-level results such as the totality and monotonicity of the simulator model are proved [24].

The same approach is used by van Tassel to embed a VHDL subset in HOL [59]. Again an abstract data type is used to represent program terms, but here a HOL package to define inductive relations is then used to derive a rule induction principle from a relational semantics. This is a more general induction than Goossens' structural induction on the abstract syntax of programs [14] and the fixed-point (or computational) induction [40] in LCF. LAMBDA version 3.2 permits only functional semantics, whereas the HOL system allows more general relational semantics.

More recently a spate of embeddings in HOL has been reported [15].

5 Conclusions

In this article we attempted to illustrate the evolution of the separation of structural and behavioural aspects in formal hardware verification. Although behavioural hardware descriptions have shortcomings in this respect, their ease of manipulation compared to operational semantics based approaches is an advantage [24]. Due to the use of the underlying logic relational hardware description (*e.g.* [32]) is especially efficient.

References

1. L Aiello, M Aiello, and R Weyhrauch. The semantics of Pascal in LCF. Memo STAN-CS-74-447, Stanford Artificial Intelligence Laboratory, Computer Science Department, Stanford University, August 1974.
2. H Barringer, G Gough, and B Monahan. Operational semantics for hardware design languages. In P Prinetto and P Camurati, eds., *Advanced Research Workshop on Correct Hardware Design Methodologies*, pages 313–334. North Holland, June 1991.
3. H Barringer, G Gough, B Monahan, and A Williams. A semantics for Core ELLA. Deliverable D2.3b, Department of Computer Science, University of Manchester, November 1992. Formal Verification Support for ELLA, IED project 4/1/1357.

4. H Barrow. Verify: A program for proving correctness of digital hardware designs. *Artificial Intelligence*, 24:437–491, 1984.

5. D Basin. Extracting circuits from constructive proofs. In *1991 Int'l Workshop on Formal Verification in VLSI Design*. January 1991.

6. W Bevier, W Hunt, Jr, J Moore, and W Young. An approach to systems verification. *J. of Automated Reasoning*, 5:411–428, 1989.

7. J Bormann, H Nusser-Wehlan, and G Venzl. Formal design in an industrial research laboratory: Lessons and perspectives. In J Staunstrup and R Sharp, eds., *Workshop on Designing Correct Circuits*, pages 193–213, Denmark, January 1992.

8. D Borrione and J L Paillet. An approach to the formal verification of VHDL descriptions. Technical Report RR 683-I-, IMAG/ARTEMIS, November 1987.

9. R Boulton, A Gordon, M Gordon, J Harrison, J Herbert, and J van Tassel. Experience with embedding hardware description languages in HOL. In V Stavridou, T Melham, and R Boute, eds., *Theorem Provers in Circuit Design: Theory, Practice and Experience*, pages 129–156. North Holland, June 1992.

10. R Boulton, M Gordon, J Herbert, and J van Tassel. The HOL verification of ELLA designs. Technical Report 199, University of Cambridge Computer Laboratory, August 1990.

11. M Breuer. General survey of design automation of digital computers. *Proceedings of the IEEE*, 54(12):1708–1721, December 1966.

12. B Brock and W Hunt, Jr. The formalization of a simple hardware description language. In L Claesen, ed., *Applied Formal Methods For Correct VLSI Design*, pages 778–792, Amsterdam, November 1989. Elsevier Science Publishers.

13. B Brock, W Hunt, Jr, and W Young. Introduction to a formally defined hardware description language. In V Stavridou, T Melham, and R Boute, eds., *Theorem Provers in Circuit Design: Theory, Practice and Experience*, pages 3–35. North Holland, June 1992.

14. R Burstall. Proving properties of programs by structural induction. *The Computer Journal*, 12(1):41–44, 1969.

15. L Claesen and M Gordon, eds. *Higher Order Logic Theorem Proving and its Applications*, Leuven, Belgium, September 1992. North Holland.

16. A Cohn. High level proof in LCF. Internal Report CSR-35-78, Department of Computer Science, University of Edinburgh, November 1978.

17. A Cohn. The notion of proof in hardware verification. *J. of Automated Reasoning*, 5(2):127–139, June 1989.

18. R Constable and D Howe. Nuprl as a general logic. In P Odifreddi, ed., *Logic and computer science*, volume 31 of *APIC studies in data processing*, pages 77–90. Academic Press, 1990.

19. I Fillippenko. VHDL verification in the state delta verification system (SDVS). In *1991 Int'l Workshop on Formal Verification in VLSI Design*, January 1991.

20. R Floyd. Assigning meanings to programs. *Proceedings of American Mathematical Society, Symposia in Applied Mathematics*, 19:19–32, 1967.

21. M Fourman and E Mayger. Formally based system design – interactive hardware scheduling. In G Musgrave and U Lauther, eds., *Int'l Conf. on VLSI*, 1989.

22. M Francis, S Finn, and E Mayger. *Reference Manual for the Lambda System*. Abstract Hardware Limited, version 3.2, November 1990.

23. K G W Goossens. Embedding a CHDDL in a proof system. In P Prinetto and P Camurati, eds., *Advanced Research Workshop on Correct Hardware Design Methodologies*, pages 359–374. North Holland, June 1991.

24. K G W Goossens. *Embedding Hardware Description Languages in Proof Systems.* PhD thesis, Laboratory for Foundations of Computer Science, Department of Computer Science, University of Edinburgh, December 1992.

25. K G W Goossens. Operational semantics based formal symbolic simulation. In L Claesen and M Gordon, eds., *Higher Order Logic Theorem Proving and Its Applications*, pages 487–506, Leuven, Belgium, September 1992. North Holland.

26. G Gopakakrishnan, R Fujimoto, V Akella, N Mani, and K Smith. Specification-driven design of custom hardware in HOP. In G Birtwistle and P Subrahmanyam, eds., *Current Trends in Hardware Verification and Automated Theorem Proving*, pages 128–170, New York, 1988. Springer Verlag.

27. A Gordon. The formal definition of a synchronous hardware description language in higher order logic. In *Int'l Conf. on Computer Design*, October 1992.

28. M Gordon, R Milner, and C Wadsworth. *Edinburgh LCF*, volume 78 of *Lecture Notes in Computer Science*. Springer Verlag, 1979.

29. M Gordon. HOL: A proof generating system for higher-order logic. In G Birtwistle and P Subrahmanyam, eds., *VLSI Specification, Verification and Synthesis*, pages 73–128, Boston, 1987. Kluwer Academic Publishers.

30. M Gordon. LCF_LSM. Technical Report 41, University of Cambridge Computer Laboratory, September 1983. Second Printing with Corrections and Additions.

31. M Gordon. Proving a computer correct. Technical Report 42, University of Cambridge Computer Laboratory, 1983.

32. M Gordon. Why higher-order logic is a good formalisation for specifying and verifying hardware. In G Milne and P Subrahmanyam, eds., *Formal Aspects of VLSI Design*, pages 153–177, Amsterdam, 1985. North Holland.

33. F K Hanna and N Daeche. Specification and verification using higher-order logic. In C Koomen and T Moto-Oka, eds., *CHDL 85: 7th Int'l Symposium on Computer Hardware and Description Languages and their Applications*, pages 418–433, Amsterdam, 1985. North Holland.

34. G Hays. Computer-aided design: Simulation of digital design logic. *IEEE Trans. on Computers*, C-18(1):1–10, January 1969.

35. W Hunt, Jr. FM8501: A verified microprocessor. Technical Report 47, Institute for Computing Science. The University of Texas at Austin, December 1985.

36. IEEE computer, December 1974. Special edition on Hardware Description Languages.

37. J Joyce. A verified compiler for a verified microprocessor. Technical Report 167, University of Cambridge Computer Laboratory, March 1989.

38. S Leinwand and T Lamdan. Design verification based on functional abstraction. In *16th Design Automation Conf.*, pages 353–359, San Diego, California, June 1979.

39. J-C Madre and J-P Billon. Proving circuit correctness using formal comparison between expected and extracted behaviour. In *Proceedings of the 25th ACM/IEEE Design Automation Conf.*, pages 205–210, 1988.

40. Z Manna, S Ness, and J Vuillemin. Inductive methods for proving properties of programs. *ACM SIGPLAN Notices*, 7:27–50, 1972.

41. T Melham. Using recursive types to reason about hardware in higher order logic. Technical Report 135, University of Cambridge Computer Laboratory, May 1988.

42. R Milner and R Weyhrauch. Proving compiler correctness in a mechanised logic. In B Meltzer and D Mitchie, eds., *Machine Intelligence*, chapter 3. Edinburgh University Press, 1972.

43. R Milner. Processes: A mathematical model of computing agents. In Rose and Shepherdson, eds., *Logic Colloquium 73: Studies in Logic and Foundations of Mathematics*, volume 80, pages 157–173. North Holland, 1973.

44. R Milner. *A Calculus of Communicating Systems*, volume 92 of *Lecture Notes in Computer Science*. Springer Verlag, 1980.

45. F Moller. The definition of CIRCAL. In L Claesen, ed., *Applied Formal Methods For Correct VLSI Design*, pages 178–187, Amsterdam, November 1989. Elsevier Science Publishers.

46. J O'Donnell. Hardware description with recursion equations. In M Barbacci and C Koomen, eds., *CHDL 87: 8th Int'l Symposium on Computer Hardware Description Languages and Their Applications*, pages 363–382. North Holland, 1987.

47. L Pierre. From a HDL description to formal proof systems: Principles and mechanization. In D Borrione and R Waxman, eds., *CHDL 91: 10th Int'l Symposium on Computer Hardware Description Languages and Their Applications*, April 1991.

48. V Pitchumani and E Stabler. A formal method for computer design verification. In *19th Design Automation Conf.*, pages 809–814, 1982.

49. G Plotkin. A structural approach to operational semantics. Technical Report FN-19, Computer Science Department, Aarhus University (DAIMI), 1981.

50. Praxis Systems plc, 20 Manvers Street, Bath BA1 1PX. *The ELLA Language Reference Manual*, issue 3.0, 1986. ELLA is now marketed by R^3 Systems.

51. D Schmidt. *Denotational Semantics, A Methodology for Language Development*. Allyn and Bacon Inc, Boston, 1986.

52. S Sokolowski. Soundness of Hoare's logic: An automated proof using LCF. *ACM Trans. on Programming Languages and Systems*, 9(1):100–120, January 1987.

53. R Spence. Progress in computer-aided circuit design. *CAD*, 1(4):19–24, 1969.

54. E Stabler. System description languages. *IEEE Trans. on Computers*, C-19(12):1160–1173, December 1970.

55. V Stavridou, J Goguen, S Elker, and S Aloneftis. FUNNEL: A CHDL with formal semantics. In P Prinetto and P Camurati, eds., *Advanced Research Workshop on Correct Hardware Design Methodologies*, pages 115–137. North Holland, June 1991.

56. V Stavridou, J Goguen, A Stevens, S Eker, S Alonefits, and K M Hobley. FUNNEL and 2OBJ: Towards and integrated hardware design environment. In V Stavridou, T Melham, and R Boute, eds., *Theorem Provers in Circuit Design: Theory, Practice and Experience*, pages 197–223. North Holland, June 1992.

57. G Umbreit. Providing a VHDL-interface for proof systems. In *EURO-DAC*, pages 698–703, September 1992. IEEE Computer Society Press.

58. J van Tassel and D Hemmendinger. Toward formal verification of VHDL specifications. In L Claesen, ed., *Applied Formal Methods For Correct VLSI Design*, pages 261–270, Amsterdam, November 1989. Elsevier Science Publishers.

59. J van Tassel. A formalisation of the VHDL simulation cycle. In L Claesen and M Gordon, eds., *Higher Order Logic Theorem Proving and Its Applications*, pages 359–374, Leuven, Belgium, September 1992. North Holland.

60. W vanCleemput. Computer hardware description languages and their applications. In *16th Design Automation Conf.*, pages 554–560, San Diego, California, June 1979.

61. L-G Wang. Hardware synthesis logic and its independence. Manuscript. Laboratory for Foundations of Computer Science, Computer Science Department, University of Edinburgh, November 1991.

62. P Wilsey. Developing a formal semantic definition of VHDL. In J Mermet, ed., *VHDL for Simulation, Synthesis and Formal Proofs of Hardware*, pages 243–256. Kluwer Academic Publishers, 1992.

Degrees of Formality in Shallow Embedding Hardware Description Languages in HOL*

Catia M. Angelo, Luc Claesen, Hugo De Man

IMEC vzw, Kapeldreef 75, B3001 Leuven, Belgium

Abstract. Theorem proving based techniques for formal hardware verification have been evolving constantly and researchers are getting able to reason about more complex issues than it was possible or practically feasible in the past. It is often the case that a model of a system is built in a formal logic and then reasoning about this model is carried out in the logic. Concern is growing on how to consistently interface a model built in a formal logic with an informal CAD environment. Researchers have been investigating how to define the formal semantics of hardware description languages so that one can formally reason about models informally dealt with in a CAD environment. At the University of Cambridge, the embedding of hardware description languages in a logic is classified in two categories: deep embedding and shallow embedding. In this paper we argue that there are degrees of formality in shallow embedding a language in a logic. The choice of the degree of formality is a trade-off between the security of the embedding and the amount and complexity of the proof effort in the logic. We also argue that the design of a language could consider this verifiability issue. There are choices in the design of a language that can make it easier to improve the degree of formality, without implying serious drawbacks for the CAD environment.

1 Introduction

To interface theorem proving frameworks with CAD environments, researchers have been formalizing the semantics of hardware description languages. The two approaches for embedding HDLs in the HOL system [1], *deep embedding* and *shallow embedding*, are described in [2]. In *deep embedding*, the abstract syntax of an HDL is represented in the HOL logic and, within the logic, semantic functions assign meanings to the programs. The syntax of the HDL is represented by a HOL type. For instance, one could define in the HOL logic a type for the syntactic class of expressions of the HDL and then define in the logic a function giving meaning to the expressions. Deep embedding allows one to reason about classes of programs because it is possible to quantify over syntactic structures [3]. For instance, one could quantify over expressions of the HDL. The type definition package developed by Tom Melham [4] provides automation and support for the user to define a class of recursive types and to make proofs involving them.

---

* This research was sponsored by CNPq (Brazilian Government) and CHEOPS Project (ESPRIT BRA "3215").

However, defining the types to represent the abstract syntax of a language and semantic functions in the logic can be very complex and time-consuming. In *shallow embedding*, semantic operators are defined in the HOL logic and an ML [5] interface interprets programs into semantic structures in the logic. The abstract syntax of the HDL is represented by a ML type. Since the production rules of the language are not modeled in the logic, one cannot reason directly about the language in the logic. However, shallow embedding a language in the logic still allows one to model and reason about a specification written in the embedded language. One can also prove theorems about the semantic operators defined in the logic. Shallow embedding is less secure than deep embedding because the interpretation of programs is outside of the logic and therefore one cannot reason about this interpretation in the logic. However, shallow embedding is simpler than deep embedding. In general, it is much easier to define ML types than HOL types[2]. The shallow embedding method in HOL has been used in [7, 8, 9, 10, 11] and the deep embedding method in HOL has been used in [12, 13].

In the shallow embedding style, the semantics of a language is needed to consistently map a specification written in that language into its meaning in the HOL logic and, once in the logic, to reason about it. The reasoning in the HOL logic is safe but the parsing of programs written in the HDL and the semantic interpretation (the assigning of meanings to the programs) are possible sources of errors because they are implemented in ML and not in the logic. The simpler the parsing and the semantic interpretation, the safer the whole system. However, the more elaborate the semantic interpretation is, the simpler the model in the logic can be and, therefore, reasoning about it can be easier.

The shallow embedding style has two informal components (the parsing and the semantic interpretation) and a formal component (the reasoning about the meaning of programs in the logic). The degree of formality of the system is defined by the relative "weights" of these components in the whole system. We mean that a component A of the embedding has a heavier "weight" than a component B of the embedding if "more issues" about the embedding are addressed by A than by B and/or if the issues addressed by A are "more complex" (according to a certain criteria) than the issues addressed by B. The choice of the degree of formality is a trade-off between the security of the embedding and the amount and complexity of the proof effort in the logic. The development of infra-structure to build models and reason about them in the HOL logic can reduce the proof effort to reason about complex issues. This kind of infra-structure makes it easier to formally address complex issues that would otherwise be addressed by an informal component of the embedding. Therefore, this kind of infra-structure makes it easier to improve the degree of formality of language embeddings, which means to move in the scale of formality closer to the logic.

The deep embedding style has one informal component (the parsing) and two formal components (the semantic interpretation and the reasoning about the meaning of programs in the logic). One might think of deep embedding as a

---

[2] In [6], standard ML types and HOL types are discussed.

particular case of embedding style that has one less informal component than the shallow embedding style has[3]. On the other hand, one might think of informal programs (based on algorithmical approaches for the verification of aspects of a specification described in a language) as an extreme case of embedding style that has no formal component (no reasoning in a logic).

The embedding of languages in a logic also has to address well-formedness issues of specifications. Whether these issues are addressed informally or in the logic also has an impact on the security of the whole system and on the complexity of the proof effort in the logic. In the shallow embedding style, well-formedness issues can be addressed informally by implementing an algorithm in ML to check properties of specifications described in the embedded language. These issues can also be addressed formally by implementing proof procedures in ML that can prove in the logic properties of specific programs described in the embedded language. These proof procedures can be based on theorems about the semantic operators defined in the logic. In contrast with the shallow embedding style, in the deep embedding style one has additionally the possibility of using predicates defined over the types representing the syntax in order to address well-formedness issues in the logic.

In this paper we present a discussion about degrees of formality in the shallow embedding style, based on the definition of the multi-rate semantics of Silage [11]. In the next section we give an informal overview of the Silage language. Then we outline the basic principles of the semantics of Silage, focusing on a particular aspect. Next we discuss the semantic model and the idea of degrees of formality. Finally, some conclusions are drawn.

2 An Overview of Silage

The Silage language [14, 15, 16, 17, 18] used as input of the CATHEDRAL [19] silicon compilers is an applicative language suitable for describing DSP algorithms represented as a data-flow graph. No assignment operators are allowed. A basic concept of the language is the notion of a signal. A signal is an infinite stream of data indexed by discrete and periodic time instants. Signals are defined by function applications. The single assignment principle allows signals to be defined only once. A definition involving signals is a set of equations about the samples of the signals involved. The operations are vectorial and the indexing is implicit. Previous samples of the signals can be referred to in a signal definition. Arrays of samples can be manipulated. There are iterators and an if construct in Silage. No statements are made about the order or concurrency of the operations and the parallelism of computation can be exploited by the synthesis tool by analyzing the data flow dependencies.

There are operators in Silage that can generate signals with periods that are different from the periods of the signals they are derived from. The period of

---

[3] This does not mean that the deep embedding style is a particular case of the shallow embedding style, which is not true.

signals are implicit in Silage programs and they are derived by the synthesis tool. One cannot explicitly define the periods of the signals.

In Silage definitions, signals or samples on the left-hand side of the equality symbol are defined in terms of the expressions on the right-hand side. The order of the definitions is irrelevant. For instance, consider the piece of Silage code below. The second definition is recursive. The first definition initializes the second one. The operator @ is used to access previous samples of a signal and the operator @@ is used to define the values of the *initialisation samples*. The second definition means that the n-th sample of the signal x is defined by the addition of the n-th sample of a signal in with the previous sample of the signal x. The period of the signal x must be equal to the period of the signal in.

```
x@@1 = 0;
x = in + x@1;
```

Tuples of the form (a,...,z) in Silage aggregate signals that may have different periods. Tuples cannot be nested. On the other hand, the aggregate constructor of the form {a,...,z} aggregates signals with the same period. Functions in Silage returning multiple signals can be called using tuples. The two pieces of code below (where the initialisations are omitted) illustrate Silage functions and mutually recursive definitions in Silage. Both functions take as argument a signal in and return a signal out. The signals a and b are internal signals. When a Silage function has more than one argument signal, it is implicit that they have the same period. Unlike the first piece of code, the second one is illformed because of cyclic data dependencies between samples.

```
func Ok (in:word) out:word =      func notOk (in:word) out:word =
a:word; b:word;                   a:word; b:word;
begin                             begin
   b = a - b@2;                      b = a - b@2;
   out = b + a@1;                    out = b + a@1;
   a = in + a@2 + b@1;               a = in + a@2 + b;
end;                              end;
```

So far, except for the tuples and function definitions, only the Silage operators that handle signals with identical periods have been considered. When a signal is defined using these operators, it is implicit that the periods of all the signals involved are the same. Next we consider the **decimate**, the **interpolate** and the **switch** functions, that handle signals with different periods.

The **decimate** function takes as argument a source signal **sourceS** and returns a list of n signals **listS**. The signal **sourceS** is decomposed into n signals with a period n times greater than the period of **sourceS**. The samples of **sourceS** are distributed among the output signals in the order they appear in **listS**.

The **interpolate** function takes as argument a list **listS** of n signals with equal periods and returns a destination signal **destS**. The samples of the signals in **listS** are interleaved generating the signal **destS** with a period **n** times smaller than the period of the signals in **listS**.

The following is an example of illformed Silage program because of inconsistencies with the periods. The first definition defines the period of the signal **out1** as half of the period of the signal **in1** and the definition of **out2** is only possible if the periods of **in1** and **out1** are equal.

```
func notOkToo (in1:word; in2:word) out1:word; out2:word =
    begin
            out1 = interpolate (in1, in2);
            out2 = in1 + out1;
    end;
```

The **switch** function takes as arguments **n** signals and returns **m** signals. The **n** argument signals are interpolated generating a temporary signal and this is decimated into **m** signals. In particular, if (**m** = 1), then **switch** is equivalent to **interpolate** and if (**n** = 1), then **switch** is equivalent to **decimate**.

In a Silage program, the periods of all signals depend on each other. One Silage definition alone does not define the period of the signals involved, but it defines a relation between them. One has to keep track of the period information in *every* relation modeling a Silage definition. Even for the operations that handle signals with equal periods, this information has to be explicit in the semantic interpretation. Most operations are only meaningful if the periods of the signals involved have some property. For instance, only signals with equal periods can be added or interpolated. Each Silage definition relates the samples and the periods of the involved signals.

3   Modeling the Period Semantics of Silage

In [11] the formal multi-rate semantics of Silage is defined relationally in the shallow embedding style. Each Silage definition is interpreted separately stating relations between the samples, the periods, and the phases of the signals. No distinction is made between the input and output signals of the DSP algorithm. In this section, we show informally the basic principles of the multi-rate semantics of Silage, with reference to the periods of the signals. We will refer to this view of Silage programs as the *period semantics* of Silage programs. This view of Silage programs will be used to illustrate the idea of degrees of formality in the shallow embedding style.

Next we introduce the basic ideas of the model of the periods of the signals in the multi-rate semantics of Silage, by refining the interpretation of some examples. In [11], Silage signals are modeled by a tuple of three components: samples (a function from time to sample values), period (the period of the signal), and the phase of the signal (the time instant when the first sample happens). Consider the following Silage definition def: $\boxed{y = x}$, where y and x are signal variables. Suppose also that **PER** is a selector defined in the HOL logic that applied to a signal returns the period component of the signal. Then the period semantics of the definition def should be:

$$[y = x]_p = ((\text{PER } y) = (\text{PER } x))$$

A signal variable is a particular case of Silage expression. Silage expressions are typed in ML. In [11], the period of every kind of Silage expression is defined. In particular, the period of a signal can be a multiple of the periods of other signals when the multi-rate Silage constructs **decimate, interpolate** and **switch** are involved. Suppose that the period of a generic Silage expression is $[e]_p$. Then a first attempt to define the period semantics of the Silage equality between two expressions $e_1 = e_2$ could be:

$$[e_1 = e_2]_p = ([e_1]_p = [e_2]_p)$$

However, there is a problem with this definition of $[e_1 = e_2]_p$. In Silage, there might be constant signals and signals obtained by interpolating constants. These signals do not have period information. We have defined $[e]_p$ such that, whenever an expression does not have period information, $[e]_p$ is zero. However, the interpretation of a Silage definition such as $y = 2$ should not define the period of y as zero. No signal in a Silage program has period equal to zero. So, we improve the definition of $[e_1 = e_2]_p$ as below, where **T** means true in the logic.

$$[e_1 = e_2]_p = (\text{if } ([e_2]_p = 0) \text{ then } \mathbf{T} \text{ else } ([e_1]_p = [e_2]_p))$$

There is still a problem with the definition of $[e_1 = e_2]_p$. Suppose that we have a definition such as $z = x + y$, where z, x and y are signal variables. Then we would have:

$$[z = x + y]_p = ([z]_p = [x + y]_p)$$

But the period of $(x + y)$ must be either the period of x or the period of y. Whatever it is, there must be a guarantee that the period of x is equal to the period of y. If the periods of every signal in Silage were declared, this could be checked statically by defining checking rules for each kind of Silage expression. Since this is not the case, whenever an expression is composed of many subexpressions, there are implicit constraints about the periods of the subexpressions. In [11] we have defined in ML the *check semantics of expressions* $[e]_c$ to model these constraints in the logic for every kind of Silage expression. So, refining the definition of $[e_1 = e_2]_p$ we have:

$$[e_1 = e_2]_p = (\text{if } ([e_2]_p = 0) \text{ then } \mathbf{T} \text{ else } (([e_1]_p = [e_2]_p) \wedge [e_2]_c))$$

In particular, $[e]_c$ and $[e]_p$ are defined in such a way that:

$$[\![x + y]\!]_c = ((\text{PER } x) = (\text{PER } y))$$
$$[\![x + y]\!]_p = (\text{PER } x)$$

Then the period semantics of $z = x + y$ is defined as follows:

$$[\![z = x + y]\!]_p = (((\text{PER } z) = (\text{PER } x)) \land ((\text{PER } x) = (\text{PER } y)))$$

The body of a Silage function is composed of a set of Silage definitions. No external or internal signal in a Silage function has period equal to zero. The arguments of a Silage function (inputs of the DSP algorithm) must have equal periods. These constraints and the set of period semantics of the definitions in the body of a Silage function define a linear system of equations with natural numbers. If a Silage function defines consistent periods for the signals, there must be a solution for this system. For instance, consider the following Silage function (where the initialisations are omitted):

```
func example (in1:word;in2:word) out:word =
a:word; b:word;
begin
    b = in1 + (a - b@2);
    out = interpolate(a,b);
    a = in2 + (a@2 + b@1);
end;
```

The period semantics of this function is shown below and it is defined in terms of the HOL functions **NOT_ZERO**, **SYNC**, and **MULTIPLE**. Given a list of natural numbers, the predicate **NOT_ZERO** holds if all elements of the list are different from zero. The predicate **SYNC** holds if all elements in a list are equal to a given element. Given two natural numbers n and m, the predicate **MULTIPLE** holds if n is a multiple of m.

```
∀ in1 in2 out. PER_example(in1,in2,out) =
(NOT_ZERO[PER in1; PER in2; PER out]) ∧
(SYNC [PER in1; PER in2] (PER in1)) ∧
(∃ a b.
    (NOT_ZERO[PER a; PER b]) ∧
    (((PER b) = (PER in1)) ∧ ((PER in1) = (PER a)) ∧ ((PER a) = (PER b))) ∧
    (((PER out) = ((PER a) DIV (LENGTH [a;b]))) ∧
    (MULTIPLE (PER a) (LENGTH [a;b])) ∧
    (SYNC [PER a;PER b] (PER a))) ∧
    ((PER a) = (PER in2)) ∧ ((PER in2) = (PER a)) ∧ ((PER a) = (PER b)))
```

4 Discussing the Model

In this section we discuss the model outlined in the previous section and the idea of degrees of formality in the shallow embedding style.

The interpretation of Silage definitions in a relational style is less elaborate than it is in a functional style because it is less context dependent. In the relational style, Silage definitions are interpreted one for one into terms of the logic in one step of compilation. In a functional style, a group of Silage definitions is interpreted all together to define the samples of a signal as a function from time instants to values. The definitions take as parameters the input signals and return the output signals. For each signal, the definitions about its initialisation samples and its algorithmical samples are considered all together to define the signal as a function. Such an approach has to handle functions that are not defined for all discrete time instants and has also to cope with the complexity of mutually recursive definitions. The interpretation of Silage programs in the functional style cannot be performed in one step of compilation and it involves more informal manipulation in ML to derive the meaning of Silage programs in the logic than the relational style does. Therefore, the choice between a relational style and a functional style has an impact on the degree of formality of the shallow embedding of Silage in HOL.

The simpler the semantics of a language in the shallow embedding style, the lighter the weight of the informal part of the embedding (the ML code) and therefore the safer is the whole system. There is no formal way to measure the complexity of a semantic interpretation and the complexity of the verification to be performed on specific interpretations. However, it is intuitive that the greater the complexity of a semantic interpretation (in ML) with respect to the complexity of the verification to be performed on specific interpretations (in the HOL logic), the worse is the quality of the embedding with respect to security. Ultimately, both the semantic interpretation and the verification should be informal (to have a system that is simpler but not much less safe than the one built in an unbalanced shallow embedding style) or the deep embedding style should be used (to have a system that is much safer than the one built in the unbalanced shallow embedding style).

Not only the way meaning is attached to programs affects the degree of formality in shallow embedding languages in HOL. Consistency issues also have to be addressed informally or formally, affecting the degree of formality. Next we discuss how one could address in the logic the well-formedness issue of consistency of the periods in Silage. Let X_p be the relational model of the period semantics of a Silage function x. For instance, the relational model of the period semantics of the Silage function **example** of the previous section is **PER_example**. The Silage function x defines consistent periods for its signals only if there are signals for which X_p holds. This is not the case, for instance, of the Silage function **notOkToo** defined in a previous section. This consistency issue can be addressed informally by an algorithm implemented in ML but it can also be addressed more formally in the logic, improving the degree of formality of the embedding. Although we have not automated the proof that a Silage program defines consistent periods for its signals, we have proved theorems about the constants defined to handle the periods of Silage expressions that allow us to prove automatically that the period semantics of a Silage function is equal to a linear system of equations about the periods. We will call this system **x_system**[4]. The proof about the consistency of the periods can be reduced to a first-order proof about natural numbers. If **x_system** can be solved informally using conventional techniques, the solutions of this system are the witnesses necessary to prove in the logic that the periods are consistent. *Addressing well-formedness issues in the logic, rather than informally, is a way to move closer to the logic in the scale of formality.* A system built in the shallow embedding style that addresses consistency issues both in its informal components (ML code) and in its formal component (HOL logic) is a hybrid verification system.

The main motivation for the definition of the multi-rate semantics of Silage was to prove the correctness of source-to-source transformations [20, 21, 22]. These transformations are used to optimize the results of the silicon compilation but they should not change the input-output behavior of the DSP algorithm specified in Silage. To prove the correctness of the transformations, one has to prove that the period semantics of a Silage function is equal to the period semantics of a new Silage function[5]. In particular, to prove the correctness of some transformations, it would be easier to have a more compact model of the periods of the signals. This simpler model could be obtained by trying to solve the **x_system** with conventional informal techniques. For instance, suppose that, for the Silage function **example** of the previous section we would have the following model:

$\forall$ in1 in2 out. PER_example'(in1,in2,out) = $\exists$ a b.
  (NOT_ZERO[PER in1; PER in2; PER out;PER a; PER b]) $\wedge$
  (SYNC [PER in1; PER in2; PER a; PER b] (2 * (PER out)))

One can prove in HOL the theorem below and then use it to prove the correctness of transformations on the Silage function **example**.

$\vdash$ $\forall$ in1 in2 out. PER_example(in1,in2,out) = PER_example'(in1,in2,out)

If the semantic interpretation were more elaborate so that **PER_example'**, rather than **PER_example**, would be the period semantics of **example**, the effort to prove the equivalence between these two models would be saved, but we would be moving away from the logic in the scale of formality. This is because the equivalence between the two models would have to be done informally by implementing an algorithm in ML. A proof procedure that formalizes the steps of this algorithm could prove in the logic, rather than informally, the equivalence between the two models in general. *In the shallow embedding style, to move simplification algorithms from the semantic interpretation to proof procedures formalizing them means to move closer to the logic in the scale of formality.*

---

[4] This system should be possible and not determined.

[5] This is necessary but not sufficient.

This move implies more safety for the system but it also implies more reasoning in the logic.

If Silage had specific constructs to define explicitly the periods of every signal in Silage, the interpretation of these constructs would be a straightforward and clean period semantics of Silage programs in the logic. The interpretation of these constructs would be the only information necessary to prove the equivalence of the periods of the input/output signals of two well-formed Silage programs. Proving the equivalence of the period semantics of two Silage programs would then be simple for any kind of transformation. As far as the periods are concerned, there would be no need to elaborate the ML interpretation to get a simpler model such as **PER_example'** or to prove intermediate theorems about the period semantics of two Silage programs in order to prove the correctness of the transformations in a straightforward way. Requiring that Silage programs have explicit period information might be a drawback for the CAD environment. However, this is not a serious drawback and explicit period information in Silage programs could reduce the proof effort to verify the correctness of transformations in Silage without deteriorating the quality of the embedding.

Whether or not there is explicit period information in Silage programs, the consistency of the periods has to be addressed somewhere, either informally in ML or in the logic. The difference is that, with the explicit period information, the check is simpler than without the explicit period information. Therefore, with the explicit period information, addressing the consistency of the periods informally, rather than in the logic, means less loss in the degree of formality than it would mean without having the explicit period information. With the explicit period information, the checking rules of the *check semantics* could be used in a straightforward way to verify informally the consistency of the periods in Silage programs, rather than be used to define in the logic what the periods of the signals are. The explicitness of the period information in Silage would also make it simple to address the issue of consistency of the periods formally in the logic without much proof effort. There are choices in the design of a language that can make it easier to move closer to the logic in the scale of formality, without implying serious drawbacks for the CAD environment.

5 Conclusions

We have discussed that there are degrees of formality in shallow embedding a language in HOL. The choice of the degree of formality is a trade-off between the security of the embedding and the amount and complexity of the proof effort in the logic. The design of a language could consider this verifiability issue. There are choices in the design of a language that can make it easier to improve the degree of formality, without implying serious drawbacks for the CAD environment. We have illustrated these ideas with the multi-rate semantics of Silage.

References

1. M. Gordon. "HOL: A Proof Generating System for Higher-Order Logic". In G. Birtwistle and P.A. Subrahmanyam, editors, *VLSI Specification, Verification and Synthesis*, pages 73–128. Kluwer Academic Publishers, 1988.

2. R. Boulton, A. Gordon, M. Gordon, J. Harrison, J. Herbert, and J. Van Tassel. "Experience with Embedding Hardware Description Languages in HOL". In V. Stavridou, T.F. Melham, and R. Boute, editors, *Proceedings of the IFIP International Conference on Theorem Provers in Circuit Design: Theory, Practice and Experience*, pages 129–156. Nijmegen, The Netherlands, North-Holland, Amsterdam, June 1992.

3. T.F. Melham. "Using Recursive Types to Reason about Hardware in Higher Order Logic". In G.J. Milne, editor, *The Fusion of Hardware Design and Verification: Proceedings of the IFIP WG 10.2 Working Conference*, pages 27–50. Glasgow, North-Holland, Amsterdam, July 1988.

4. T.F. Melham. "Automating Recursive Type Definitions in Higher Order Logic". In G. Birtwistle and P.A. Subrahmanyam, editors, *Current Trends in Hardware Verification and Automated Theorem Proving*, pages 341–386. Springer-Verlag, 1989.

5. G. Cousineau, M. Gordon, G. Huet, R. Milner, L. Paulson, and C. Wadsworth. *The ML Handbook*. INRIA, France, 1986.

6. E.L. Gunter. "Why We Can't Have SML Style Datatype Declarations in HOL". In L. Claesen and M. Gordon, editors, *Proceedings of the IFIP International Workshop on Higher Order Logic Theorem Proving and its Applications - HOL-92*, pages 561–568. IMEC, Leuven, Belgium, Elsevier Science Publishers B. V. (North-Holland), Amsterdam, September 1992.

7. R. Boulton, M. Gordon, J. Herbert, and J. Van Tassel. "The HOL Verification of ELLA Designs". In *Proceedings of the ACM/SIGDA International Workshop in Formal Methods in VLSI Design*. Miami, FL, January 1991.

8. R. Boulton. *A HOL Semantics for a Subset of ELLA*. Technical Report 254, University of Cambridge Computer Laboratory, April 1992.

9. A.D. Gordon. *A Mechanised Definition of Silage in HOL*. Technical Report 287, University of Cambridge Computer Laboratory, February 1993.

10. A.D. Gordon. "The Formal Definition of a Synchronous Hardware-description Language in Higher Order Logic". In *ICCD92: 1992 IEEE International Conference on Computer Design: VLSI in Computers & Processors*, pages 531–534. Cambridge, Massachusetts, IEEE Computer Society Press, October 1992.

11. C.M. Angelo, L. Claesen, and H. De Man. "The Formal Semantics Definition of a Multi-Rate DSP Specification Language in HOL". In L. Claesen and M. Gordon, editors, *Proceedings of the IFIP International Workshop on Higher Order Logic Theorem Proving and its Applications - HOL-92*, pages 375–394. IMEC, Leuven, Belgium, Elsevier Science Publishers B. V. (North-Holland), Amsterdam, September 1992.

12. J. Van Tassel. *A Formalisation of the VHDL Simulation Cycle*. Technical Report 249, University of Cambridge Computer Laboratory, March 1992.

13. J. Van Tassel. "A Formalisation of the VHDL Simulation Cycle". In L. Claesen and M. Gordon, editors, *Proceedings of the IFIP International Workshop on Higher Order Logic Theorem Proving and its Applications - HOL-92*, pages 359–374. IMEC, Leuven, Belgium, Elsevier Science Publishers B. V. (North-Holland), Amsterdam, September 1992.

14. P.N. Hilfinger. "Silage, a High-level Language and Silicon Compiler for Digital Signal Processing". In *Proceedings of the IEEE 1985 Custom Integrated Circuits Conference - CICC-85*, pages 213–216. Portland, OR, May 1985.

15. P.N. Hilfinger. *Silage Reference Manual*, December 1987.

16. D. Genin, P.N. Hilfinger, J. Rabaey, C. Scheers, and H. De Man. "DSP Specification Using the Silage Language". In *Proceedings of the IEEE International Conference on Accoustics, Speech and Signal Processing*, pages 1057–1060. Albuquerque, NM, April 1990.

17. L. Nachtergaele. *A Silage Tutorial*. IMEC, Leuven, Belgium, May 1990.

18. L. Nachtergaele. *User Manual for the S2C Silage to C Compiler*. IMEC, Leuven, Belgium, May 1990.

19. H. De Man, J. Rabaey, P. Six, and L. Claesen. "Cathedral-II: a Silicon Compiler for Digital Signal Processing". *IEEE Design & Test of Computers*, 3(6):73–85, December 1986.

20. P. Lippens. *Defining Control Flow from an Applicative Specification*. Technical report, Philips Research Laboratories, Eindhoven, December 1988.

21. I. Verbauwhede. *VLSI Design Methodologies for Application-specific Cryptographic and Algebraic Systems*. PhD thesis, Katholieke Universiteit Leuven - IMEC, Leuven, Belgium, 1991.

22. J. Vanhoof. *Multi-rate Expansion for CATHEDRAL-II/III. A tutorial*. Technical report, IMEC, Leuven, Belgium, October 1992.

A Functional Approach for Formalizing Regular Hardware Structures[*]

Dirk Eisenbiegler[1], Klaus Schneider[1] and Ramayya Kumar[2]

[1] Universität Karlsruhe, Institut für Rechnerentwurf und Fehlertoleranz,
(Prof. D. Schmid), P.O. Box 6980, 76128 Karlsruhe, Germany,
e-mail:schneide@ira.uka.de
[2] Forschungszentrum Informatik, Haid-und-Neustraße 10-14, 76131 Karlsruhe, Germany,
e-mail:kumar@fzi.de

Abstract. An approach for formalizing hardware behaviour is presented which is based on a small functional programming language called primitive ML (PML). Since the basic constructs of PML are simply typed λ-terms, PML lends itself both to simulation and verification. The semantics of PML is formally embedded in higher-order logic.

The formalization scheme is based on PML-functions that allow hardware descriptions from the logical level up to the algorithmic level. Besides descriptions of real circuits, abstract forms of hardware descriptions can also be dealt with in PML. The main emphasis is thereby put on regular hardware structures which are described by means of primitive recursion. PML-descriptions can easily be converted to syntactic structures, called hardware formulae, which can then be verified by the MEPHISTO system.

1 Introduction

Embedding hardware description languages (HDLs) in a logic or some calculi is essential for verification. The semantics of such embedded HDLs, which correspond to certain formulae in the underlying formal framework, can then be used to

- verify certain properties of an implementation,
- prove equivalences of two or more implementations and
- perform correct HDL-to-HDL translations.

Existing HDLs such as VHDL and ELLA are very powerful and complex. The expressive power is an advantage for circuit design, but their semantics are not formally defined due to their complexity. Formalizing the semantics of a HDL means bridging the gap between a high level design language and the simpler elements of the logic or a particular calculus.

Functional HDLs such as ELLA are rather close to logic. Those elements which consist exclusively of λ-terms can be converted to higher-order logic almost unchanged. In contrast to functional languages, procedural languages are more difficult to be formalized due to their operational semantics. For every basic instruction, it

[*] This work has been partly financed by a german national grant, project Automated System Design, SFB No.358.

must be described, how the execution of the instruction changes the global state. The effect of compound instructions must be derived from the effect of the basic instructions and the control structures.

There are several research projects about formalizing existing HDLs in higher-order logic [BGGH92, BGHT91, CaGM86] or other calculi [Hunt86, BoPS92]. In contrast to these projects, the starting-point of the approach presented in this paper is not a given HDL. Instead, a simple functional language called primitive ML (PML) is built on top of the logic such that its semantics is given right away. PML will be the basis for hardware descriptions. It can be regarded as a common sublanguage of HOL and ML (figure 1) — its syntax is similar to ML and every construct has a

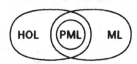

Fig. 1. Relation between HOL, ML and PML.

corresponding representation in HOL. The embedding of PML has been done in a *shallow* manner [BGGH92]. However, both PML-terms and the corresponding HOL-terms are simply typed λ-terms and there are only very small syntactical differences which could simply be overcome by pretty-printing.

Relational descriptions are frequently used for formalizing circuits, i.e. input and output signals need not be distinguished and signals can affect each other in an arbitrary manner. However, relational circuit descriptions can be ambiguous or even contradictory. In contrast, circuits are described in PML in a functional manner, i.e. circuits are represented by functions mapping input signals onto output signals.

PML cannot be viewed as a hardware description language such as VHDL, but as a general purpose programming language we use for describing hardware. In contrast to VHDL, PML has no hardware specific syntactic elements such as signals, interfaces and timing declarations. PML programmes can merely describe primitive recursive and μ-recursive functions and these functions can be regarded as a representation of the corresponding hardware.

In our hardware formalization scheme, we will describe two kinds of PML-functions: ones that represent single real circuits and others that describe sets of real circuits. Functions representing exactly one real circuit are called *concrete circuits* and functions describing a set of real circuits are called *abstract circuits*. Abstract circuits do not represent real circuits, but correspond to a scheme for describing regular hardware structures. Concrete circuits can be derived from abstract circuits by type instantiation and variable substitution.

In this paper, only the formalization of circuits is described. PML can be viewed as a more abstract layer for MEPHISTO [KuSK93, ScKK93c]. The verification of descriptions using PML is achieved by converting them into formulae which can be handled by MEPHISTO.

After having given a description of PML in section 2, concrete and abstract circuits are formalized in section 3 and 4, respectively. Finally, we briefly discuss the use of PML in simulation and verification in section 5.

2 Syntax and Semantics of PML

PML is a sublanguage of ML (figure 1). The syntax is similar, but there are less syntactical constructs — for example there are no exceptions and no side effects. Furthermore, data types in PML correspond to HOL–style data types and therefore they are weaker than ML data types [Gunt92]. Although PML has only a small number of basic elements, arbitrary μ–recursive functions can be expressed by PML functions, i.e. PML is Turing–complete.

2.1 Data Types and Primitive Recursion

In PML only HOL–style data types are allowed. For a detailed explanation of data types in HOL and the mechanism for defining them see [Melh88]. The syntax of a PML data type declaration is as follows:

```
primitive_datatype string ;
```

The parameter *string* defines the data type by giving the name of the type, the names of the constructors and the types of their arguments. The syntax is the same as in the HOL function define_type, e.g.

```
primitive_datatype " bool = T | F " ;
primitive_datatype " num  = Zero | Suc of num " ;
```

The semantics of HOL–style data types is described by a theorem which states that the primitive recursion over this type is unambiguous. In HOL, this theorem is derived whenever a new data type is introduced by define_type. A data type declaration in PML also introduces a basic function named PRIMREC_type, which is generated automatically. PRIMREC_type is derived from the semantics of the data type and it can be used for expressing primitive recursion over that data type. For example, the data types *num* and *bool* lead to the functions PRIMREC_bool and PRIMREC_num, respectively. They have the following semantics:

PRIMREC_bool T $a\ b$ = a
PRIMREC_bool F $a\ b$ = b

PRIMREC_num Zero $a\ f$ = a
PRIMREC_num (Suc n) $a\ f$ = $f\ n$ (PRIMREC_num $n\ a\ f$)

Arbitrary primitive recursive functions can be expressed by constant definitions based on PRIMREC–functions.

2.2 Derived Functions

As in ML, functions and constants can be added by the language constructs fun and val, respectively. However, function and constant definitions in PML both correspond to constant definitions in HOL. Therefore, function and constant definitions of PML are less powerful than those in ML. The restrictions are:

- There must be only one equation within a fun or val definition, e.g.
  fun is_zero 0 = T | is_zero(Suc n) = F; is not a valid PML definition.

- The parameters on the left hand side of the equation may only be variables and paired variables, e.g. val (Suc n) = y; is not allowed.
- The expressions on the right hand side are built up by function applications (f a) and λ-abstractions (fn x => a). The only basic functions are PRIMREC-functions and WHILE (WHILE will be introduced later in section 2.5).
- The function being defined must not appear on the right hand side of the equation, e.g. fun odd n = PRIMREC_num n F (fn a => fn b => not(odd a)); is not a valid PML definition. Recursion can always be expressed by equivalent definitions which use PRIMREC-functions and WHILE.
- There is no exception handling.
- (case...of...) has not yet been implemented.

2.3 Predefined Data Types

Some data types are already defined in order to support pretty-printing for them. For example, it is possible to write 2 instead of Suc(Suc Zero). However, there is no pretty-printing for user defined types. The predefined data types are:[1]

```
primitive_datatype  " bool = T | F " ;
primitive_datatype  " prod = Comma of 'a # 'b " ;
primitive_datatype  " list = Nil | Cons of 'a # list " ;
primitive_datatype  " num  = Zero | Suc of num " ;
```

The following syntactic sugar refers to the predefined types. They can all be put down to expressions based on data type constructors and PRIMREC-functions.

- (a, b) may be used instead of (Comma a b)
- (fn (x, y) => $p[x, y]$) may be used instead of
  (fn z => PRIMREC_prod z (fn x => fn y => $p[x, y]$))
- (let val x = p in $q[x]$ end) may be used instead of ((fn x => $q[x]$) p).
- numerals 0, 1, 2,... may be used instead of Zero, (Suc Zero), (Suc(Suc Zero)),...
- [], [a], [a, b],... may be used instead of Nil, (Cons a Nil), (Cons a (Cons b Nil)),...

2.4 Example

We illustrate the use of the language constructs by a traffic light controller. First, a new data type named *state* is defined, which represents the states of the traffic light.

```
primitive_datatype  " state = Green | Yellow | Red " ;
```

This type declaration automatically introduces the function PRIMREC_state with the following semantics:

PRIMREC_state Green a b c = a
PRIMREC_state Yellow a b c = b
PRIMREC_state Red a b c = c

---

[1] In PML, type variables are expressed by 'a, 'b, 'c, etc. whereas the corresponding type variables in HOL are expressed by α, β, γ, etc. It should be noted that the polymorphism in PML corresponds to that of HOL only.

A constant named init containing the initial state Red can be defined as:

```
val init = Red;
```

The function next takes a state as parameter and calculates the successor state. In this simple traffic light controller, the state changes from red directly to green, but changes from green to red via yellow.

```
fun next x = PRIMREC_state x Yellow Red Green;
```

Figure 2 shows the entire programme and the corresponding HOL–formula describing its semantics.

PML–Program	Semantics
```primitive_datatype "state = Green | Yellow | Red"; val init = Red; fun next x = PRIMREC_state x Yellow Red Green;```	$(\forall a\, b\, c.\, \exists_1 g.$ $\quad (g\, \text{Green} = a) \wedge (g\, \text{Yellow} = b) \wedge$ $\quad (g\, \text{Red} = c)) \wedge$ $(\forall a\, b\, c.$ $\quad (\text{PRIMREC\_state Green}\, a\, b\, c = a) \wedge$ $\quad (\text{PRIMREC\_state Yellow}\, a\, b\, c = b) \wedge$ $\quad (\text{PRIMREC\_state Red}\, a\, b\, c = c)) \wedge$ $(\text{init} = \text{Red}) \wedge$ $(\forall x.\, \text{next}\, x =$ $\quad \text{PRIMREC\_state}\, x\, \text{Yellow Red Green})$

Fig. 2. Traffic light programme.

2.5   μ–Recursion

According to Church's Thesis, there are several equivalent schemes for describing computable functions. μ–recursive functions are one means for describing computable functions. With the elements described until now, only primitive recursive functions can be described, while μ–recursive functions cannot. Primitive recursive functions are sufficient for describing hardware *implementations*, but they are too weak for formalizing algorithmic *specifications*. Previous work such as the approaches followed by the Boyer–Moore community, are limited to primitive recursive specifications that cannot express all kinds of algorithms.

Unlike primitive recursive functions, μ–recursive functions need not be total. In ML it is possible that the evaluation of a function application does not terminate. The equations of an ML function definition can be considered as a constant specification in HOL where the specified function need not be described unambiguously, i.e. nothing can be said about the value of a function application where the evaluation does not terminate. In contrast to ML, the result of a PML function always has an explicitly defined value, even if the function application does not terminate. In this case, the value of the function is explicitly defined to be the constant Undefined, otherwise the result is (Defined y) for a certain y. The data type *partial* is used for describing values of μ–recursive functions:

```
primitive_datatype  " partial = Defined of 'a | Undefined " ;
```

The corresponding function PRIMREC_partial has the following semantics:

PRIMREC_partial (Defined x) f a = f x
PRIMREC_partial Undefined f a = a

The function WHILE is the basis for μ-recursion in PML and can be used to create loops. Given functions f and g and a parameter x, it iterates f until a value x is reached with g $x = $ F.

⊢ iota f = PRIMREC_bool ($\exists_1$ f) (Defined(ε f)) Undefined
⊢ terminates(f, n) = (f n) $\wedge$ ($\forall m.$ $m < n$ $\Rightarrow$ $\neg(f$ m))
⊢ mu f = iota($\lambda m.$ terminates(f, m))
⊢ power f n x =
   PRIMREC_num n (Defined x) ($\lambda a\, b.$ PRIMREC_partial b f Undefined)
⊢ WHILE g f x =
   PRIMREC_partial
     (mu($\lambda n.$ PRIMREC_partial (power f n x)
            ($\lambda y.$ PRIMREC_bool (g y) F T) F))
     ($\lambda n.$ power f n x)
   Undefined

The semantics of WHILE is described using four auxiliary constants: iota, terminates, mu and power. The function iota resembles the Hilbert operator. But in contrast to the Hilbert operator its value is (Defined y) in case the predicate specifies a *unique* value and Undefined if it does not. The predicate terminates(f, n) states that n is the smallest number such that (f n) becomes true. mu is a formalization of the μ-operator where mu f = Undefined corresponds to $\mu(f)\uparrow$ [2] and mu f = Defined k corresponds to $\mu(f) = k$. The function power computes the multiple application of a function, i.e. (power f n x) computes $f^n(x)$. The expression (WHILE g f x) calculates $f^{\mu(\lambda n. g(f^n(x))=\mathrm{F})}(x)$. [3]

2.6 Example

The traffic light example of section 2.4 is extended by a μ-recursive function red_time. red_time is calculating the next time when the traffic light becomes red. The traffic light is described by a function $f_{num \rightarrow state}$, that is assigning a state to every time. For a given function $f_{num \rightarrow state}$ and a time t_{num}, the function red_time calculates the smallest $n \geq t$ with f n = Red.

Figure 3 shows an implementation in PML in comparison with an implementation in ML. The implementations in PML and ML are not really equivalent, since their types differ. In PML the result of red_time has the type (*num partial*) whereas in ML it is *num*.

---

[2] $\mu(f)$ is the smallest element of $\{f(x) = \mathrm{T} | x \in N\}$ and $\mu(f)\uparrow$ denotes that $\{f(x) = \mathrm{T} | x \in N\}$ is empty.

[3] Extending primitive recursive functions by WHILE is equivalent to the extension by the μ-operator, i.e. both extensions lead to computable functions. However, WHILE-expressions can be evaluated more efficiently by an interpreter than expressions using the μ-operator.

PML–Implementation | ML–Implementation

```
fun is_not_red x =
   PRIMREC_state x T T F;

fun red_time f t =
   WHILE
      (fn n => is_not_red(f n))
      (fn n => Defined(Suc n))
      t;
```

```
fun red_time f t =
   if ((f t) = Red) then
      t
   else
      (red_time f (Suc t));
```

Fig. 3. Implementations of red_time.

3   Concrete Circuits

The functions described in this section are called concrete circuits since each of them corresponds to exactly one real circuit. A concrete circuit is represented by a function that assigns an input signal onto an output signal. Hardware structures are build up by function definitions. Expressing a structure by a function definition is possible, if and only if the structure does not have cycles. Since structures of sequential circuits essentially have cycles, sequential circuits are represented by a triple consisting of a combinational transient circuit, a combinational output circuit and an initial state. Instead of combining sequential circuits directly, their transient circuits, output circuits and initial states are combined.

3.1   Combinational Circuits

Individual signals of combinational circuits have type *bool*. Other than individual signals can be obtained by pairing individual signals. Combinational circuits are represented by functions assigning an input signal to an output signal. Figure 4 shows a 1–bit fulladder implemented by the circuits and, or and xor.

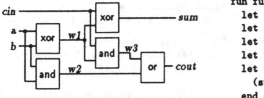

```
fun fulladder (cin,(a,b)) =
   let val w1 = xor(a,b) in
   let val w2 = and(b,a) in
   let val sum = xor(cin,w1) in
   let val w3 = and(cin,w1) in
   let val cout = or(w3,w2) in
      (sum,cout)
   end end end end end;
```

Fig. 4. Structure of a 1–bit fulladder and the PML representation.

3.2   Sequential Circuits

The signals of sequential circuits are time dependent. Their type is $num \rightarrow \gamma$ where the type *num* represents the discrete time and γ is the type of a time independent signal. Sequential circuits map time dependent input signals onto time dependent output signals, thus they have the following type: $(num \to \alpha) \to (num \to \beta)$.

The description style used for combinational circuits does not allow cycles which are necessary for sequential circuits. In order to use this scheme also for sequential circuits, we define a sequential circuit by a triple (f, g, q) consisting of a combinational transient circuit f, a combinational output circuit g and an initial state q. Thus, sequential structures can be expressed by interconnecting combinational circuits. Since structures of Mealy machines might lead to zero–delay–cycles, in this paper only Moore circuits will be considered (see figure 5).

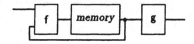

Fig. 5. Scheme of a Moore circuit.

A function named makeseq is introduced, which computes a sequential circuit for a given triple (f, g, q). makeseq can be defined by the equations below. The function definition given by these equations does not have the form of a PML function definition. It can rather be regarded as a ML–style function definition or a constant specification in HOL. These equations describe the desired properties of the intended PML function in a clearer manner.

$$\text{makeseq } (f, g, q) \, a \, 0 \quad = \quad f \, q$$
$$\text{makeseq } (f, g, q) \, a \, (\text{Suc } t) = \quad \text{makeseq } (f, g, g(a\, t, q)) \, (\lambda t. \, a(\text{Suc } t)) \, t$$

The corresponding implementation in PML:

```
fun makeseq (f,g,q) a t =
    f (PRIMREC_num t q (fn n => fn r => g(a n,r)));
```

It shall be demonstrated, how structures of sequential circuits can be described in PML by function definitions. Figure 6 shows an example for a structure consisting

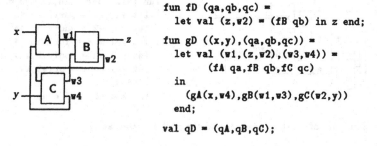

```
fun fD (qa,qb,qc) =
    let val (z,w2) = (fB qb) in z end;

fun gD ((x,y),(qa,qb,qc)) =
    let val (w1,(z,w2),(w3,w4)) =
        (fA qa,fB qb,fC qc)
    in
        (gA(x,w4),gB(w1,w3),gC(w2,y))
    end;

val qD = (qA,qB,qC);
```

Fig. 6. Structure of a sequential circuit and the PML representation.

of three sequential circuits A, B and C. The circuits A, B and C are represented by (fA, gA, qA), (fB, gB, qB) and (fC, gC, qC). The entire circuit is called D and its triple (fD, gD, qD) can directly be extracted from the structure.

4   Abstract Circuits

In the previous section, functions were used for describing single real circuits. In contrast to concrete circuits, abstract circuits represent sets of (concrete) circuits and are therefore more powerful than concrete circuits. Similar to concrete circuits, abstract circuits can also be represented by PML–functions. Abstract circuits can be polymorphic and allow parameters which have types that are not restricted to pairs (e.g. lists and trees may be used for instance). Concrete circuits can be obtained from abstract circuits by type instantiation and variable substitution.

The function mux is an example for an abstract circuit:

```
fun mux((s:bool),(a:'a),(b:'a)) = PRIMREC_bool s b a;
```

Concrete circuits can be derived from mux by instantiating the type variable α (α is expressed by 'a within the PML syntax). Figure 7 shows two instances of mux.

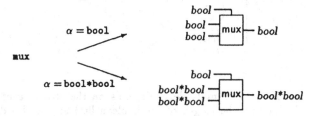

Fig. 7. Instances of a polymorphic 2:1–multiplexer.

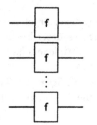

Abstract circuits are used for formalizing regular hardware structures which can be expressed by means of primitive recursion. In general, regular circuit structures lead to regular signal structures. If for example, a structure is described, that consists of n combinational circuits connected in parallel, then it would be appropriate to use the type list for grouping together the input and the output signals.

The structure of the input signal determines the structure of the circuit and the structure of the output signal. Grouping signals together by recursive types such as *list* is flexible, since the structure of the signals and especially the number of the individual signals depends on the value.

Fig. 8.: Regular circuit.

Other types than *list* can be used for grouping signals. In the next example, signals are grouped together by a list and a binary tree.

```
primitive_datatype "list = Nil | Cons of 'a # list";
primitive_datatype "btree = Bleaf of 'a | Bnode of btree # btree";
```

The semantics of the corresponding PRIMREC–functions is:

PRIMREC_list Nil a f $=$ a
PRIMREC_list (Cons x y) a f $=$ f x y (PRIMREC_list y a f)
PRIMREC_btree (Bleaf x) f g $=$ f x
PRIMREC_btree (Bnode x y) f g $=$
   g x y (PRIMREC_btree x f g) (PRIMREC_btree y f g)

The input of the 2^n:1–multiplexer consists of a group of data inputs and an address signal for selecting one of the data inputs. The data input signals have an arbitrary type α and they are grouped together as (α btree). The address signals are represented by a list of booleans (see figure 8).

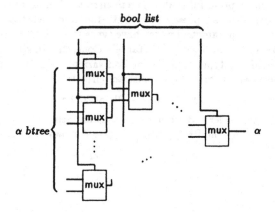

Fig. 9. 2^n:1–multiplexer.

The structure of the 2^n:1–multiplexer depends on the structure of the input signals (i.e. it depends on the length of the boolean list) and it also depends on the shape of the binary tree. The PML function representing the 2^n:1–multiplexer function is total and so the 2^n:1–multiplexer has to be designed for arbitrary lists and arbitrary binary trees, even though after instantiation the binary tree has a constant depth which is equal to the length of the list.

Figure 10 illustrates, how the the structure of the 2^n:1–multiplexer bmux is defined. For a given structure of the input signals, a circuit structure describes how the 2^n:1–multiplexer can recursively be put down to other 2^n:1–multiplexers having 'smaller' input structures in the sense of a canonical term–ordering.

The following equations give a formal definition of the description in figure 10. The equations correspond to the circuit structures of the figure in a one–to–one manner.

$$
\begin{aligned}
\mathsf{bmux}(x, \mathsf{Bleaf}\ a) &= a \\
\mathsf{bmux}(\mathsf{Nil}, \mathsf{Bnode}\ b\ c) &= \mathsf{bmux}(\mathsf{Nil}, b) \\
\mathsf{bmux}(\mathsf{Cons}\ h\ t, \mathsf{Bnode}\ b\ c) &= \mathsf{mux}(h, \mathsf{bmux}(t, b), \mathsf{bmux}(t, c))
\end{aligned}
$$

Obviously bmux is a primitive recursive function, but these equations cannot directly be used for the PML implementation. To implement bmux in PML, the definition has to be transformed: the interlocking primitive recursions over *list* and *btree* have to be broken up:

$$
\begin{aligned}
\mathsf{bmux}(x, \mathsf{Bleaf}\ a) &= a \\
\mathsf{bmux}(\mathsf{Nil}, \mathsf{Bnode}\ b\ c) &= \mathsf{bmux}(\mathsf{Nil}, b) \\
\mathsf{bmux}(\mathsf{Cons}\ h\ t, \mathsf{Bnode}\ b\ c) &= \mathsf{mux}(h, \mathsf{bmux}(t, b), \mathsf{bmux}(t, c))
\end{aligned}
$$

$\Downarrow$

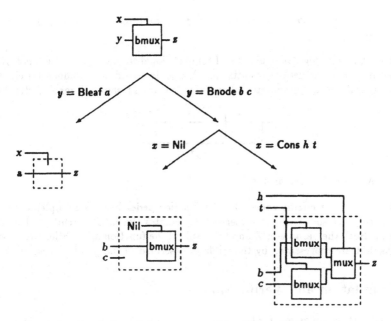

Fig. 10. Recursive description of the structure of the 2^n:1-multiplexer.

$$
\begin{aligned}
\mathsf{bmux}(\mathsf{Nil}, \mathsf{Bleaf}\,a) \quad &= \quad a \\
\mathsf{bmux}(\mathsf{Nil}, \mathsf{Bnode}\,b\,c) \quad &= \quad \mathsf{bmux}(\mathsf{Nil}, b) \\[4pt]
\mathsf{bmux}(\mathsf{Cons}\,h\,t, \mathsf{Bleaf}\,a) \quad &= \quad a \\
\mathsf{bmux}(\mathsf{Cons}\,h\,t, \mathsf{Bnode}\,b\,c) &= \quad \mathsf{mux}(h, \mathsf{bmux}(t, b), \mathsf{bmux}(t, c)) \\
&\Downarrow \\
\mathsf{bmux}(\mathsf{Nil}, y) \quad &= \quad \mathsf{PRIMREC\_btree}\; y\; (\lambda\,a.\,a)\;(\lambda\,a\,b\,v\,w.\,v) \\[4pt]
\mathsf{bmux}(\mathsf{Cons}\,h\,t, y) &= \quad \mathsf{PRIMREC\_btree}\; y\; (\lambda\,a.\,a) \\
&\qquad\quad (\lambda\,a\,b\,v\,w.\,\mathsf{mux}(h, \mathsf{bmux}(t, v), \mathsf{bmux}(t, w))) \\
&\Downarrow \\
\mathsf{bmux}(x, y) \quad &= \quad \mathsf{PRIMREC\_list}\; x \\
&\qquad (\lambda\,s.\,\mathsf{PRIMREC\_btree}\; s\; (\lambda\,a.\,a)\;(\lambda\,a\,b\,v\,w.\,v)) \\
&\qquad (\lambda\,h\,t\,r\,s. \\
&\qquad\quad \mathsf{PRIMREC\_btree}\; s\; (\lambda\,a.\,a)\;(\lambda\,b\,c\,v\,w.\,\mathsf{mux}(h, r\,b, r\,c))\,) \\
&\qquad y
\end{aligned}
$$

The corresponding implementation in PML is:

```
fun bmux (x,(y:'a btree)) =
  PRIMREC_list x
    (fn s =>
      PRIMREC_btree s (fn a => a)
        (fn a => fn b => fn v => fn w => v) )
    (fn h => fn t => fn r => fn s =>
      PRIMREC_btree s (fn a => a)
```

```
            (fn b => fn c => fn v => fn w => mux(h,r b,r c)) )
    y;
```

Up to now, merely regular structures of combinational circuits were considered. Regular structures of sequential circuits can also be described since sequential circuits can be put down to combinational circuits. Figure 11 shows a regular structure

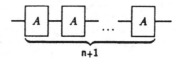

Fig. 11. A series of Moore circuits.

based on a Moore circuit called A. The function series has been implemented in PML. It takes A and n as parameters and calculates the entire circuit as shown in figure 11. Both the parameter A and the result of the function application are Moore circuits that are represented by the triple as described in the previous section.

5 Simulation and Verification

5.1 An Interpreter for PML

PML programmes can simply be executed by a ML interpreter, however, the ML environment has to be extended by some functions and data types. The type declaration construct primitive_datatype has to be implemented as a ML–function, the predefined data types of PML have to be declared and the function WHILE has to be implemented.

As all PML programmes are also ML programmes and the extended ML interpreter still accepts ML programmes, it is not tested whether the input is a PML programme or more general, a ML programme.

5.2 Simulation Tools

Some general tools for simulating circuits have been implemented in the extended ML interpreter. These tools are not PML functions, but they take PML functions which describe circuits as arguments. Moreover, they display values of output signals during the simulation of combinational and sequential circuits.

For this reason, output functions called *type_to_string* have been implemented for all predefined data types. These output functions convert a value of a certain PML type to a string. When a new data type is added, a corresponding output function should also be implemented. Output functions are used as parameters of the following simulation tools.

A function called function_table has been implemented for combinational circuits for performing the simulation and displaying the results as a table.

Sequential circuits that are represented by triples (f, g, q) could be simulated using the function makeseq. However, if the output is considered over a period, the use of make_seq would be very inefficient because for every single output, the calculation would start from the beginning. The simulation function for sequential circuits that has been implemented does not have this disadvantage. The circuit is simulated only once until the last point of time of the considered period is reached. The parameters for a sequential simulation are: the circuit represented by (f, g, q), a time dependent input signal, a condition for terminating the simulation and an output function for converting the circuits output to a string.

5.3 Verification

A function called extend_theory_by_pml is implemented for converting a PML programme to HOL. Some tools are provided for reasoning about PML functions. They are concerned with: extending constant abbreviations, evaluation and induction.

For concrete circuits and some classes of abstract circuits, a more direct approach for verification is used. The PML-terms can be converted into certain formulae, called hardware–formulae [ScKK93c], which can be automatically verified within the MEPHISTO system [KuSK93] (see figure 12). Thus PML descriptions can also be used as a front–end specification language within this verification framework.

```
fun fulladder (cin,(a,b)) =                  ∀cin a b sum cout.
    let val w1 = xor(a,b) in                     fulladder(cin, a, b, sum, cout) ⇔
    let val w2 = and(b,a) in                      ∃w1 w2 w3.
    let val sum = xor(cin,w1) in      →              xor(a, b, w1) ∧
    let val w3 = and(cin,w1) in                      and(b, a, w2) ∧
    let val cout = or(w3,w2) in                      xor(cin, w1, sum) ∧
      (sum,cout)                                     and(cin, w1, w3) ∧
    end end end end end;                             or(w3, w2, cout)
```

Fig. 12. Converting PML circuit descriptions to hardware–formulae.

6 Conclusion and Future Work

We have presented a general purpose programming language that is formally embedded in higher–order logic and we have also demonstrated, how this language can be used for formalizing both combinational and sequential circuits. The main emphasis has been put on demonstrating how regularity can be expressed by means of primitive recursion.

PML is a very simple language and writing PML programmes can be rather tough since all function definitions have to be broken up to the primitive recursion and μ–recursion constructs. It is intended to improve the applicability of PML by adding a more comfortable ML–style mechanism for expressing recursive functions.

In future research it shall be analyzed, how PML descriptions of circuits can be used for hardware design. PML descriptions shall be used in several fields: verification, simulation, symbolic simulation, synthesis and optimization (HDL–to–HDL transformations).

References

[BGGH92] R. Boulton, A. Gordon, M. Gordon, J. Herbert, and J. van Tassel. Experiences with Embedding hardware description languages in HOL. In V. Stavridou, T.F. Melham, and R. Boute, editors, *Conference on Theorem Provers in Circuit Design*, IFIP Transactions A-10, pages 129–156. North-Holland, 1992.

[BGHT91] R. Boulton, M. Gordon, J. Herbert, and J. van Tassel. The HOL verification of ELLA designs. In *International Workshop on Formal Methods in VLSI Design*, January 1991.

[BoPS92] D. Borrione, L. Pierre, and A. Salem. Formal verification of VHDL descriptions in Boyer-Moore: First results. In J. Mermet, editor, *VDHL for Simulation, Synthesis and Formal Proofs of Hardware*, pages 227–243. Kluwer Academic Press, 1992.

[CaGM86] A. Camilleri, M.J.C. Gordon, and Th. Melham. Hardware verification using higher order logic. In D. Borrione, editor, *From HDL Descriptions to Guaranteed Correct Circuit Designs*, pages 41–66. North-Holland, 1986.

[Cami88] J. Camilleri. Executing behavioural definitions in higher order logic. Technical Report 140, University of Cambridge Computer Laboratory, 1988.

[Cami91] J. Camilleri. Symbolic compilation and execution of programs by proof: a case study in HOL. Technical Report 240, University of Cambridge Computer Laboratory, 1991.

[Gunt92] E. Gunter. Why we can't have SML-style datatype declarations in HOL. In L.J.M. Claesen and M.J.C. Gordon, editors, *Higher Order Logic Theorem Proving and its Applications*, volume A-20 of *IFIP Transactions*, pages 561–568, Leuven, Belgium, 1992. North-Holland.

[Hunt86] W.A. Hunt. The mechanical verification of a microprocessor design. In D. Borrione, editor, *From HDL Descriptions to Guaranteed Correct Circuit Designs*, pages 89–132. North-Holland, 1986.

[Jone87] S.L.P. Jones. *The Implementation of Functional Programming Languages*. Prentice Hall, 1987.

[KuSK93] R. Kumar, K. Schneider, and Th. Kropf. Structuring and automating hardware proofs in a higher-order theorem-proving environment. *Journal of Formal Methods in System Design*, 2(2):165–223, 1993.

[Melh88] T. F. Melham. Automating recursive type definitions in higher order logic. Technical Report 146, University of Cambridge Computer Laboratory, September 1988.

[ScKK93c] K. Schneider, R. Kumar, and Th. Kropf. Eliminating higher-order quanitifers to obtain decision procedures for hardware verification. In *International Workshop on Higher-Order Logic Theorem Proving and its Applications*, Vancouver, Canada, August 1993.

[Shee88] M. Sheeran. Retiming and slowdown in Ruby. In G.J. Milne, editor, *The fusion of Hardware Design and Verification*, pages 289–308, Glasgow, 1988. North Holland.

A Proof Development System for the HOL Theorem Prover

Laurent Théry

University of Cambridge Computer Laboratory*
and
INRIA Sophia-Antipolis

Abstract. In this paper, we present a system to improve the interaction between HOL and the user when doing proofs.

1 Introduction

Learning how to use any theorem prover requires a rather important investment. The user has not only to assimilate the specificities of the prover in terms of the logic and the definition mechanisms but also to digest a command language with its own syntax and nomenclatures. In that respect, HOL is not an exception. On the contrary, the constant addition of new theories and new tools makes it even harder to handle.

Our aim in building a proof development system is to simplify the use of HOL by providing a user-friendly environment for doing proofs. From generic ideas about interface of theorem provers given in [10], some specific tools have been implemented for HOL. In this paper, we present these different tools starting from the theory level, through the proof level and finally to the tactic level.

2 Theories

HOL has the notion of *theory* to structure and organise the proof activity. A theory regroups the definitions of some new constructors and the theorems that can be derived from these definitions. A theory also imports other theories by declaring them as parents. Large examples as [7] may contain more than 50 theories. Therefore, visualising the dependences and the contents of these theories becomes an important issue for the comprehension of the environment in which proofs are done.

In the proof development system, a tool has been developed to display the hierarchy of the theories in the form of a graph. In the graph, every theory is represented as a node and an arrow is drawn between two theories if one is a parent of the other. The Figure 1 presents an example of such a graph for the initial HOL theories. The dependences between theories can then be investigated by splitting different subcomponents of the graph in subviews. Some standard browsing operations are also available to inspect the contents of theories.

* Supported by SERC grant GR/G 33837 and a grant from DSTO Australia

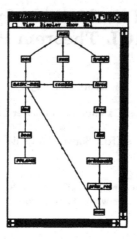

Fig. 1. The theory graph for the HOL theory

3 The proof

3.1 Controlling the construction

We limit our interest in the construction of goal oriented proofs: a proof starts with an initial goal, then commands (tactics) are applied to break this goal into several presumably simpler ones till trivial facts are obtained. Each step of the proof construction is then characterised by

- the selection of a subgoal among the remaining ones;
- the construction of a tactic to be applied on the selected subgoal.

As advocated in [8], the choice of the subgoal to attack is important to avoid rigid proof strategies. It is also necessary to add an "undo" mechanism to these two basic operations. The standard solution of undoing by restoring some previous states is here not satisfactory. As the user may work simultaneously on different subgoals, it is necessary to implement a notion of undo *local* to a subgoal, which does not affect steps on other subgoals.

In order to provide a natural user-interface for all these operations (subgoal selection, tactic construction, undoing), the proof of a proposition is transformed into the construction of a tactic that proves the proposition. To explain how this can be done, we first introduce the notion of *incomplete tactic* and then define the mapping between a proof state and an incomplete tactic.

An incomplete tactic is a tactic where place holders can be used at any location as a value. Textually these place holders are represented by the type expected at this location surrounded by ⟨⟩. From this definition it follows that REPEAT ⟨*tactic*⟩ THEN EXISTS_TAC ⟨*term*⟩ is an incomplete tactic. It has two place holders, the first stands for a tactic, and the second for a term.

Mapping a proof state into an incomplete tactic is based on the tactical THENL. This tactical groups a sequence of tactics in one tactic. Given a subgoal σ, if the application of Tactic creates two new subgoals σ_1 and σ_2 and Tactic1 proves the first subgoal and Tactic2 proves the second one, then the tactic Tactic THENL [Tactic1;Tactic2] proves the subgoal σ. With this tactical, the mapping of a proof state into an incomplete tactic is simply done by composing the tactics that lead to this state with THENL tacticals and representing the unsolved subgoals with ⟨tactic⟩ place holders. For example, suppose that the steps to prove σ are first to apply Tactic on the initial goal, then Tactic2 on the second subgoal and finally Tactic1 on the first one. The corresponding incomplete tactics are ⟨tactic⟩ for the initial state, then Tactic THENL [⟨tactic⟩; ⟨tactic⟩] after the application of Tactic, Tactic THENL [⟨tactic⟩; Tactic2] after the application of Tactic2 and finally Tactic THEN [Tactic1; Tactic2]. All the operations of the proof construction can be explained in terms of edition on the corresponding incomplete tactics:

- Goal selection: Selection of a place holder ⟨tactic⟩.
- Tactic construction: Replacement of the selected place holder by a tactic.
- Local undoing: Replacement of a tactic by a place holder ⟨tactic⟩.

Every time a place holder is replaced by a tactic, the system has to check the coherence of the replacement. If the tactic solves the corresponding subgoal, the modification is accepted, if not a THENL is automatically appended to the tactic with as many ⟨tactic⟩ as the generated subgoals. For example, in the proof of σ, the replacement of the initial ⟨tactic⟩ by Tactic automatically generates Tactic THENL [⟨tactic⟩; ⟨tactic⟩] as the application of Tactic creates two new subgoals. A proof is finally completed when its corresponding incomplete tactic has no place holders left.

4 Construction of a new proof step

Once a subgoal is selected, the construction of a new proof step is achieved by building a tactic to be applied on this subgoal. The conventional way of doing this is by typing the tactic as a sequence of characters. An alternative and interesting possibility is to use the syntax and the type structure of the command language to help this construction. Such approaches have been originally developed for program construction (see for example Centaur [4] and Cornell Synthesizer [6]). The basic idea is to guide the construction with a menu that provides a list of possible constructions for any edited expression.

4.1 Guided edition

Each item of the menu represents a scheme of construction called *pattern*. Patterns are basically ML expressions where place holders can be used. Given an edited expression, a pattern is possible if its type is compatible with the type of the expression. The Figure 2 shows an example of such a menu. The tactic section is highlighted, only items from this section can be selected: the sub-expression that is edited is a tactic.

The content of the menu can be interactively extended. With the help of the type checker, any ML function is automatically expanded into a pattern. For example, in order to insert the tactical **THEN** in the tactic section, it is sufficient to enter the term **THEN** and the system transforms it automatically into the incomplete tactic ⟨*tactic*⟩ **THEN** ⟨*tactic*⟩.

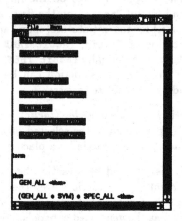

Fig. 2. Guided edition with a menu

4.2 Using the goal

The construction of a proof step is relative to a certain subgoal. In addition to the menu, a selection on a goal can provide four different types of argument for this construction: a term, an occurrence, a theorem and a tactic.

Providing a term Given a selection on a goal, a term can be obtained by simply pasting the selected subterm into the tactic text. However, it is necessary to ensure that this argument will be accepted by the system, ie that it can be type checked. The fact that the subterm is taken from a well-typed expression is not sufficient. For example, if "$\forall x \neg(x + 1 = x)$" is well typed, its subterm "x" cannot be type checked by the system. In order to ensure this property, the type information is automatically added on every variable, so "x" is copied as "$x : num$".

Providing an occurrence Occurrences are used as arguments to denote a location in a goal rather than a value. The notion of occurrence in goals is not native in HOL, but a notion of occurrence in terms, called *director string*, has already been defined by David Shepherd in his conversion library [9]. A director string is a sequence built from an alphabet of three letters f, a and b. Each letter represents a movement from a term to one of its immediate subterms:

- f designates the left part of an application;
- a designates the right part of an application;
- b designates the body of an abstraction.

To build the director string associated to a subterm, one simply concatenates in a string the successive movements to reach the subterm starting from the root. For example, given the term "$\forall x \neg (x+1 = x)$", the director string corresponding to "1" is 'abafaa'.

To extend the notion of occurrence from terms to goals, the following ML type is defined:

```
type occ = GOAL_OCC of string |
           ASM_OCC of ((goal -> term) # string) ;;
```

GOAL_OCC takes a director string to represent an occurrence in the conclusion. ASM_OCC takes a pair to represent an occurrence in an assumption. The first element of this pair is a function that, given a goal, returns the selected assumption. The second element is a director string on this assumption. Having a function for selecting the assumption makes the method to refer to assumptions more flexible. The current implementation generates occurrences referring to assumptions by value.

Providing a theorem From a boolean term A, it is possible to generate the theorem $A \vdash A$. The main interest in producing such a theorem is when the selection is on an assumption. This allows assumptions to be used as "standard" theorems.

Providing a tactic The generation of a tactic from a selection follows the method of *proof by pointing* developed in [1]. The basic idea is to give a semantics to the selection in term of elimination of logical connectives. For example, let us consider the case of a disjunction with the following goal:

$$\Gamma \text{ ?- } A \vee B$$

where Γ is a list of assumptions and A, B arbitrary formulas. A possible way of proving this goal is to show that A is true. Another one is to show that B is true. This is represented in HOL by two different tactics, namely DISJ1_TAC and DISJ2_TAC:

$$\text{DISJ1\_TAC} : \frac{\Gamma \text{ ?- } A \vee B}{\Gamma \text{ ?- } A} \qquad\qquad \text{DISJ2\_TAC} : \frac{\Gamma \text{ ?- } A \vee B}{\Gamma \text{ ?- } B}$$

A closer look at these two tactics shows that their purpose is to define which branch of the ∨ to keep. It is then possible to unify these two tactics in a single one, FINGER_TAC, which takes an occurrence as the argument. The behaviour of this tactic on a disjunction can be defined as follows:

$$\text{FINGER\_TAC}: \quad \frac{\Gamma\ ?\text{-}\ \boxed{A}\ \vee\ B}{\Gamma\ ?\text{-}\ A} \qquad\qquad \frac{\Gamma\ ?\text{-}\ A\ \vee\ \boxed{B}}{\Gamma\ ?\text{-}\ B}$$

If the left part of an ∨ is selected, then FINGER_TAC applies DISJ1_TAC, if it is the right part, it applies DISJ2_TAC.

Similar behaviours can be defined for the other logical connectives. The complete list is given in Figure 3. It is then possible to compose these basic behaviours to obtain a tactic FINGER_TAC, that has two arguments: an occurrence and a list of terms. This list provides the successive witnesses for the existential quantifiers and the values to specialise the universal quantifiers. To show how this composition works, let us consider the following goal:

$$?\text{-}\ \exists x. \forall y. (P\,y) \Rightarrow ((Q\,x\,y) \wedge (R\,x\,y))$$

A selection on the term $Q\,x\,y$ produces the following incomplete tactic:

$$\text{FINGER\_TAC (GOAL\_OCC 'ababafa') } [\langle term\rangle]$$

where (GOAL_OCC 'ababafa') denotes the occurrence of $Q\,x\,y$ in the goal and ⟨term⟩ is a place holder for the witness corresponding to the existentially quantified variable x. Suppose that this place holder is replaced by a term "t", the application of the resulting tactic gives:

$$\text{FINGER\_TAC (GOAL\_OCC 'ababafa') } [\text{"t"}]: \quad \frac{?\text{-}\ \exists x. \forall y. (P\,y) \Rightarrow (\boxed{(Q\,x\,y)} \wedge (R\,x\,y))}{P\,y\ ?\text{-}\ (Q\,t\,y) \qquad P\,y\ ?\text{-}\ (R\,t\,y)}$$

In order to reach the selected subterm, this tactic firsts gives "t" as witness for "x", then generalises the ∀, discharges the implication and finally splits the conjunction into two subgoals.

In the proof development system for HOL, the goal corresponding to the current proof step is displayed in the Goal window. The Figure 4 gives an example of such a window. In this window, a structured selection is attached to the mouse designation. Sending arguments to the new proof step is provided by the ⟨Send⟩ button that transforms the selected subterm into an argument of the expected type as defined above.

5 Other tools

5.1 Rewriting tool

One of the main issues when building a proof development system is to provide some tools to ensure a maximal re-usability of already proved theorems. Solving

$$\frac{\Gamma \,?\!\!-\, \boxed{A} \wedge B}{\Gamma \,?\!\!-\, A \qquad \Gamma \,?\!\!-\, B} \qquad\qquad \frac{\Gamma \,?\!\!-\, A \wedge \boxed{B}}{\Gamma \,?\!\!-\, B \qquad \Gamma \,?\!\!-\, A}$$

$$\frac{\ldots \boxed{A} \wedge B \ldots \,?\!\!-\, C}{\ldots A \ldots \,?\!\!-\, C} \qquad\qquad \frac{\ldots A \wedge \boxed{B} \ldots \,?\!\!-\, C}{\ldots B \ldots \,?\!\!-\, C}$$

$$\frac{\Gamma \,?\!\!-\, \boxed{A} \vee B}{\Gamma \,?\!\!-\, A} \qquad\qquad \frac{\Gamma \,?\!\!-\, A \vee \boxed{B}}{\Gamma \,?\!\!-\, B}$$

$$\frac{\ldots \boxed{A} \vee B \ldots \,?\!\!-\, C}{\ldots A \ldots \,?\!\!-\, C \qquad \ldots B \ldots \,?\!\!-\, C} \qquad \frac{\ldots A \vee \boxed{B} \ldots \,?\!\!-\, C}{\ldots B \ldots \,?\!\!-\, C \qquad \ldots A \ldots \,?\!\!-\, C}$$

$$\frac{\Gamma \,?\!\!-\, \boxed{A} \Rightarrow B}{A, \Gamma \,?\!\!-\, B} \qquad\qquad \frac{\Gamma \,?\!\!-\, A \Rightarrow \boxed{B}}{A, \Gamma \,?\!\!-\, B}$$

$$\frac{\ldots \boxed{A} \Rightarrow B \ldots \,?\!\!-\, C}{\ldots \,?\!\!-\, A \qquad \ldots B \ldots \,?\!\!-\, C} \qquad \frac{\ldots A \Rightarrow \boxed{B} \ldots \,?\!\!-\, C}{\ldots B \ldots \,?\!\!-\, C \qquad \ldots \,?\!\!-\, A}$$

$$"t" \; \frac{\Gamma \,?\!\!-\, \exists x . \boxed{P\,x}}{\Gamma \,?\!\!-\, P\,t} \qquad \frac{\ldots \exists x . \boxed{P\,x} \ldots \,?\!\!-\, C}{\ldots P\,c \ldots \,?\!\!-\, C}$$

$$\frac{\Gamma \,?\!\!-\, \forall x . \boxed{P\,x}}{\Gamma \,?\!\!-\, P\,c} \qquad "t" \; \frac{\ldots \forall x . \boxed{P\,x} \ldots \,?\!\!-\, C}{\ldots P\,t \ldots \,?\!\!-\, C}$$

$$\frac{\Gamma \,?\!\!-\, \forall \boxed{x} . P\,x}{}$$

Induction Scheme on the type of x

where

c is a new constant
t an arbitrary term.

Fig. 3. Providing a tactic: **FINGER_TAC**

Fig. 4. The Goal window

this problem in all its generality is still an open question. More pragmatically in our system we propose a solution for some extremely useful cases. The rewriting operation is one of these cases.

Rewriting is a technique that implements the natural principle of substituting equals for equals. In the HOL system, rewriting is one of the most useful tactics. It is used not only to expand definitions but also as a way of doing symbolic computations. The construction of a tool to help rewriting first requires the generation of a specific database for rewritings from the existing theorems and then the implementation of an efficient algorithm on this database that, given a term, retrieves all the possible rewritings that can be applied on that term.

In order to produce a relevant database of rewritings from theorems, different standard techniques can be applied (see [3] for example). A first implementation of the tool has been built that handles conditional rewriting. Every equality that is reachable going through $\forall$, $\wedge$ and the right part of $\Rightarrow$ is considered as a rewriting equation. For example from the theorem

$$\forall x\,y\,z.\,(x * y) * z = x * (y * z) \wedge (0 < z) \Rightarrow (x * z < y * z) = (x < y)$$

both $(x*y)*z = x*(y*z)$ and $(x*z < y*z) = (x < y)$ are considered as rewriting equations, the second one being conditional. A rewriting equation $A = B$ can produce two rewriting rules: A is rewritten into B $(A \longrightarrow B)$ and B is rewritten into A $(B \longrightarrow A)$. In order to discard irrelevant rewriting rules, a rewriting rule must fulfill the two following conditions to be included in the database of rewritings:

- The left-hand side of the rewriting rule is not a variable.
- The left-hand side of the rewriting rule must contain the variables of the right-hand side: no new variable is introduced by the rewriting.

Applying this to the previous rewriting equations produces three rewriting rules for the database: two from the first one $(x*y)*z \longrightarrow x*(y*z)$ and $x*(y*z) \longrightarrow (x*y)*z$ as both sides contain the same variables, and only one from the second one $(x*z < y*z) \longrightarrow (x < y)$.

Given a term A and in order to find all the rewritings that can be applied on A, it is necessary to test if the pattern B matches A for each rewriting rule $B \longrightarrow C$. Instead of doing the matching test separately for each rule, all the patterns are grouped in a discrimination net [5]. The possible rewritings are then found with only one matching test. A possibility of limiting the search space is also given to the user by defining in which theories the search for rewritings has to be performed. The composition of these two techniques, the discrimination net and the limitation of the search space, has revealed to be sufficient to get a reasonable reaction time on practical examples.

The activation of the Rewriting tool works as follows. First of all, the theories in which the user wants to perform the search have to be selected with the graph interactor. Then each time the $\boxed{\text{Rewriting?}}$ button is pressed in the Goal window, a request is sent in order to find all the possible rewritings for the selected term. The answers are then displayed in the Retrieval window. The Figure 5 shows the result of a search for the term "$1 + x$" in the theory for arithmetic. A reader familiar with the HOL system may recognise that the first answer corresponds to the application of a rewriting rule generated from the theorem SUC_ONE_ADD: $\forall n.\,(SUC\,n) = 1 + n$, and the second to the application of a rewriting rule generated from the theorem ADD_SYM: $\forall m\,n.\,m + n = n + m$. A substitution is performed by selecting one of the items of the menu returned by the search of rewritings and pressing the $\boxed{\text{Do it}}$ button.

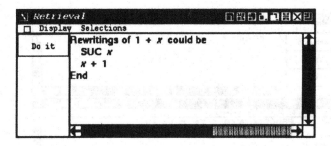

Fig. 5. The retrieval window for rewriting

5.2 Edition of the subgoal

The rewriting tool has been developed to guide local transformations. Given a term A, it is possible to get hints from the system to transform A into another term B. An alternative is to assert first the value B into which A has to be transformed. This operation can be justified by the following tactic:

$$"B"\ \dfrac{\Gamma\ ?\text{-}\ P[A]}{\Gamma\ ?\text{-}\ A = B \qquad \Gamma\ ?\text{-}\ P[B]}$$

This way of proceeding has several advantages. First, it structures the proof script in a natural way by grouping all the transformations relative to a particular sub-expression in one subgoal. Secondly, in some cases, the resulting equality $A = B$ may be checked automatically by some decision procedures. For example when reasoning about arithmetic expressions, Richard Boulton's arithmetic decision procedure [2] can be used in combination with this technique to provide a rather power tool.

In the proof development system, the above tactic is implemented by an edition of the goal. In the Goal window, selecting a subterm A and pressing the ⌐Edit⌐ button send the value of A to an editor where it can be edited into B. After this edition, a predefined decision procedure is called on $A = B$ in order to try to prove it automatically. If this procedure succeeds, the above tactic generates only one new goal where the substitution has been performed, if not it generates the two expected subgoals.

6 The proof development system in action

Now that all the different tools have been described, it is possible to show how they combine with each other to provide an environment to do proofs. For this, we are going to give the complete proof of the proposition **ADD0**: $\forall m.\ m + 0 = m$ inside this new environment.

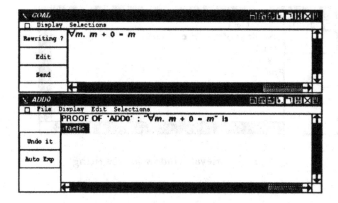

When we start the proof of **ADD0**, the tactic window in which we are going to construct the proof contains a place holder ⟨*tactic*⟩ and the Goal window shows the initial proposition corresponding to this place holder.

The first step of the proof is an induction on m. For doing this, it is sufficient to select the variable m using the technique of proof by pointing. This replaces the place holder by a **FINGER_TAC** tactic:

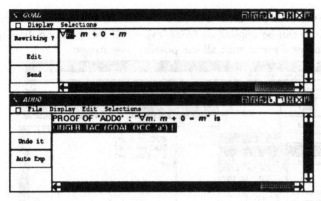

As a place holder has been replaced by a valid tactic, the system checks automatically the consistency of the change and adds a continuation with two new place holders, one for the base case and one for the induction case:

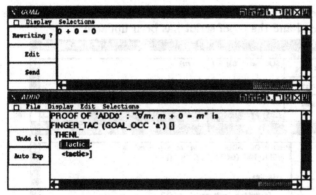

The first place holder is selected, so the Goal window displays the corresponding subgoal. As there are two place holders, we are free to start proving one or the other. In order to attack the induction case first, we simply select the second place holder with the mouse:

For this subgoal, we can send a request to the Rewriting tool to find which transformations can be applied on $(SUC\,m) + 0$. This operation displays in the Retrieval window a menu with all the possible rewritings:

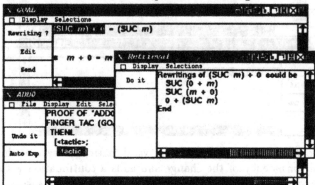

The second rewriting is the interesting one. After its application, we get the expected subgoal and the proof script has been updated:

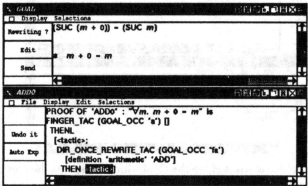

The new subgoal is a simple consequence of the assumption and can simply be proved using the tactic ASM_REWRITE_TAC □. After this, only the base case remains to be proved. For this goal, we use the Rewriting tool on the subterm 0+0 one more time:

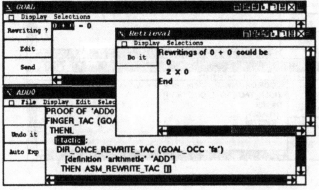

Applying this final rewriting solves the base case. The resulting tactic contains no place holder: the proof has been completed.

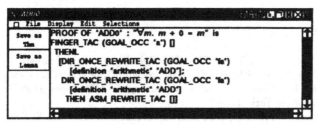

7 Conclusions and future works

The main motivation for building the proof development system presented in this paper was to ease doing proofs in HOL. The guided edition, the proof by pointing and the retrieval of theorems are among the tools that go in that direction. The guided edition provides a natural way of constructing a tactic by respecting the syntax and the type structure of the command language. The proof by pointing covers most of the interaction with logical connectives that structure the proofs. Finally, the Rewriting tool gives a way to use theorems without having to learn their actual nomenclature.

A more surprising result is that the nature of the user-interface affects the way proofs are done. For example, occurrences provide a natural and precise control on the proof construction. This can be used to define useful notions such as the local rewriting. If it is possible to use occurrences with the standard interface of HOL, it requires too much skill from the user (the computation of a path from a location in a term). So, alternative and often more complicated solutions are developed in order to produce the same result. In the interface, the ability to produce an occurrence from a selection is then an issue that may change the way proofs are done.

One of the possible extension of the system is to cover the integrality of the interaction with HOL. So far, the main focus has been on proof construction. A natural idea is to embed not only proof construction but also theory construction. The expected benefit would be an environment that has a clear idea of the dependences in the HOL system, dependences between theories and dependences between theorems. For example, given a modification in a definition, the system should first detect which theorems could be affected by the change and then provide an environment to replay the proofs of these theorems in order to update them.

A second possible extension is the addition of a tool to post-process proofs. The final tactic produced at the end of a proof is the result of an interaction, it cannot be considered as an acceptable final result. Before this, some transformations may be needed to get a more robust or a more efficient tactic. In particular, explicit references to contextual information should be removed. If this kind of activity seems difficult to automatise, having a tool that provides some help for these transformations would already be an important improvement.

References

1. Y. Bertot, G. Kahn, L. Théry, "Proof by Pointing", to be published.
2. R. J. Boulton, "The arithmetic decision procedure", in the HOL system.
3. R. S. Boyer, J S. More, "A Computational Logic", ACM Monograph Series, Academic Press, New York, 1979.
4. "The Centaur 1.3 Manual", I. Jacobs, *ed.*, available from INRIA-Sophia-Antipolis, January 1993.
5. E. Charniak, C. K. Riesbeck, D. V. McDermott, J. R. Meehan, "Artificial intelligence programming", Lawrence Erlbaum Associates Publishers, New Jersey 1987.
6. T. W. Reps, T. Teitelbaum, "The Synthesizer generator: a system for constructing language-based editors", Springer-Verlag, 1988.
7. B. T. Graham, "The SECD Microprocessor, A Verification Case Study", Kluwer Academic Publishers, Boston, 1992.
8. S. Kalvala, "Developing an Interface for HOL", in *Proceedings of the '91 International Workshop on the HOL Theorem Proving System and its Applications*, Davis, Cal., IEEE Computer Society Press, August 1991.
9. D. Shepherd, "The convert library", in the HOL system.
10. L. Théry, Y. Bertot, G. Kahn, "Real Theorem Provers Deserve Real User-Interfaces", in *Proceedings of the Fifth ACM SIGSOFT Symposium on Software Development Environments*, Tyson's Corner, Va, USA, Software Engineering Notes, Vol. 17, no. 5, ACM Press, 1992.

A HOL Package for Reasoning about Relations Defined by Mutual Induction

Rachel E. O. Roxas

The Australian National University, Canberra, Australia **

Abstract. The work of Melham [3] which provided tools built on the HOL system to enable reasoning with inductively defined relations is extended to handle mutually inductive definitions. This paper presents this extension and some examples of its use.

1 Introduction

A technique of defining relations using their *inductive definitions* in the HOL system was introduced by Melham [3]. The package provides theorem-proving tools for reasoning about these relations. Case studies, which include the definitions and proofs of properties of an operational semantics, combinatory logic and rudimentary process algebra, using these tools were presented by Camilleri and Melham [2].

Unfortunately, though, their work does not handle mutual definition. Therefore, we describe here an extension of their implementation to handle mutually inductive definitions. The succeeding discussions follow closely their presentation, highlighting the additions, differences and similarities. The range of problems is enlarged when mutual induction is introduced.

2 Inductive Definitions

An inductive definition of a relation consists of a set of rules. For instance, the set of even natural numbers can be inductively defined by the following inference rules:

$$(1) \quad Even\ 0 \qquad\qquad (2) \quad \frac{Even\ n}{Even\ (n+2)}$$

Rules of the first form are called *axioms* and rules of the second form are called *implicative rules*.[3] Both of these forms of inference rules, axioms and implicative rules, assert membership in the relation *Even*. (Predicates are unary relations.)

In general, an m-place relation R is defined inductively using a set of *inference rules* of the form:

$$\frac{R(t_1^1, ..., t_m^1)\ \ ...\ \ R(t_1^k, ..., t_m^k)}{R(t_1, ..., t_m)} \quad S_1,\ ...\ ,S_j$$

** Present address: University of the Philippines at Los Baños, Laguna, Philippines

[3] Implicative rules will be called *rules* hereon. Future reference to implicative rules vis-a-vis inference rules will be explicitly specified.

where the terms above the line are called *premises* and the one below the line is the *conclusion*. The rule can also have side conditions $S_1, ..., S_j$ $(j \geq 0)$ which may be arbitrary propositions not involving the relation R. The superscript k $(k \geq 0)$ is the number of assertions for the relation R in the premise. If $k = 0$, then there are no premises, and the inference rule is expressed as:

$$R(t_1, ..., t_m)$$

and is an axiom. If $k > 0$ then the inference rule is an implicative rule. An inference rule such as the ones described above which assert membership in the relation being defined, in this case relation R.

The relation being defined by this collection of rules is taken to be the *least relation closed under the rules* specified. This means that the rules used to define the relation are exhaustive; that is, no other membership assertions hold for this relation that cannot be derived from the rules. From the definitions, a theorem called the *case analysis theorem* can be derived for the relation defined; it enables the proofs of the properties of the relations to be constructed using finite sequences of applications of the rules. Each of these are described in detail in the following sections. See [1], [2] and [3] for more details about the theory of inductive definitions.

A syntactic convention will be introduced to simplify the notations. Let A denote the premise set of a rule which consists of an arbitrary number of terms that asserts the relation being defined with term parameter vector of length m; let C be terms of the conclusion of the rule which are assertions of membership in the relation being defined with term parameters; and let S be the set of side conditions of the rule; such that the rule above can be expressed as follows:

$$\frac{A}{C} \quad S$$

If the premise set A is empty, then the inference rule is expressed as an axiom C with the set of side conditions S.

2.1 Inference Rules

For the *Even* example above, Melham's package automatically proves the following theorems which state that the required rules hold of the predicate *Even*:

$$\vdash Even\ 0$$
$$\vdash \forall n.\ Even\ n \supset Even\ (n+1)$$

In general, the definitions derived for the rules of inference have the following form: [4]

$$\vdash \forall \bar{v}.\ (\exists \bar{z}.\ \hat{A}) \supset C$$

---

[4] Strictly, Melham's package moves the quantification of variables in $\bar{v}$ past the implication if they do not appear in $\hat{A}$

where $\bar{v}$ represents the set of variables contained in the terms in the parameters of the conclusion C, and $\bar{z}$ represents the set of variables contained in the premises A that are not contained in the conclusion, and $\hat{A}$ denotes the conjunct of all the elements of the premise set A. Both $\bar{v}$ and $\bar{z}$ have arbitrary sizes. For the purpose of simplicity, the side conditions of the rules are placed in the assumption list and will not be shown in the succeeding discussions.

Mutually inductive definitions can also be defined using a set of inference rules. Consider the following inductive definition of even and odd numbers. The rules are as follows:

$$(1)\ Even\ 0 \quad (2)\ \dfrac{Odd\ n}{Even\ (n+1)} \quad (3)\ Odd\ 1 \quad (4)\ \dfrac{Even\ n}{Odd\ (n+1)}$$

Inference rules (1) and (2) assert membership in the relation $Even$, while (3) and (4) assert membership in the relation Odd. Inference rules (1) and (3) are axioms.

In general, the rules to describe the inductive definition of n mutually inductive relations $R_1, R_2, ..., R_n$ are of the following form (for all i, where $0 < i \leq n$):

$$\dfrac{A_1\ ...\ A_n}{C_i}\ S_1, ..., S_j$$

Each A_k (where $0 < k \leq n$) represents a set containing an arbitrary number of assertions of the relation R_k in the premise, and C_i be the conclusion of the rule which asserts the relation R_i with term parameters.

For the even-odd example above, the following theorems which state that the required rules hold of the predicates $Even$ and Odd should be proven:

$$\vdash\ Even\ 0$$
$$\vdash\ \forall n.\ Odd\ n \supset Even\ (n+1)$$
$$\vdash\ Odd\ 1$$
$$\vdash\ \forall n.\ Even\ n \supset Odd\ (n+1)$$

In general, the above theorems are of the following form:

$$\vdash\ \forall \bar{v}.\ (\exists \bar{z}.\ \hat{A}_1 \wedge ... \wedge \hat{A}_n) \supset C_i$$

where $\bar{v}$, $\bar{z}$ and $\hat{A}_j$ (where $0 < j \leq n$) are as previously defined.

2.2 Induction Theorems

In this section, we describe a recipe for generating an induction theorem from any set of rules that define several relations. The induction theorem produced by the procedure for our even-odd example is as follows:

$$\vdash\ \forall P_1\ P_2.\ (P_1\ 0 \wedge (\forall n.\ P_2\ n \supset P_1(n+1)) \wedge$$
$$P_2\ 1 \wedge (\forall n.\ P_1\ n \supset P_2(n+1))) \supset$$
$$(\forall n.\ Even\ n \supset P_1\ n) \wedge (\forall n.\ Odd\ n \supset P_2\ n)$$

This theorem is based on the premise that the inductively defined relations *Even* and *Odd* are least relations that satisfy the rules. So, if there are relations, say P_1 and P_2, which also satisfy the same rules that *Even* and *Odd* do (a condition captured by the antecedent of the theorem above) then *Even* and *Odd* are subsets of them. That is:

$$\forall n.\ Even\ n \supset P_1\ n$$
$$\forall n.\ Odd\ n \supset P_2\ n$$

which are the conjuncts of the conclusion.

For n mutually inductive relations, say $R_1, ..., R_n$, the induction theorem appears as follows:

$$\vdash \forall P_1 ... P_n.\ (IndRules\ P_1 \wedge\ ...\ \wedge\ IndRules\ P_n) \supset$$
$$(\forall \overline{x_1}.\ R_1(\overline{x_1}) \supset P_1(\overline{x_1})) \wedge\ ...\ \wedge$$
$$(\forall \overline{x_n}.\ R_n(\overline{x_n}) \supset P_n(\overline{x_n}))$$

where each predicate P_i is generated and is of type R_i, and each $\overline{x_i}$ stands for a variable vector of length m_i, the arity of the relation R_i. Each ($IndRules\ P_i$) is directly computed from the axioms and rules asserting membership in the relation R_i, is the conjunct of the following terms, for axioms and implicative rules respectively, for all i $(0 < i \leq n)$:

$$\forall \overline{v_i}.\ C_i\ [R_i/P_i]$$

$$\forall \overline{v_i}.\ (\exists\ \overline{z_i}.\ \hat{A}_1\ [R_1/P_1]\ \wedge\ ...\ \wedge\ \hat{A}_n\ [R_n/P_n]) \supset\ C_i\ [R_i/P_i]$$

where $\overline{v_i}$ represents the set of variables contained in the terms in the parameters of the conclusion, $\overline{z_i}$ represents the set of variables in the terms in the parameters of the antecedent but not in the conclusion, $[R_i/P_i]$ means to instantiate all occurrences of R_i in the term (or set of terms) to which it has been applied by the corresponding P_i.

A stronger induction theorem can be generated based on the induction theorem described above. For instance, a stronger inductive theorem for the even-odd example is:

$$\vdash \forall P_1 P_2.(P_1\ 0 \wedge (\forall n.\ Odd\ n\ \wedge P_2\ n \supset P_1(n+1)) \wedge$$
$$P_2\ 1 \wedge (\forall n.\ Even\ n \wedge P_1\ n \supset P_2(n+1))) \supset$$
$$(\forall n.\ Even\ n \supset P_1\ n) \wedge (\forall n.\ Odd\ n \supset P_2\ n)$$

A stronger induction theorem is derived by weakening the implicative terms in ($IndRules\ P_i$) by adding terms into their antecedents, as shown below:

$$\forall \overline{v_i}.\ (\exists \overline{z_i}.(\hat{A}_1\ \wedge\ \hat{A}_1\ [R_1/P_1]) \wedge\ ...\ \wedge$$
$$(\hat{A}_n\ \wedge\ \hat{A}_n\ [R_n/P_n])) \supset\ (C_i\ [R_i/P_i])$$

where $\overline{v_i}, \overline{z_i}, \hat{A}_i$ and $[R_i/P_i]$ (for all i, where $0 < i \leq n$) are as previously defined.

2.3 Case Analysis Theorems

For the even-odd example, the cases theorems for the relations *Even* and *Odd* are:

$$\vdash \forall n.\ Even\ n\ =\ (n=0)\ \vee\ (\exists n'.\ (n=n'+1)\ \wedge\ Odd\ n')$$
$$\vdash \forall n.\ Odd\ n\ =\ (n=1)\ \vee\ (\exists n'.\ (n=n'+1)\ \wedge\ Even\ n')$$

A case analysis theorem is proven for each relation being defined. The theorem for *Even* states that the general membership assertion *Even n* for relation *Even* holds when either it is derivable by the axiom:

$$Even\ 0$$

such that $n = 0$; or it is derivable by the implicative rule:

$$\frac{Odd\ n}{Even\ (n+1)}$$

where there is a value n' such that $(n = n' + 1)$ and $Odd\ n'$. Each disjunct in the right hand side of the equation is computed from the inference rule that asserts membership in the considered relation and contains assertions about the variables in the parameter vector of the relation. The theorem for *Odd* is derived in a similar fashion.

In general, if we consider a mutually inductive definition with n relations using a set of inference rules, say *Rules*, the case analysis theorems will have the following format:

$$\forall \overline{x_i}.\ R_i(\overline{x_i})\ =\ \bigvee\ \{\ caseR_i(rule)\ |\ rule\ \in\ Rules\ \}$$

where $\overline{x_i}$ is a variable vector which has a length equivalent to the arity of R_i; and where $caseR_i(rule)$ reduces the term to $false$ when the rule does not assert membership in the relation R_i, and otherwise, derives a formula to capture the behavior of the rule. These are shown in the following:

Inference Rules asserting $R_i,\ (0 < i \le n)$	$caseR_k$ where $(k = i)$	$caseR_k$ where $(k \ne i)$
$R_i(t_{i1}, ..., t_{im_i})$ **AXIOMS**	$\exists \overline{y_i}.\ (x_{i1}\ =\ t_{i1}\ [\overline{w_i}/\overline{y_i}])\ \wedge\ ...$ $\wedge\ (x_{im_i} = t_{im_i}[\overline{w_i}/\overline{y_i}])$	$false$
$\dfrac{A_1\ ...\ A_n}{R_i(t_{i1}, ..., t_{im_i})}$ IMPLICATIVE RULES	$\exists \overline{y_i}.\ \exists \overline{z_i}.\ (x_{i1}\ =\ t_{i1}\ [\overline{w_i}/\overline{y_i}])\ \wedge\ ...$ $\wedge\ (x_{im_i} = t_{im_i}[\overline{w_i}/\overline{y_i}])$ $\wedge\ (\forall j.\ A_j\ [\overline{w_i}/\overline{y_i}])$	$false$

where $\overline{w_i}$ is the set of variables that are contained in both the conclusion of the rule and the variable vector $\overline{x_i}$ of the relation, and:

$$\overline{y_i} = \{ \ gen \ w \mid w \in w_{ij} \ where \ 0 < j \leq (card \ \overline{w_i}) \ \}$$
$$[\overline{w_i}/\overline{y_i}] = [w_{i1}/y_{i1}] \ o \ [w_{i2}/y_{i2}] \ o \ ... \ o \ [w_{i(card \ \overline{w_i})}/y_{i(card \ \overline{w_i})}]$$

and $\overline{z_i}$ is the set of variables that are in the premise of the corresponding rule that asserts membership in the relation R_i but not in its conclusion.

3 Implementation in Higher Order Logic

Additions and modifications were made to the functions implemented by Camilleri and Melham [2]. Some of their internal functions were imported to the new system. There are four parts of the system. First is the specification of definitions. Second is the proof of the existence of the smallest relations satisfying the rules in the specification. This consists of a theorem specifying that the relations satisfy the rules and another theorem capturing the rule induction for the relations. The third is the proof of the stronger induction theorem. And the fourth is the derivation of the theorems expressing the various cases of the relations.

We will use the even-odd example to illustrate the use of these functions.

3.1 Definitions

The function implemented to inductively define mutually recursive relations is:

```
mutrec_new_inductive_definition :
string ->                              definition name
(bool # term # term list) list ->      information about relations being defined
(term list # term) list ->            inference rules
(thm list) # thm                       result
```

The first parameter identifies the name at which the result is to be saved. The second is a list of tuples specifyinging properties of the relations being defined. The next parameter supplies the list of inference rules to define the relations, each of which is of the form:

$$([premises \ and \ side \ conditions], conclusion)$$

The *result* is automatically proven from what states the existence of the smallest relations that satisfies the rules; and consists of the **rules**, inductive theorems of the relations **inds** and the inductive theorem for all the relations **ind**.

The specification of the even-odd example is as follows:

```
let rules,ind =
    let EVEN = "EVEN:num->bool" in
        let ODD = "ODD :num->bool" in
            mutrec_new_inductive_definition
                'evenodd'
                [(false,"^EVEN n",□);
                 (false,"^ODD  n",□)]
                [ □,   "^EVEN 0"              ;

                 [         "^ODD n"
                 %---------------------------- % ],
                           "^EVEN (n+1)"       ;

                 □,        "^ODD 1"            ;

                 [         "^EVEN n"
                 %---------------------------- % ],
                           "^ODD (n+1)"        ];;
```

The *result* consists of a list of theorems and a theorem. The list of theorems generated expresses the rules of the inductively defined relations:

```
rules = [ ⊢ EVEN 0;
          ⊢ ∀n. ODD n ⊃ EVEN(n + 1);
          ⊢ ODD 1;
          ⊢ ∀n. EVEN n ⊃ ODD(n + 1) ]
```

The induction theorems proven are expressed as:

```
ind = ⊢ ∀P P'. (P 0 ∧ (∀n. P' n ⊃ P(n + 1)) ∧
               P' 1 ∧ (∀n. P n ⊃ P'(n + 1)))
           ⊃ ((∀n. EVEN n ⊃ P n) ∧ (∀n. ODD n ⊃ P' n))
```

3.2 Stronger induction theorem

A stronger induction theorem can be proved using the following derived inference rule:

```
mutrec_derive_stronger_ind_thm :
(rules #                          rules generated by definition
ind) ->                           inductive theorem generated by definition
thm                               stronger induction theorem
```

The stronger induction theorem generated for the even-odd example is as follows:

```
mutrec_derive_stronger_ind_thm(rules, ind);;

⊢ ∀P P'. P 0 ∧ (∀n. ODD n ∧ P' n ⊃  P(n + 1)) ∧
         P' 1 ∧ (∀n. EVEN n ∧ P n ⊃  P'(n + 1))
         ⊃  (∀n. EVEN n ⊃  P n) ∧ (∀n. ODD n ⊃  P' n)
```

3.3   Case Analysis Theorems

A useful tool in reasoning with inductively defined relations by a set of rules is when an exhaustive case analysis theorem is proven using these rules and the induction theorem formalized by the definition. The function to derive this was implemented by Camilleri and Melham [2], and was modified to handle mutually recursive relations. The inference rule implemented to derive the case analysis theorems for the defined relations has the following syntax:

```
mutrec_case_analysis_thm :
(rules #            rules generated by definition
ind) ->             inductive theorem generated by definition
(thm)list           case analysis theorems for the relations
```

The corresponding derived cases theorem was proven for the even-odd example:

```
( ⊢ ∀n. EVEN n = (n = 0) ∨ ((∃n'. n = n' + 1) ∧ ODD  n'),

  ⊢ ∀n. ODD  n = (n = 1) ∨ ((∃n'. n = n' + 1) ∧ EVEN n') )
```

4   Another Example: Operational Semantics

In a programming language, the rules for the evaluation of its constructs can be mutually recursive.[5] Example programming languages were tested using several mutually recursive constructs, but for the purpose of illustration in this paper, we consider a smaller language with the following types of constructs:

$$Expr \quad := \quad Constant\, n \quad | \quad Var\, v \quad | \quad FCall\, f$$
$$Comm \quad := \quad Null \quad | \quad v := Expr \quad | \quad Comm; Comm$$

where the body of the function f consists of ($Comm$; $return\ Expr$).

---

[5] Representation and implementation of mutually recursive types is an altogether different area of study.

The following inference rules describe the operational semantic definition of this language:

$$(Constant\ n), (E, s) \Rightarrow (n, s)$$

$$(Var\ i), (E, s) \Rightarrow (get\_val\ i\ s), s$$

$$\frac{(get\_comm\ E\ f), (E', s1) \Rightarrow s2 \quad (get\_expr\ E\ f), (E', s2) \Rightarrow s3}{(FCall\ f), (E, s1) \Rightarrow (n, s3)}$$

$$Null, (E, s) \Rightarrow s$$

$$\frac{e, (E, s1) \Rightarrow (n, s2)}{(i := e), (E, s1) \Rightarrow (put\_val\ i\ n\ s2)}$$

$$\frac{c1, (E, s) \Rightarrow s1 \quad c2, (E, s1) \Rightarrow s2}{(c1; c2), (E, s) \Rightarrow s2}$$

(where E' is a modified environment of E to handle scoping properties which will not be discussed here in detail).

Transition relations $\Rightarrow$ for expressions and commands are represented as TR_Expr and TR_Comm, respectively, and are defined inductively. Let:

$$TR\_Expr : Expr \to Env \to state \to num \to state \to bool$$
$$TR\_Comm : Comm \to Env \to state \to state \to bool$$

The inference rules are as follows:

$$TR\_Expr\ (Const\ n)\ E\ s\ n\ s$$

$$TR\_Expr\ (Var\ i)\ E\ s\ (get\_val\ i\ s)\ s$$

$$\frac{TR\_Comm\ (get\_comm\ E\ f)\ E'\ s_1\ s_2 \quad TR\_Expr\ (get\_expr\ E\ f)\ E'\ s_2\ n\ s_3}{TR\_Expr\ (FCall\ f)\ E\ s_1\ n\ s_3}$$

$$TR\_Comm\ Null\ E\ s\ s$$

$$\frac{TR\_Expr\ e\ E\ s_1\ n\ s_2}{TR\_Comm\ (i := e)\ E\ s_1\ (put\_val\ i\ n\ s_2)}$$

$$\frac{TR\_Comm\ c_1\ E\ s\ s_1 \quad TR\_Comm\ c_2\ E\ s_1\ s_2}{TR\_Comm\ (c_1; c_2)\ E\ s\ s_2}$$

The definition specification is as follows:

```
let rules,ind =
    let TR_Expr = "TR_Expr: Expr->Env->state->num->state->bool" in
    let TR_Comm = "TR_Comm: Comm->Env->state->state->bool" in
        mutrec_new_inductive_definition
            'opsem'
            [(false,"^TR_Expr e E s1 n s2",□);
             (false,"^TR_Comm c E s1   s2",□)]
            [□,     "^TR_Expr (Constant n) E s n s"          ;

              □,     "^TR_Expr (Var i) E s (get_val i s) s"  ;

              [  "^TR_Comm (get_comm E f) E' s1 s2";
                 "^TR_Expr (get_expr E f) E' s2 n s3"
               % ---------------------------------------------- % ],
                 "^TR_Expr (FCall f) E s1 n s3"               ;

              □,         "^TR_Comm Null E s s"                ;

              [  "^TR_Expr e E s1 n s2"
               % ---------------------------------------------- % ],
                 "^TR_Comm (i := e) E s1 (put_val i n s2)"    ;

              [  "^TR_Comm c1 E s s1";  "^TR_Comm c2 E s1 s2"
               % ---------------------------------------------- % ],
                 "^TR_Comm (c1;c2) E s s2"                    ];;
```

The definition above yields the following list of rule theorems:

```
rules =
[ ⊢ ∀i E s. TR_Expr (Var i) E s (get_val i s) s;
  ⊢ ∀n E s. TR_Expr (Constant n) E s n s;
  ⊢ ∀E f s1 n s3. (∃s2.  TR_Comm (get_comm(E f)) E' s1 s2 ∧
                         TR_Expr (get_expr(E f)) E' s2 n s3)
                     ⊃ TR_Expr (FCall f) E s1 n s3;
  ⊢ ∀E s.    TR_Comm Null E s s;
  ⊢ ∀e E s1 n s2. TR_Expr e E s1 n s2
               ⊃ (∀i. TR_Comm (i := e) E s1 (put_val i n s2));
  ⊢ ∀c1 E s c2 s2. (∃s1. TR_Comm c1 E s s1 ∧ TR_Comm c2 E s1 s2)
               ⊃ TR_Comm (c1;c2) E s s2]
: thm list
```

The induction theorem generated is expressed:

```
ind =
⊢ ∀P P'.
   (∀i E s. P (Var i) E s (get_val i s) s) ∧
   (∀n E s. P (Constant n) E s n s) ∧
   (∀E f s1 n s3.
     (∃s2. P' (get_comm(E f)) E' s1 s2 ∧
           P (get_expr(E f)) E' s2 n s3) ⊃ P (FCall f) E s1 n s3) ∧
   (∀E s. P' Null E s s) ∧
   (∀e E s1 n s2.
     P e E s1 n s2 ⊃ (∀i. P' (i := e) E s1 (put_val i n s2))) ∧
   (∀c1 E s c2 s2.
     (∃s1. P' c1 E s s1 ∧ P' c2 E s1 s2) ⊃ P' (c1;c2) E s s2) ⊃
   (∀e E s1 n s2. TR_Expr e E s1 n s2 ⊃ P e E s1 n s2) ∧
   (∀c E s1 s2.   TR_Comm c E s1   s2 ⊃ P' c E s1   s2)
```

The stronger induction theorem generated is as follows:

```
⊢ ∀P P'.
   (∀i E s. P (Var i) E s (get_val i s) s) ∧
   (∀n E s. P (Constant n) E s n s) ∧
   (∀E f s1 n s3. (∃s2. TR_Comm (get_comm(E f)) E' s1 s2 ∧
                        P' (get_comm(E f)) E' s1 s2 ∧
                        TR_Expr (get_expr(E f)) E' s2 n s3 ∧
                        P (get_expr(E f)) E' s2 n s3)
                  ⊃ P (FCall f) E s1 n s3) ∧
   (∀E s. P' Null E s s) ∧
   (∀e E s1 n s2. TR_Expr e E s1 n s2 ∧ P e E s1 n s2
                  ⊃ (∀i. P' (i := e) E s1 (put_val i n s2))) ∧
   (∀c1 E s c2 s2. (∃s1. TR_Comm c1 E s s1  ∧ P' c1 E s s1 ∧
                        TR_Comm c2 E s1 s2 ∧ P' c2 E s1 s2)
                  ⊃ P' (c1;c2) E s s2) ⊃
   (∀e E s1 n s2. TR_Expr e E s1 n s2 ⊃ P e E s1 n s2) ∧
   (∀c E s1 s2.   TR_Comm c E s1   s2 ⊃ P' c E s1 s2)
```

The derived cases theorems [6] for *TR_Expr* and *TR_Comm* are as follows:

```
⊢ ∀e E s1 n s2.
   TR_Expr e E s1 n s2 =
       (∃i. (e  = Var i) ∧ (n  = get_val i s) ∧ (s2 = s1)) ∨
       (    (e = Constant n) ∧ (s2 = s1)) ∨
       (∃f s2'. (e  = FCall f) ∧
                TR_Comm (get_comm (E f)) E' s1 s2' ∧
                TR_Expr (get_expr (E f)) E' s2' n s3)
```

```
⊢ ∀c E s1 s2.
    TR_Comm c E s1 s2 =
    (c = Null) ∧ (s2 = s1)  ∨
    (∃e n s2' i. (c = (i := e)) ∧
                 (s2 = put_val i n s2') ∧ TR_Expr e E s1 n s2') ∨
    (∃c1 c2 s1'. (c = (c1;c2)) ∧
                 TR_Comm c1 E s1 s1' ∧ TR_Comm c2 E s1' s2)
```

The automatic derivation of the cases theorem for the operational semantics definition of a programming language mechanizes the approach discussed
by Roxas and Newey [4] which manually computes for the corresponding complementary rules of the inference rules to obtain the equality desired as above.
This automation ensures the preservation of the consistency of the programming language definition, as well as, assists in the automated reasoning of the
language.

5 Conclusions

This paper presented an extension to the work of Camilleri and Melham which
provided automated tools built on the HOL system to enable reasoning with inductive defined relations. The extension caters to mutually inductive definitions.
The facility also offers a mechanism to automate the generation of the definition
rules, induction theorem (and stronger induction theorem) and the various cases
theorems. The functions implemented were illustrated using a numerical example (even and odd numbers) and a structural one (definition of the semantics of
a programming language using operational semantics).

Acknowledgements. Special thanks to the researchers of the inductive definition package, T. Melham and J. Camilleri, and to my academic supervisor, M.
Newey, for their valued help and advice. This research was assisted by support
from DSTO, Australia.

References

1. P. Aczel. An introduction to inductive definitions. In J. Barwise, editor, *Handbook
   of Mathematical Logic*, pages 739–782. North-Holland, 1977.
2. J. Camilleri and T. F. Melham. Reasoning with inductively defined relations in the
   HOL theorem prover. Technical Report 265, Computer Laboratory, University of
   Cambridge, August 1992.
3. T. F. Melham. A package for inductive relation definitions in HOL. In M. Archer,
   J. Joyce, K. Levitt, and P. Windley, editors, *1991 International Workshop on the
   HOL Theorem Proving System & its Applications*, pages 350–357. IEEE Computer
   Society Press, 1992.
4. R. E. O. Roxas and M. C. Newey. Proofs of program transformation. In K. Levitt
   M. Archer, J. Joyce and P. Windley, editors, *1991 International Workshop on the
   HOL Theorem Proving System & Its Applications*, pages 223–230. IEEE Computer
   Society Press, 1992.

---

[6] The equations derived using the $caseR_k$ function discussed in section 2.3 are simplified such that entries of the form $(\exists x. (x_1 = x) \wedge (x_2 = x)...)$ where x is free in the
rest of the term are rewritten as $(x_1 = x_2)$.

A Broader Class of Trees for Recursive Type Definitions for HOL

Elsa L. Gunter

AT&T Bell Laboratories, Rm.#2A-432
600 Mountain Ave., Murray Hill, N.J. 07974, USA
elsa@research.att.com

Abstract. In this paper we describe the construction in HOL of the inductive type of arbitrarily branching labeled trees. Such a type is characterized by an initiality theorem similar to that for finitely branching labeled trees. We discuss how to use this type to extend the system of simple recursive type specifications automatically definable in HOL to ones including a limited class of functional arguments. The work discussed here is a part of a larger project to expand the recursive types package of HOL which is nearing completion. All work described in this paper has been completed.

1   A Broader Class of Recursive Type Definitions

The work described in this paper forms the foundation of a project to expand the class of recursive type specifications for which HOL is capable of automatically defining the types specified and proving the initiality theorem, which acts as an axiomatization for the defined types. The full class of specifications the project aims to handle are those BNF-style specification of the form

$$rty_1 ::= C_{1,1} ty_{1,1,1} \cdots ty_{1,1,k_{1,1}} \mid \cdots \mid C_{1,m_1} ty_{1,m_1,1} \cdots ty_{1,m_1,k_{1,m_1}}$$

$$\vdots$$

$$rty_n ::= C_{n,1} ty_{n,1,1} \cdots ty_{n,1,k_{n,1}} \mid \cdots \mid C_{n,m_n} ty_{n,m_n,1} \cdots ty_{n,m_n,k_{n,m_n}}$$

where each type description $ty_{i,j,k}$ must be *admissible*, as defined below (and where we can show that every type specified is well-founded, or in essence, has a base case).

Definition: A type description ty is *admissible* (in a given mutually recursive type specification) if it satisfies one of the following three conditions:

- ty is an existing type.
- ty is rty_i for some i, $1 \leq i \leq n$
- ty is of the form $ety \rightarrow tyd$ where ety is an existing type and tyd is an admissible type description.

It is also possible to extend the notion of admissibility to include occurrences of certain kinds of type constructors, but the precise definition of this case is quite complicated and we omit it here.

This project is composed of three major aspects. The first is the development of a theory of a broader class of (broader) trees in HOL to form the basic building blocks for all other types defined by specifications of the kind described above. The second is the construction of types for a simplified subclass of specifications. The third aspect is the translation between the full class of specifications and the simplified subclass.

The specifications of the simplified subclass are those of the form

$$rty ::= C_1 ty_{1,1} \ldots ty_{1,k_1} \mid \ldots \mid C_n ty_{n,1} \ldots ty_{n,k_n}$$

where $ty_{i,j}$ is either an existing type, or of the form $ty \rightarrow rty$ for some existing type ty. To see that this simplified class is already a generalization of specifications that are currently handled, notice that any type rty is isomorphic to the type one $\rightarrow rty$.

The first two aspects of the project have been completed and are discussed in this paper in some detail. The third aspect will only be discussed briefly in the concluding section.

2 Broadening Trees

The current system for automatic definition of recursive types from specifications is founded upon the type of finitely branching labeled trees of finite height. (See [2]). This type will not do as the foundation for the class of specifications which we are attempting to handle. To see this, consider the following specification:

$$\text{tree} ::= \text{Leaf integer} \mid \text{Node integer (num} \rightarrow \text{tree)}$$

and the following specific tree of that type:

$$\text{flat} = \text{Node (neg (INT 1)) } (\lambda n.\text{Leaf (INT} n))$$

Any finitely branching tree of finite height can have only finitely many leaves, but this tree has a countably infinite set of leaves (one for each natural number). The natural structure to use to model this specification is the collection of arbitrarily branching trees. To be a bit more precise, we will create a polymorphic type of tree, parametrized by a type α, which will act as the indexing set for the branching of trees, and by a type β for labeling the nodes. The type tree described above could be modeled using num for the branching indexing set and integer for the labeling set.

2.1 Partial Functions

In defining our type of broader trees, and providing a succinct "axiomatization" for it, we will want a theory of *partial functions* between two types. As an example of why we would expect to use partial functions in defining this new type of trees, consider the collection of all trees labeled by strings and having at most continuum many subtrees at each node (that is, each node has no more subtrees than there are real numbers). Then, the collection of subtrees at any node of a tree of this type is a set of trees, again of this type, indexed by a subset of the reals. That is, it is a partial function from the reals to trees of this type.

HOL is equipped with a notion of total function, but no built-in notion of partial function. So the question is, how do we best encode the notion of a partial function, given only total functions. To answer this, we define a type constructor lift that is the solution to the specification

$$\alpha \text{ lift} ::= \text{lift } \alpha \mid \text{undefined}$$

The type α lift is characterized by the initiality theorem

$$\forall fe. \ \exists! fn. \ (\forall x. \ fn(\text{lift } x) = f \ x) \land (fn \ \text{undefined} = e).$$

The constructor lift is one-to-one and has a left inverse lower.

By a partial function from α to β we mean a function from α to β lift. Total functions are injected into partial functions in the obvious manner by the constant lift_fun : $(\alpha \rightarrow \beta) \rightarrow (\alpha \rightarrow \beta \text{ lift})$. The constant

$$\text{lift\_compose} : (\beta \rightarrow \gamma \text{ lift}) \rightarrow (\alpha \rightarrow \beta \text{ lift}) \rightarrow (\alpha \rightarrow \gamma \text{ lift})$$

is the expected composition of partial functions. The domain of definition of a partial function (*i.e.* the set of values on which the partial function takes on a value other than undefined) is given by the predicate

$$\text{part\_fun\_domain} : (\alpha \rightarrow \beta \text{ lift}) \rightarrow \alpha \rightarrow \text{ bool}.$$

The range of a partial function is given by

$$\text{part\_fun\_range} : (\alpha \rightarrow \beta \text{ lift}) \rightarrow \beta \rightarrow \text{ bool}.$$

2.2 Broad Trees

Using the type lift, and the constants associated with it, we can now say more precisely what we mean by the type of α-branching β-labeled trees. We mean any polymorphic type $(\alpha, \beta)\tau$ such that the following initiality theorem holds:

$$\exists Node : \beta \rightarrow (\alpha \rightarrow (\alpha, \beta)\tau \text{ lift}) \rightarrow (\alpha, \beta)\tau.$$
$$\forall Cases. \exists! f. \forall \ label \ subtrees.$$
$$f(Node \ label \ subtrees) = \tag{1}$$
$$Cases \ (\text{lift\_compose} \ (\text{lift\_fun} \ f) \ subtrees) \ label \ subtrees$$

The value *subtrees* is the indexed set of immediate subtrees of our tree, and *label* is the label at the root node. The term lift_compose (lift_fun f) *subtrees* is the set of recursive values of f on the immediate subtrees of the tree, indexed in the same manner over α as the subtrees are. At first glance, this may strike some as a peculiar statement of initiality since there is no obvious separate base case. To see that there is in fact a base case, note that if *subtrees* is the everywhere undefined function, then so is lift_compose (lift_fun f) *subtrees*, and hence there are no recursive calls for such trees.

Before we begin the discussion of the construction of such trees, let us note a couple of facts about them. Firstly, it is inevitable that we will have to find a model for such trees, if we wish to be able to handle all (well-founded) specifications of the sort described in Section 1, since the type of α-branching β-labeled trees can be described by

$$\text{bonsai} ::= \text{bonsai\_NODE } \beta \ (\alpha \rightarrow \text{bonsai lift})$$

This is an allowable specification, and as we noted above, it has a base case. Therefore, what ever mechanism we devise for handling our class of specifications, it will have to be able to generate a model for this type of α-branching β-labeled trees.

Secondly, these trees behave a bit differently than the finitely branching trees used as the foundation for the current recursive type package in HOL. Not only are these trees potentially infinitely branching, but they potentially have infinite height. It is possible to define a function over these trees using the initiality theorem (1) which returns twice the height of the tree, if the tree has finite height, and returns thirteen otherwise. (See Appendix A.1 for details of the example.) When this function is applied to a tree of infinite height (which can exist), there is no finite subtree for which this function will return the same result. Moreover, no finite number of unwindings of the recursive equation will allow us to eliminate the recursion and directly compute the answer. In the domain theoretic sense, this function is not the limit of its finite approximates. This is radically different behavior than is had by the finitely branching trees. There, every function given by a primitive recursive equation, when applied to a specific tree, can be directly computed by unwinding the recursion as many times as the height of the tree. This fact is central to the proof of existence of functions given by primitive recursion. We do not have any such fact available to us, and hence we shall have to take a different route entirely to show existence.

To begin the construction of our broader trees, we will first build the unlabeled kind. An α-branching unlabeled tree is represented by the set (described by a predicate) of its finite branches (where a branch always starts at the root). A finite branch is a list of αs, describing which index was selected at each height. A set of finite branches is the branch set of an α-branching unlabeled tree provided that the empty list (or trivial branch) is in the set and, if a branch is in the set, then all prefixes of the branch are again branches and thus are in the set. Hence, we have the following definition:

$\forall$ *branch_set* : α list $\rightarrow$ bool.
  Is_unlabeled_tree *branch_set* =
    branch_set [] $\land$
    ($\forall b_1\ b_2.\ branch\_set$ (APPEND $b_1\ b_2$) $\implies branch\_set\ b_1$)

A labeling of an unlabeled tree is a partial function mapping the nodes of the tree to labels. A node is given by the path (or branch) from the root to it. Thus, if the type of the labels is β, a labeling is a partial function from α list to β whose domain of definition is an unlabeled tree.

$\forall l : \alpha$ list $\rightarrow \beta$ lift. Is_labeling l = Is_unlabeled_tree(part_fun_domain l)

At this point, we can define a type of α-branching β-labeled trees, by identifying them with partial functions that are labelings. We will call this type (α, β)broad_tree. Thus we have

$\exists rep : (\alpha, \beta)$broad_tree $\rightarrow (\alpha$ list $\rightarrow \beta$ lift). TYPE_DEFINITION Is_labeling rep

In the usual manner, we can define the representation and abstraction functions giving the isomorphism between the new type (α, β)broad_tree and the set described by Is_labeling:

($\forall a.$ broad_tree_ABS (broad_tree_REP a) = a) $\land$
($\forall r.$ Is_labeling r = broad_tree_REP (broad_tree_ABS r) = r)

For more information on defining new types in HOL, see [1, 2].

The next thing we want to do is define a function broad_tree_NODE that will behave as a constructor for terms of type (α, β)broad_tree. Given a function *subtrees* supplying the subtrees of a tree, and a label *label* for the root of the tree, how do we reconstruct the tree? To determine this, we need to know what the set of branches of the new tree is, and what the new labeling is. Each branch of the new tree is either the trivial branch (given by the empty list) or is an element of type α followed by a branch of the tree indexed by that element. Thus, if *subtrees* x = lift t and b is a branch of t, then $x :: b$ is a branch of broad_tree_NODE *label subtrees*. A labeling on this set of branches returns the root label *label* on the trivial branch, and on a nontrivial branch returns the label of the node found at the end of the branch of the subtree indexed by the head of the given branch, where the branch of the subtree is described by the tail of the given branch. Hence, if *subtrees* x = lift t and the branch l of t is labeled by c, then c is the labeling of $x :: l$ in the tree broad_tree_NODE b *subtrees*.

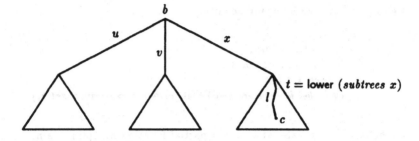

Formally, the definition of broad_tree_NODE is

$\forall(label:\beta)\ (subtrees:\alpha \rightarrow (\alpha,\beta)broad\_tree).$
$\quad$ broad_tree_NODE *label subtrees* =
$\qquad$ broad_tree_ABS($\lambda l.\ (l = [])\Rightarrow$ (lift *label*)
$\qquad\qquad\qquad\qquad$ | (($subtrees$ (HD l) = undefined) $\Rightarrow$ undefined
$\qquad\qquad\qquad\qquad$ | (broad_tree_REP (lower ($subtrees$ (HD l))) (TL l))))

At this point, it would seem that we are almost done. All we have to do is prove that the initiality theorem holds for broad_tree_NODE, right? Unfortunately, life is not so simple. The initiality theorem (1) is false for broad_tree_NODE. Both existence and uniqueness fail. (What we have built so far is not the initial algebra, but the final algebra, instead.) For an example of this failure, see Appendix A.2. Since uniqueness fails and induction implies uniqueness, induction must fail also.

2.3 Bonsai

To remedy the failing of induction, we do the standard construction, basically the same construction that allowed us to build the natural numbers from the type of individuals. Let Is_bonsai be the intersection of all predicates on (α,β)broad_tree that are closed under broad_tree_NODE. More precisely,

$\forall tr.$ Is_bonsai $tr =$
$\quad (\forall P.\ (\forall subtrees\ label.\ (\forall sbtr.\ \text{part\_fun\_range}\ subtrees\ sbtr \Longrightarrow P\ sbtr) \Longrightarrow$
$\qquad\qquad P\ (\text{broad\_tree\_NODE}\ label\ subtrees)) \Longrightarrow$
$\qquad P\ tr)$

In the same way as with the definition of the naturals, we immediately get an induction principle from this construct:

$\forall P.\ (\forall subtrees\ label.\ (\forall sbtr.\ \text{part\_fun\_range}\ subtrees\ sbtr \Longrightarrow P\ sbtr) \Longrightarrow$
$\qquad P\ (\text{broad\_tree\_NODE}\ label\ subtrees)) \Longrightarrow$ $\qquad\qquad\qquad\qquad$ (2)
$\qquad (\forall tr.\ \text{Is\_bonsai}\ tr \Longrightarrow P\ tr)$

Using the predicate Is_bonsai, we can now introduce a new type (α,β)bonsai that is in one-to-one correspondence with the set described by Is_bonsai. (The name bonsai was chosen, in part not to conflict with the names of existing types of trees in HOL, but also in part because the set of bonsai is the subset of broad_tree consisting precisely of those trees having only finite branches. That is, only the broad_trees with short branches are bonsai.) We can also pull over broad_tree_NODE to the type bonsai to get a node constructor bonsai_NODE:

$$\forall label\ subtrees.\ \textsf{bonsai\_NODE}\ label\ subtrees =$$
$$\textsf{bonsai\_ABS}(\textsf{broad\_tree\_NODE}$$
$$label$$
$$(\textsf{lift\_compose}\ (\textsf{lift\_fun}\ \textsf{bonsai\_REP})\ subtrees))$$

It follows easily from this definition that bonsai_NODE is one-to-one, since broad_tree_NODE is.

It is not so immediate, however, that bonsai_NODE is onto. What we know immediately is only that every broad_tree satisfying Is_bonsai is a node, all of whose subtrees are broad_trees. We don't know off hand that the subtrees of a tree in Is_bonsai are again in Is_bonsai. This fact is proved using the induction principle (2) for Is_bonsai. From the fact that all subtrees of a tree in Is_bonsai are again in Is_bonsai, it follows straight-forwardly that bonsai_NODE is onto. Using the induction principle for Is_bonsai, together with the fact that bonsai_NODE is onto, we can then derive the following induction principle for the type bonsai:

$$\forall P.\ (\forall subtrees\ label.\ (\forall sbtr.\ \textsf{part\_fun\_range}\ subtrees\ sbtr \Longrightarrow P\ sbtr) \Longrightarrow$$
$$P\ (\textsf{bonsai\_NODE}\ label\ subtrees)) \Longrightarrow \qquad\qquad (3)$$
$$(\forall tr.\ P\ tr)$$

2.4 Proving Initiality

Now we are back to trying to prove the initiality theorem (1) again. This time, we know we have an induction principle for the type bonsai and the constructor bonsai_NODE. The uniqueness of functions defined by structural induction over bonsai_NODE follows immediately from our induction principle. Therefore, all we need to show is existence. In previous work with finite trees, the existence of functions defined by structural induction was shown using the heights of the trees. In essence, it could be shown that for a tree of height n, it sufficed to unwind the recursion n times to be able to compute the value of the function on the tree without further recursive calls. As discussed above, this approach will not work in our setting because our trees will not in general have finite height. Another approach must be sought.

The approach we take to demonstrating the existence of such functions is a rather set-theoretic approach: to show such functions exist, we demonstrate a graph which is the graph of such a function. By a graph we mean a relation $g : (\alpha, \beta)\textsf{bonsai} \to \gamma \to \textsf{bool}$. Given a case function

$$Cases : (\alpha \to \gamma\ \textsf{lift}) \to \beta \to (\alpha \to (\alpha, \beta)\textsf{bonsai}\ \textsf{lift}) \to \gamma$$

we need to find a relation on $(\alpha, \beta)\textsf{bonsai}$ and γ that is closed under $Cases$ and that is functional. We define what it means for a relation to be closed under a case function as follows:

$\forall Cases\ fun\_rel.$ rel_is_case_closed $Cases\ fun\_rel =$
$(\forall subtrees\ label\ rec\_fun.$
$\quad (\forall x.((subtrees\ x = \text{undefined}) = (rec\_fun\ x = \text{undefined}))\wedge$
$\quad (\neg(subtrees\ x = \text{undefined}) \Longrightarrow$
$\quad fun\_rel\ (\text{lower}(subtrees\ x))\ (\text{lower}\ (rec\_fun\ x)))) \Longrightarrow$
$\quad fun\_rel\ (\text{bonsai\_NODE}\ label\ subtrees)\ (Cases\ rec\_fun\ label\ subtrees))$

The graph we are looking for is the smallest graph that is closed under *Cases*. That is, it is the intersection of all graphs that are closed under *Cases*.

$\forall Cases\ tr\ z.$ smallest_bonsai_fun_rel $Cases\ tr\ z =$
$\quad (\forall fun\_rel.$ rel_is_case_closed $Cases\ fun\_rel \Longrightarrow fun\_rel\ tr\ z)$

As was the case with our definition of ls_bonsai, it follows fairly immediately that smallest_bonsai_fun_rel *Cases* satisfies rel_is_case_closed *Cases*, and that we have the following induction principle:

$$\forall Cases\ fun\_rel.\ \text{rel\_is\_case\_closed}\ Cases\ fun\_rel \Longrightarrow \qquad (4)$$
$$(\forall tr\ z.\ \text{smallest\_bonsai\_fun\_rel}\ Cases\ tr\ z \Longrightarrow fun\_rel\ tr\ z)$$

The fact that smallest_bonsai_fun_rel *Cases* satisfies rel_is_case_closed *Cases* gets us that the function f described by the graph smallest_bonsai_fun_rel *Cases* satisfies the existence half of the initiality theorem, namely that

$\forall subtrees\ label.\ f(Node\ label\ subtrees) =$
$\quad Cases\ (\text{lift\_compose}\ (\text{lift\_fun}\ f)\ subtrees)\ label\ subtrees$

assuming we know that smallest_bonsai_fun_rel *Cases* describes a function.

To prove that smallest_bonsai_fun_rel *Cases* describes a function, we need to show two things. We need to show that it describes a partial function:

$\forall tr\ z_1\ z_1.\ (\text{smallest\_bonsai\_fun\_rel}\ Cases\ tr\ z_1 \wedge$
$\quad \text{smallest\_bonsai\_fun\_rel}\ Cases\ tr\ z_2) \Longrightarrow (z_1 = z_2)$

and we need to show that it is total in its first argument:

$\forall tr.\ \exists z.\ \text{smallest\_bonsai\_fun\_rel}\ Cases\ tr\ z$

This latter fact follows by induction on bonsai (3) using the fact that smallest_bonsai_fun_rel *Cases* is closed under *Cases*. The former fact is a bit more involved. Its proof used both induction on bonsai (3) and the induction principle for smallest_bonsai_fun_rel (4). This is the last of the pieces required to get us the initiality theorem we have been seeking:

$\forall Cases.\ \exists! f.\ \forall subtrees\ label.\ f(\text{bonsai\_NODE}\ label\ subtrees) =$
$\quad Cases\ (\text{lift\_compose}\ (\text{lift\_fun}\ f)\ subtrees)\ label\ subtrees$

3 A Broader Class of Simple Recursive Types

The next step toward supporting our broader class of recursive type definitions is to handle the simple recursive case. In this section we show how to solve recursive type specifications of the form

$$rty \ ::= \ C_1 ty_{1,1} \ldots ty_{1,k_1} \mid \ldots \mid C_n ty_{n,1} \ldots ty_{n,k_n}$$

where $ty_{i,j}$ is either an existing type, or of the form $ty \rightarrow rty$ for some existing type ty. Given such a specification, a solution for it is a new type rty, constructors $C_i : ty_{i,1} \rightarrow \ldots \rightarrow ty_{i,k_i} \rightarrow rty$ and an initiality theorem analogous to (but more complicated than) the one for bonsai. Thus, we need to identify a type in which we can build a model for our specification, define a predicate on that type identifying the elements of the model, and introduce a new type that is isomorphic to the model. Then we need to define the constructors C_i and we need to prove the initiality theorem. In the description that follows, we will often resort to giving examples for each of these steps, rather than giving a completely rigorous description.

3.1 Building the Type

The background type that we are going to use to solve the specification is (σ, τ)bonsai, for some branching type σ and some labeling type τ. The branching type σ and the labeling type τ are each a sum type having one component for each case in the specification. The contribution of a particular case $C_i ty_{i,1} \ldots ty_{i,k_i}$ to the branching type is sum of all $ety_{i,j}$ where $ty_{i,j} = ety_{i,j} \rightarrow rty$, or one if none such exist. The contribution to the labeling type is the product of each $ty_{i,j}$ that is an existing type, if there are any, and one elsewise. For example, the specification

toto::=A bool num | B ($\alpha \rightarrow$ toto) (ind $\rightarrow$ toto) | C | D bool (num $\rightarrow$ toto)

is modeled using the background type

(one + (α + ind) + one + num, (bool × num) + one + one + bool) bonsai

Suppose that $tr = $ bonsai_NODE $label \ subtrees$ is a bonsai of the background type. Then $label$ is uniquely in one of the summands, say the i'th summand, of the labeling type. The predicate that describes the subset of the background type that models the specification is true of tr provided that the domain of definition of $subtrees$ is either exactly the i'th summand of the branching type, if the i'th case has a type argument of the form $ty_{i,j} = ety_{i,j} \rightarrow rty$, or is empty otherwise.

Thus, for our example this becomes

$$\lambda tr.\forall label\ subtrees.(tr = \text{bonsai\_NODE}\ label\ subtrees) \implies$$
$$(\text{ISL}\ label \land (\forall x.\text{part\_fun\_domain}\ subtrees\ x = \text{F})) \lor$$
$$(\text{ISR}\ label \land \text{ISL}(\text{OUTR}\ label)) \land$$
$$(\forall x.\ \text{part\_fun\_domain}\ subtrees\ x = \text{ISR}\ x \land \text{ISL}(\text{OUTR}\ x))) \lor$$
$$(\text{ISR}\ label \land \text{ISR}(\text{OUTR}\ label) \land \text{ISL}(\text{OUTR}(\text{OUTR}\ label))) \land$$
$$(\forall x.\ \text{part\_fun\_domain}\ subtrees\ x = \text{F})) \lor$$
$$(\text{ISR}\ label \land \text{ISR}(\text{OUTR}\ label) \land \text{ISR}(\text{OUTR}(\text{OUTR}\ label))) \land$$
$$(\forall x.\ \text{part\_fun\_domain}\ subtrees\ x =$$
$$\text{ISR}\ x \land \text{ISR}(\text{OUTR}\ x) \land \text{ISR}(\text{OUTR}(\text{OUTR}\ x))))$$

Using new_type_definition with this predicate gets us the type that we need to solve the specification.

3.2 Making the Constructors

The next phase is making the constructors for the type. Each constructor needs to make a bonsai and then abstract it to the new type. It makes the bonsai using bonsai_NODE, and thus it needs to make a label and an indexed collection of subtrees. The label is simply the product of all the arguments to the constructor of pre-existing type (or one if there are none) injected into the corresponding summand of the label type. The indexed collection of subtrees is rather more complicated. The i'th constructor sends all summands of the branching type, except the i'th to undefined. If it has no argument types of the form $ty_{i,j} = ety_{i,j} \to rty$, then it sends everything of branching type to undefined. Otherwise, the i'th summand of the branching type is a sum of all $ety_{i,j}$ such that $ty_{i,j} = ety_{i,j} \to rty$. A summand of the i'th summand of the branching type of type $ety_{i,j}$ is sent to the bonsai representation of the corresponding rty value. For example, the second constructor for the type specification given above is

$$\text{B} = \lambda(f_1 : \alpha \to \alpha\ \text{toto})(f_2 : \text{ind} \to \alpha\ \text{toto}).$$
$$\text{toto\_ABS}(\ \text{bonsai\_NODE}$$
$$(\text{INR}(\text{INL}\ \text{one}))$$
$$(\lambda z.(\ \text{ISR}\ z \land \text{ISL}(\text{OUTR}\ z)) \Rightarrow$$
$$((\text{ISL}(\text{OUTL}(\text{OUTR}\ z))) \Rightarrow$$
$$(\text{toto\_REP}(f_1(\text{OUTL}(\text{OUTL}(\text{OUTR}\ z)))))$$
$$|\ (\text{toto\_REP}(f_2(\text{OUTR}(\text{OUTL}(\text{OUTR}\ z)))))))$$
$$|\ \text{undefined}))$$

Given these definitions for our constructors we show that no element is in the range of two distinct constructors and that every is element is in the range of one of the constructors.

3.3 Deriving Initiality

We prove initiality roughly in the following manner. Assume we have case functions for each of the cases of the specification (that is, build variables of the right types to be such case functions). Using these case functions we can build a case function over **bonsai** as follows. Given an indexed collection of recursive values *rec_fun*, a label *label*, and an indexed collection of subtrees *subtrees*, for each constructor, test whether there exist arguments such that bonsai_NODE *label subtrees* is the representation of the definition of the constructor applied to those arguments. If it is, return the appropriate case function applied to the appropriate arguments. If it fails to match any constructor, return @x.T. For example, consider the specification

$$\text{tutu}::=\text{A bool} \mid \text{B (num} \rightarrow \text{tutu)}$$

The case function we get then is

$$
\begin{aligned}
&\lambda \ rec\_fun \ label \ subtrees. \\
&\quad (\exists x.\text{bonsai\_NODE} \ label \ subtrees = \\
&\quad \quad \text{bonsai\_NODE (INL } x) \ (\lambda z.\text{undefined})) \Rightarrow \\
&\quad (A\_case(\text{OUTL } label)) \\
&\quad \mid ((\exists g.\text{bonsai\_NODE} \ label \ subtrees = \\
&\quad \quad \text{bonsai\_NODE (INR one) } (\lambda z.(\text{ISR } z) \Rightarrow \\
&\quad \quad \quad\quad\quad\quad\quad\quad\quad\quad (\text{tutu\_REP}(g(\text{OUTR } z))) \\
&\quad \quad \quad\quad\quad\quad\quad\quad\quad\quad \mid \text{undefined})) \Rightarrow \\
&\quad (B\_case \ (\lambda x.rec\_fun(\text{INR } x)) \ (\lambda x.\text{tutu\_ABS}(subtrees(\text{INR } x)))) \\
&\quad \mid @x.\text{T})
\end{aligned}
$$

If h is the function over **bonsai** given by the case function so generated, then $f = \lambda x.h(rty\_\text{REP} \ x)$ is the function over the new recursive type given by the case functions over that type. We use the fact that the constructors have distinct images to demonstrate that the desired equations hold for f. We use the fact that every element is in the image of some constructor together with the uniqueness of h to show the uniqueness of the function satisfying the equations of the initiality theorem. The resultant initiality theorem for the tutu specification is

$$
\begin{aligned}
\forall A\_case \ B\_case. \ \exists! f. (\forall x. \ f(\text{A } x) = A\_case \ x) \ \wedge \\
(\forall g. \ f(\text{B } g) = B\_case \ (f \circ g) \ g)
\end{aligned}
$$

4 Future Work

As was mentioned in Section 1, this work is a part of a project to support a class of mutually recursive specifications with nestings of type constructors. The general approach we take involves translating the specifications into progressively simpler forms, preserving information necessary to translate solutions back. First we eliminate the nested type constructors in favor of larger mutually recursive specifications having no such nestings. For an example of this, consider the specification we gave for **bonsai**:

$$\text{bonsai} ::= \text{bonsai\_NODE} \; \beta \; (\alpha \rightarrow \text{bonsai lift})$$

The type **bonsai** being specified has an occurrence within the recursive type constructor **lift**. Using the specification for **lift** (which we can reconstruct from the initiality theorem for **lift**) we can convert this specification into the following form:

$$\text{bonsai}' ::= \text{bonsai\_NODE}' \; \beta \; (\alpha \rightarrow \text{bonsai\_lift})$$
$$\text{bonsai\_lift} ::= \text{lift}' \; \text{bonsai} \; | \; \text{undefined}'$$

The type **bonsai'** is isomorphic to the type **bonsai** and the type **bonsai_lift** is isomorphic to the type **bonsai lift**.

The next translation transforms the mutually recursive specification into a simple recursive specification. This translation is basically the same as the one previously outlined by Thomas Melham in [3]. This translation does not yield an isomorphism, and so a second predicate must be defined on the type returned by the simple recursive specification culling out those terms that represent well-formed terms of the mutually recursive types. One way in which we differ slightly from the description given by Melham is that we must translate each occurrence of the recursive type *rty* as an argument to a constructor into an occurrence of **one** $\rightarrow$ *rty*. Perhaps one subtlety in this translation that was not discussed in [3] was determining whether the mutually recursive specification is well-founded, and the computation of witnesses for each type. It is possible for one or more of the types being defined in a mutually recursive specification to have no base case by itself, and yet for the system to be well-founded. This is the case, for example, with the translated specification for **bonsai**.

It should be noted, that although we haved talked in terms of making various definitions at various intermediate stages along the path to finding a solution to a specification, in reality we want to avoid actually introducing multiple intermediate types and constructors, and thus in our actual package we deal with the definitions of the objects instead of actually introducing these objects. It is rather important for purposes of both space and time efficiency not to introduce formally definitionally the intermediate stages.

Once the package for defining types from the specifications described in Section 1 is completed, there will still need to be a package developed for the flexible definition of functions over these types. The tools for proving results by induction over such types will also need to be extended.

5 Acknowledgements

It has come to my attention that David Shepherd of INMOS is carrying out a independent project parallel to the one discussed in this paper, and in particular, that he has developed a comparable theory of arbitrarily branching labeled trees as a foundation.

A Infinite Trees

A.1 An infinitely Tall bonsai (with no infinite branches)

Let us assume that we have a type bonsai described by

$$\text{bonsai} ::= \text{bonsai\_NODE } \beta \ (\alpha \to \text{bonsai lift})$$

with the initiality theorem

$$\forall Cases. \ \exists! f. \ \forall subtrees \ label. \ f(\text{bonsai\_NODE } label \ subtrees) =$$
$$Cases \ (\text{lift\_compose (lift\_fun } f) \ subtrees) \ label \ subtrees$$

By primitive recursion on the natural numbers, we can define

$$\text{mk\_tree } 0 = \text{bonsai\_NODE one } (\lambda x. \text{ undefined}) \ \wedge$$
$$\forall n. \ \text{mk\_tree } (\text{SUC } n) = \text{bonsai\_NODE one } (\lambda m.\text{lift}(\text{mk\_tree } n))$$

Using these trees, we can then define

$$\text{tall} = \text{bonsai\_NODE one (lift\_fun mk\_tree)}$$

Given a partial function from α to num, we can define the least upper bound of the function by

$\forall p. \ \text{lub } p =$
$\quad (\forall x. \ p \ x = \text{undefined}) \Rightarrow \text{lift } 0$
$\quad | \ (\exists n.(\forall x.(p \ x \neq \text{undefined}) \Longrightarrow (\text{lower}(p \ x) \leq n)) \wedge (\exists x. \ p \ x = \text{lift } n)) \Rightarrow$
$\qquad \text{lift}(@n.(\forall x.(p \ x \neq \text{undefined}) \Longrightarrow (\text{lower}(p \ x) \leq n)) \wedge (\exists x. \ p \ x = \text{lift} n))$
$\quad | \ \text{undefined}$

Now consider the following case function for trees:

$\quad Cases \ rec\_fun \ label \ subtrees =$
$\qquad (\exists x. \ rec\_fun \ x = \text{lift } 13) \Rightarrow 13$
$\qquad | \ (\text{lub } rec\_fun = \text{undefined}) \Rightarrow 13 \ | \ (2 + \text{lower}(\text{lub } rec\_fun))$

If we define our DoubleHeight function by

$$\text{DoubleHeight (bonsai\_NODE } label \ subtrees) =$$
$$Cases(\text{lift\_compose (lift\_fun } f) \ subtrees) \ label \ subtrees$$

then we can show that DoubleHeight tall $= 13$.

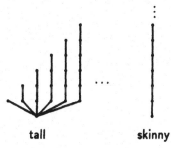

tall   skinny

A.2 broad_tree is not Initial

Notice that the function λl : one list. one satisfies Is_labeling. Let

$$\text{skinny} = \text{broad\_tree\_ABS}(\lambda l : \text{one list. one})$$

Then we can show the following:

$$\text{skinny} = \text{broad\_tree\_NODE one } \lambda x : \text{one. skinny}$$

That is, every immediate subtree of skinny (there is only one) is equal to skinny. Consider the following two case functions over (one, one)broad_tree:

$$\text{U } rec\_fun \text{ label subtrees} = ((rec\_fun \text{ one} = \text{undefined}) \vee \text{lower}(rec\_fun \text{ one}))$$

and

$$\text{E } rec\_fun \text{ label subtrees} = ((rec\_fun \text{ one} = \text{undefined}) \vee \neg\text{lower}(rec\_fun \text{ one}))$$

Then we have both that

$$\forall label \text{ } subtrees.(\lambda tr.\text{T})(Node \text{ } label \text{ } subtrees) =$$
$$\text{U}(\text{lift\_compose}(\text{lift\_fun }(\lambda tr.\text{T})) \text{ } subtrees) \text{ } label \text{ } subtrees$$

and

$$\forall label \text{ } subtrees.(\lambda tr.tr \neq \text{skinny})(Node \text{ } label \text{ } subtrees) =$$
$$\text{U}(\text{lift\_compose}(\text{lift\_fun }(\lambda tr.tr \neq \text{skinny})) \text{ } subtrees) \text{ } label \text{ } subtrees$$

Therefore, there is no unique function defined by U. The situation is even worse with E. There we can prove

$$\forall f. \text{ } f(\text{broad\_tree\_NODE one } \lambda x : \text{one. skinny}) \neq$$
$$\text{E}(\text{lift\_compose}(\text{lift\_fun } f) \text{ } \lambda x : \text{one. skinny}) \text{ one } \lambda x : \text{one. skinny}$$

References

1. M. J. C. Gordon. *The HOL System.* Cambridge Research Centre, SRI International, and DSTO Australia, 1989.
2. T. F. Melham. Automating recursive type definitions in higher order logic. In G. Birtwistle and P. A. Subrahmanyam, editors, *Current Trends in Hardware Verification and Automated Theorem Proving*, pages 341 – 386. Springer-Verlag, 1989.
3. T.F. Melham. Email correspondence. info-hol email, 26 April 1992.

Some Theorems We Should Prove

David Lorge Parnas

Telecommunications Research Institute of Ontario (TRIO)
Communications Research Laboratory
Department of Electrical and Computer Engineering
McMaster University, Hamilton, Ontario, Canada L8S 4K1

ABSTRACT

Mathematical techniques can be used to produce precise, provably complete documentation for computer systems. However, such documents are highly detailed; oversights and other errors are quite common. To detect the "early" errors in a document, one must attempt to prove certain simple theorems. This paper gives some examples of such theorems.

1 Introduction

In [PM4], we have shown how the contents of key computer systems documents can be defined in terms of mathematical functions and relations. We also reminded our readers that (1) functions and relations can be viewed as sets of ordered pairs, (2) sets can be characterised by predicates and described by logical expressions, (3) predicates can be represented in a more readable way using multidimensional (tabular) expressions whose components are logical expressions and terms, and (4) the meaning of these tables can be defined by rules for translating those tables into more conventional expressions. A complete discussion of these tabular expressions can be found in [Pa5]. The most recent illustration of their use can be found in [PMI3].

Our efforts have very pragmatic goals. We are not trying to provide mathematical proofs of program correctness; our goals are much more mundane. We wish to use mathematical methods to improve the quality of documentation in software systems. We believe, and have demonstrated using both practical and "academic" examples, ([HKPS1, PAM6, PMI3]), that we can provide mathematically precise documents that can be read by both programmers and properly prepared users.

Although we are not working on program verification *per se*, we believe that the ability to provide readable mathematical documentation is a prerequisite for regular practical use of mathematical methods in software development. It does no good to prove that a piece of software satisfies a specification, if that specification cannot be read, understood, and criticised by potential users or their representatives.

Although we are not trying to prove programs correct, we do have a need for theorem provers. The formulae in our tabular expressions must satisfy certain mathematical conditions. When we have used these tables in practice (e.g. [PAM6]), we have found that the documents submitted for review often fail to satisfy those conditions; as a result, the reviewers spent much too much of their time and energy checking for simple, application-independent, properties. This distracted us from the more difficult, safety relevant, issues and we felt that the preliminary checking should be done by a computer. Tools that check these tables must prove theorems, but theorems that are different from those that arise in program verification. The purpose of this paper is to formulate, but not prove, examples of those theorems. We would like to know which theorem provers or theorem proving support systems, are best able to deal with this type of theorem.

2 An Introductory Example

The example below describes a function in terms of a single real variable, x, applying a previously defined function, denoted by "$\sqrt{}$", which represents a function that is defined on a domain containing only non-negative real-numbers. The value of the function is a pair with two elements named y and z. The intent is to describe a function whose domain includes all real values. Each column of the table describes the value of the function in the subset of the function's domain that is characterised by the predicate expression in the column header. Each row in the table corresponds to an element of the pair; the rows are identified by the labels "y" and "z"...

	$x < 0$	$x = 0$	$x > 0$
			H_1
y	$x + 2$	$x + 4.21$	$5.4 + \sqrt{x}$
z	$5 + \sqrt{-x}$	x-4	x

H_2 G

Figure 1: Arbitrary function of a single real represented by "x"

The first theorem that must be proven is given in Figure 2. It confirms our intent that the domain of the function described includes all real numbers.

$$(\forall x, (x < 0 \vee x = 0 \vee x > 0))$$

Figure 2: Domain Coverage Theorem for Figure 1

In addition, we wish to make sure that each column deals with a disjoint subset of the domain. This can be expressed by the theorem in Figure 3,

$$(\forall x, \neg((x < 0 \wedge x = 0) \vee (x < 0 \wedge x > 0) \vee (x > 0 \wedge x = 0)))$$

Figure 3: Disjoint Domains Theorem for Figure 1

Finally, since the function applied in this table is a partial function, we want to prove that there is a defined value for every application of such a function in each column. Below, we introduce notation for referring to a predicate that characteriseses the domain of an application of a function, **domain( )**, so that we can motivate the following two (very trivial) theorems...

$$x < 0 \Rightarrow \mathbf{domain}(\sqrt{-x})$$

Figure 4: Definedness theorem for Column 1 of Figure 1

$$x > 0 \Rightarrow \mathbf{domain}(\sqrt{x})$$

Figure 5: Definedness theorem for Column 3 of Figure 1

Each of these theorems is "obviously true", but they should be checked routinely when preparing these tabular descriptions of functions. "Proving" them requires knowing the definitions of each of the relations and functions that appear as well as knowing the characterisation of the domain of any partial functions. The functions used in these examples are familiar functions, but, in practice, designers define unfamiliar functions for their applications. Thus, it must be possible to add new functions and relations to the "vocabulary" of the prover. The users of these tables cannot be assumed to be mathematically sophisticated, or even rigorous. Thus, we would like the theorems to be formulated and verified automatically wherever possible.

3 More Advanced Examples

The example in Section 2 illustrates the meaning of our tabular expressions and the way that "theorems" are derived from such tables. In more advanced examples we want to prove the same general theorems, but the expressions become more complex. The primary source of new problems is the use of quantification over finite sets.

3.1 Array Search Example

The example below describes programs that deal with an array, B, with indices 1... N. Like many others, we treat such arrays as partial functions whose domain consists of the integers 1 ... N. The value of the array (partial function) is not defined for other values.

Figure 6 specifies the behaviour of a program that must search the array B, looking for an element whose value is the same as the value of the program variable x[1]. To describe the behaviour of this program completely, we must distinguish two cases depending on whether or not there is such an element. The table describes the required properties of the *final* values of j and present (denoted by "j' " and "present' ") in both cases. In the first row, we state a predicate that j' must satisfy. Note that if the value of x cannot be found in the array, any value of j will satisfy the specification. In the next row, we provide a term whose value gives the required value of present'. We further indicate that the variables x and B should not change (by writing "NC(x, B)"). [2]

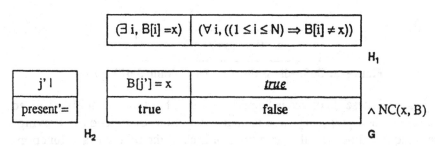

$(\exists\, i, B[i] = x)$	$(\forall\, i, ((1 \leq i \leq N) \Rightarrow B[i] \neq x))$	H_1

j' |	$B[j'] = x$	*true*	
present'=	true	false	$\wedge\, NC(x, B)$

H_2 G

Figure 6: Relational Description of a program that searches B for the value of x

---

[1] In these tables, *true* and *false* are predicate values, while true and, false represent the values of program variables. "|" is read "such that" and indicates that the value of the variable must satisfy a predicate given in the appropriate column.

[2] NC(x,B) is our abbreviation for x'='x ∧ 'B = B'

For Figure 6 to be proper, the two columns must be mutually exclusive. Further, we would like the domain of the function described to be the universe. This means that we would want the formulae in Figures 7 and 8 to evaluate to *true* for any array B. . .

$$(\exists\, i,\, B[i] = x) \vee (\forall\, i,\, ((1 \leq i \leq N) \Rightarrow B[i] \neq x))$$

Figure 7: Domain Coverage Theorem for Figure 6

$$\neg((\exists\, i,\, B[i] = x) \wedge (\forall\, i,\, ((1 \leq i \leq N) \Rightarrow B[i] \neq x))$$

Figure 8: Disjoint Domains Theorem for Figure 6

It should be noted that the theorems in Figures 7 and 8 are not as obvious as they might appear. In the logic that we use (described in [Pa2]), if both "=" and "≠" denote primitive relations, (i.e. one is not defined to be the complement of the other) they are *not* complementary. If i is not in the index set of B, both $B[i] = x$ and $B[i] \neq x$ will be *false*. Thus one cannot apply the standard transformation in this case. We can simplify both the expressions, and the proofs, if we do not include "≠" in the set of primitive relations and define "≠" to denote the complement of "=". If we do that, all occurrences of the formulae: "$(\forall\, i,\, ((1 \leq i \leq N) \Rightarrow B[i] \neq x))$", can be replaced by "$(\forall\, i,\, B[i] \neq x)$". Alternatively, we could replace "$B[i] \neq x$" by "$\neg(B[i] = x)$" and get the same simplifications. The simplified theorems are shown in Figures 9 and 10. . .

$$(\exists\, i,\, B[i] = x) \vee (\forall\, i,\, \neg(B[i] = x))$$

Figure 9: Simplified Domain Coverage Theorem for Figure 6

$$\neg((\exists\, i,\, B[i] = x) \wedge (\forall\, i,\, \neg(B[i] = x))$$

Figure 10: Simplified Disjoint Domains Theorem for Figure 6

Column 1 of Figure 6 requires that one prove that there will be a value of j' that satisfies the condition specified. This gives rise to the theorem of Figure 11. Since the expressions in the second row are constants, we need not state the corresponding theorems.

$$(\exists\, i,(\, B[i] =x) \Rightarrow \text{nonempty}(\{j'\,|\, B[j'] = x\}))$$

Figure 11: Definedness Theorem for Column 1 of Figure 6.

3.2 Searching for a palindrome of length, n, in an array, A, with index set 1...N

Figure 12 shows the use of some additional mathematical functions. Although we could have avoided it, we use of floor function and integer division in the expression "$\lfloor n+2\rfloor$". The Domain Coverage Theorem and the Disjoint Domain Theorem for Figure 12 are trivial as the header for column 2 is explicitly given as the complement of the header for column 1, The definedness theorem *should* be easy, but checking it is likely to be forgotten in practice. It is stated in Figure 13.

	$(\exists\, l, (\forall\, i, 0 \le i < \lfloor n+2\rfloor \Rightarrow$ $A[l+i] = A[l+n-1-i]))$	$\neg(\exists\, l, (\forall\, i, 0 \le i < \lfloor n+2\rfloor \Rightarrow$ $A[l+i] = A[l+n-1-i]))$
$l'\,|$	$(\forall\, i, 0 \le i < \lfloor n+2\rfloor \Rightarrow$ $A[l'+i] = A[l'+n-1-i])$	*true*
present'=	*true*	*false*

H_1

H_2 $\wedge NC(n,A)$ G

Figure 12: Find a Palindrome of length l in A[1:n]

$$(\exists\, l, (\forall\, i, 0 \le i < \lfloor n+2\rfloor \Rightarrow A[l+i] = A[l+n-1-i])) \Rightarrow$$
$$\text{nonempty}(\{l'|\, (\forall\, i, 0 \le i < \lfloor n+2\rfloor \Rightarrow A[l'+i] = A[l'+n-1-i])\})$$

Figure 13: Definedness Theorem for Column 1 of Figure 12.

3.3 Looking for the longest palindrome in A [1:N], N>0

The program described by Figure 12 could be used in looking for the longest palindrome that can be found in an array. n' is to designate the length of this palindrome and l' will indicate a location where a palindrome of that length can be found. A specification of such a program is given in Figure 14.

Note that maxel(x), where x is a non-empty set of integers, is a function whose value is the largest value in x. Even though we do not need to distinguish cases, the table format is still useful. Because, this table has only one column, the Domain Coverage Theorem is as trivial as one can get. However, the Definedness Theorem could be interesting because it depends on our recognising that there will always be palindromes of length 1 in any array of non-zero length.

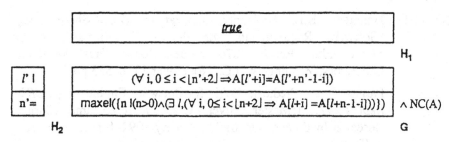

Figure 14: Finding the longest Palindrome in a non-empty array.

$$N > 0 \Rightarrow \text{nonempty}(\{n \mid (\exists \, l, (\forall \, i, 0 \le i < \lfloor n \div 2 \rfloor \Rightarrow A[l+i] = A[l+n-1-i]))\})$$

Figure 15: Definedness Theorem for Figure 12.

4 CONCLUDING OBSERVATIONS

Mathematicians reading this paper will find the theorems posed trivial; they are certainly shallow when compared to the theorems that mathematicians prove in published papers. However, they are more difficult than the majority of the theorems that arose in the documentation and inspection of the Darlington Nuclear Plant Shutdown Systems discussed in [PAM6]. Most of the theorems that we had to check, were similar in nature to the ones discussed in Section 2. The scrupulously careful inspection resulted in about 40 kg. of such trivial tables. If these theorems can be proven automatically by today's theorem proving programs, we should be using those programs. If these theorems still require human intervention, perhaps the developers of theorem proving programs would like to turn their attention to this type of theorem.

5 ACKNOWLEDGEMENTS

This work was supported by the Government of Ontario, through TRIO, and by the Government of Canada through NSERC's Research Grant programme. I am grateful to Dennis Peters, John Rushby, and Mandayam Srivas for their careful checking of an earlier draft of this paper.

References

[HKPS1] Heninger, K.L., Kallander, J., Parnas, D.L., Shore, J.E., "Software Requirements for the A-7E Aircraft", NRL Memorandum Report 3876, United States Naval Research Laboratory, Washington D.C., November 1978, 523 pp.

[Pa2] Parnas, D.L., "Predicate Logic for Software Engineering", *CRL Report 241*, McMaster University, TRIO (Telecommunications Research Institute of Ontario), February 1992, 8 pgs. To appear in IEEE Transactions on Software Engineering.

[PMI3] Parnas, D.L., Madey, J., Iglewski, M., "Formal Documentation of Well-Structured Programs", *CRL Report 259*, McMaster University, TRIO (Telecommunications Research Institute of Ontario), September 1992, 37 pgs.

[PM4] Parnas, D.L., Madey, J., "Functional Documentation for Computer Systems Engineering (Version 2)", CRL Report 237, McMaster University, Hamilton Canada, TRIO (Telecommunications Research Institute of Ontario), September 1991, 14 pgs.

[Pa5] Parnas, D.L., "Tabular Representation of Relations", *CRL Report 260*, McMaster University, TRIO (Telecommunications Research Institute of Ontario), October 1992, 12 pgs.

[PAM6] Parnas, D.L., Asmis, G.J.K., Madey, J., "Assessment of Safety-Critical Software in Nuclear Power Plants", *Nuclear Safety*. vol. 32, no. 2, April-June 1991, pgs. 189-198.

Using PVS to Prove Some Theorems of David Parnas

John Rushby and Mandayam Srivas
Computer Science Laboratory
SRI International
Menlo Park CA 94025 USA

June 9, 1993

Abstract

David Parnas [13] describes some theorems representative of those encountered in support of certification of software for the Darlington nuclear reactor. We describe the verification of these theorems using PVS.

1 Introduction

David Parnas [13] describes some theorems representative of those encountered in support of certification of software for the Darlington nuclear reactor. After noting that these illustrative theorems "will appear trivial to mathematicians," Parnas [13] writes "if these theorems can be proven automatically by today's theorem proving programs, we should be using those programs. If these theorems still require human intervention, perhaps the developers of theorem proving programs would like to turn their attention to this type of theorem." In this note, we describe the transcription of these theorems into the language of PVS and their mechanized verification using the PVS proof checker [10]. PVS is not an automatic theorem prover; it is primarily intended for the proof of hard theorems such as those that arise in verification of interesting algorithms [11]. Consequently, PVS provides rather powerful automation of the lower levels of deduction (for example, decision procedures for ground arithmetic), but is generally guided by a human operator. We find this arrangement to be the most productive for kinds of theorems that arise in the problem domains we have studied.

However, PVS proof steps can be composed into larger steps that we call "strategies" (these are akin to the "tacticals" of LCF-like systems). It is straightforward to construct strategies that can deal automatically with various classes of elementary theorems. One such strategy is the "tcc-strategy" built-in to PVS for the purpose of automatically discharging many of the "type-correctness conditions" (tcc's) that arise during typechecking. This strategy iteratively expands explicit definitions, skolemizes, and performs propositional and arithmetic simplification and simple heuristic instantiation. It is able to prove all but one of the theorems posed by Parnas. The one that it fails to prove requires something close to "insight" and can be completed with just a couple of additional user-supplied steps.

Implicit in Parnas' "challenge" to prove these theorems automatically is the expectation that the theorem and supporting specification text submitted to the prover should be in a form close to that employed by Parnas. In the following section, we describe the specification text submitted to PVS and also explain the theorem proving mechanisms used to prove each theorem.

2 The Theorems and Their Proofs

The theorems suggested by Parnas appear as figures in [13]; not all the figures are theorems, however—some are specification (expressed as tables) whose various well-formedness constraints give rise to the theorems. In this section we describe each of the theorems in a separate subsection, labeling each with the figure number from [13] of the corresponding theorem.

2.1 The Theorem of Figure 2

For real number x, prove that

$$(\forall x, (x < 0 \lor x = 0 \lor x > 0)).$$

In PVS this is represented as

```
fig2: THEOREM FORALL (x: real): x < 0 OR x = 0 OR x > 0
```

The proof follows by skolemization and ground arithmetic.

The PVS prover includes decision procedures for linear arithmetic over both integers and reals [14–17]. That means it can decide *any* true ground (variable-free) formula of arithmetic involving the propositional connectives, equality over uninterpreted functions symbols, the arithmetic relations $<, >, \leq, \geq$, and the operators of addition, subtraction, multiplication and division, but with the latter two restricted to the linear case (i.e., one argument must be a literal constant). Theorem provers that lack automation of arithmetic can be tedious to use, since even ostensibly non-numerical specifications can often generate numerous proof obligations concerning (usually trivial) arithmetic facts.

2.2 The Theorem of Figure 3

Here the requirement is to prove

$$(\forall x, \neg((x < 0 \land x = 0) \lor (x < 0 \land x > 0) \lor (x > 0 \land x = 0))).$$

In PVS, this becomes

```
fig3: THEOREM
   FORALL (x: real): NOT ((x < 0 AND x = 0)
                       OR (x < 0 AND x > 0)
                       OR (x > 0 AND x = 0))
```

and again the proof follows directly from skolemization and ground arithmetic.

2.3 The Theorem of Figure 4

Here we are required to prove

$$x < 0 \Rightarrow \text{domain}(\sqrt{-x}).$$

Parnas employs a partial term logic [12] in order to accommodate partial functions such as square root,[1] whereas PVS uses classical logic and total functions. However, PVS provides predicate and dependent types that can be used to constrain the domains of what would otherwise be partial functions. Thus, for example, division is a total function on the domain that excludes 0 in its second argument position.

Here, we define the predicate nonneg_real? to be true of just the non-negative reals. Given a predicate p in PVS, the parenthesized construction (p) is a type-expression that denotes the subtype of the domain of p that satisfies p. Thus, in particular, (nonneg_real?) denotes the subtype of the reals consisting of just the non-negative reals. We then introduce nonneg_real as a synonym for this subtype. Next, we specify the square root function sqrt to be a function that takes a nonneg_real as its argument and returns one as its value.[2] We do not supply an interpretation for the sqrt function since none is needed for the proofs considered here.

```
nonneg_real?(x: real): bool = (x >= 0)
nonneg_real: TYPE = (nonneg_real?) CONTAINING 0
sqrt: [nonneg_real -> nonneg_real] % (could return a pair if required)
```

The clause CONTAINING 0 is provided in order to discharge the tcc that will require the nonemptiness of this subtype to be demonstrated.

Given this machinery, we can state the required theorem as

```
x: VAR real
fig4: THEOREM x < 0 => nonneg_real?(-x)
```

Notice that, unlike the earlier theorems that used local bindings for x and explicit quantification, here we use a global declaration and implicit quantification (all formulas in PVS are automatically closed by universally quantifying their free variables). This is purely a syntactic convenience and has no semantic consequences. The proof follows by skolemizing, expanding the definition of nonneg_real, and ground arithmetic reasoning.

Notice that another way to generate the theorem of interest would be to propose an expression such as

---

[1] The logic used is essentially that of Beeson [2] (more accessible references are [1, Chapter 6, Section 1] and [3, Section 5]), but for some reason these references are not cited by Parnas. For a general account of "free logics" (of which Beeson's is an example) see [4], and for a brief discussion of their application to partial term logics see [18, volume I, chapter 2, section 2]. PX [8] is a computational logic based on these ideas, while a (higher-order) logic of this kind is mechanized in the IMPS system [5, 6], and variants have been proposed for other specification languages [9]. Gumb [7, Chapter 5] uses a free logic to express facts about execution-time errors in programs.

[2] Alternatively, by a trivial modification, we could specify it to return a pair consisting of a nonneg_real and a nonpos_real.

```
    variant: THEOREM x<0 => sqrt(-x) = sqrt(-x)
```

The typechecker would then automatically generate a theorem identical to fig4 above as a tcc necessary to ensure type-correctness of the expression sqrt(-x).

2.4 The Theorem of Figure 5

This is a trivial variation on the previous example:

$$x > 0 \Rightarrow \mathbf{domain}(\sqrt{x}).$$

The PVS specification is

```
    fig5: THEOREM x > 0 => nonneg_real?(x)
```

and the proof proceeds just as before.

2.5 The Theorem of Figure 7

Here we are required to prove, for any array B indexed by the integers $1 \ldots N$,

$$(\exists i, B[i] = x) \vee (\forall i, ((1 \le i \le N) \Rightarrow B[i] \neq x)).$$

N is presumably some fixed constant, but it is not clear whether B is intended to be an arbitrary constant or a variable. Since skolemization will reduce the latter to the former, it doesn't matter which we choose here, so we have used a constant. Parnas treats B as a partial function with domain $1 \ldots N$, whose value is not defined for other indices. In PVS, we define $1 \ldots N$ as a subtype (called index) of the integers and introduce B as a *total* function on this domain. The range type of B is not specified by Parnas, so we introduce an uninterpreted type T to serve this purpose.

```
    index: TYPE = {i: int | 1<= i & i <= N} CONTAINING 1
    T: TYPE
    B: [index -> T]
```

We can then state the required theorem as:

```
    x: VAR T
    fig7: THEOREM (EXISTS (i: index): B(i) = x)
                OR (FORALL (i: index): B(i) /= x)
```

Notice that we have overloaded x to be both a variable of type T and one of type real (defined earlier). The PVS typechecker disambiguates these names in context by virtue of their types. Notice that because we specify the quantified variables i to be of type index, we do not need the range qualification used in Parnas' form of the theorem.

The proof of this theorem follows by skolemization, propositional simplification, and heuristic instantiation.

2.6 The Theorem of Figure 8

This is a trivial variant on the previous example (it follows by De Morgan's rule).

$$(\exists i, B[i] = x) \vee (\forall i, ((1 \leq i \leq N) \Rightarrow B[i] \neq x)).$$

The PVS version is obvious

```
fig8: THEOREM NOT ((EXISTS (i: index): B(i) = x)
                   AND (FORALL (i: index): B(i) /= x))
```

and its proof is the same as the previous one.

We do not dicuss the proofs of Figures 9 and 10 since they are merely simplified instances of the previous two.

2.7 The Theorem of Figure 11

Here we are required to prove

$$(\exists i, B[i] = x) \Rightarrow \mathbf{nonempty}(\{j' | B[j'] = x\}).$$

The PVS version is an almost exact transliteration:

```
fig11: THEOREM (EXISTS (i: index): B(i) = x)
               => nonempty?({j: index | B(j) = x})
```

Sets are identified with predicates in PVS,[3] and the set expression {j: index | B(j) = x} is simply a syntactic variation on the predicate definition (LAMBDA (j: index): B(j)=x). The function nonempty? is from the PVS prelude (i.e., built-in) theory called sets, which supplies definitions for various set-theoretic constructs. The ones relevant here are:

```
sets [T: TYPE] : THEORY
  BEGIN
  member(x:T, a:set): bool = a(x)
  empty?(a:set): bool = (FORALL (x:T)) NOT member(x,a))
  nonempty?(a:set): bool = NOT empty?(a)
  END sets
```

The proof of fig11 is obtained by expanding the definitions of nonempty?, empty?, and member, followed by skolemization, propositional simplification, and heuristic instantiation.

2.8 The Theorem of Figure 13

Here we are required to prove

$$(\exists l, (\forall i, 0 \leq i < \lfloor n \div 2 \rfloor \Rightarrow A[l + i] = A[l + n - 1 - i])) \Rightarrow$$
$$\mathbf{nonempty}(\{l' | (\forall i, 0 \leq i < \lfloor n \div 2 \rfloor \Rightarrow A[l' + i] = A[l' + n - 1 - l])\})$$

---

[3]PVS is a simply-typed higher-order logic, so the axiom of comprehension (identifying sets with predicates) is sound.

The idea is that the array A (indexed by $1 \ldots N$) has a palindrome of length n starting at position l if $A[l + i] = A[l + n - 1 - i]$ for all i from 0 up to (but not including) $\lfloor n \div 2 \rfloor$. The theorem states that if there exists such an l, then the set of such ls is not empty.

We could state this directly in PVS, but to save some typing and to provide a clearer specification, we prefer to define an auxiliary predicate: palindrome?(1, n, A) will be true when there is a palindrome starting at position 1, of length n in the array A. Then we have:

```
A: VAR [index -> T]
n: VAR posnat
1: VAR index
palindrome?(1, n, A): bool
fig13: THEOREM (EXISTS 1: palindrome?(1, n, A))
                => nonempty?({1 | palindrome?(1, n, A)})
```

Notice that we have given a signature, but no interpretation for the predicate palindrome?. This is because the truth of the theorem fig13 is independent of the interpretation of this predicate (Fig11 has a similar form). The theorem is proved in exactly the same way as the previous two.

2.9 The Theorem of Figure 14

This theorem is given as

$$n > 1 \Rightarrow \textbf{nonempty}(\{n | \exists l, (\forall i, 0 \leq i < \lfloor n \div 2 \rfloor \Rightarrow A[l + i] = A[l + n - 1 - i])\}).$$

This formula contains both a free occurrence of n and a bound occurrence. We can make no sense of the free occurrence. The label given by Parnas to the theorem suggests that it is intended to assert that the set of lengths of palindromes in the nonempty array A is nonempty. This suggests that the antecedent to the implication should be $N > 0$ rather than $n > 1$.

With this interpretation, we can state the theorem in PVS as

```
fig14: THEOREM nonempty?({n | (EXISTS 1: palindrome?(1, n, A))})
```

Notice that we do not need to impose the condition N > 0 because this is embedded in the type (posnat) specified for N.

In order to evaluate this theorem, we do need to give an interpretation to the predicate palindrome?:

```
palindrome?(1, n, A): bool =
     (FORALL (i: nat): i < floor(n/2) => A(1+i) = A(1+n-1-i))
```

This in turn requires a definition for the function floor:

```
floor(q: real): int = epsilon({n: int | n <= q
                        AND (FORALL (m: int): m <= q => m <= n)})
floorprop: LEMMA floor(x) <= x
                        AND (FORALL (m: int): m <= x IMPLIES m <= floor(x))
```

The definition is given in terms of Hilbert's ε (choice) operator and is followed by a lemma which states that the floor of a real number x is the largest integer less than or equal to x. The definition and lemma come from the PVS prelude; we do not explain the definition or the proof of the lemma here, since their level of sophistication is out of keeping with the elementary level of the rest of this note. All the PVS user needs to understand is the property stated in floorprop.

The tcc-strategy built-in to PVS is unable to prove the theorem fig14 automatically. It is necessary for the user to supply some "insight": namely, that the reason this theorem is true is that palindromes of length 1 occur at every position. Thus, to finish the proof after the tcc-strategy has done its work, it is necessary for the user to suggest two instantiations (one for the length n and another for the starting position 1) and to invoke floorprop.[4]

Although it looks as though we have completed the assignment suggested by Parnas, a few loose ends remain. Typechecking a PVS specification can lead to the generation of proof obligations called tccs that must be discharged before the specification is considered type-correct. In the present case, most of the tccs are trivial[5] and are discharged automatically by the tcc-strategy that also disposes of the main theorems suggested by Parnas. However, the declaration of palindrome? generates some tccs that inspection shows to be unprovable (in fact, false)—for example:

```
palindrome?_TCC1:
  OBLIGATION (FORALL (i: nat), 1, n:
    i < floor(n / 2) IMPLIES 1 + i > 0 AND 1 + i <= N)
```

The issue here is that the definition of palindrome? requires accessing the values of the array A at index positions 1+i and 1+n-1-i where 1 is known to be of type index, n is a posnat, and 0 <= i < floor(n/2). Because PVS is a logic of total functions, it requires us to prove that all accesses to A will be within its domain—this is the content of the tcc palindrome_TCC1 shown above. Unfortunately, it is easy to see that some of these accesses could actually be outside the domain of A and that the tcc is consequently false. The way to repair this deficiency is to realize that once we have chosen a value for 1 (the starting position of the putative palindrome), the length of the longest palindrome that can lie within the array is restricted to N+1-1. Thus not all combinations of 1 and n are valid arguments to palindrome? and we need to restrict its domain appropriately. This is done by means of a *dependent type* declaration—a type that depends on the *value* of some term appearing earlier in the specification. Here the appropriate specification is the following.

---

[4]Notice that this theorem remains true (and becomes stronger) if the existential quantifier is replaced by a universal one. This variant also has an easier proof, since the starting position 1 is skolemized and we do not need to supply an instantiation for it.

[5]For example, the declaration of the type index requires that we show $0 < 1$ and $1 \leq N$.

```
palindrome?(l, (n | 1+n <= N+1), A): bool =
   (FORALL (i: nat): i < floor(n / 2) => A(1 + i) = A(1 + n - 1 - i))
```

Notice that the type of the second argument is now restricted to the range 1 ...N+1-1. With this adjustment, the tcc becomes

```
palindrome?_TCC1:
   OBLIGATION (FORALL (i: nat), l, (n: posnat | 1 + n <= N + 1):
      i < floor(n / 2) IMPLIES 1 + i > 0 AND 1 + i <= N)
```

which is (easily) provable. With more precise type constraints on the predicate palindrome?, we need to adjust the statements of fig13 and fig14 so that they remain type-correct:

```
fig13: THEOREM (EXISTS l, (n|1+n<=N+1): palindrome?(l, n, A))
    => nonempty?({l | (EXISTS (n|1+n<=N+1): palindrome?(l, n, A))})

fig14: THEOREM nonempty?({n|(FORALL (l|1+n<=N+1): palindrome?(l, n, A))})
```

This correction to the specification has no effect on the proofs of these theorems (it just makes their statements valid in the logic of PVS).

It might seem that Parnas' partial term logic has avoided this complication. In one sense it has: his specification has a meaning even if accesses can occur outside the domain of the array. However, when we come to prove a theorem such as fig14 it is not enough to know that the expression palindrome?(l, n, A) has *some* meaning—we need to evaluate its *actual* meaning in order to decide the truth of the theorem. That will require evaluating the expression A(1+i) = A(1+n-1-i) for all combinations of the variables concerned. In order to consider the truth of this equality, we need to know whether the expressions on either side are defined or not (partial term logics use Kleene equality, which is true if both its arguments are undefined, false if just one of them is, and behaves like ordinary equality if both are defined). Thus, in formally evaluating this expression as it appears in fig14, a theorem prover based on a partial term logic would probably pose definedness lemmas identical to the tcc shown earlier.[6]

One other feature of PVS we should mention before closing: PVS provides a LaTeX-printer that can be customized by simple tables to recreate the preferred notation of a particular applicaiton area. The following two-line incantation causes it to employ the symbols used by Parnas for the floor and division functions.

```
floor 1   1 {\lfloor #1 \rfloor}
/     key 1 {\div}
```

The result of LaTeX-printing the definition of palindrome? is shown below.

$$\text{palindrome?}(l, (n \mid l + n \leq N + 1), A) : \text{bool}$$
$$= (\forall (i: \text{nat}): i < \lfloor n \div 2 \rfloor \Rightarrow A(l + i) = A(l + n - 1 - i))$$

---

[6] It would be interesting to try this in a system such as IMPS that uses a partial term logic.

3 Conclusions

We appreciate this opportunity to demonstrate the effectiveness of a modern theorem prover on problems representative of those that arise in the software engineering and verification methodology developed by David Parnas. All but one of the theorems posed by Parnas were proved automatically by the tcc strategy of PVS. The entire development (transcription of the theorems into PVS and their proof) took about an hour—essentially the time required to interpret the theorems and type them in. We speculate that proof of somewhat harder theorems could also be largely automated: it has been our experience that any given development tends to generate many theorems of a similar form. An expert could therefore develop an appropriate PVS strategy for the class of theorems generated by a particular development. However, we do not consider it productive to attempt the construction of automatic proof procedures for genuinely hard theorems—our experience is that it is better and faster for a skilled and knowledgeable user to guide the proof than to attempt heroic automation.

One question not raised by the theorems examined here is what to do when an automated proof attempt fails. This can happen for two reasons: the theorem may be true but the automated procedures are inadequate to prove it, or the theorem may be false. In our experience, it requires some skill to distinguish between these cases, and it may not be easy to develop that skill when relying on automated proof procedures.

Finally, although the purpose of this note has been to demonstrate theorem proving, we hope the reader may also find something of value and interest in style of specification supported by PVS. In particular, we believe that predicate and dependent subtyping (and the attendant requirement to prove theorems in order to decide type-correctness) provide very effective solutions to issues (such as partial functions) that pose considerable difficulties in other treatments. Also, although we can appreciate some of the attractions of the tabular approach to specification, we suggest that those who practice this approach should also consider whether direct expression in a logic such as that of PVS might not be advantageous in some circumstances. Tables are an effective way to present control-dominated constructions, but simple functions and algorithms may be more perspicuous when presented in logic. The very strong typechecking and the guarantees of conservative extension provided by a system such as PVS automatically take care of some of the checks that generate proof obligations in the tabular approach.

References

[1] Michael J. Beeson. *Foundations of Constructive Mathematics.* Ergebnisse der Mathematik und ihrer Grenzgebiete; 3. Folge · Band 6. Springer Verlag, 1985.

[2] Michael J. Beeson. Proving programs and programming proofs. In *International Congress on Logic, Methodology and Philosophy of Science VII,* pages 51–82, Amsterdam, 1986. North-Holland. Proceedings of a meeting held at Salzburg, Austria, in July, 1983.

[3] Michael J. Beeson. Towards a computation system based on set theory. *Theoretical Computer Science*, 60:297–340, 1988.

[4] Ermanno Bencivenga. Free logics. In Dov M. Gabbay and Franz Guenthner, editors, *Handbook of Philosophical Logic–Volume III: Alternatives to Classical Logic*, volume 166 of *Synthese Library*, chapter III.6, pages 373–426. D. Reidel Publishing Company, Dordrecht, Holland, 1985.

[5] William M. Farmer. A partial functions version of Church's simple theory of types. *Journal of Symbolic Logic*, 55(3):1269–1291, September 1990.

[6] William M. Farmer, Joshua D. Guttman, and F. Javier Thayer. IMPS: An interactive mathematical proof system. In Mark E. Stickel, editor, *10th International Conference on Automated Deduction (CADE)*, pages 653–654, Kaiserslautern, Germany, July 1990. Volume 449 of *Lecture Notes in Computer Science*, Springer Verlag.

[7] Raymond D. Gumb. *Programming Logics: An Introduction to Verification and Semantics*. John Wiley and Sons, New York, NY, 1989.

[8] Susumu Hayashi and Hiroshi Nakano. *PX: A Computational Logic*. Foundations of Computing. MIT Press, Cambridge, MA, 1988.

[9] C. A. Middelburg and G. R. Renardel de Lavalette. LPF and MPL_ω—a logical comparison of VDM SL and COLD-K. In S. Prehn and W. J. Toetenel, editors, *VDM '91: Formal Software Development Methods*, pages 279–308, Noordwijkerhout, The Netherlands, October 1991. Volume 551 of *Lecture Notes in Computer Science*, Springer Verlag. Volume 1: Conference Contributions.

[10] S. Owre, J. M. Rushby, and N. Shankar. PVS: A prototype verification system. In Deepak Kapur, editor, *11th International Conference on Automated Deduction (CADE)*, pages 748–752, Saratoga, NY, June 1992. Volume 607 of *Lecture Notes in Artificial Intelligence*, Springer Verlag.

[11] Sam Owre, John Rushby, Natarajan Shankar, and Friedrich von Henke. Formal verification for fault-tolerant architectures: Some lessons learned. In J. C. P. Woodcock and P. G. Larsen, editors, *FME '93: Industrial-Strength Formal Methods*, pages 482–500, Odense, Denmark, April 1993. Volume 670 of *Lecture Notes in Computer Science*, Springer Verlag.

[12] David Lorge Parnas. Predicate logic for software engineering. Technical Report TRIO-CRL-241, Telecommunications Research Institute of Ontario (TRIO), Faculty of Engineering, McMaster University, Hamilton, Ontario, Canada, February 1992.

[13] David Lorge Parnas. Some theorem we should prove. Technical report, Telecommunications Research Institute of Ontario (TRIO), Faculty of Engineering, McMaster University, Hamilton, Ontario, Canada, June 1993.

[14] Robert E. Shostak. On the SUP-INF method for proving Presburger formulas. *Journal of the ACM*, 24(4):529–543, October 1977.

[15] Robert E. Shostak. An algorithm for reasoning about equality. *Communications of the ACM*, 21(7):583–585, July 1978.

[16] Robert E. Shostak. A practical decision procedure for arithmetic with function symbols. *Journal of the ACM*, 26(2):351–360, April 1979.

[17] Robert E. Shostak. Deciding combinations of theories. *Journal of the ACM*, 31(1):1–12, January 1984.

[18] A. S. Troelstra and D. van Dalen. *Constructivism in Mathematics: An Introduction*, volume 121 and volume 123 of *Studies in Logic and the Foundations of Mathematics*. North-Holland, Amsterdam, Holland, 1988. In two volumes.

Extending the HOL Theorem Prover with a Computer Algebra System to Reason About the Reals

John Harrison[*] & Laurent Théry[**]

University of Cambridge Computer Laboratory

Abstract. In this paper we describe an environment for reasoning about the reals which combines the rigour of a theorem prover with the power of a computer algebra system.

1 Introduction

Computer theorem provers are a topic of research interest in their own right. However much of their popularity stems from their application in computer-aided verification, i.e. proving that designs of electronic or computer systems, programs, protocols and crypto-systems satisfy certain properties.

Such proofs, as compared with the proofs one finds in mathematics books, usually involve less sophisticated central ideas, but contain far more technical details and therefore tend to be much more difficult for humans to write or check without making mistakes. Hence it is appealing to let computers help.

Some fundamental mathematical theories, such as arithmetic, are usually required for such proofs, and in particular the real numbers are useful in several fields:

- The verification of floating-point hardware. It is often desirable to specify the intended behaviour of a floating-point unit with reference to the real numbers which the bit-strings approximate.
- The verification of systems incorporating linear components, such as vehicle braking mechanisms or nuclear reactor controllers.
- Extension of theorem-proving methods into the realm of day-to-day applied mathematics as undertaken by engineers and scientists.

It is mainly the last application which will concern us here. Computer Algebra Systems (CASs) are very popular for such applications as multi-precision arithmetic, differentiation and integration, and the expansion of functions in power series. However they have two drawbacks.

Firstly, they tend to have no concept of a logical language in which to state precise mathematical theorems. Usually, they accept an algebraic expression

[*] Supported by the Science and Engineering Research Council, UK.
[**] Supported by SERC grant GR/G 33837 and a grant from DSTO Australia.

and return another purportedly equal to it. They may not even mean equality in the true sense; there may be restrictions, or 'equality' may mean some weaker notion such as equality as meromorphic functions, e.g. $(x^2 - 1)/(x - 1) = x + 1$. By contrast, theorem provers are fundamentally based on logic, and offer good facilities for the management and organization of large proofs.

Secondly, even where the result from a CAS has some precise and unambiguous meaning, it may be false! CASs tend to simplify expressions fairly aggressively, taking shortcuts which are not always rigorously justified. In any case, they implement some extremely complicated algorithms to get their results, and it would be surprising if there were no bugs. For example Maple [6] evaluates $\int_{-1}^{1} \sqrt{x^2} \, dx$ to 0. Mathematica [21] gets this integral right and returns 1, but it returns 0 when given $\int_{-1}^{1} \frac{1}{\sqrt{x^2}} \, dx$, forgetting the singularity at 0. Computer theorem provers can offer much greater reliability; this is particularly so with strictly foundational systems like HOL.

However there is one obvious advantage that CASs have over theorem provers: they have many powerful decision procedures and heuristics which make them effective to use. To emulate these in a computer theorem prover would be an enormous task, and the burden of producing a logical proof would doubtless slow down the algorithms a great deal.

In this paper, we present an approach that combines the best of the two worlds by a careful exploitation of a link between HOL and a CAS. In the next section we outline the theoretical background to the formal development of integral calculus in HOL. We then consider the technical aspects of the link and how it might be useful. In the final section we draw the threads together by showing how we can evaluate certain trigonometric integrals in an entirely rigorous manner.

2 The reals in HOL

HOL is strictly definitional: rather than the existence of a new mathematical structure being asserted, the structure is *defined* in terms of existing structures, and the characterizing 'axioms' are derived by formal proof. For example, the real numbers have been constructed in HOL using a version of Dedekind's construction; a brief overview is given in [10], which sketches the procedure and discusses some significant parts of elementary analysis, including differentiation. Here we concentrate on the extension of this work to include integration, which we will use in a later example.

A consequence of the definitional approach is that we must be particularly careful about the way we define mathematical notions. In some cases, the appropriate definitions are uncontroversial. However many areas of mathematics offer a range of subtly different approaches. Integration is a particularly difficult case (its history is traced in [7] and [19]). Many people think of integration as the opposite of differentiation. Undergraduate mathematics courses usually present the Riemann integral. At a more advanced level, Lebesgue theory seems dominant; consider the following quote from [4]

It has long been clear that anyone who uses the integral calculus in the course of his work, whether it be in pure or applied mathematics, should normally interpret integration in the Lebesgue sense. A few simple principles then govern the manipulation of expressions containing integrals.

We shall consider these notions in turn and explain our selection of the Kurzweil-Henstock gauge integral. For our purposes it is particularly important to get clear the relationship between differentiation and integration. Ideally we would like the Fundamental Theorem of Calculus

$$\int_a^b f'(x)dx = f(b) - f(a)$$

to be true whenever f is differentiable at all points of the interval $[a, b]$ and its derivative at x is $f'(x)$.

The Newton Integral Newton actually *defined* integration as the reverse of differentiation. Integrating f means finding a function which when differentiated, gives f (called an *antiderivative*). Therefore the Fundamental Theorem is trivially true for the Newton Integral.

Newton's approach however has certain obvious defects as a formalization of the notion of the area under a curve. It is not too hard to prove that all derivatives are *Darboux continuous*, that is, they attain all intermediate values. Consequently, a simple step function:

$$f(x) = \begin{cases} 0 \text{ if } x < 1 \\ 1 \text{ if } x \geq 1 \end{cases}$$

which intuitively has a perfectly well-defined area, does not have a Newton integral.

The Riemann Integral The Riemann integral defines the area under a curve in terms of the areas of strips bounded by the curve, as the width of the strips tends to zero. It handles the step function but has other defects. Infinite limits have to be written as limiting cases of other integrals in various ad-hoc ways. The integral does not have convenient convergence properties: limits of sequences of integrable functions can fail to be integrable. And, particularly relevant to the present work, the Fundamental Theorem of Calculus fails to hold (the example given below for the Lebesgue integral also serves for the Riemann).

The Lebesgue Integral The Lebesgue integral is superior to the Riemann integral in a number of important respects. It accommodates infinite limits without any ad-hoc devices, and obeys some useful convergence theorems. Any (directly) Riemann integrable function is also Lebesgue integrable, and some functions which have no Riemann integral nonetheless have a Lebesgue integral, the classic example being the indicator function of the rationals:

$$f(x) = \begin{cases} 1 \text{ if } x \in \mathbb{Q} \\ 0 \text{ if } x \notin \mathbb{Q} \end{cases}$$

One feature that the Lebesgue integral shares with the Riemann integral is that the Fundamental Theorem of Calculus is *still* not generally true. The following counterexample was given in Lebesgue's thesis:

$$f(x) = \begin{cases} 0 & \text{if } x = 0 \\ x^2 sin(1/x^2) & \text{if } x \neq 0 \end{cases}$$

This is an inevitable consequence of the fact that the Lebesgue integral, in common with the Riemann integral, is an *absolute* integral, meaning that whenever f is integrable, so is $|f|$.

Other Integrals Various integrals have been proposed which extend the Lebesgue integral and for which the Fundamental Theorem is true. The first was Denjoy integral [15] which, starting with the Lebesgue integral, constructs a sequence of integrals by a process of transfinite recursion which Denjoy called *'totalisation'*. A very simple characterization of the Denjoy integral was given by Perron [3], but it is not constructive and the development of the theory uses theorems about the Lebesgue integral.

The Kurzweil-Henstock Gauge Integral Surprisingly recently it was observed that a simple modification of the Riemann limit process could give an integral equivalent to the Denjoy and Perron integrals. This seems to have first been made explicit by Kurzweil [17], but its later development, in particular the proof of Lebesgue-type convergence theorems, was mainly due to Henstock [11]. It is known as the 'Kurzweil-Henstock gauge integral' or simply 'gauge integral'. In the following, we give a sketch of the definition of this integral following the terminology given in [18] and note some results that have already been proved in HOL. A fuller introduction may be found in the undergraduate textbook [8] or the definitive [12].

The limiting process involved in the gauge integral seems rather obscure at first sight, but the intuition can be seen quite clearly if we consider integrating a derivative. Suppose f is differentiable for all x lying between a and b. Then given any such x and any $\epsilon > 0$, we know that there exists a $\delta > 0$ such that whenever $|y - x| < \delta$

$$\left| \frac{f(y) - f(x)}{y - x} - f'(x) \right| < \epsilon$$

For some fixed ϵ, this δ can be considered as a function of x which always returns a strictly positive real number, i.e. a *gauge*.

Consider now splitting the interval $[a, b]$ into a *tagged division*, i.e. a finite sequence of non-overlapping intervals, each interval $[x_i, x_{i+1}]$ containing some nominated point t_i called its *tag*. We shall say that a division is δ-fine (or fine with respect to a gauge δ) if for each interval in the division:

$$[x_i, x_{i+1}] \subseteq (t_i - \delta(t_i), t_i + \delta(t_i))$$

It is an important property that a δ-fine division exists for any gauge. This has been proved in HOL using Bolzano's technique of bisection, as detailed in [10].

Now we say that f has *gauge integral* I on the interval $[a, b]$ if for any $\epsilon > 0$, there is a gauge δ such that for any δ-fine division, the usual Riemann-type sum approaches I closer than ϵ:

$$\left| \sum_{i=0}^{n} t_i(x_{i+1} - x_i) - I \right| < \epsilon$$

The property that any derivative has a gauge integral has been proved in HOL starting from the basic definition. It has also been proved that the Fundamental Theorem holds.

As hinted earlier, the gauge integral is nonabsolute, but it has all the attractive convergence properties of the Lebesgue integral. There is quite a simple relationship between the two integrals: f has a Lebesgue integral precisely when both f and $|f|$ have a gauge integral. A more surprising connection is that dropping the insistence that the tag actually be a member of the corresponding interval gives exactly the Lebesgue integral, but without requiring any of the usual measure-theoretic machinery.

3 How to use a CAS inside a theorem prover?

How we use a CAS inside a theorem prover depends on the degree of confidence about the accuracy of the answers given by the CAS. Roughly there are three possible attitudes: to trust the CAS completely, to trust it partially, or to not trust it at all.

Complete trust Trusting the CAS completely means that given a request A, if the CAS's answer is B, then the theorem prover has to accept the theorem $\vdash A = B$. Implementing this solution in HOL violates the usual strict principle of deriving all theorems from the axioms. However a technique given in [9] can be applied to get around this problem. This technique consists in defining a constant CAS logically equivalent to false. With this constant, it is then possible to produce theorems of the form $CAS \vdash A = B$ which are (trivially) true. The first advantage of using this technique is that the generated theorems have a natural reading: "providing the CAS is correct, A equals B". The second advantage is that every theorem that uses such a theorem in its proof will automatically inherit the assumption CAS. This means that there is a simple criterion to differentiate 'purely true' theorems from theorems proved with the help of the CAS.

Partial trust The notion of partial trust comes from efficiency considerations. Algorithms in CASs can be described as optimized algebraic manipulations on ad-hoc data structures. Trying to implement the same algorithms in a prover where the notions of genericity and consistency are preponderant usually entails an important degradation of performance. This degradation may jeopardize the interactive use of the theorem prover where a reasonable reaction time is mandatory. A typical example of such a situation can be found in arithmetic with HOL's REDUCE library. This library provides an automatic procedure to compute ground arithmetic terms. Given a term, it returns a theorem establishing the equality between this term and its reduced form. In order to produce such a theorem, it uses the primitive inference rules of Peano arithmetic. This means that on practical examples, the procedure tends to be rather slow. For example it takes more than 10 seconds to obtain the theorem $\vdash 2^{10}/2^9 = 2$ from the term $2^{10}/2^9$. By contrast, most of the CASs return this result in less than 0.1 second.

The technique of partially trusting the CAS consists in accepting the result of the CAS during the interactive proof of the proposition, but finally when the theorem has to be recorded, *checking* all the results that have been computed by the CAS. The benefit of such a technique is to combine an efficient interaction with the theorem prover with the security of the result. The implementation of such a technique is a direct application of the lazy approach developed by Richard Boulton [2]. This approach gives a natural framework for delaying the proofs of some subgoals. The justification of these subgoals can then be postponed and handled by a batch process where the efficiency requirements are less stringent.

No trust at all The main drawback of the previous approach is that as noted above, the CAS may sometimes generate wrong answers. The benefit of speeding up the interaction using the CAS can then be overwhelmed by the time spent on trying to prove unprovable subgoals. The only way to avoid this kind of situation is never to trust the CAS's answers. Even in this case, having a CAS can be very useful, since there are many situations where finding a solution of a problem is computationally more complex than verifying it. Then we can delegate the 'finding' part of the procedure to the CAS and automate the 'checking' part in the prover with an efficient procedure. The CAS is then used as an oracle; one might regard it as a proof planner, though in general these guide the detailed structure of the proof, rather than just selecting the starting point.

Conclusion The first solution is convenient to give a simple integration between HOL and a CAS without actually generating false theorems. Proofs using the CAS are marked in a clear and simple way. However it is not secure because a determined user may create any theorem with the assumption CAS. This sort of pragmatic approach has already been used to delegate some low-level parts of hardware proofs to other tools [20]. The second is primarily useful for simple arithmetic, where a wrong answer is almost inconceivable and the insistence on carrying the proof through in the theorem prover at all would be questioned by many. The third alternative is the one we are most interested in here. It has the merit of only using the CAS as a guide; the proof itself is still done rigorously in the usual manner. In real analysis, these situations occur quite frequently for problems like integration or factorization. The example given at this end of the paper illustrates this idea.

4 Implementation Issues

Connecting a theorem prover with other systems is not a new idea. Some experiments have already been done in linking HOL with other theorem provers (see [1] and [16] for example). Our application differs slightly from these previous experiments. First of all, the relation between HOL and the CAS is a clear master/slave relation: the use of the CAS is limited to some precise tasks (mainly algebraic simplifications) within an overall proof. Another difference is that the requests sent to the CAS only concern standard algebraic expressions. The translation from the HOL term representation to the CAS one is then straightforward and involves mostly minor syntactic modifications.

The actual implementation of the link between HOL and the CAS follows the lines given in [5] and can be depicted by the following drawing:

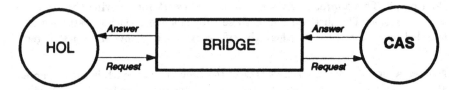

The organization involves three different processes: HOL, the CAS and a bridge. The communication between them uses broadcasting messages. HOL and the CAS send and receive messages with string representing formulas in their own syntax. For example, $(\sin x + \cos x)'$ is represented as `diff(sin(x)+cos(x),x)` in Maple, as `D[Sin[x]+Cos[x],x]` in Mathematica but as `deriv x.(sin x)+(cos x)` in HOL. The role of the bridge is to perform the data coercion automatically. Communicating by broadcasting makes the prover completely independent of the CAS: HOL just requests an answer from an oracle. This model allows us not only to experiment with connecting different CASs without any perturbation on the prover side, and furthermore it actually transforms the CAS into a server. Thus it is possible to share this server between several HOL sessions by connecting these sessions to the same CAS.

From the prover's point of view, the access to the CAS is represented by a single function `call_CAS` whose type is `term -> string -> term`. Its first argument is the term to transform, the second argument represents the method to apply. In the current implementation, only two methods are available. `SIMPLIFY` gives the answer as if the term had been typed at the top level of the CAS. `FACTORIZE` tries to factorize the term. With the function `call_CAS`, it is then possible to compute some arithmetic expressions via the CAS:

```
# call_CAS "(((FACT 5) EXP 2) - 1) MOD (3 EXP 2)" 'SIMPLIFY';;
"8" : term
Run time: 0.1s

# call_CAS "(x*x)+(7*x)+12" 'FACTORIZE';;
"(x + 3) * (x + 4)" : term
Run time: 0.1s
```

5 An example

In order to illustrate how the combined system can be used, we explain the construction of a *secure* procedure to automatically integrate polynomial expressions in cosine and sine. This procedure is composed of five different steps. The steps 2, 4 and 5 are proof steps, i.e. steps that produce a theorem. The steps 1 and 3 are CAS steps, i.e. steps that produce only a result. This procedure is secure in the sense that as described in Section 3 it never trusts the CAS. In order to have a concrete example to explain what every step is supposed to do, we are going to apply this procedure to compute the following integral:

$$\int_0^t \sin^3 u \; du$$

The first step of the procedure is to call the CAS to get the actual value of this integral. With our example, we get the result:

$$\int_0^t \sin^3 u \; du = -\frac{1}{3} \sin^2 t \cos t - \frac{2}{3} \cos t + \frac{2}{3}$$

If we can *prove* that the derivative of the right part of this equality is in fact $\sin^3 x$, we get this result as a theorem by applying the Fundamental Theorem of Calculus (see Section 2). Step 2 consists in computing this derivative inside the prover. This is done with an automated conversion which composes the standard theorems about the differentiation of sums, products, cosine and sine. Step 2 on our example gives the theorem:

$$\vdash (-\frac{1}{3} \sin^2 t \cos t - \frac{2}{3} \cos t + \frac{2}{3})' = -\frac{1}{3} (2 \sin t \cos t \cos t - \sin^3 t) + \frac{2}{3} \sin t$$

In order to prove that this derivative is in fact $\sin^3 t$, the remaining steps have to establish the following theorem:

$$\vdash -\frac{1}{3} (2 \sin t \cos t \cos t - \sin^3 t) + \frac{2}{3} \sin t - \sin^3 t = 0$$

In order to do this, we transform this expression in a polynomial by replacing $\sin t$ by x and $\cos t$ by y, and then factorize it with the help of the CAS. This is Step 3 and we get the result:

$$-\frac{1}{3} (2 \, x \, y \, y - x^3) + \frac{2}{3} x - x^3 = -\frac{2}{3} x (y^2 + x^2 - 1)$$

In Step 4, this result is proved automatically inside the prover. The procedure expands out products and collects together monomials of the same degree. The coefficients should then all be zero; this is checked by a simple extension of the REDUCE library to rational arithmetic. Applied to the example, we get:

$$\vdash -\frac{1}{3}\left(2\,x\,y\,y - x^3\right) + \frac{2}{3}x - x^3 = -\frac{2}{3}x\left(y^2 + x^2 - 1\right)$$

Finally as x is $\sin t$ and y is $\cos t$ we have $y^2 + x^2 - 1 = 0$. Furthermore, such a factor will occur in any trigonometric identity of this form[1], so our procedure is complete. From this, we deduce in Step 5 the following theorem:

$$\vdash -\frac{1}{3}\left(2\sin t\cos t\cos t - \sin^3 t\right) + \frac{2}{3}\sin t = \sin^3 t$$

By composition of this theorem, the theorem of Step 2 and the Fundamental Theorem of Calculus, we get the expected theorem:

$$\vdash \int_0^t \sin^3 u\ du = -\frac{1}{3}\sin^2 t\cos t - \frac{2}{3}\cos t + \frac{2}{3}$$

Although this theorem has been completely derived from the definitions, the use of the CAS has allowed us to obtain it from a simple set of tools, namely a procedure to compute derivatives and a decision procedure about the equality of two polynomials. This procedure can then be applied to compute any polynomial integral. Figure 1 gives some results on different examples in term of run-time and number of primitive inferences used to integrate them.

Value	Run time	Intermediate theorems generated
$\int_0^t \sin u\ du$	6.3s	1901
$\int_0^t \sin u\cos u\ du$	7.8s	2276
$\int_0^t \sin^2 u\ du$	20.4s	6442
$\int_0^t \sin^3 u\ du$	24.6s	7707
$\int_0^t \sin^4 u\ du$	49.4s	16527
$\int_0^t \sin^5 u\ du$	70.4s	25387
$\int_0^t \sin^6 u\ du$	110.4s	40814
$\int_0^t \sin^7 u\ du$	301.4s	125093
$\int_0^t \sin^8 u\ du$	2345.2s	1019832

Fig. 1. Integration Benchmark for Sun4/ Allegro Common Lisp

---

[1] This result is a simple consequence of the Hilbert Nullstellensatz since $x^2 + y^2 - 1$ is still irreducible over $\mathbb{C}[x, y]$.

6 Conclusions and Future Work

The example of Section 5 shows the immediate benefit of having a CAS inside a prover. Some operations like integration and factorization become available inside the prover without jeopardizing the security of the result. The performance, as can be seen in the last integrations of Figure 1, is still below acceptable standard. Performing differentiation and collecting monomials are a small part of this; what dominates is doing rational arithmetic. It is our intention to implement a binary representation inside the HOL logic, which would speed up arithmetic on large numbers enormously without sacrificing security. It is also our intention to tackle more realistic examples where the proof management of the theorem prover and the heuristics of the CAS can be fully exploited.

From the implementation's point of view, the link between HOL and a CAS has mainly been with the Maple system [6]. But it already seems probable that having only one CAS is not sufficient as there is not one best and universal CAS but rather different systems which specialize in different areas. In order to make the best features of these different CASs available inside HOL, a possible extension is to handle multiple connection via an auction mechanism as in [14]. In that system, all the requests are sent simultaneously to different CASs and then the auction mechanism is applied to select the best of all the answers. This extension and an alternative connection with the AXIOM system [13] are under examination.

References

1. M. Archer, G. Fink, L. Yang, *Linking Other Theorem Provers to HOL Using PM: Proof Manager*, Proceedings of the 1992 International Workshop on the HOL Theorem Proving System and its Applications, North-Holland 1993.
2. R. J. Boulton, *A Lazy Approach to Fully-Expansive Theorem Proving*, Proceedings of the 1992 International Workshop on the HOL Theorem Proving System and its Applications, North-Holland 1993.
3. P. Bullen, *Non-absolute integrals: A survey*, Real Analysis Exchange, vol. 5 pp. 195-259, 1980.
4. J. C. Burkill, *The Lebesgue Integral*, Cambridge Tracts in Mathematics and Mathematical Physics no. 44, Cambridge University Press 1965.
5. D. Clément, F. Montagnac, V. Prunet, *Integrated Software Components: a Paradigm for Control Integration*, Proceedings of the European Symposium on Software Development Environments and CASE Technology, Königwinter, Springer-Verlag LNCS 509, June 1991.
6. B. W. Char, K. O. Geddes, G. H. Gonnet, B. L. Leong, M. B. Monagan, S. M. Watt, *A Tutorial Introduction to Maple V*, Springer-Verlag, 1992.
7. D. van Dalen, A. F. Monna, *Sets and Integration: an Outline of the Development*, Wolters-Noordhoff 1972.
8. J. DePree, C. Swartz, *Introduction to Real Analysis*, Wiley 1988.
9. M. Gordon, Private Communication.
10. J. R. Harrison, *Constructing the Real Numbers in HOL*, Proceedings of the 1992 International Workshop on the HOL Theorem Proving System and its Applications, North-Holland 1993.

11. R. Henstock, *A Riemann-type integral of Lebesgue power*, Canadian Journal of Mathematics vol. 20 pp. 79-87, 1968.

12. R. Henstock, *The General Theory of Integration*, Clarendon Press, Oxford 1991.

13. R. D. Jenks, R. S. Sutor, *AXIOM: the Scientific Computation System*, Springer-Verlag 1992.

14. N. Kajler, *Cas/Pi: A Portable and Extensible Interface for Computer Algebra Systems*, Proceedings of ISSAC '92, ACM 1992.

15. A. Kechris, *The Complexity of Antidifferentiation, Denjoy Totalization, and Hyperarithmetic Reals*, Proceedings of the International Conference of Mathematicians, Berkeley 1986.

16. R. Kumar, T. Kropf, K. Schneider, *Integrating a First-Order Automatic Prover in the HOL environment*, Proceeding of the 1991 International Workshop on the HOL Theorem Prover System and its Applications, IEEE Computer Society Press 1991.

17. J. Kurzweil, *Generalized Ordinary Differential Equations and Continuous Dependence on a Parameter*, Czechoslovak Mathematics Journal vol. 7(82), pp. 418-446, 1958.

18. E. J. McShane, *A Unified Theory of Integration*, American Mathematical Monthly 80 pp. 349-357, 1973.

19. I. N. Pesin, *Classical and modern integration theories*, Academic Press 1970.

20. C. Seger, J. J. Joyce, *A Two-level Formal Verification Methodology using HOL and COSMOS*, Technical Report 91-10, Department of Computer Science, University of British Columbia, 1991.

21. S. Wolfram, *Mathematica, A System for Doing Mathematics by Computer*, Addison-Wesley 1988.

The HOL-Voss System:
Model-Checking inside a General-Purpose
Theorem-Prover

Jeffrey Joyce and Carl Seger

Integrated Systems Design Laboratory
Department of Computer Science
University of British Columbia

Abstract. We have extended the HOL theorem-prover with an efficient
implementation of symbolic trajectory evaluation. Using this extension
we can obtain verification results for models of digital hardware – usually
with much less effort than would be required using a conventional interactive theorem-proving approach. We illustrate the use of this extension
with three examples, namely, the formal verification of a 32-bit adder,
an 8-bit by 8-bit multiplier and the MAJORLOGIC block of the Viper
microprocessor.

1 Introduction

Symbolic trajectory evaluation is a form of model-checking that can be used to
check assertions expressed in a simple temporal logic with respect to the behaviour of a finite-state machine [7]. In principle, symbolic trajectory evaluation
could be implemented entirely within the HOL system [4] using only built-in
proof procedures. (This is sometimes referred to as a "secure implementation".)
However we believe that the implementation of symbolic trajectory evaluation
in this manner would likely be too inefficient to be useful for the kinds of practical applications that interest us. Instead, we have taken a less orthodox but
more practical approach that involves the integration of HOL with a very efficient implementation of symbolic trajectory evaluation provided by a separate
verification tool called Voss [8]. We refer to this extension of HOL, together with
supporting infrastructure, as the HOL-Voss system. A more theoretical presentation of our hybrid approach to formal hardware verification is given in [9].
A discussion of how our approach might be integrated with conventional CAD
practice may be found in [6]. In this paper, we focus mainly on the thesis that
many verification problems that might be regarded as non-trivial, difficult or
even practically impossible for a pure HOL-based approach to formal hardware
verification can be solved automatically or almost automatically with symbolic
trajectory evaluation

2   HCL - A Simple Specification Language

Although one can begin with a "HOL-style" specification as we will demonstrate
later in this paper, we normally prefer to use a simple specification language
called HCL (Higher-level Constraint Language) to express the desired verification result at the start of a proof effort. We find the specification of a verification
result in terms of HCL operators to be (at least) as readable as a HOL-style specification. We prefer to use HCL as a starting point rather than starting with a
HOL-style specification because this often reduces the proof effort to the application of a single tactic, namely, VOSS_TAC, thus avoiding the additional effort
required to reduce a HOL-style specification of a verification problem into a goal
that can be solved by VOSS_TAC. In this section, we provide a very brief summary
of HCL focusing specifically on a specific set of HCL operators, is, node_vec,
is_vec, unconstrained, when, during, and and.

The notion of a "trajectory" is central to both a description of HCL and,
more generally, an understanding of our formalization of symbolic trajectory
evaluation in HOL. A trajectory is an infinite sequence of states. A state is
mapping from strings to Boolean values. A finite-state machine in our approach
is set of trajectories. We use the type abbreviations,

```
state = string->bool

trajectory = num->state

fsm = trajectory->bool
```

for the representation of states, trajectories and finite-state machines in our
formalization of symbolic trajectory evaluation.

HCL is a language for specifying constraints on trajectories. These constraints
are specified with respect to a finite interval of time – where a single unit of time
corresponds to a single state transistion. We use HCL expressions to specify the
behaviour of hardware – each trajectory state represents a mapping from nodes
(actually, the names of nodes) to Boolean values. For example, the HCL term
('x' is T) during (10,35) expresses the constraint that the value of node
'x' is T from time 10 up to, but not including, time 35 for a given trajectory.

((s:string) is (b:bool)):state→bool
  – yields a predicate on single states that checks whether the value of the
  node named by the string s in a given state is equal to the specified value b.

(node_vec (n:num) (s:string)):(string)list
  – yields a list of node names given the name of a vector as a string s and its
  size n; for example, node_vec 3 'a' generates the string list
  ['a.0';'a.1';'a.2'].

`((sl:(string)list) is_vec (bl:(bool)list)):state→bool`
  – yields a predicate on single states that checks whether all of the values of
  a vector of nodes named by a list of strings **sl** in a given state are equal to
  the specified values **bl**.
  node_vec and **is_vec** are often used together; for example, the expression
  `((node_vec 3 'a') is_vec [T;F;F])` would be true in a state where `'a.0'`,
  `'a.1'` and `'a.2'` are equal to T, F and F respectively.

`unconstrained:trajectory→bool`
  – is true for any given trajectory.

`((p:state→bool) during (n:num,m:num)):trajectory→bool`
  – yields a predicate on trajectories that checks whether the predicate **p** is
  true in each state of a finite sub-sequence of a given trajectory – in particular, from state **n** up to but not including state **m**.

`((p:trajectory→bool) when (b:bool)):trajectory→bool`
  – yields a predicate on a trajectory that is true if the predicate **p** is true for
  the given trajectory or **b** is false.

`((p:trajectory→bool) and (q:trajectory→bool)):trajectory→bool`
  – is a "lifted" version of logical conjunction for predicates on trajectories.

`(foralln. (p:(bool)list→bool)):bool`
  – is true if the predicate **p** is true for all Boolean lists of length n where n is
  a fixed value.

In addition to the above operators, the examples in this paper use a predicate
HCL,

$$
\vdash_{def} \text{HCL fsm (ante,cons)} =
$$
$$
(\forall tj. \text{ fsm } tj \Rightarrow (\text{ante } tj \Rightarrow \text{ cons } tj)) \wedge
$$
$$
(\exists tj. \text{ fsm } tj \wedge \text{ante fsm})
$$

that expresses the general form of a conclusion that can be solved directly by
VOSS_TAC. The parameters **ante** (called the "antecedent") and **cons** (called the
"consequent") are both predicates on trajectories. The predicate **HCL** expresses
the condition that, for a given set of trajectories **fsm**, any trajectory that satisfies
the antecedent **ante** must also satisfy the consequent **cons**. This predicate also
checks that the antecedent is satisfiable for at least one trajectory of the finite-
state machine.

3 A 32-bit Adder

In this section we use the example of verifying the implementation of a 32-bit
adder to illustrate how HOL-Voss may be used to reason about models of digital
hardware.

The exercise of verifying the implementation of a digital circuit using HOL-Voss begins with a description of this implementation. For example, Fig. 1 shows the description of an implementation of the 32-bit adder in a language called NET [1]. Various tools (outside the scope of our formal approach) would then be used to compile this description into a behavioural model. The result of this compilation step is a file called **adder32.exe** which contains the representation of a behavioural model for the 32-bit adder.

```
(macro adder (a b cin cout sum)
  (local loc1 loc2 loc3)
  (xor a b loc1)
  (nand2 a b loc2)
  (xor loc1 cin sum)
  (nand2 loc1 cin loc3)
  (nand2 loc2 loc3 cout))

(macro adder32 (a b sum cout)
  (local loc)
  (repeat i 0 31 (adder a.i b.i loc.i loc.(+ i 1) sum.i))
  (connect loc.0 gnd)
  (connect loc.32 cout))

(node a b sum cout)

(adder32 a b sum cout)
```

Fig. 1. NET description of a 32-bit adder.

Once a behavioural model for the 32-bit adder has been generated, the user may begin an interactive HOL-Voss session and evaluate the HOL-Voss command,

```
new_fsm_specification 'adder32';;
```

to introduce a new constant **adder32** and specify this constant in terms of the behavioural model stored externally in the file **adder32.exe**. This constant denotes the characteristic function of a set of trajectories. The introduction of a new constant in this manner also has the effect of specifying that this constant satisfies the property of being both "suffix-closed" and "non-trivial". The property of being suffix-closed means that every state of the finite-state machine is a possible initial state. The property of being non-trivial means that there is at least one trajectory in the set characterized by the new constant. After evaluating the above HOL-Voss command, the current theory will contain an axiom,

```
⊢ SuffixClosed adder32 ∧ NonTrivial adder32
```

specifying that the behavioural model of the 32-bit adder is suffix-closed and non-trivial.

We can use HOL-Voss to show that the behavioural model now represented in this HOL-Voss session by the constant **adder32** satisfies the property that, if the inputs are held stable for at least one hundred time units, the value of the output of this 32-bit adder will be equal to the sum of the values of its two 32-bit inputs. The condition that the inputs must be held stable for at least one hundred time units is expressed by an HCL term,

```
(((node_vec 32 'a') is_vec a) during (0,100)) and
(((node_vec 32 'b') is_vec b) during (0,100))
```

where the variables **a** and **b** represent lists of Boolean values. The desired effect, that the output of this 32-bit adder will be equal to the sum of the values of its two 32-bit inputs, is expressed by another HCL term,

```
((node_vec 32 'sum') is_vec (sized 32 (a bvplus b))) during (100,101))
```

where the function **bvplus** is used in the above verification result to denote the addition of the unsigned binary representation of two numbers and the function **sized** is used to convert the result of the addition operation into a Boolean list with exactly thirty-two elements.

Both of the above HCL terms are predicates on trajectories: any trajectory of the set characterized by **adder4** that satisfies the first of these two HCL terms (the antecedent) must also the second HCL term (the consequent). These two HCL terms are combined using the predicate **HCL** to express the desired verification result as the conclusion,

```
forall32 a b.
  HCL adder32 (
    ((((node_vec 32 'a') is_vec a) during (0,100)) and
    (((node_vec 32 'b') is_vec b) during (0,100))),
    (((node_vec 32 'sum') is_vec (sized 32 (a bvplus b)))
     during (100,101))))
```

of a goal that can be solved directly with an application of **VOSS_TAC**.

We remark, parenthetically, that it would be necessary in this case to give the Voss system some additional instructions (not shown here) on how to order the sixty-four Boolean variables in this problem. While variable ordering has no logical significance, it is generally necessary for non-trivial problems to specify a variable ordering to constrain the size of the BDD's (Binary Decision Diagrams) used by the Voss system to represent this verification problem.

4   HCL-style verses HOL-style Specifications

The reader may wonder how the HCL-style specification of the verification result for the 32-bit adder compares to the style of specification often used in other HOL-based examples of formal hardware verification. For example, a HOL-style specification of the 32-bit adder would typically be specified in terms of the relationship expressed by the predicate ADDER32,

```
⊢def ADDER32 (a,b,sum) =
     ∀t n m.
       (∀i. i < 100 ⇒  (a (t+i) = sized 32 (num2bv n))) ∧
       (∀i. i < 100 ⇒  (b (t+i) = sized 32 (num2bv m)))
       ⇒
       (sum (t+100) = sized 32 (num2bv (n + m)))
```

where "for all times t", if the inputs are held stable for one hundred time units beginning at time t, the value of the output at time t+100 will be the sum of the two inputs. This HOL-style specification would appear to be "stronger" than the HCL-style specification: when interpreted literally, the HCL-style specification appears to have a temporal scope limited to the first one hundred instants of time. Nevertheless, as we will explain in this section, the temporal scope of HCL-style specification of the verification result extends infinitely along every trajectory of the finite-state machine – and thus, in this respect, it is at least as strong as the HOL-style specification of this result. More generally, we will sketch how the HCL-specification of the verification result can be transformed, if so desired, into a HOL-style specification. While we do not think that it is necessarily useful to transform HCL-style specifications into a HOL-style specification (as mentioned before, we find HCL-style to be as readable as HOL-style specifications), we provide a sketch of this transformation as evidence that verification results obtained automatically by VOSS_TAC have the potential to be at least as strong as verifications results expressed in terms of a more conventional HOL-style specification.

The temporal scope of the verification result for the 32-bit adder does indeed extend infinitely along every trajectory of the finite-state machine. This is a consequence of the fact that adder32 is specified to be suffix-closed. As explained earlier, this means that every state of adder32 is a possible initial state. Therefore, it is sufficient to show that a desired verification result holds for all intervals that begin at time 0 to show that it holds for all intervals. The fact that this verification result holds "for all times", can be expressed explicitly in a theorem such as,

```
⊢ forall32 a b.
    ∀t.
      HCL adder32 (
        ((((node_vec 32 'a') is_vec a) during (t,t+100)) and
        (((node_vec 32 'b') is_vec b) during (t,t+100))),
        (((node_vec 32 'sum') is_vec (sized 32 (a bvplus b)))
          during (t+100,t+101))))
```

which can be derived very easily from the verification result obtained by means
of **VOSS_TAC** and the fact that **adder32** is suffix-closed.

The verification result for the 32-bit adder could be further transformed into
a "more abstract" form using proven lemmas about **bvplus** and **sized** as well
as a function, **num2bv**, that converts numbers into bit-vectors.

```
⊢ ∀n m.
    let a = sized 32 (num2bv n) in
    let b = sized 32 (num2bv m) in
    let sum = sized 32 (num2bv (n + m)) in
    ∀t. HCL adder32 (
        ((((node_vec 32 'a') is_vec a) during (t,t+100)) and
        (((node_vec 32 'b') is_vec b) during (t,t+100))),
        (((node_vec 32 'sum') is_vec sum) during (t+100,t+101))))
```

It is even possible, if so desired, to apply several more transformations to
derive a HOL-style version of the verification result. Using the function
get_trace_of_node_vec to extract a sequence of node vector values (called a
trace) from a trajectory we can obtain the following HOL-style version of the
verification result,

```
⊢ ∀tj.
    adder32 tj
    ⇒
    let a = get_trace_of_node_vec tj ('a',32) in
    let b = get_trace_of_node_vec tj ('b',32) in
    let sum = get_trace_of_node_vec tj ('sum',32) in
    ADDER32 (a,b,sum)
```

where the previously defined predicate **ADDER32** is used to specify the desired
relationship between sequences extracted from a trajectory for the node vectors
'a', 'b' and 'sum'.

5 An 8-bit by 8-bit multiplier

We now consider the use of HOL-Voss to reason about the behavioural model of
an 8-bit by 8-bit multiplier. This example demonstrates how symbolic trajectory
evaluation can be used to reason about the behaviour of a sequential circuit over
several clock cycles.

Figure 2 shows a RTL (Register Transfer Level) view of a simple implementation of an 8-bit by 8-bit multiplication circuit. When the **load** line is set high and the circuit is clocked through a full cycle, the two 8-bit inputs **ain** and **bin** are loaded into two internal 8-bit registers, **A** and **B**. There is also an internal 16-bit register called OUT which will be set to zero during a load cycle. When the **load** line is reset low and the circuit is clocked through a compute cycle, the circuit performs a single step in a multiplication algorithm. If the least significant bit of the A register is low, then the current contents of OUT will be simply shifted one position "left" where the least significant bit appears at the left end of the representation. Otherwise, if the least significant bit of the A register is high, the 8-bit contents of the B register are added to bits 8-15 of the contents of OUT. The result of this addition is then shifted one position left and stored back into the OUT register. A single phase clock is used in this circuit. The clock line **phi** is set high for the first half of a clock cycle and then low for the second half of the cycle. We have used a clock cycle with a length of 200 time units in this example.

A behavioural model of the 8-bit by 8-bit multiplier can be generated from a description of this circuit in NET. (The NET description is too large to be included in this paper.) To show that this model implements a multiplier, we would enter a HOL-Voss session and use **new_fsm_specification** to introduce a new constant, **mult8_8**, to denote the behavioural model generated from the NET description.

We then define a predicate, **cycles**,

```
⊢_def (cycles 0 phi = unconstrained) ∧
      (cycles (SUC n) phi =
        ((phi is T) during (n*200,(n*200)+100)) and
        ((phi is F) during ((n*200)+100,(n*200)+200)) and
        cycles n phi)
```

which will be used to describe the behaviour of the single phase clock when specifying a desired verification result for the multiplier.

Next, we define a predicate **mult8_8_ante** that expresses a set of constraints on inputs to the multiplier. This predicate is parameterized by the number of clock cycles, c and the values, **ain** and **bin**, to be loaded into registers A and B.

```
⊢_def mult8_8_ante (c,ain,bin) =
      ((cycles (c+1) 'phi') and
      (('load' is T) during (0,200)) and
      (('load' is F) during (200,(c+1)*200)) and
      (((node_vec 8 'ain') is_vec ain) during (0,200)) and
      (((node_vec 8 'bin') is_vec bin) during (0,200)))
```

We can now prove that the 8-bit by 8-bit multiplier correctly computes the product of its inputs, under certain timing constraints, for all possible inputs. This is easily accomplished in a one-tactic proof by applying **VOSS_TAC** directly to a goal with the following conclusion:

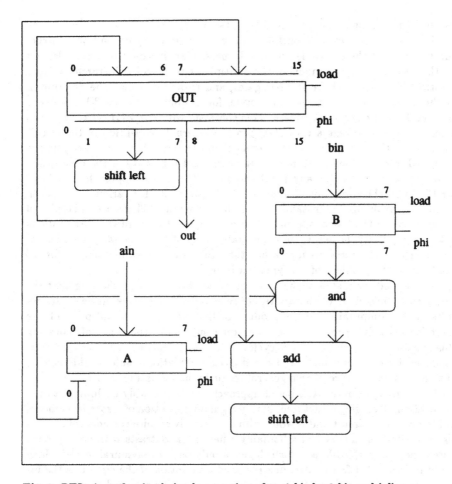

Fig. 2. RTL view of a simple implementation of an 8-bit by 8-bit multiplier.

```
forall8 ain bin.
  let out = sized 16 (ain bvmult bin) in
HCL mult8_8 (
  mult8_8_ante (5,ain,bin),
  (((node_vec 16 'out') is_vec out) during ((5*200)+50,(5*200)+100)))
```

It is interesting to compare the one-tactic proof of the multiplier in the HOL-Voss system with the amount of proof effort that would probably be required to verify a model of the 8-bit by 8-bit multiplier using a pure HOL-based approach.

It is very likely that a pure HOL-based approach would be organized into two main steps. The first step of the proof would show that the behaviour of the multiplier circuit during a single arbitrary cycle is correct with respect to a specification of the single-cycle behaviour of the multiplier. The second step of the proof would show that this single-cycle behaviour repeated five times results in the computation of the product of the two inputs.

The first of the two steps outlined above would probably depend on a "structural" model of the multiplier circuit. In particular, it would probably depend on the specification of a model of the multiplier circuit as the hierarchical composition of registers, shifters, AND-gates, and an adder — and the description of these components in an iterative style, for example, that the 8-bit adder is the iteration of eigth 1-bit adders. This contrasts with our HOL-Voss approach where hierarchy or regularity (if any) in the external description of the circuit (in NET, VHDL or some other language) disappears when the description is compiled into a behavioural model. Hierarchy and regularity in the description of a circuit is needed (in any practical) pure HOL-based approach as a basis for the hierarchical organization of the proof effort and for various verification strategies involving, for instance, proofs by induction. (There are examples of when a pure HOL-based approach has been applied to hardware descriptions without regularity, however, these generally amount to a *tour-de-force* exercise in interactive theorem-proving rather than the practical application of formal methods to the problem of verifying hardware.)

The second main step of the proof outlined above, that is, showing that the single-cycle behaviour of the multiplier repeated for five consecutive cycles results in the computation of the produce of the two inputs, would probably be established by first proving a more general result about the result of repeating this single-cycle behaviour for n cycles. This more general result would be established by induction on n and the desired verification result would then be obtained as an instance of this general result by instantiating n to 5.

In summary, a pure HOL-based approach would probably be limited to the verification of a highly structured, highly regular description of a gate-level model of the multiplier circuit and it would almost certainly require considerable expertise and effort on the part of a human verifier. This contrasts with an approach based the use of HOL-Voss which depends only on a behavioural model which could be generated from a flat, unstructured description of the circuit as the interconnection of switches. While some expertise may be required to write down a HCL-style specification of the desired verification (but no more than the expertise required to write down a HOL-style specification), the actual proof effort is limited to the application of a single tactic.

6   The Viper microprocessor MAJORLOGIC circuit

In the late 1980's A. Cohn used the HOL system to verify aspects of a commercial microprocessor called Viper. One version of this proof involved using HOL to (partially) establish a relationship between a specification of the implementation of Viper at the level of ten major functional units (called "blocks") and a specification of its programming level behaviour [3]. In this section we demonstrate how HOL-Voss could be used to verify implementations of (at least some of) these major blocks with respect to the behavioural specifications for these block as given in Cohn's report. The example in this section concerns the combinational logic for the MAJOR block which is one of the simpler blocks in the block level implementation of Viper.

The combinational logic of the MAJOR block is given the name MAJOR-LOGIC in Cohn's report. This circuit has eight inputs **stop** (1 bit), **call** (1 bit), **timeoutbar** (1 bit), **nextmnbar** (1 bit), **advance** (1 bit), **reset** (1 bit), **cond** (4 bits) and **major** (4 bits). There are two outputs: **intreset** (1 bit) and **nextmajor** (4 bits).

We have created a NET description for an implementation of this combinational circuit and compiled this description into a behavioural model. (The NET description is too large to be included in this paper.) We have re-used the behavioural specification given by Cohn for this block of combinational logic – which, for the convenience of the reader, we have included here:

```
⊢def ILLEGAL_MAJOR major =
        let majorval = VAL4 major in
        (majorval = 5) OR
        ((majorval = 8) OR
        ((majorval = 9) OR
        ((majorval = 12) OR
        ((majorval = 13) OR (majorval = 14) OR (majorval = 15)))))

⊢def INTRESET (timeoutbar,major) =
        (ILLEGAL_MAJOR major) OR (NOT timeoutbar)

⊢def NEXT_MAJOR (stop,call,cond,major) =
        let majorval = VAL4 major in
        (majorval = 0) →   cond |
        ((majorval = 1) →   cond |
        ((majorval = 2) →   (WORD4 1) |
        ((majorval = 3) →   (WORD4 4) |
        ((majorval = 4) →   (stop →   (WORD4 8) | (WORD4 1)) |
        ((majorval = 6) →   (WORD4 4) |
        ((majorval = 7) →   (call →   (WORD4 3) | (WORD4 4)) |
        (WORD4 1)))))))

⊢def FIND_MAJOR
        (reset,intreset,advance,nextmain,stop,call,cond,major) =
        (reset →   (WORD4 2) |
        (intreset →   (WORD4 8) |
        ((advance AND nextmain) →   NEXT_MAJOR (stop,call,cond,major) |
        major)))

⊢def MAJORLOGIC
        (stop,call,timeoutbar,nextmnbar,advance,reset,cond,major) =
        let intreset = INTRESET (timeoutbar,major) in
        let nextmajor =
            FIND_MAJOR
                (reset,intreset,advance,NOT nextmnbar,stop,call,cond,major) in
        (nextmajor,NOT intreset)
```

We have made just two slight modifications to the original versions of these definitions. First, we have re-defined **VAL4** and **WORD4** in terms of bit-vector functions which are built into HOL-Voss (instead of using the WORDn library):

```
⊢_def VAL4 v = bv2num (sized 4 v)

⊢_def WORD4 v = sized 4 (num2bv v)
```

Second, we have made a slight modification to the definition **NEXT_MAJOR** such that the value **WORD4 1** is used as a "don't care" value instead of **ARB** (which denotes an arbitrary value).

The predicate **majorlogic_ante** is defined to express a set of constraints on inputs to this combinational circuit:

```
⊢_def majorlogic_ante
       (stop,call,timeoutbar,nextmnbar,advance,reset,cond,major) =
       ((('stop' is stop) during (0,200)) and
        (('call' is call) during (0,200)) and
        (('timeoutbar' is timeoutbar) during (0,200)) and
        (('nextmnbar' is nextmnbar) during (0,200)) and
        (('advance' is advance) during (0,200)) and
        (('reset' is reset) during (0,200)) and
        (((node_vec 4 'cond') is_vec cond) during (0,200)) and
        (((node_vec 4 'major') is_vec major) during (0,200)))
```

The predicate **majorlogic_cons** is defined to express a set of constraints that are expected to be satisfied by two outputs, **nextmajor** and **intresetbar**, of this circuit. The function **MAJORLOGIC** is used directly in the definition of **majorlogic_cons**:

```
⊢_def majorlogic_cons
       (stop,call,timeoutbar,nextmnbar,advance,reset,cond,major) =
       let (nextmajor,intresetbar) =
       MAJORLOGIC
          (stop,call,timeoutbar,nextmnbar,advance,reset,cond,major) in
       ((((node_vec 4 'nextmajor') is_vec nextmajor)
        during (100,200)) and
        (('intresetbar' is intresetbar) during (100,200)))
```

Finally, we combine the antecedent expressed by **majorlogic_ante** and the consequent expressed by **majorlogic_cons** to set up a goal with the following conclusion:

```
∀stop call timeoutbar nextmnbar advance reset.
  forall4 cond major.
    HCL majorlogic
      (majorlogic_ante
        (stop,call,timeoutbar,nextmnbar,advance,reset,cond,major),
       majorlogic_cons
        (stop,call,timeoutbar,nextmnbar,advance,reset,cond,major))
```

The above conclusion can be solved by a short sequence of tactics that terminates with **VOSS_TAC**. We suspect that the other nine functional blocks that form the basis of Cohn's proof effort could be verified with similar ease.

7 Future Development

The HOL-Voss system represents our first attempt at building a prototype verification tool that combines interactive theorem-proving with symbolic trajectory evaluation. We are also investigating several other approaches to building verification tools to support our hybrid verification methodology.

One such approach is the enhancement of the Voss system as a stand-alone verification tool. Much of the logical infrastructure for Voss is developed using the HOL system. This includes, for instance, the development of our bit-vector library, e.g., functions such as **bvplus**, **bvmult**, **size**, **num2bv** and **bv2num**. However, an ordinary user of the Voss system would be sheltered from the complexity of the HOL system by the simpler logical framework of the Voss system.

Another possibility is the development of a HOL-like system where, in addition to the current set of primitive inference rules, there would be a primitive inference rule that would simplify proposition terms (i.e., terms composed from Boolean variables and the constants $\neg$ and $\lor$) to a canonical form. For efficiency reasons, this inference rule would be implemented using BDD techniques. This extension would provide a basis for building a symbolic trajectory evaluator (as well as other kinds of model-checkers) into a HOL-like system.

Another direction of future work is the development of a new back-end for the Voss system. Currently, the Voss system can only be applied to behavioural models generated from the combination of circuit descriptions in languages such as NET and VHDL and a switch level model of circuit behaviour. We are currently developing a new back-end that would accept behavioural models generated from higher levels of abstraction such as VHDL simulation models. Moreover, symbolic trajectory evaluation is not necessarily tied to models of digital hardware; we would also like to generate behavioural models from notations such as Statecharts [5] which may be used to describe software as well as hardware.

8 Conclusions and Future Research

Symbolic trajectory evaluation is an extremely powerful technique for verifying models of digital hardware. As we have demonstrated in this paper, many verification problems that might be regarded as non-trivial, difficult or even practically impossible for a pure HOL-based approach to formal hardware verification can be solved automatically or almost automatically with symbolic trajectory evaluation.

Nevertheless, there are practical limits to the power of symbolic trajectory evaluation. For example, we have shown in this paper how symbolic trajectory evaluation can be used to verify an 8-bit by 8-bit multiplier – however, it has been shown that this could not be used directly to obtain the same result for significantly larger multiplier, for instance, a 32-bit by 32-bit version [2]. In general, symbolic trajectory evaluation is limited by the number the size of BDD's generated during the verification process. This is a function of both the number of Boolean variables and the inherent complexity of the Boolean functions represented by the BDD's.

This difficulty leads us to consider the development of tools, techniques and methodologies for partitioning verification problems which exceed the capacity of symbolic trajectory evaluation into smaller problems that can be handled by symbolic trajectory evaluation. In this case, interactive theorem-proving would be used mostly as a technique for managing proof results and combining them into larger results.

Acknowledgements

This research was supported by operating grants OGPO 109688 and OGPO 046196 from the Natural Sciences Research Council of Canada, by research contract 92-DJ-295 from the Semiconductor Research Corporation, and by fellowships from the Advanced Systems Institute.

References

1. Randal E. Bryant, "NET – simple network description language", Carnegie-Mellon University, on-line documentation.
2. R.E. Bryant, "On the Complexity of VLSI Implementations and Graph Representations of Boolean Functions with Applications to Integer Multiplication", *IEEE Transactions on Computers*, Vol. C-40, No. 2, February 1991.
3. Avra Cohn, "Correctness Properties of the Viper Block Model: The Second Level", in: G. Birtwistle and P. Subrahmanyam, eds., *Current Trends in Hardware Verification and Automated Theorem Proving*, Springer-Verlag, 1989, pp. 1-91. Also Report No. 134, Computer Laboratory, Cambridge University, May 1988.
4. M. J. C. Gordon and T. F. Melham (eds.,), *Introduction to HOL: a theorem proving environment for higher order logic*, Cambridge University Press, 1993.
5. David Harel *et al.*,, "STATEMATE: A Working Environment for the Development of Complex Reactive Systems", IEEE Transactions on Software Engineering, Vol. 16, No. 4, April 1990, pp. 403-413.
6. Jeffrey J. Joyce and Carl-Johan H. Seger, "Linking BDD-Based Symbolic Evaluation to Interactive Theorem-Proving", Proceedings of the 30th Design Automation Conference, IEEE Computer Press, June 14-18, 1993.
7. C-J. Seger and R. E. Bryant, "Formal Verification of Digital Circuits by Symbolic Evaluation of Partially-Ordered Trajectories", in preparation.
8. C-J. Seger, "Voss — A Formal Verification System (User's Guide)", Department of Computer Science, University of British Columbia, in preparation.
9. Carl-Johan H. Seger and Jeffrey J. Joyce, "A Mathematically Precise Two-Level Formal Hardware Verification Methodology", Department of Computer Science, University of British Columbia, April 1993.

Linking Higher Order Logic to a VLSI CAD System*

Juin-Yeu Lu and Shiu-Kai Chin

Dept. of Electrical & Computer Engineering, Syracuse University, Syracuse NY

Abstract. We show how hardware implementation descriptions written in Higher Order Logic (HOL) are transformed into VLSI layouts. The intent is to link formally verified designs to physical implementations. Once HOL structural descriptions are transformed to structural descriptions and parameterized cell generators in a VLSI CAD system, the tools associated with the VLSI CAD system can be used such as standard cell libraries, automatic placement and routing, and simulation. The cell generator programs derived from verified HOL descriptions provide a means by which verified and parameterized HOL designs can augment a cell library within a CAD system. The particular VLSI CAD system used here is the GDT system from Mentor Graphics. The GDT system includes a CMOS standard cell library, a macrocell synthesis tool which includes automatic placement and routing, and the Lsim multilevel simulator. GDT uses a structural language called L. HOL structural descriptions are first mapped to L, then layouts corresponding to the L descriptions are created using the standard cell library and the layout generation tools. We illustrate this design process by creating the layout for an 8-bit serial/parallel multiplier which has been verified in HOL.

1 Introduction

HOL has been widely used to formally verify hardware descriptions, [1] - [5], [7, 8, 9, 11]. Once a HOL description is verified, we would like to translate the HOL description to a form which is usable by a VLSI CAD system. To do this transformation, we must understand how the components of a HOL structural description are related to objects and functions in the CAD system.

In HOL, hardware is described by using predicates, universally and existentially quantified variables, and recursion. HOL objects are related to CAD objects like cell modules, I/O ports, signals, and module generators. This paper describes how we make such linkages. Specifically, we implemented this link to the Mentor Graphics GDT (Generator Development Tools) system, [12]. GDT supports the design of VLSI circuits at multiple levels of description. The focus of this work is the mapping of structural descriptions in HOL to structural descriptions in GDT, and recursive definitions in HOL module generators in GDT. GDT structural descriptions use the L language, [12]. The transformations preserve hierarchical descriptions in HOL which means that designs need not be

---

* Supported by NY State Center for Advanced Technology at Syracuse University.

flattened before being transformed. This allows us to use the module generation capability of GDT. Figure 1 illustrates the link between HOL and the GDT system.

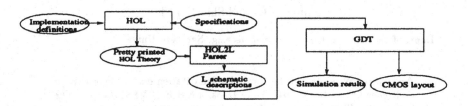

Fig. 1. Linking HOL to the GDT system.

The type of link described here is similar to the link made between the HOL-based MEPHISTO verification system and the CADENCE system [14]. The main difference is our use of GDT's parameterized cell generator capability in L. Our view is to preserve as much hierarchy as is possible when making the translation from HOL to GDT to avoid costly cell generation within HOL. The preservation of hierarchy is significant. Previous approaches flattened HOL netlists so that recursive descriptions in HOL could not be added to the library of module generators in the CAD system. This results in a significant loss of utility since module generators support parameterized design synthesis.

The translation from HOL specifications to GDT structural descriptions consists of four phases: source analysis, dependency graph generation, mapping and code generation. The source analysis phase translates the source into an intermediate representation. This information is used to create the dependency graph of a design. Mapping assigns each primitive component within the graph to a specific cell (a GDT standard cell). Finally, the code generation phase translates the graph into an L program along with cell attributes that link with a runtime library.

In translating specifications to implementations, technology mapping plays an important role. This connects primitive components between the two systems. To ensure a correct mapping, we start at the gate level where the GDT standard cells we use have been validated against the specifications described in HOL using exhaustive simulation. Since the standard cells are simple functions, this approach is practical. As there is no semantic definition of L, we have not as yet verified the compiler from HOL to L.

The organization of the rest of this paper is as follows. In Section 2, we review how relational descriptions are used to describe hardware. Section 3 gives an overview of the GDT system. Section 4 describes the L language and how a schematic cell module is created using the L language. Section 5 shows how HOL structural descriptions are compiled to the corresponding L module generators. Section 6 shows a complete example of a parameterized *n-bit* serial/parallel multiplier. In this example, the recursive description of the multiplier is presented and compiled into the L description, then the layout is shown. Furthermore, a simulation is performed based on the initial conditions required for correct behavior as described by the proved correctness theorem, [4, 6].

2 Relational Descriptions

Relational hardware descriptions use predicates as described in [2]. The predicates describe relationships which hold among its parameters. A hardware block is described by a predicate whose parameters are the external input and output lines. These input/output parameters are typically universally quantified variables. Modules having n external lines are specified as n-ary predicates. For example, a device as shown in Figure 2a can be specified as $D(i1,...,in,o1,...,om)$, where $i1,...,in,o1,...,om$ are external lines. Internal lines within a device are hidden by existentially quantifying internal variables which do not appear as parameters of the predicate. The connection of two components is represented by conjunction "$\wedge$". Considering the example of Figure 2b, the relation is specified as $sm(i,o) = \exists l1\ l2.COM(i,l2,o,l1) \wedge M(l1,l2)$.

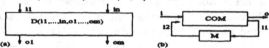

Fig. 2. Relational descriptions.

2.1 Recursion

Recursion is used to describe a circuit which is constructed from a regular interconnection of identical components. A primitive recursion consists of two parts – the base case and the recursive case. The base case specifies the primitive building block of a circuit, while the recursive case shows how primitive building blocks are cascaded. To illustrate the idea, consider the n-bit ripple adder in Figure 3.

$\vdash$ ($\forall$ $i1$ $i2$ cin sum $cout$. FADDERN 0 $i1$ $i2$ cin sum $cout$ = FADD1($i1$ 0,$i2$ 0,cin,sum 0,$cout$)) $\wedge$ ($\forall$ n $i1$ $i2$ cin sum $cout$. FADDERN($n+1$) $i1$ $i2$ cin sum $cout$ = ($\exists$ $l1$.FADDERN n $i1$ $i2$ cin sum $l1$ $\wedge$ FADD1($i1(n+1),i2(n+1),l1,sum(n+1),cout$)))

Fig. 3. An n-bit ripple adder.

The recursion in Figure 3 is an example of a simple recursion which corresponds to the repeated use of a single hardware block which itself is not described recursively, e.g. *FADD1*. Currently, layouts for any these types of descriptions are generated automatically.

We do not automatically generate layouts for descriptions with nested recursions. In these cases, it is possible to automatically generate schematic netlists. Within the netlist will be blocks corresponding to uninstantiated cells which are the hardware blocks generated by the nested recursions. In these cases, the user must instruct GDT to instantiate the cells to create a layout.

3   The GDT System

The GDT system is a design environment for VLSI circuits at multiple levels of abstraction. There are six basic tools, the L language, the L database, the Lsim simulator, the Led graphics editor, the M language, and cell library components. The L database and the Lsim simulator form the major framework that supports design description, modeling and simulation. The Led editor, the L language, and cell library components are the tools used to describe and model designs in the L database. Figure 4 is a block diagram of the GDT environment.

L is a structural description language that supports various levels of design descriptions. These descriptions range from block diagrams to circuit layouts. The Led graphics editor accesses the L database directly. It can edit a block diagram, a schematic, and a layout simultaneously. The L database has electrical connectivity and geometric information. The L database is created by the L compiler from L programs or by the Led graphics editor. Through the L database, models at different levels of abstraction are connected, e.g. a gate-level model connected to a layout can be simulated. The Lsim simulator provides a multi-

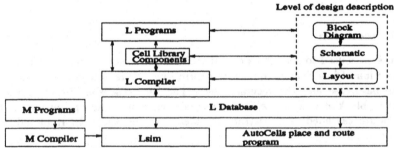

Fig. 4. The GDT environment.

level simulation environment from high-level system descriptions to the low-level circuit descriptions. High-level descriptions use the M language. The M language, which is a superset of C, is a modeling language used to describe the behavioral models of circuit blocks in terms of information flowing between blocks and information stored within a block. Mc is the M modeling language compiler, where object code produced by the Mc compiler is simulated by the Lsim simulator. To create a model, one may use a standard cell generator from the GDT library or an existing generator developed by the designer.

Within GDT, M descriptions are simulated and compared to simulations of L structures for partial verification. By using verified HOL descriptions, we hope to eliminate the need for M simulations for functional verification of macrocells. Also, verified and parameterized HOL descriptions add to the library of cell generators in GDT.

Finally, layouts are created by the AutoCells place and route program to automatically place and route a standard cell implementation.

4　L Language

In GDT, the VLSI circuit design process has three levels of description: a functional specification; a structural circuit description; and a mask layout. The structural description describes how a circuit is constructed by interconnecting instances of its building blocks. L supports various levels of structural descriptions. At the high level, it supports electrical and geometric primitives such as wires, terminals, transistors, and cell hierarchies. At the low level, it supports physical layer primitives such as metal, poly, n-well, etc. One uses the L compiler to transform a higher-level description to a lower-level (router netlist format) description in which the cell hierarchy is flattened. Such router format descriptions are used to automatically route a circuit.

The L language has traditional program constructs such as variables, logic and arithmetic expressions, and control expressions such as *IF ELSE* and *WHILE* expressions. The control and iteration constructs allow the creation of cell *generators*, i.e. parameterized L programs which create designs from iterative arrays of previously defined cells.

Cells are the basic hierarchical unit defined in the L language and entered in the L database. A design is represented in terms of cells and interconnections of cells. L supports various levels of cell types such as schematic, layout, icon, etc.

4.1　Schematic Descriptions

A schematic representation ignores information related to physical layers such as poly, metal, etc. Schematic cells are represented in terms of the connectivity of their building blocks where a building block is an instance of an existing cell. The following shows how a schematic cell is described. For simplicity, we present the simplified syntax in BNF, where all key words consists of capital characters and symbols are quoted.

A schematic cell may have I/O ports that are used to pass signals in and out of the cell to other cells. There are three types of terminals: *IN*, *OUT*, and *INOUT*. In addition, one may declare an array of terminals with subscripts. This is used to systematically produce ports for a cell generator where its ports are indexed.

terminal_declaration: terminal_type terminal_name
terminal_type : IN | OUT | INOUT
terminal_name : name;
name : identifier | identifier '[' subscript ']'
subscript : identifier | number

An instance of a cell is an instantiation of a previously defined cell. Cell instances can be included as components of cell definitions. If a cell definition is modified, then all the instances referring to it inherit the change. Cell instances are declared using *INST*.

A net describes how L objects are connected. Legal L objects are terminals and instance terminals. *WIRE* is used to connect two terminals of blocks. *SIG* is used to attach a signal to a terminal. All objects with the same signal name are netted together through *SIGs*.

```
cell_instance : INST existing_cell_name inst_name ';'
existing_cell_name: identifier
inst_name : name
net : WIRE object TO object ';' | SIG object '"' signal_name '"'
signal_name: identifier
object : terminal | instance_terminal
instance_terminal: inst_name '.' terminal_name
```

In the following example, we make a full-adder from instances of its building blocks. Two half-adders and an *OR* gate are connected. The instances of the half-adder are *HADD1[0]* and *HADD1[1]*, and *or[0]* is the instance of the *OR* gate. The connections between two terminals are done through the use of *WIREs*. Terminals are connected by attaching the terminals to a same signal name. For example, instance terminals *HADD1[0].sum* and *HADD1[1].i1* have the same signal *l1* attached to them; thus, they are connected.

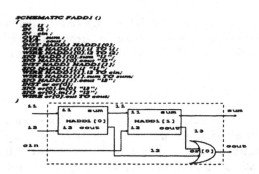

Fig. 5. FADD1().

There are two kinds of conditional control statements, *if_statements* and *while_statements*. With these statements and the net expressions, one may wire terminals conditionally. See Section 5.3 for details.

```
cond_statement : if_statement ';' | while_statement ';'
if_statement : IF '(' bool_exp ')' '{' stmt1 '}' | IF '(' bool_exp ')' '{' stmt1 '}' ELSE '{' stmt2 '}'
stmt1 : statements
stmt2 : statements
while_statement: WHILE '(' bool_exp ')' '{' statements '}'
```

5 Compiling HOL to L

This section describes how circuits expressed as relations are translated to the corresponding L expression. In Section 5.2 we describe how L descriptions of non-recursive HOL definitions are created. In Section 5.3 we describe how iterative L cell generators are created from recursive HOL definitions.

5.1 Inputs and Outputs

In HOL relational descriptions, no distinction is made between inputs and outputs. In an L module the direction (input or output) of a port has to be explicitly specified. To automate the translation from relational definitions to L descriptions, the user must specify which parameters are inputs and which parameters are outputs. This is done as part of the HOL description. Four HOL types are introduced: *variable* (which is of type string), *port_width*, *port*, and *module_port*.

variable:string
port_width = pin | bus num
port = in_port variable port_width | out_port variable port_width
module_port = module_name # (port)list

A port is either a single pin or a bus (multiple pins). The direction of a port is specified by the keyword *in_port* or *out_port*. For example, the input-output information and relational definition of a full adder are shown below.

⊢ *cell_port_def = ['FADD1', [in_port 'i1' pin;in_port 'i2' pin;in_port 'cin' pin;out_port 'sum' pin;*
out_port 'cout' pin]]
⊢ ∀*i1 i2 b cin sum cout. FADD1(i1,i2,cin,sum,cout)=(∃l1 l2 l3. HADD1(i1,i2,l1,l2) ∧*
HADD1(l1,cin,sum,l3)∧ or(l3,l2,cout))

In this example, *i1*, *i2* and *cin* are inputs with port width 1, while *sum* and *cout* are outputs with port width 1.

5.2 Translating Relational Definitions

A relational definition has the following form.

$$\forall i_i...i_n o_1...o_m.P(i_i,...,i_n,o_1,...,o_m) = \exists l_1...l_k.P_1(par\_list_1)\wedge P_2(par\_list_2)\wedge...\wedge P_p(par\_list_p)$$

P is the predicate being defined and predicates P_i are instances of existing predicates. Each variable in the parameter list par_list_i is universally or existentially quantified. Predicates being defined always appear on the left of the "=". Components of the definition appear on the right. Relational definitions coupled with the cell-port definitions are translated as follows.

1. Each predicate P_i in the right hand side of a relation is renamed as a distinct instance of an existing cell definition. The distinct instances are created by attaching *[num]* to each cell name where *num* is an integer.
2. Universally quantified variables of P (the predicate being defined) are declared as an I/O terminals with a terminal type *IN* or *OUT*.

3. Each existentially quantified variable becomes a signal.

4. Parameters of component predicates P_i are instance terminals.

5. If a variable of a component P_i is universally quantified, then the corresponding instance terminal is netted to the corresponding I/O terminal by the L construct *WIRE*.

6. If a variable of P_i is existentially quantified - i.e. has previously been declared as a signal, then the instance terminal of P_i is attached to the signal by using the construct *SIG*.

5.3 Translating Recursive Definitions

The objective here is to create an iterative L program, i.e. a parameterized module generator, from a recursive HOL definition. We have assumed that recursive HOL definitions are recursions over the natural numbers so there is a natural correspondence with cell indicies. Recursive definitions based on the natural numbers have two predicates – one for the base case and the other for the recursive case. The base case describes the primitive building block of a cell module generator. The recursive case describes how a primitive building block is attached to an existing one to increase the size of a cell module by one. In this work, we have limited ourselves to one level of recursion which has the form shown below.

$$(\forall i_1...i_r o_1...o_m.P\ 0\ i_1...i_r o_1...o_m = B(i_1,...,i_r,o_1,...,o_m))$$
$$(\forall n i_1...i_r o_1...o_m.\ P\ (n+1)\ i_1...i_r o_1...o_m = \exists l_1...l_k.P\ n\ in\_list\ out\_list \wedge B(par\_list))$$

The identifier *in_list* denotes input ports with free occurrences of $i_1,...,i_r$, and the identifier *out_list* denotes output ports with free occurrences of $o_1, ..., o_m$. The identifier *par_list* represents ports with free occurrences of $i_1,...,i_r$, $o_1, ..., o_m$. In a term, existentially quantified variables are substituted for free occurrences of universally quantified variables. In the recursive part of a definition, existentially quantified variables indicate how two primitive building blocks are netted together. This enables us to translate a recursive definition to a L cell which is created iteratively from its primitive building block. In addition, the cell size is parameterized.

Before describing the translation scheme, we introduce some notation. P_0 denotes the term $P\ 0\ i_1...i_r o_1...o_m$, i.e. the base case definition of the module being defined. P_{n+1} denotes the term $P\ (n+1)\ i_1...i_r o_1...o_m$, i.e. the recursive case definition of the module being defined. B_0 represents an instance of an exiting predicate B used in the base case definition P_0. P_n and B_{n+1} denote P $n\ in\_list\ out\_list$ and $B(par\_list)$ respectively. B_{n+1} is an instance of a previously defined predicate used in the definition of P_{n+1}. P_n is the module generated after n iterations which is used in the description of the module generated by $n+1$ iterations.

Creating the corresponding L module generator from a HOL recursive definition is based on interpreting the context in which a variable appears. The context is determined by

1. How the variable is quantified, i.e. universally or existentially, and
2. Where in the definition the variable occurs, i.e. in P_0, B_0, P_n, P_{n+1} or B_{n+1}.

Also necessary to create a cell generator is a means by which a cell instance terminal name, e.g. $FADD1[0].cin$, is determined based on the port position of a variable in its corresponding predicate. To help backtrack to terminal names from variable positions we define two abstract functions *pos* and *port*. Given a parameter name v and a predicate $p(w,...,v,...)$, the function application $pos(v,p)$ returns the position of v in the parameter list of p. Given a predicate $p(w,...,v,...)$ and a position $n=pos(v,p)$, the function application $port(n,p)$ returns the corresponding parameter name v.

To create an L module generator program from a recursive HOL description and a cell port definition list, the universally quantified pin names are declared as inputs or outputs, then the definition is scanned from left to right. The instance terminal names are determined and wired together using the rules in Table 1. Note that if the variable corresponds to a bus, then it is indexed.

Table 1. Case analysis for variables.

	Var	U/E	Var Occurrence			Iteration	Interconnection
			B_0	P_n	B_{n+1}	$0 \leq i \leq m, 1 \leq m$	
Case1	x	U	√	√		$i = 0$	WIRE B[0].t' TO x where $pos(x,P_n)=k$, $port(k,P_0)=t$ $pos(t,B_0)=k'$, $port(k',B)=t'$
Case2	x	U	√	√	√	$0 \leq i \leq m$	WIREB[i].t TO x (or $x[i]$) where $pos(x,B_0)=k$, $pos(x,B_{n+1})=k$ $port(k,B)=t$
Case3	x	U	√		√	$i = m$	WIRE B[m].t TO x where $pos(x,B_0)=k$, $pos(x,B_{n+1})=k$ $port(k,B)=t$
Case4	y	E		√	√	$i \neq 0$	WIRE B[i-1].t' TO B[i].t'' where $pos(y,P_n)=k$, $port(k,P_0)=t$ $pos(t,B_0)=k'$, $port(k',B)=t'$ $pos(y,B_{n+1})=k''$, $port(k'',B)=t''$

Since the translation scheme is somewhat intricate, we illustrate the process using the example of creating the L generator for the ripple adder shown in Figure 3. The port definitions for $FADDERN$ and $FADD1$ along with the definition of $FADDERN$ are shown below. The parameterized L program which generates the ripple adder and an explanation of its function follow.

```
⊢ cell_port_def= ['FADD1', [in_port 'i1' pin;in_port 'i2' pin; in_port 'cin' pin;out_port 'sum' pin;
out_port 'cout' pin]; ['FADDERN',[in_port 'i1' (bus 1);in_port 'i2' (bus 1); in_port 'cin' pin;
out_port 'sum' (bus 1);out_port 'cout'pin]]
⊢(∀i1 i2 cin sum cout. FADDERN 0 i1 i2 cin sum cout = FADD1(i1 0,i2 0,cin,sum 0,cout))
wedge (∀n i1 i2 cin sum cout. FADDERN(n+1) i1 i2 cin sum cout = (∃l1.FADDERN n i1 i2
cin sum l1 wedge FADD1(i1(n+1),i2(n+1),l1,sum(n+1),cout)))
```

The first line, $SCHEMATIC\ FADDERN\ (INT\ hol\_n = 1)$, is derived from the name of the recursively defined predicate. Since we have assumed that the recursion is based on the natural numbers, the parameter of $FADDERN$ is an L

```
SCHEMATIC FADDERN (INT hol_l_n = 1)
{ IN  cin;
  OUT  cout;
  INT  hol_l_i;
  hol_l_i = 0;
  WHILE(hol_l_i<hol_l_n) {
    IN  i1[hol_l_i];
    IN  i2[hol_l_i];
    OUT  sum[hol_l_i];
    INST FADD1 FADD1[hol_l_i];
    IF(hol_l_i == 0){
      WIRE FADD1[0].cin TO cin;
    }
    WIRE FADD1[hol_l_i].i1 TO i1[hol_l_i];
    WIRE FADD1[hol_l_i].i2 TO i2[hol_l_i];
    WIRE FADD1[hol_l_i].sum TO sum[hol_l_i];
    IF(hol_l_i != 0){
      WIRE FADD1[hol_l_i-1].cout TO
              FADD1[hol_l_i].cin;
    }
    IF(hol_l_i == hol_l_n - 1){
      WIRE FADD1[hol_l_i].cout TO cout;
    }
    hol_l_i++;
  }
}
```

integer *hol_l_n* whose default value is 1. The rest of the generator is based on a case analysis of the context in which variables in the recursive definition occur. As shown in Table 1 there are four variable contexts.

Case 1 is illustrated by variable *cin*. *cin* is universally quantified and is a primary input or output. It appears only in the definitions B_0 and P_n, and appears only in the first iteration.

Case 2 is illustrated by variables *i1, i2*, and *sum*. These variables are indexed by the recursion variable n. They appear in all iterations and in the base case and recursive case definitions. Since they are universally quantified they are primary inputs and outputs of the module.

Case 3 is illustrated by variable *cout*. *cout* is universally quantified so it is a primary input or output. It appears only in the base case cell B_0 and as a parameter of B_{n+1}. It only is used in the last iteration of the implementation.

Case 4 pertains to internal connections which are existentially quantified variables. This case is illustrated by variable *l1*. Since internal connections are not seen by the user, they will not occur as parameters in the base case B_0 nor in P_{n+1}. Internal variables will occur in P_n and B_{n+1}. What must be determined are the instance terminal names which are *WIREed* together by the internal variable.

6 An n-bit Serial/Parallel Multiplier Example

In this section, we show a complete example of an *n-bit* serial/parallel multiplier. First, the verified HOL structural description of an *n-bit* multiplier is given. Second, we present the correctness theorem of the multiplier. Third, we translate it into the corresponding L description. Finally, we use the L description with the GDT tools to automatically produce the layout, and perform the simulation based on the initial condition given by the correctness theorem [4, 6]. In this example, the cell size is set to 8. The block diagram of an 8-bit serial/parallel multiplier is shown in Figure 6.

6.1 HOL Description

The *n-bit* multiplicand is loaded in parallel. The multiplier bits come in as a bit-serial stream least-significant bit first. Below are the definitions for the com-

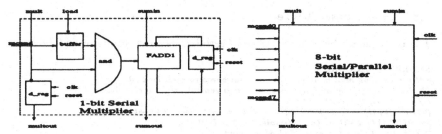

Fig. 6. Block diagram of a bit-serial/parallel multiplier

ponents of the multiplier. *HADD1* is a half adder. *FADD1* is a full adder. *SAD-DER1* is a bit-serial adder. *BIT_WORD_PROD* is an *AND*-gate with a latch to hold one bit of the multiplicand. *WORDPRODCKT1* is *BIT_WORD_PROD* with a D flip-flop to hold one bit of the multiplier. *SMULT1* is the basic building block of the bit-serial/parallel multiplier with no hidden wires. *SMULT* is the same as *SMULT1* except the *co* and *pprod* lines are hidden. *SMULTN* is the parameterized bit-serial/parallel multiplier.

⊢ *cell_port_def* = ['HADD1',[in_port 'i1' pin;in_port 'i2' pin;out_port 'sum' pin;out_port 'cout' pin];
'FADD1',[in_port 'i1' pin;in_port 'i2' pin;in_port 'cin' pin; out_port 'sum' pin;out_port 'cout' pin];
'SADDER1', [in_port 'clk' pin;in_port 'i1' pin;in_port 'i2' pin;out_port 'co' pin;out_port 'sum' pin; in_port 'reset' pin];
'BIT_WORD_PROD',[in_port 'word' pin;in_port 'bit_stream' pin;out_port 'pprod' pin; in_port 'load' pin];
'WORDPRODCKT1',[in_port 'mcand' pin;in_port 'mult' pin;out_port 'multout' pin;out_port 'pprod'pin;in_port 'load' pin;in_port 'reset' pin;in_port 'clk' pin];
'SMULT1', [in_port 'mcand' pin;in_port 'mult' pin;out_port 'multout' pin;in_port 'sumin' pin;out_port 'sumout' pin;out_port 'pprod' pin; out_port 'co' pin;in_port 'reset' pin;in_port 'load' pin;in_port 'clk' pin];
'SMULT',[in_port 'mcand' pin;in_port 'mult' pin;out_port 'multout' pin;in_port 'sumin' pin;out_port 'sumout' pin;in_port 'reset' pin;in_port 'load' pin;in_port 'clk' pin];
'SMULTN',[in_port 'mcand' (bus 1);in_port 'mult' pin;out_port 'multout' pin;in_port 'sumin' pin;out_port 'sumout' pin;in_port 'reset' pin;in_port 'load' pin;in_port 'clk' pin]]
⊢ ∀i1 i2 sum cout. HADD1(i1,i2,sum,cout) = (∃t1. xor(i1,i2,sum) ∧ nand(i1,i2,t1) ∧ inv(t1,cout))
⊢ ∀a b cin sum cout. FADD1(i1,i2,cin,sum,cout) =
(∃l1 l2 l3.HADD1(i1,i2,l1,l2) ∧ HADD1(l1,cin,sum,l3) ∧ or(l3,l2,cout))
⊢ ∀clk i1 i2 co sum reset.SADDER1(clk,i1,i2,co,sum,reset) =
(∃cin. d_reg(co,clk,reset,cin) ∧ FADD1(i1,i2,cin,sum,co))
⊢ ∀word bit_stream pprod load. BIT_WORD_PROD(word,bit_stream,pprod,load) =
(∃b1. and(b1,bit_stream,pprod) ∧ buffer(word,b1,load))
⊢ ∀mcand mult multout pprod load reset clk.
WORDPRODCKT1(mcand,mult,multout,pprod,load,reset,clk) =
BIT_WORD_PROD(mcand,mult,pprod,load)∧d_reg(mult,clk,reset,multout)
⊢ ∀ mcand mult multout sumin sumout pprod co reset load clk.
SMULT1(mcand,mult,multout,sumin,sumout,pprod,co,reset,load,clk) =
WORDPRODCKT1(mcand,mult,multout,pprod,load,reset,clk) ∧
SADDER1(clk,pprod,sumin,co,sumout,reset)
⊢ ∀mcand mult multout sumin sumout reset load clk.
SMULT(mcand,mult,multout,sumin,sumout,reset,load,clk) =
(∃ co pprod.SMULT1(mcand,mult,multout,sumin,sumout,pprod,co,reset,load,clk))
⊢ (∀mcand mult multout sumin sumout reset load clk = SMULT(mcand 0 mcand mult multout sumin sumout reset load clk = SMULT(mcand 0,mult,multout,sumin,sumout,reset,load,clk)) ∧
(∀n mcand mult multout sumin sumout reset load clk. SMULTN(n+1)mcand mult multout sumin sumout reset load clk =
(∃L_multout L_sumout. SMULTN n mcand mult L_multout sumin L_sumout reset load clk ∧
SMULT(mcand(n+1),L_multout,multout,L_sumout,sumout,reset,load,clk)))

6.2 Correctness Theorem

The correctness theorem for the *n-bit* serial/parallel multiplier is as follows.

```
SMULT_CKT1_CORRECT
⊢ ∀ n k t mcand mult multout sumin sumout pprod co reset load clk.
SMULTN_CKT1 n mcand mult multout sumin sumout pprod co reset load clk ⊃
(reset t = Hi) ∧ (load t = Hi) ∧
(∀t1.(clk(t+t1) = Hi_Lo)∧(reset((t+1)+t1) = Lo)∧ (load((t+1)+t1) = Lo) ∧
(mult((t+1)+(t1+k)) = Lo) ∧ (sumin(t+t1) = Lo))
⊃ ((WORDVAL1 n t mcand) * (VAL k t mult) =
(((n = 0) ∧ k > 0) → VAL k t sumout | (((k = 0) ∧ n > 0) → VAL n t sumout |
(((n = 0) ∧ (k = 0)) → BV(sumout t) | VAL(n+k+1))t sumout))))
```

Theorem *SMULT_CKT1_CORRECT* says that if the initial conditions in the antecedent are satisfied, e.g. *load* and *reset* occur at time *t*, etc., then the product of the values of the multiplier and multiplicand bits equals the value of the bit-serial stream, *sumout*. *SMULT_CKT1* is not in a form which is directly translatable to layout. *SMULT_CKT2* shown below is. Theorem *SMULTN_CKT1_CKT2_EQV* as shown shows *SMULTN_CKT2* is equivalent to *SMULT_CKT1*, and Theorem *SMULTN_CKT2_SMULTN_EQV_multout_sumout* shows *SMULTN* having the same input-output relation as *SMULTN_CKT2*. Thus, *SMULTN* has the same arithmetic properties as stated in Theorem *SMULT_CKT1_CORRECT*.

```
SMULTN_CKT2
⊢ (∀mcand mult multout sumin sumout pprod co reset load clk.
SMULTN_CKT2 0 mcand mult multout sumin sumout pprod co reset load clk =
SMULT1(mcand 0,mult,multout,sumin,sumout,pprod 0,co 0,reset,load,clk)) ∧
(∀n mcand mult multout sumin sumout pprod co reset load clk.
SMULTN_CKT2 (n+1) mcand mult multout sumin sumout pprod co reset load clk = (∃L_multout
L_sumout.SMULTN_CKT2 n mcand mult L_multout sumin L_sumout pprod co reset load clk ∧
SMULT1(mcand(n+1),L_multout,multout,L_sumout,sumout,pprod(n+1),co(n+1),reset,load,clk)))
SMULTN_CKT1_CKT2_EQV
⊢ (∀n mcand mult multout sumin sumout pprod co reset load clk.
SMULTN_CKT1 n mcand mult multout sumin sumout pprod co reset load clk =
SMULTN_CKT2 n mcand mult multout sumin sumout pprod co reset load clk
```

6.3 L Description

In Figure 7, the *n-bit* serial multiplier as shown in the previous section is translated into the L description and the word length is set to 8.

6.4 Simulation Results and the CMOS Layout

The 8-bit serial/parallel multiplier layout was simulated under the initial condition as given in Theorem *SMULT_CKT1_CORRECT*. In Figure 8 we show an example simulation and the CMOS layout derived from the HOL logic. Note that the simulation results are consistent with the correctness theorem except for some glitches. The glitches are due to static hazards, i.e. glitches which occur when a function changes from one minterm to another in separate implicants. Static hazards are eliminated by introducing another prime implicant which covers two adjacent minterms or by latching the output.

Fig. 7. L description of an n-bit serial/parallel multiplier

7 Conclusions

We have shown how hardware descriptions in higher-order logic are linked to the GDT system. Some styles of HOL structural descriptions, including parameterized recursive descriptions, can be transformed to corresponding L implementation descriptions. Notably, this includes parameterized cell generators. This means verified parameterized HOL descriptions can augment the cell generator library of the GDT system making designs in HOL accessible to VLSI designers.

It is possible to make links to other design technologies such as field programmable gate arrays, (FPGAs). In addition to the work presented here, we have constructed a compiler that transforms the L router format to the Xilinx Net Format (XNF)[13].

References

1. Mike Gordon, "A proof Generating System for Higher-Order Logic," in VLSI Specification, Verification and Synthesis, edited by Graham Birtwistle and P.A. Subrahmanyam, Kluwer, 1987.
2. Mike Gordon, "Why higher-order logic is a good formalism for specifying and verifying hardware," Formal Aspects of VLSI Design, edited by G. Milne and P.A. Subrahmanyam, North Holland, 1986.

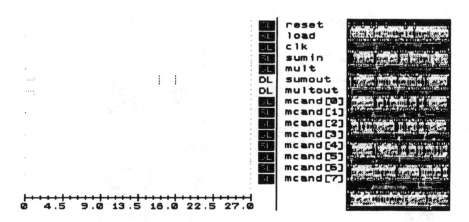

Fig. 8. Simulation timing diagram & CMOS layout of an 8-bit s/p multiplier

3. Shiu-Kai Chin and Edward P. Stabler, "Synthesis of Arithmetic Hardware Using Hardware Metafunctions," IEEE Trans. Computer-Aided Design, Vol. 9, No. 8, August 1990.

4. Shiu-Kai Chin and Juin-Yeu Lu, The Mechanical Verification and Synthesis of Parameterized Serial/Parallel Multiplier, CASE Center Tech. Report No. 9140, Syracuse University, March 1991.

5. Shiu-Kai Chin and Graham Birtwistle, "Implementing and Verifying Finite-State Machines Using Types in Higher-Order-Logic," Proc. 1991 Int. Tutorial Workshop on the HOL Theorem Proving System and its Applications, IEEE Comp. Soc. Press, 1992.

6. Juin-Yeu Lu and Shiu-Kai Chin, Linking HOL to a VLSI CAD System, CASE Center Tech. Report, Syracuse University, April 1993.

7. Albert John Camilleri, Executing Behavioural Definitions in Higher Order Logic, Tech. Report No. 140, University of Cambridge Computer Lab., Feb. 1988.

8. A. J. Cohn, "A Proof of Correctness of the VIPER Microprocessor: The First Level," in VLSI Specification, Verification and Synthesis, edited by Graham Birtwistle and P.A. Subrahmanyam, Kluwer, 1988.

9. J. Joyce, "Formal Verification and Implementation of a Microprocessor," in VLSI Specification, Verification and Synthesis, edited by Graham Birtwistle and P.A. Subrahmanyam, Kluwer, 1988.

10. Thomas F. Melham, Automating Recursive Type Definitions in Higher Order Logic, Tech. Report No. 146, University of Cambridge Computer Lab., Jan. 1989.

11. Mandayam Srivas, Mark Bickford, "Formal Verification of a Pipelined Microprocessor," IEEE Software, Sep. 1990.

12. Mentor Graphics Corporation, GDT Manuals V.5, CA, 1990.

13. Xilinx Inc., The Programmable Gate Array Data Bool, CA, 1989.

14. Thomas Kropf, Ramayya Kumar, Klaus Schneider, "Embedding Hardware Verification within a Commercial Design Framework," Advanced Research Working Conference on Correct Hardware Design Methodologies (CHARME93), Lecture Notes in Computer Science, 1993.

Alternative Proof Procedures for Finite-State Machines in Higher-Order Logic*

Klaus Schneider,[1] Ramayya Kumar,[2] and Thomas Kropf[1]

[1] Universität Karlsruhe, Institut für Rechnerentwurf und Fehlertoleranz,
(Prof. Dr. D. Schmid), P.O. Box 6980, 76128 Karlsruhe, Germany,
e-mail:{schneide|kropf}@ira.uka.de
[2] Forschungszentrum Informatik, Haid-und-Neustraße 10-14, 76131 Karlsruhe, Germany,
e-mail:kumar@fzi.de

Abstract. Verification of digital circuits in higher-order logic often requires the proof of temporal propositional logic formulae. The implementation of decision procedures for this logic or finite-state machines is however not very easy within the HOL system, since it requires the proof of certain fixpoint theorems and a creation of a new theory based on it. The main contribution of this paper is to give some alternative proof procedures so that proof tactics can be developed for directly solving these goals. These proof procedures can be classified into two categories. Firstly, a set of easily implementable proof methods which do not use knowledge of fixpoint theorems are given. Since these methods are incomplete, the second category exploits an external program for computing fixpoint lemmata which can then be easily proved in HOL.

1 Introduction

The approaches to hardware-verification which are based on the verification of properties of finite-state machines can be fully automated. Recently, efficient methods have been implemented [BuCL91, CoBM89, BCMD90] which are on the threshold of industrial use. On the other hand, verification at higher levels of abstraction requires higher-order logic, which cannot be automated sufficiently. However, parts of digital circuits can nevertheless be verified very elegantly by finite-state machine approaches. Moreover, in [KuSK93] first proof paradigms for the verification of circuits at register transfer level have been formulated and in [ScKK93c] these paradigms have been extended to decision procedures. The transformations which have been used in [ScKK93c] yield in subgoals which resemble classical finite-state machine descriptions. These subgoals, which we call 'finite-state machine goals' (FSM-goals) are defined as follows:

Definition 1.1 (FSM-goals) *Given the variables* $q_1, \ldots, q_l, i_1, \ldots, i_m$ *of type* $\mathbb{N} \rightarrow \mathbb{B}$, *the quantifier-free formulae* $\Omega_1(\vec{p}), \ldots, \Omega_l(\vec{p}), \Phi(\vec{p})$ *with variables* $p_1, \ldots, p_{l+m}$ *of*

* This work has been partly financed by a german national grant, project Automated System Design, SFB No.358.

type $\mathbb{B}$, *and a boolean tuple* $\vec{\omega}_0 \in \mathbb{B}^l$, *the following goals are called universal FSM-goals*[1]:

$$\underbrace{\forall t. [\vec{q}(0) \leftrightarrow \vec{\omega}_0] \wedge \left[\vec{q}(t+1) \leftrightarrow \vec{\Omega}(\vec{q}(t), \vec{i}(t))\right]}_{=: \Gamma_{\vec{q}, \vec{i}}} \vdash \forall t. \Phi(\vec{q}(t), \vec{i}(t)).$$

Moreover, goals of the form $\Gamma_{\vec{q}, \vec{i}} \vdash \exists t. \Phi(\vec{q}(t), \vec{i}(t))$ *are called existential FSM-goals.*

States of the underlying finite-state machine described by the assumption $\Gamma_{\vec{q}, \vec{i}}$ of a FSM-goal are given as boolean-valued tuples $\vec{q}(t)$ which depend on the time t (modelled by natural numbers). The machines have an initial state given by the tuple $\vec{\omega}_0$ and a transition function which has been given implicitly by the tuple of formulae $\vec{\Omega}(\vec{q}(t), \vec{i}(t))$. Under these state transition conditions which form the assumption $\Gamma_{\vec{q}, \vec{i}}$ of the goals, it has to be proved that a property $\Phi(\vec{q}(t), \vec{i}(t))$ holds for each time t or at a certain point of time. As we will concentrate on universal FSM-goals in this paper, we will neglect the supplement 'universal' in most cases.

In order to prove FSM-goals, one can make use of already existing model checkers by integrating them in interactive theorem proving tools [JoSe93]. This has a lot of advantages, as on the one hand the higher-order facilities of HOL can be used, and on the other hand one also benefits from the efficiency of already existing model checkers. However, this means that one has to trust the external model checker. As model checkers are complex programs, they however can contain errors. Moreover, model checkers can handle only a limited class of formulae, i.e. they can only check the validity of a propositional formula in each state of a finite-state machine. If on the other hand the state traversal would be integrated in HOL, then the validity of higher-order formulae can be proved during the state traversal. Such formulae can result from the interactions between generic data paths (described using abstract data types) and controllers.

In order to implement decision procedures for FSM-goals, knowledge about finite-state machines is required, e.g. the fact that there is only a finite number of reachable states. Hence, a formal treatment in HOL can be done by the implementation of a theory about finite-state systems [Loew92]. This, however, is itself not an easy task and it shifts the reasoning about those formulae onto a meta-level.

In this paper, we will therefore discuss several approaches for automating proofs of FSM-goals which neither require the implementation of a special theory nor the trust in external programs. Not all of the discussed approaches are *complete*, i.e. they are not able to prove all valid FSM-goals, although in practice these approaches turned out to be quite useful. The outline of this paper is as follows: the next section briefly describes the concepts of the various proof procedures. Section 3 to 6 discuss the proof procedures for universal FSM-goals in detail and section 7 indicates a simple decision procedure for existential FSM-goals. The paper concludes with some experimental results and a summary.

---

[1] By the equivalence of tuples of formulae $\vec{\varphi} \leftrightarrow \vec{\psi}$, we mean the conjunction of the equivalences of the corresponding components $\bigwedge\limits_{j=0}^{n} (\varphi_j \leftrightarrow \psi_j)$.

2 Overview of the Proof Methods

Before we discuss the proof procedures, we recapitulate some theoretical results proved in [ScKK93c] which indicate a first decision procedure for universal FSM-goals. The formulae $\vec{\Theta}_j^{0,\vec{i}}$ and $\vec{\Theta}_j^{t,\vec{i}}$ represent the states which are reachable in j steps from the initial state with the inputs $\vec{i}(0), \ldots, \vec{i}(j)$ and from the state $\vec{q}(t)$ with the inputs $\vec{i}(t), \ldots, \vec{i}(t+j)$, respectively.

Lemma 2.1 (Instance Lemma) *Defining the tuples of formulae $\vec{\Theta}_j^{0,\vec{i}}$ and $\vec{\Theta}_j^{t,\vec{i}}$ recursively by $\vec{\Theta}_0^{0,\vec{i}} := \vec{\omega}_0$, $\vec{\Theta}_{j+1}^{0,\vec{i}} := \vec{\Omega}(\vec{\Theta}_j^{0,\vec{i}}, \vec{i}(j))$ and $\vec{\Theta}_0^{t,\vec{i}} := \vec{q}(t)$, $\vec{\Theta}_{j+1}^{t,\vec{i}} := \vec{\Omega}(\vec{\Theta}_j^{t,\vec{i}}, \vec{i}(t+j))$, the following holds: $\Gamma_{\vec{q},\vec{i}} \vdash \forall t.\Phi(\vec{q}(t), \vec{i}(t))$ iff $\vdash \Phi(\vec{\Theta}_j^{0,\vec{i}}, \vec{i}(j))$ holds for all $j \in \mathbb{N}$.*

The above lemma was used in [ScKK93c] to establish the decidability of FSM-goals. This decidability result is analogous to the pumping lemma for regular expressions and is stated below:

Theorem 2.1 (Decidability) *Given a FSM-goal $\Gamma_{\vec{q},\vec{i}} \vdash \forall t.\Phi(\vec{q}(t), \vec{i}(t))$ and the formulae $\vec{\Theta}_j^{0,\vec{i}}$ and $\vec{\Theta}_j^{t,\vec{i}}$ as defined in Lemma 2.1, there is a number n_0, such that $\Gamma_{\vec{q},\vec{i}} \vdash \forall t.\Phi(\vec{q}(t), \vec{i}(t))$ holds iff $\vdash \Phi(\vec{\Theta}_j^{0,\vec{i}}, \vec{i}(j))$ holds for all canonical terms $j < n_0$. Moreover n_0, which is called the sequential depth of $\Gamma_{\vec{q},\vec{i}}$, can be effectively computed, thus the validity of the considered goals is decidable.*

The above decidability result can be visualized as a proof tactic (shown in figure 1). An implementation of this tactic in HOL according to the proof given in [ScKK93c]

$$\frac{\Gamma_{\vec{q},\vec{i}} \vdash \forall t.\Phi(\vec{q}(t), \vec{i}(t))}{\vdash \bigwedge_{j=0}^{n_0} \Phi(\vec{\Theta}_j^{0,\vec{i}}, \vec{i}(j))}$$

Fig. 1.: SEQ_TAC

is however complicated, since this proof uses the quotient set of an equivalence relation built on the cartesian product between all possible propositional models and the formulae $\vec{\Theta}_j^{0,\vec{i}}$. This motivated us to implement alternative proof methods like *propositional abstraction* (PL), *first-order logic* (FOL), *temporal induction* (TEMP_INDUCT), and *state-enumeration* (STATE_ENUM) (see figure 2). A symbolic form of the state-enumeration method leads then to a simpler implementation of SEQ_TAC.

Propositional abstraction is a trivial means of establishing an automated proof procedure, and should be used whenever possible, since this method is the most efficient one. *First-order logic* is an extension of propositional logic, which allows to prove more FSM-goals, but it turns out that not all FSM-theorems can be proved without further tactics either.

Another extension of propositional logic is obtained by the approach called *temporal induction*. The basic idea of this approach is the following: as time is modelled by natural numbers and induction is a classical proof tactic for the natural numbers, it seems to be quite obvious to use the induction tactic for the solution of this problem. Although the simple application of the induction tactic is insufficient in theory, it turned out to be quite useful in practice. We will also state necessary conditions which guarantee that a goal can be proved by this approach.

The last presented method, called *state-enumeration*, uses implicit knowledge about

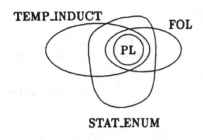

TEMP_INDUCT

FOL

PL

STAT_ENUM

Fig. 2.: Power of the approaches.

finite-state machines in such a way, that it first computes the reachable states and then proves that the property which has to be proved holds in each reachable state. It has to be noted, that for the implementation of this tactic no explicit knowledge in a finite-state machine theory has to be provided and that the computation of the reachable states is not done by a HOL-tactic, but by an external program. When these states have been computed, a lemma is formally proved in HOL which states that at each time the current state is one of the computed reachable states[2]. The advantage of this technique is that time critical computations can be given to fast external programs without sacrifying the philosophy of HOL.

3 Propositional Abstraction and Rewriting

A first and quite simple method to prove FSM-goals is to prove the resulting formula by propositional means.

Lemma 3.1 (Propositional Abstraction) *If the outermost universal quantifiers of a higher-order formula φ are removed and the atomic and quantified subformulae are replaced by propositional variables, the resulting formula ψ is called the propositional abstraction of φ. If ψ is a tautology, φ is a theorem of higher-order logic as well, otherwise nothing can be said about the validity of φ.*

It is clear, that this method will not be able to solve much of the goals without extensions, but it is very efficient, so that we should use it whenever possible. However, it has to be noted, that all circuits without delays can be proved by this approach.

In order to enhance the efficiency, we add a further tactic: Equivalences enforce normally a lot of case distinctions in a tautology checker, such that often a combinatorial explosion takes place. We circumvent this problem by using the antecedent of the sequents as a term rewriting system and try to prove only the rewritten formula by propositional abstraction. This is justified by the following lemma:

Lemma 3.2 (Rewrite Lemma)
Given a quantifier-free formula $\Upsilon(\alpha_1, \ldots, \alpha_n)$ with atomic formulae $\alpha_1, \ldots, \alpha_n$ and some other formulae $\varphi_1, \ldots, \varphi_n$ which do not depend on the α_j's, the following holds[3] (Υ is allowed to contain more atoms than the α_j's and not all α_j's have to occur in Υ):

$$\bigwedge_{j=1}^{n} (\alpha_j \leftrightarrow \varphi_j) \vdash \Upsilon(\alpha_1, \ldots, \alpha_n) \text{ holds iff } \vdash \Upsilon(\varphi_1, \ldots, \varphi_n) \text{ holds}$$

---

[2] This methodology resembles the approach that we have taken for integrating automated first-order provers in HOL [KuSK93]. First, the external prover is used to compute a proof with a list of instantiations which are then used to validate the proof in HOL.

[3] It has not been stated that $\vdash \left[\left(\bigwedge_{j=1}^{n} (\alpha_j \leftrightarrow \varphi_j) \right) \to \Upsilon(\alpha_1, \ldots, \alpha_n) \right] \leftrightarrow \Upsilon(\varphi_1, \ldots, \varphi_n)$ is a theorem. This is not the case since the simple example $[(a \leftrightarrow b) \to a] \leftrightarrow b$ is not a tautology (interpret $a := T, b := F$).

$$\bigvee_{i=1}^{m} \bigwedge_{j=1}^{n} (\alpha_j \leftrightarrow \varphi_{i,j}) \vdash \Upsilon(\alpha_1, \ldots, \alpha_n) \text{ holds iff } \vdash \bigwedge_{i=1}^{m} \Upsilon(\varphi_{i,1}, \ldots, \varphi_{i,n}) \text{ holds.}$$

Proof. The proof of the first proposition is split up into two implications:

'$\Leftarrow$:' Suppose, $\vdash \Upsilon(\varphi_1, \ldots, \varphi_n)$ is a theorem, then we conclude with modus ponens with the simple theorem $\bigwedge_{i=1}^{n}(\alpha_i \leftrightarrow \varphi_i) \vdash \Upsilon(\varphi_1, \ldots, \varphi_n) \rightarrow \Upsilon(\alpha_1, \ldots, \alpha_n)$, that $\bigwedge_{i=1}^{n}(\alpha_i \leftrightarrow \varphi_i) \vdash \Upsilon(\alpha_1, \ldots, \alpha_n)$ is a theorem, too.

'$\Rightarrow$:' We show the equivalent statement: If $\vdash \Upsilon(\varphi_1, \ldots, \varphi_n)$ is not a theorem, then $\bigwedge_{i=1}^{n}(\alpha_i \leftrightarrow \varphi_i) \vdash \Upsilon(\alpha_1, \ldots, \alpha_n)$ cannot be a theorem either. Suppose $\vdash \Upsilon(\varphi_1, \ldots, \varphi_n)$ is not a theorem, thus there is an interpretation I_1 of the variables of the φ_i's such that $\mathcal{VAL}^{I_1}(\Upsilon(\varphi_1, \ldots, \varphi_n)) = \mathsf{false}$. We define a new interpretation I_2 by assigning $I_2(p) := I_1(p)$ for each variable p occurring in the φ_i's and $I_2(\alpha_i) := \mathcal{VAL}^{I_1}(\varphi_i)$ for $i \in \{1, \ldots, n\}$. Then it is straightforward to show $\mathcal{VAL}^{I_2}(\bigwedge_{i=1}^{n}(\alpha_i \leftrightarrow \varphi_i)) = \mathsf{true}$ and $\mathcal{VAL}^{I_2}(\Upsilon(\alpha_1, \ldots, \alpha_n)) = \mathcal{VAL}^{I_1}(\Upsilon(\varphi_1, \ldots, \varphi_n)) = \mathsf{false}$, thus $\bigwedge_{i=1}^{n}(\alpha_i \leftrightarrow \varphi_i) \vdash \Upsilon(\alpha_1, \ldots, \alpha_n)$ cannot be a theorem.

The proof of the second proposition follows from the first one by simple sequent calculus rules.

After having rewritten the goal $\Gamma_{\vec{q},\vec{\imath}} \vdash \forall t.\Phi(\vec{q}(t), \vec{\imath}(t))$ with $\Gamma_{\vec{q},\vec{\imath}}$, we can neglect $\Gamma_{\vec{q},\vec{\imath}}$ totally for the propositional abstraction. Propositional abstraction can prove not only combinatorial circuits, as shown by the example below:

Example 3.1 *Consider the following proof goal:*

$$\begin{bmatrix} a(0) \leftrightarrow F, \\ a(1) \leftrightarrow F, \\ \forall t.a(t+2) \leftrightarrow \neg a(t) \wedge \neg a(t+1) \end{bmatrix} \vdash \forall t.a(t+1) \rightarrow \neg a(t+2)$$

First, the rewrite lemma is used to obtain the goal $\vdash \forall t.a(t+1) \rightarrow \neg(\neg a(t) \wedge \neg a(t+1))$ which is then abstracted to the tautology $\vdash p \rightarrow \neg(\neg q \wedge \neg p)$.

Nevertheless, propositional abstraction is very limited, e.g. existential FSM-goals like the following can not be proved with it:

$$\boxed{(out(0) \leftrightarrow F) \wedge \forall t.(out(t+1) \leftrightarrow T) \vdash \exists t.(\neg out(t) \wedge out(t+1))}$$

4 First-Order Logic

First-order proof techniques are also an incomplete method for proving FSM-goals, if they are directly applied on the goals, since first-order provers have no knowledge about the natural numbers, i.e. the prover does not know that the induction axiom holds. As a consequence, first-order proof procedures instantiate arbitrary terms $a, b, \ldots$, although the induction axiom tells us that each natural number can be written as a canonical term of the form: $0, SUC(0), SUC(SUC(0)), \ldots$.

A goal which is not provable by without induction, is e.g. the following:

$$\boxed{(out(0) \leftrightarrow F) \wedge \forall t.(out(t+1) \leftrightarrow out(t)) \vdash \forall t.(out(t) \leftrightarrow F)}$$

Nevertheless, first-order proof procedures are able to make required copies of instantiations of quantified formulae and therefore, they are able to solve more goals than the propositional abstraction method. For example, the last goal of the previous section could not be proved with propositional abstraction, but can easily be proved by the first-order prover FAUST [ScKK92b, ScKK93a].

However, it cannot be stated that first-order theorem proving methods are not strong enough to solve FSM-problems, but the problems have to be presented in another way. There are transformations which translate FSM-goals into formulae which are really first-order, hence first-order provers can be made complete for FSM-goals with additional tactics[4]. Automated first-order provers like FAUST [ScKK93a] are then able to solve *all* of these goals. First-order proof procedures are moreover very flexible, since the goals have to fulfill only certain syntactical restrictions, i.e. they have to be first-order.

5 Temporal Induction

Normally, induction over natural numbers is performed by the well-known proof tactic given on the left side of the box below.

$$
\frac{\Gamma \vdash \forall t.\Phi(t)}{\Gamma \vdash \Phi(0) \mid \Gamma \vdash \forall t.\Phi(t) \to \Phi(t+1)} \qquad \frac{\Gamma \vdash \forall t.\Phi(t)}{\Gamma \vdash \Phi(0) \mid \Gamma \vdash \forall t.\Phi(t+1)}
$$

The first subgoal is called the induction base and the second one is called the induction step. One might also suggest the weaker modification of the tactic on the right side of the above box which we call a *case distinction* over the natural numbers. Clearly, this tactic is also sound and its subgoals are easier to prove since the induction hypothesis is omitted and the length of the formula is therefore smaller. However, the case distinction tactic is not as powerful as the induction tactic (even for FSM-goals) as can be shown by the following example:

$$
\begin{bmatrix} \forall t.(p(0) \leftrightarrow T) \land (p(t+1) \leftrightarrow q(t)); \\ \forall t.(q(0) \leftrightarrow F) \land (q(t+1) \leftrightarrow p(t)) \end{bmatrix} \vdash \forall t.p(t) \not\leftrightarrow q(t)
$$

If the case distinction tactic is applied, rewriting the second subgoal with its assumptions leads immediately back to the original goal, thus we end up in an infinite loop. On the other hand, the proof can be done with one induction step, rewriting and propositional abstraction. For this reason, we concentrate on the more powerful induction rule.

Sometimes the application of one induction tactic is not sufficient for the proof. Then the induction tactic has to be applied again on the induction step of the first application. Therefore, the following subgoals are obtained:

- $\Gamma \vdash \Phi(0) \to \Phi(1)$
- $\Gamma \vdash \forall t.\, [\Phi(t) \to \Phi(t+1)] \to [\Phi(t+1) \to \Phi(t+2)]$

---

[4] For example, if a finite type 'state' would be introduced, then each FSM-goal can be translated in a form which uses variables of this type.

If a lot of induction rules have to be applied, the repeated use of the induction tactic would create induction steps with an exponentially growing size. Fortunately, $([a \rightarrow b] \rightarrow [b \rightarrow c]) \leftrightarrow (b \rightarrow c)$ is a tautology, thus we can replace the induction step above by $\forall t. [\Phi(t+1) \rightarrow \Phi(t+2)]$ and obtain therefore the same subgoals as if we would have applied the case distinction rule. *Thus, having once applied the induction tactic, further induction tactic applications are equivalent to the application of the case distinction tactic.* Using this fact, the length of the subgoals grows only linearly and is nearly constant.

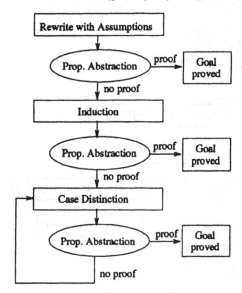

Fig. 3.: Temporal Induction

The general proof strategy is given in figure 4. First, the rewritten succedent of the goal is abstracted to propositional logic and an attempt is undertaken to prove it by a tautology checker. If this succeeds, a proof is obtained, otherwise the induction tactic has to be applied. The induction base is proved by propositional abstraction and an attempt is undertaken to prove the induction step also by propositional means. If this fails, case distinctions will be used with propositional abstraction to prove the remaining goal until a proof can be found.

Example 5.1 *Consider for example the following proof goal:*

$$\left[\begin{array}{l} \forall t. e(t) \leftrightarrow F; \\ \forall t. (q_1(0) \leftrightarrow F) \wedge (q_1(t+1) \leftrightarrow e(t)); \\ \forall t. (q_2(0) \leftrightarrow F) \wedge (q_2(t+1) \leftrightarrow q_1(t)); \\ \forall t. (q_3(0) \leftrightarrow F) \wedge (q_3(t+1) \leftrightarrow q_2(t)); \\ \forall t. a(t) \leftrightarrow q_3(t) \end{array} \right] \vdash \forall t. \neg a(t)$$

$$\underbrace{}_{=: \Gamma}$$

In the proof tree in figure 4, rewrite rules have been applied on nodes with just one successor node. After the first rewrite rule, the normal induction tactic has been applied, then two case distinctions are used for the proof. The rightmost leaf is proven by propositional abstraction $(p_1 \rightarrow F)$.

During the temporal induction procedure, subgoals of the form $\vdash \forall t. \Phi(\vec{\Theta}_n^{t,\vec{i}}, \vec{i}(t+n)) \rightarrow \Phi(\vec{\Theta}_{n+1}^{t,\vec{i}}, \vec{i}(t+n+1))$ are created. Even if the given FSM-goal is valid, it is not guaranteed, that these goals are valid, too:

Lemma 5.1 (Incompleteness of temporal induction)
There is a valid universal FSM-goal $\Gamma_{\vec{q},\vec{i}} \vdash \forall t. \Phi(\vec{q}(t), \vec{i}(t))$ such that for each natural

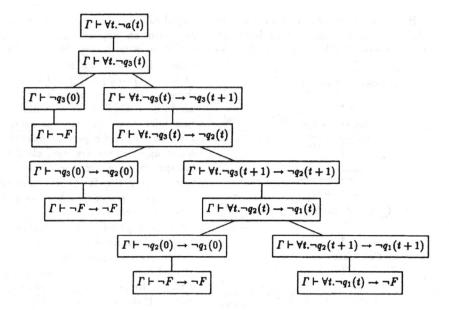

Fig. 4. A proof tree using temporal induction.

number $n \in \mathbb{N}$, $\vdash \forall t.\Phi(\vec{\theta}_n^{t,\vec{\imath}}, \vec{\imath}(t+n)) \rightarrow \Phi(\vec{\theta}_{n+1}^{t,\vec{\imath}}, \vec{\imath}(t+n+1))$ *is not a tautology.*
Thus this theorem cannot be proved by temporal induction.

Proof. Consider for example the following goal $\mathcal{G}_1$:

$$
\left[
\begin{array}{l}
(p(0) \leftrightarrow F) \wedge \forall t.p(t+1) \leftrightarrow p(t); \\
(q(0) \leftrightarrow F) \wedge \\
\quad \forall t.q(t+1) \leftrightarrow p(t) \Rightarrow [q(t) \leftrightarrow \neg(e_0(t) \vee e_1(t))] \\
\quad\quad\quad\quad\quad\quad\quad | \; [q(t) \leftrightarrow \neg(e_0(t) \wedge e_1(t))]
\end{array}
\right] \vdash \forall t.\neg(p(t) \wedge q(t))
$$

The formulae of the antecedent describe the state transition diagram given in figure 5. The initial state is 00 and it can be easily seen that the state 11 could never be reached. This can however not be proved by temporal induction, since the induction step would state that given any state different from 11, then the next state will also be different from 11. But this is not true, for example consider the state 10. It is different from state 11, but 11 is a successor state of it, thus the induction step is false and can therefore never be proved.

In general, it can be stated that all theorems can be proved this way, which do not contain a cycle consisting of non-reachable states in the state transition diagram as in these formulae, there is a finite number n of iteration steps such that each nth successor state of a non-reachable state is a reachable state. If on the other hand cycles of non-reachable states occur, then it might nevertheless be possible to prove the sequent with this method.

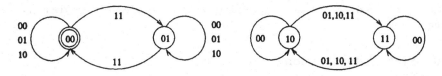

Fig. 5. State Transition diagram of a formula not provable by induction.

Furthermore, it can be easily observed, that if the FSM-goal is not a theorem, then the proof procedure will detect this after some finite time, since the procedure then successively tries to prove the goals $\vdash \varPhi(\vec{\Theta}_j^{0,\vec{\imath}}, \vec{\imath}(j))$. However, after lemma 2.1, there has to be at least one natural number n such that $\not\vdash \varPhi(\vec{\Theta}_n^{0,\vec{\imath}}, \vec{\imath}(n))$. Thus the procedure may only loop in case the given FSM-goal is indeed a theorem.

One might try to remove this incompleteness by the following *multiple induction rule*:

$$
\frac{\Gamma \vdash \forall t.\varPhi(t)}{\Gamma \vdash \bigwedge_{j=0}^{n} \varPhi(j) \quad \middle| \quad \Gamma \vdash \forall t. \left(\bigwedge_{j=0}^{n} \varPhi(t+j) \right) \rightarrow \varPhi(t+n+1)}
$$

The correctness of the tactic can be established by the simple induction tactic as given in the previous section[5]. The completeness of this tactic can however not be guaranteed:

Lemma 5.2 (Incompleteness of multiple temporal induction)
There is a theorem $\Gamma_{\vec{q},\vec{\imath}} \vdash \forall t.\varPhi(\vec{q}(t), \vec{\imath}(t))$ such that for each natural number $n \in \mathbb{N}$,

$$
\vdash \forall t. \left(\bigwedge_{j=0}^{n} \varPhi(\vec{\Theta}_j^{t,\vec{\imath}}, \vec{\imath}(t+j)) \right) \rightarrow \varPhi(\vec{\Theta}_{n+1}^{t,\vec{\imath}}, \vec{\imath}(t+n+1)) \ \text{is not a tautology.}
$$

Proof. Consider again again $\mathcal{G}_1$ and figure 5. The induction step of the multiple induction rule states here that given a state sequence of length n where all states are different from the state 11, all successor states have to be different from 11. However, this is not the case, since for each $n \in \mathbb{N}$ one can construct a state sequence of length n consisting only of the state 10, which however has 11 as a successor state.

Thus the multiple temporal induction tactic is also incomplete, i.e. there are valid FSM-goals, which can not be proved this way. On the other hand, we can prove that there is always a proof for each valid FSM-goal which uses exactly one induction tactic application. In this proof, a given FSM-goal $\Gamma_{\vec{q},\vec{\imath}} \vdash \forall t.\varPhi(\vec{q}(t), \vec{\imath}(t))$, has to be strengthened, i.e. a goal $\Gamma_{\vec{q},\vec{\imath}} \vdash \forall t.\varPsi(t)$ has to be found such that $\Gamma_{\vec{q},\vec{\imath}} \vdash (\forall t.\varPsi(t)) \rightarrow (\forall t.\varPhi(\vec{q}(t), \vec{\imath}(t)))$ holds. Moreover, $\Gamma_{\vec{q},\vec{\imath}} \vdash \forall t.\varPsi(t)$ must be provable by the application

---

[5] This can be seen if the simple induction tactic is applied on the goal $\Gamma \vdash \forall t. \bigwedge_{j=0}^{n} \varPhi(t+j)$ which is equivalent to $\Gamma \vdash \forall t.\varPhi(t)$ (the induction step has to be simplified).

of one induction tactic. Two examples for these 'intermediate goals' are given in the next section.

6 State Enumeration

Having a closer look on FSM-goals, the task is *not* to show that *each* tuple $\vec{b} \in \mathbb{B}^{|\vec{q}|+|\vec{i}|}$ of boolean values satisfies the right hand side $\Phi(\vec{b})$, but only those tuples $\vec{b} \in \mathbb{B}^{|\vec{q}|+|\vec{i}|}$ which encode states *reachable* from the initial state encoded by $\vec{\omega}_0$. We outline two complete methods which make use of this fact by enumerating the reachable states. The first method computes the reachable states explicitly while the second procedure computes them symbolically. If the reachable states are once computed, a simple propositional proof procedure is sufficient to check whether the property to be proved holds in each state or not.

6.1 Explicit State Enumeration

It is trivial that there is only a finite number of state tuples since the whole set $\mathbb{B}^{|\vec{q}|}$, which is a superset of it, is finite. By defining[6] $S_0 := \{\vec{\omega}_0\}$, $S_{j+1} := \{\mathcal{VAL}^I(\vec{\Omega}(\vec{x}, \vec{y})) \mid \vec{x} \in \mathbb{B}^{|\vec{q}|}, \vec{y} \in \mathbb{B}^{|\vec{i}|}\}$ with an arbitrary propositional interpretation I, one can easily prove that there is a finite number $n_0 \in \mathbb{N}$[7] such that $S_{n_0} = S_{n_0+1}$[8]. Now, let S_{n_0} be $\{\vec{\omega}_0, \ldots, \vec{\omega}_{|S_{n_0}|-1}\}$, then we prove the following two lemmas:

$$L_1 : \Gamma_{\vec{q}, \vec{i}} \vdash \forall t. \bigvee_{j=0}^{|S_{n_0}|-1} (\vec{q}(t) \leftrightarrow \vec{\omega}_j)$$

$$L_2 : \forall t. \bigvee_{j=0}^{|S_{n_0}|-1} (\vec{q}(t) \leftrightarrow \vec{\omega}_j) \vdash \forall t. \Phi(\vec{q}(t), \vec{i}(t))$$

L_1 states that the reachable states are $\vec{\omega}_0, \ldots, \vec{\omega}_{|S_{n_0}|-1}$ and L_2 states that the property which is to be proved holds if $\vec{q}(t)$ is one of those reachable states. In order to prove L_1, first an induction tactic is applied. Then the induction base is simply solved by rewriting it with the assumptions. The induction step is first rewritten by $\Gamma_{\vec{q}, \vec{i}}$ and the the rewrite lemma 3.2 is applied. A schematic proof tree of L_1 is given in figure 6. The remaining goal can then be proved by propositional abstraction.

In order to prove L_2, again the rewrite lemma is used and then propositional abstraction is sufficient for the proof of the remaining subgoal. Therefore, the lemmata L_1 and L_2 have been transformed into the subgoals L_1' and L_2' given below. L_1' and L_2' can both be proved by propositional abstraction.

$$L_1' : \vdash \forall t. \bigwedge_{k=0}^{|S_{n_0}|-1} \bigvee_{j=0}^{|S_{n_0}|-1} (\vec{\Omega}(\vec{\omega}_k, \vec{i}(t)) \leftrightarrow \vec{\omega}_j)$$

$$L_2' : \vdash \forall t. \bigwedge_{j=0}^{|S_{n_0}|-1} \Phi(\vec{\omega}_j, \vec{i}(t))$$

---

[6] Moreover, we conclude: $(1+j) \leq |S_j| \leq \sum_{k=0}^{j} \left(2^{|\vec{i}|}\right)^k \leq 2^{(j+1)|\vec{i}|}$ for $j \in \{1, \ldots, n_0\}$.

[7] This number n_0 is the sequential depth which is also used in the tactic SEQ_TAC.

[8] If $\mathfrak{S}$ is the set of all propositional interpretations, one could alternatively define $S_{j+1} := \{\mathcal{VAL}^I(\vec{\Theta}_j^{0, \vec{i}}) \mid I \in \mathfrak{S}\}$.

$$\Gamma_{\vec{q},\vec{i}} \vdash \forall t. \bigvee_j(\vec{q}(t) \leftrightarrow \vec{\omega}_j)$$

$$\Gamma_{\vec{q},\vec{i}} \vdash \bigvee_j(\vec{q}(0) \leftrightarrow \vec{\omega}_j) \qquad \Gamma_{\vec{q},\vec{i}}, \bigvee_j(\vec{q}(t) \leftrightarrow \vec{\omega}_j) \vdash \bigvee_j(\vec{q}(t+1) \leftrightarrow \vec{\omega}_j)$$

$$\Gamma_{\vec{q},\vec{i}} \vdash \bigvee_j(\vec{\omega}_0 \leftrightarrow \vec{\omega}_j) \qquad \bigvee_j(\vec{q}(t) \leftrightarrow \vec{\omega}_j) \vdash \bigvee_j(\vec{\Omega}(\vec{q}(t),\vec{i}(t)) \leftrightarrow \vec{\omega}_j)$$

$$\vdash \bigwedge_k \bigvee_j(\vec{\Omega}(\vec{\omega}_k,\vec{i}(t)) \leftrightarrow \vec{\omega}_j)$$

Fig. 6. The proof tree of L_1.

Having proven L_1 and L_2, the desired goal can be derived by transitivity of deduction. In the example given in figure 5, the lemmata look like this:

L_1: $\Gamma_{\vec{q},\vec{i}} \vdash \forall t.\, [(p(t) \leftrightarrow F) \wedge (q(t) \leftrightarrow F)] \vee [(p(t) \leftrightarrow F) \wedge (q(t) \leftrightarrow T)]$

L_2: $\forall t.\, [(p(t) \leftrightarrow F) \wedge (q(t) \leftrightarrow F)] \vee [(p(t) \leftrightarrow F) \wedge (q(t) \leftrightarrow T)] \vdash \forall t.\neg(p(t) \wedge q(t))$

It has to be noted that the computation of the reachable states $\{\vec{\omega}_0, \ldots, \vec{\omega}_{|S_{*_0}|-1}\}$ can be done by efficient programs outside HOL and the final proof of lemma L_1 is then done formally in HOL.

6.2 Symbolic State Enumeration

If the number of reachable states is huge, the previous procedure has to prove large lemmata. In order to decrease the size of these lemmata, the same procedure can be used with a symbolic encoding of the reachable states. The lemmata in this symbolical form look as follows:

$$SL_1 : \Gamma_{\vec{q},\vec{i}} \vdash \forall t.\exists \vec{\vartheta}. \bigvee_{j=0}^{n_0}(\vec{q}(t) \leftrightarrow \vec{\Theta}_j^{0,\vec{\vartheta}}) \qquad SL_2 : \forall t.\exists \vec{\vartheta}. \bigvee_{j=0}^{n_0}(\vec{q}(t) \leftrightarrow \vec{\Theta}_j^{0,\vec{\vartheta}}) \vdash \forall t.\Phi(\vec{q}(t),\vec{i}(t))$$

Lemma SL_1 states that if $\vec{q}(0), \vec{q}(1), \ldots$ is the state sequence of a given input sequence $\vec{i}(0), \vec{i}(1), \ldots$ then, for each computation step t, there is a finite input sequence $\vec{\vartheta}(0), \ldots, \vec{\vartheta}(n_0)$, such that $\vec{q}(t)$ equals to one of the state tuples $\vec{\Theta}_0^{0,\vec{\vartheta}}, \ldots, \vec{\Theta}_{n_0}^{0,\vec{\vartheta}}$. Thus, if it has to be checked if a property holds for each infinite state sequence $\vec{q}(0), \vec{q}(1), \ldots$, one simply has to check if it holds for each of the finite number of state tuples $\vec{\Theta}_0^{0,\vec{\vartheta}}, \ldots, \vec{\Theta}_{n_0}^{0,\vec{\vartheta}}$. In the example given in figure 5, the lemmata look like this:

$SL_1 : \Gamma_{\vec{q},\vec{i}} \vdash \forall t.\exists \vartheta_0 \vartheta_1.((p(t) \leftrightarrow F) \wedge (q(t) \leftrightarrow F)) \vee$
$\qquad\qquad ((p(t) \leftrightarrow F) \wedge (q(t) \leftrightarrow \neg\neg(\vartheta_0(0) \wedge \vartheta_1(0))))$

$SL_1 : \forall t.\exists \vartheta_0 \vartheta_1.((p(t) \leftrightarrow F) \wedge (q(t) \leftrightarrow F)) \vee$
$\qquad\qquad ((p(t) \leftrightarrow F) \wedge (q(t) \leftrightarrow \neg\neg(\vartheta_0(0) \wedge \vartheta_1(0)))) \vdash \forall t.\neg(p(t) \wedge q(t))$

These lemmata can be proved similar to the explicit ones in the previous subsection. Similar transformations as outlined there lead to the following two subgoals, where $\vec{\vartheta}$ and $\vec{\xi}$ are new variables:

$$SL'_1 : \vdash \forall \vec{\xi}.\forall t. \bigwedge_{k=0}^{n_0} \exists \vec{\vartheta}. \bigvee_{j=0}^{n_0} (\vec{\Omega}(\vec{\Theta}_k^{0,\vec{\xi}}, \vec{i}(t)) \leftrightarrow \vec{\Theta}_j^{0,\vec{\vartheta}}) \qquad \qquad SL'_2 : \vdash \forall \vec{\vartheta}.\forall t. \bigwedge_{j=0}^{n_0} \Phi(\vec{\Theta}_j^{0,\vec{\vartheta}}, \vec{i}(t))$$

SL'_2 can be proved by propositional abstraction, the proof of SL'_1 is more difficult: After removing the universal quantifiers and the conjunctions, the subgoals $\exists \vec{\vartheta}. \bigvee_{j=0}^{n_0} (\vec{\Omega}(\vec{\Theta}_k^{0,\vec{\xi}}, \vec{i}(t)) \leftrightarrow \vec{\Theta}_j^{0,\vec{\vartheta}})$ can be easily solved for $k < n_0$ by instantiating $\lambda x.(x = k) \Rightarrow \vec{i}(t) \mid \vec{\xi}(x)$ for $\vec{\vartheta}$ and rewriting (recall the definition of $\vec{\Theta}_j^{0,\vec{\vartheta}}$ in lemma 2.1). For $k = n_0$ two further tactics are required: First, one makes use of the fact that the new input sequence $\vec{\vartheta}$ is finite, i.e. $\vec{\vartheta}$ is only applied on $0, \ldots, n_0$. Therefore the quantification can be reduced to a quantification over booleans according to the following tactic:

$$\frac{\exists f : \alpha \rightarrow \beta.\Phi[f(c_0), \ldots, f(c_n)]}{\exists b_0 \ldots b_n : \beta.\Phi[b_0, \ldots, b_n]} \qquad \text{where } c_0, \ldots, c_n \text{ are variable-free terms}$$

Moreover, it is straightforward to eliminate quantification over a finite type τ (such as the type $\mathbb{B}$) which has the constants $c_0, \ldots, c_n$ according to the following theorems:

$$\vdash (\forall x : \tau.\Phi[x]) \leftrightarrow (\bigwedge_{i=0}^{n} \Phi[c_i]) \qquad \qquad \vdash (\exists x : \tau.\Phi[x]) \leftrightarrow (\bigvee_{i=0}^{n} \Phi[c_i])$$

After this, the remaining subgoal can also be proved by propositional abstraction. It should be noted, that the lemma SL'_2 is what has been proposed by SEQ_TAC in the decision procedure in [ScKK93c] (up to α-conversion). Therefore, symbolic state enumeration is an implementation of SEQ_TAC.

7 Existential FSM-goals

Existential FSM-goals $\Gamma_{\vec{q},\vec{i}} \vdash \exists t.\Phi(\vec{q}(t), \vec{i}(t))$ are much easier to prove: Given that n_0 is the sequential depth of $\Gamma_{\vec{q},\vec{i}}$, one only has to 'rewrite' the goal n_0 times with the theorem $\vdash [\exists t : \mathbb{N}.\Psi(t)] \leftrightarrow [\Psi(0) \lor \exists t.\Psi(t+1)]$. This can be visualized as follows:

$$\frac{\Gamma_{\vec{q},\vec{i}} \vdash \exists t.\Phi(\vec{q}(t), \vec{i}(t))}{\Gamma_{\vec{q},\vec{i}} \vdash \left(\bigvee_{j=0}^{n_0} \Phi(\vec{q}(j), \vec{i}(j)) \right) \lor \left(\exists t.\Phi(\vec{q}(t+n_0+1), \vec{i}(t+n_0+1)) \right)}$$

After this, the resulting subgoal is rewritten with the assumption and then the $\exists$-quantified formula can be neglected as well as the assumption. As a result, the goal $\vdash \bigvee_{j=0}^{n_0} \Phi(\vec{\Theta}_j^{0,\vec{i}}, \vec{i}(j))$ is obtained which can again be solved by propositional abstraction.

Otherwise it can be stated, that the goal is definitely not a theorem. Moreover, it should be noted, that existential FSM-goals do never require induction and can therefore always be proven by first-order techniques.

8 Experimental Results

All presented proof tactics have been implemented in HOL90. Table 1 lists the proof times and the number of (safe/unsafe) generated theorems of the temporal induction tactic TEMP_TAC, the explicit state enumeration tactic STATE_ENUM, and the first-order tactic FAUST_TAC. It can be seen that though the temporal induction tactic is incomplete, the tactic was able to prove the correctness of all example circuits. The runtimes of the temporal induction tactic behave well in comparison to the state enumeration tactic. The runtimes of FAUST_TAC are much better than the other ones, but it has to be noted that no validation has been given in HOL.

circuit	goal-type	TEMP_TAC [sec]	[thms]	STATE_ENUM [sec]	[thms]	FAUST_TAC [sec]
ADD2	$\leftrightarrow$	2.57	465/540	0.31	133/147	0.05
BCD_CORRECT	$\rightarrow$	0.50	113/166	1.28	127/544	0.09
C_COUNT	$\rightarrow$	3.03	603/973	3.56	187/1721	0.35
DETECT11	$\rightarrow$	0.04	17/27	0.25	93/163	0.03
DETECT110	$\rightarrow$	0.88	176/271	1.47	141/654	0.13
DMUX	$\rightarrow$	0.49	152/180	0.57	139/214	0.11
JK1	$\leftrightarrow$	0.32	100/164	0.53	117/289	0.10
JK2	$\leftrightarrow$	0.29	100/154	0.51	119/295	0.06
MUX	$\leftrightarrow$	0.34	109/124	0.70	25/31	0.09
PAR_SER	$\rightarrow$	1.31	277/368	10.23	344/2215	??
SPARITY	$\leftrightarrow$	1.66	382/743	1.73	261/770	??
RESET_REG	$\leftrightarrow$	0.27	95/142	0.51	138/263	0.12
SAMPLER	$\rightarrow$	1.10	139/247	31.20	339/10430	0.34
SADDER1_2	$\leftrightarrow$	0.61	124/222	3.22	252/1008	??
SADDER1_3	$\leftrightarrow$	0.42	84/165	4.97	296/1340	??
SADDER2_3	$\leftrightarrow$	0.78	116/225	15.96	436/3691	??
SREG4	$\rightarrow$	2.24	459/809	44.4	415/9689	0.41
TRC_CORRECT	$\rightarrow$	1.03	260/487	6.91	332/2373	??

Table 1. Run times of various circuits

9 Summary and Future Work

In [ScKK93c] we have shown how proof goals which occur in hardware correctness proofs can be transformed to FSM-goals and we have outlined a decision procedure called SEQ_TAC. However, the implementation of SEQ_TAC along the proof of the given decidability theorem is not straightforward and therefore we presented in this paper alternative proof procedures for FSM-goals in HOL. Two classes of proof tactics have been presented: simple methods which are more efficient, but not able to prove each valid FSM-goal and more sophisticated methods which can prove each valid FSM-goal. It is not easy to compare the proof methods with each other. For example, the number of propositional variables in the propositional abstractions is an adequate measure, e.g. the lemmata L'_1 and L'_2 have $O(|\vec{i}|)$ and the symbolical forms SL'_1 and SL'_2 have $O(n_0 \cdot |\vec{i}|)$ variables. However, the length of the formulae is also very important, and it turns out that the length of the lemmata L_1 and L_2 can be exponentially larger than in the symbolical forms SL_1 and SL_2. Moreover, it seems to be reasonable that for circuits whose sequential depth is large in relation to the number of reachable states, the explicit state enumeration technique is more efficient than the symbolical one. Temporal induction is moreover more efficient than the state enumeration techniques as less computations have to be performed, but as already stated, it is possible that no proof will be found that way.

In our future work in this area, we want to extend the methods for dealing with generic circuits. Moreover, we investigate for the relations to symbolic model checking and are integrating time abstraction mechanisms in the MEPHISTO-framework.

References

[BCMD90] J.R. Burch, E.M. Clarke, K.L. McMilian, D.L. Dill, and L.J. Hwang. Symbolic model checking: 10^{20} states and beyond. In *5th Annual Symposium on Logic in Computer Science*, 1990.

[BuCL91] J.R. Burch, E.M. Clarke, and D. E. Long. Representing circuits more efficiently in symbolic model checking. In *28th Design Automation Conference*, pages 403–407, 1991.

[CoBM89] O. Coudert, C. Berthet, and J.C. Madre. Verification of synchronous sequential machines based on symbolic execution. In *Workshop on Automatic Verification Methods for Finite State Systems*, pages 365–373, Grenoble, June 1989.

[JoSe93] J. Joyce and C. H. Seger. Linking BDD-Based Symbolic Evalutation to Interactive Theorem-Proving. In *Proceedings of the 30th Design Automation Conference*, Dallas, Texas, 1993.

[KuSK93] R. Kumar, K. Schneider, and Th. Kropf. Structuring and automating hardware proofs in a higher-order theorem-proving environment. *Journal of Formal Methods in System Design*, 2(2):165–223, 1993.

[Loew92] P. Loewenstein. A formal theory of simulations between infinite automata. In L.J.M. Claesen and M.J.C. Gordon, editors, *Higher Order Logic Theorem Proving and its Applications*, volume A-20 of *IFIP Transactions*, pages 227–246, Leuven, Belgium, 1992. North-Holland.

[ScKK92b] K. Schneider, R. Kumar, and Th. Kropf. Efficient representation and computation of tableau proofs. In L.J.M. Claesen and M.J.C. Gordon, editors, *Higher Order Logic Theorem Proving and its Applications*, volume A-20 of *IFIP Transactions*, pages 39–58, Leuven, Belgium, 1992. North-Holland.

[ScKK93a] K. Schneider, R. Kumar, and Th. Kropf. Hardware verification with first-order BDD's. In *Conference on Computer Hardware Description Languages*, 1993.

[ScKK93c] K. Schneider, R. Kumar, and Th. Kropf. Eliminating higher-order quanitifers to obtain decision procedures for hardware verification. In *International Workshop on Higher-Order Logic Theorem Proving and its Applications*, Vancouver, Canada, August 1993.

A Formalization of Abstraction in LAMBDA *

Anthony McIsaac

Laboratory for the Foundations of Computer Science
The King's Buildings, University of Edinburgh
Edinburgh EH9 3JZ, Scotland

Abstract. In a mixed approach to system verification using theorem provers with an interface to specialized model-checking tools, it may be necessary to simplify models by considering abstract versions of them. We report on work in progress that aims to develop support within LAMBDA for a systematic approach to abstraction. We give a formalization in LAMBDA of a notion of abstraction for transition systems; the abstract systems have two sorts of transition, and are related to specifications in modal process logic. We prove that formulae in the modal mu-calculus are satisfied in an abstract version of a model only if they are satisfied in the model itself. We illustrate how the proof of an inductive step in the verification of a satisfaction relation for an infinite model can be reduced to the verification of a satisfaction relation for a very small finite model.

1 Introduction

General-purpose theorem provers such as those based on higher-order logic have a place in supporting system design and verification, not least because of the wide range of properties that can be expressed in them. But, even with modern sophisticated theorem provers, formal verification is a cumbersome business. On the other hand, for properties expressed in particular formalisms, efficient automated model-checking methods have recently been implemented in tools such as the Concurrency Workbench [5] or those based on BDDs [3]. There is therefore strong reason to investigate approaches that combine the best of both worlds, the flexibility and interactive capabilities of theorem provers with the efficiency of automated specialized packages. Seger and Joyce [12, 8] have constructed a sound interface between the HOL theorem prover and the Voss symbolic trajectory evaluation system, and shown that the hybrid system can be used to good effect.

There is some middle ground between the use of a higher-order logic theorem prover to show that some verification problem expressed in a familiar language can be reduced to a model-checking task expressed in a special formalism, and the passing of this task to a specialized tool. The model may be infinite, or too enormous for the tool. In this case it will be necessary to reduce the model-checking task to questions about simpler models. One technique that can be

* This research was supported by the SERC grant GR/F 31199 'Formal System Design Tools'

used to this end is *abstraction*. This is a matter of replacing a model M with an abstract version A that contains only some of the information in M, but no information that is not in M, so that, roughly speaking, any statement that is true for A is true for M. The use of one such notion of abstraction has been explored by Clarke, Grumberg and Long [6]; we propose a somewhat more general notion. In Section 2, we investigate how this notion can be formalized in higher-order logic, and establish some theorems that can be used in applications of the technique.

We would like to explore an approach to model checking in which information about a model could be used in only two ways: to prove that one model is an abstraction of another (this is to be done within a theorem prover, making use of one's understanding of the structure of the model, and should be as straightforward as possible), and direct checking that a formula holds in a model (this is to be done by an automated tool with an interface to the theorem prover, and may involve combinatorial complexities, but the model must be of manageable size, and certainly finite). An alternative approach would be to use the theorem prover to simulate reasoning in some logic for the model, as is done for example in [11], but we feel that our approach makes a clearer division between different aspects of the task. In section 3, we test this approach on a very simple example.

This paper describes work in progress that aims to develop a systematic, formalized approach to abstraction, as part of a mixed approach to system design and verification, following ideas suggested by Mike Fourman. All the work has been carried out using the higher-order logic theorem prover LAMBDA[2] [7]. We report on how much we have formally proved in LAMBDA, but the simulated typescript in the text is not an exact transcript of our sessions.

2 Abstract Transition Systems

2.1 Formalization of Transition Systems

Our aim is to reduce the problem of checking whether a formula ϕ holds in some model M to that of checking whether some formula ψ holds in some simpler model A; the latter task must be fully automatable, and A must be finite.

The models M here will be taken to be labelled transition systems. A labelled transition system consists of a set S of states, a set K of labels, and a subset T of $S \times S \times K$, where $(x, y, k) \in T$ means that there is a transition from x to y with label k. The language we choose for the formulae ϕ is the modal mu-calculus [9], which can express a wide range of properties using its fixed-point operators. Formulations in other, more immediately comprehensible, logics such as CTL or CTL* [4], can be translated into the modal mu-calculus. Also, systems specified, for example, as Petri nets or in CCS have transition systems associated with them, so we are remaining reasonably general.

In our formalization of the satisfaction relation in LAMBDA, a model has the polymorphic type ('a -> om) * (('a * 'a * 'b) -> om). Here the states

---

[2] LAMBDA is a product of Abstract Hardware Limited

have type 'a and the labels have type 'b. The second component of the model thus defines a subset of $S \times S \times K$ (om is the type of truth values). The first component is necessary because the states might form only a proper subset of the objects of type 'a: this component is the characteristic function of the set of states. The values of the second component at triples (x, y, k), where x and y are not both states, are of no significance. Our formalization contains provisos of the form 'x is a state' at various points; these will be ignored in this account, and from now on we take it that all objects of type 'a are states, and a model has the type ('a * 'a * 'b) -> om. We have chosen to assume that all objects of type 'b are labels.

Given a type 'letters' for propositional variables, our datatype definition for formulae in the modal mu-calculus, with labels of type 'b, is:

```
datatype 'b muformula = Var of letters |
                        Tt |Ff |
                        Both of 'b muformula * 'b muformula |
                        Either of 'b muformula * 'b muformula |
                        Box of ('b -> om) * 'b muformula |
                        Diamond of ('b -> om) * 'b muformula |
                        Nu of letters * 'b muformula |
                        Mu of letters * 'b muformula;
```

Thus sets of labels, not just single labels, appear in the 'box' and 'diamond' formulae. Also, we only consider positive formulae, with no negations. We remark that any closed formula in Kozen's full modal mu-calculus with negations [9] can be rewritten as a positive formula by moving negations inwards, since any occurrence of a variable Z in $\mu Z.F$ must come within the scope of an even number of negations.

It is straightforward to encode the standard definition of the satisfaction relation in LAMBDA as a predicate 'sat', where sat m x f v means that the formula f is satisfied at the state x in the model m, for the valuation v, where a valuation assigns to each propositional variable a set of states. We give the definition in ordinary mathematical language. S is the set of states, and $v[E/c]$ is the valuation which agrees with v except at c, where it has the value E.

sat m x (Var c) $v = x \in v(c)$
sat m x Tt $v = $ TRUE
sat m x Ff $v = $ FALSE
sat m x (Both (f_1, f_2)) $v = $ sat m x f_1 $v \wedge$ sat m x f_2 v
sat m x (Either (f_1, f_2)) $v = $ sat m x f_1 $v \vee$ sat m x f_2 v
sat m x (Box (p, f) $= \forall y, k.\ m(x, y, k) \wedge p(k) \Rightarrow$ sat m y f v
sat m x (Diamond) $(p, f) = \exists y, k.\ m(x, y, k) \wedge p(k) \wedge$ sat m y f v
sat m x (Nu (c, f)) $= \exists E \subseteq S.\ x \in E \wedge \forall y \in E.$ sat m y f $v[E/c]$
sat m x (Mu (c, f)) $= \forall E \subseteq S.\ (\forall y.$ sat m y f $v[E/c] \Rightarrow y \in E) \Rightarrow x \in E$

Bradfield [1] has developed a tool that implements a tableau method [2] for checking whether a formula holds at some set of states in a model; the method applies to infinite models, and cannot be fully automated. We now introduce a way of forming a simpler abstract version of a model, and interpreting formulae with respect to this abstract version, so that checking of the satisfaction relation can be fully automated, and the satisfaction of a formula in the abstract model guarantees its satisfaction in the original model.

2.2 Forming Abstractions

The idea is to partition the set of states of the original model into subsets, and to let these subsets be the states of the abstract model. For two such subsets U and V, and a label k, there are two ways we might interpret the notion of there being a transition from U to V with label k: as 'for all states in U, there is a transition to some state in V with label k', or as 'for some state in U, there is a transition to some state in V with label k'. Clarke, Grumberg and Long [6] take an approach using just the second notion. We may want to use either of these notions, so there will be two sorts of transition in the abstract model.

Such systems with two sorts of arrows have been studied by Larsen and Thomsen [10] in their development of a theory of modal process logic. The states of their systems are intended to be interpreted as specifications of processes; there is a 'must' arrow with label k between states S and T if every process satisfying the specification S can perform the action a and thereby make a transition to some process satisfying the specification T, and there is a 'may' arrow with label k between S and T if there is some process satisfying S that can perform k and thereby make a transition to some process satisfying T. In the *modal transition systems* defined by Larsen and Thomsen, if there is a 'must' arrow between two states, then there is also a 'may' arrow between these states.

The motivation for modal process logic is the need to allow for loose specifications ('don't care'). We will select those aspects of the theory that suit our particular purposes, which have arisen more from the point of view of verification. In merging states to form a simpler model, we lose information ('don't know'); there may also be information we don't need, so don't include in the abstract model ('can't be bothered to know'). We will call our transition systems with two sorts of arrows *abstract transition systems*. The only formal difference between these and the modal transition systems of Larsen and Thomsen is that we do not insist that if there must be a transition from x to y with label k, then there may be a transition from x to y with label k. The only way that this could not be the case is for x to represent an empty set of states in some more concrete model. Of course, x could then never be reached from any other state, so there would be no need to include it in the model. However, for the sake of uniformity for parameterized models, it might be convenient to allow for non-existent states.

So we define an *abstract transition system* as consisting of a set S of states, a set K of labels, and two subsets T_1 and T_2 of $S \times S \times K$. Thus an abstract model has type $(('a * 'a * 'b) \rightarrow om) * (('a * 'a * 'b) \rightarrow om)$. Informally, for an abstract model (m_1, m_2), $m_1(x, y, k)$ means that there must be a possible transition from x to y with label k; and $m_2(x, y, k)$ means that there is permitted to be such a transition — this statement actually conveys no information, but NOT $m_2(x, y, k)$ does put a restriction on the model, namely that there are definitely no possible transitions from x to y with label k.

Bearing in mind the interpretation of an abstract transition system as an abstraction of a concrete system as above, we can define a predicate 'abssat' for satisfaction in abstract models, by replacing 'sat' with 'abssat' throughout the definition of sat, except in the case of the 'box' and 'diamond' formulae, where the definitions are:

$$\text{abssat } (a_1, a_2) \; x \; (\text{Box } (p, f)) \; v = \forall y, k. \; a_2(x, y, k) \wedge p(k) \Rightarrow \text{abssat } (a_1, a_2) \; y \; f \; v$$

$$\text{abssat } (a_1, a_2) \; x \; (\text{Diamond } (p, f)) \; v = \exists y, k. \; a_1(x, y, k) \wedge p(k) \wedge \text{abssat } (a_1, a_2) \; y \; f \; v$$

An ordinary transition system can be converted to an abstract one by duplicating the set of transitions. Thus we define a function

```
fun makeabstract m = (m,m);
```

For two abstract transition systems A_1 and A_2, we can define a notion of A_2 being more abstract than A_1, that is, the information in A_2 being a subset of the information in A_1. There are at least three ways that this could occur:

(i) Some of the states in A_1 could be merged in A_2. For example, if there must be a transition in A_1 from x_1 to y with label k, and there must not be a transition from x_2 to y with label k, and the states of A_2 are the same as those in A_1 except that x_1 and x_2 are merged to a single state x, then all one can say about transitions in A_2 from x to y with label k is that there may or may not be such a transition.

(ii) The value of the first component of the model at some (x, y, k) is changed from TRUE in A_1 to FALSE in A_2. This relaxes the requirement that the transition be possible (without insisting that it be impossible).

(iii) The value of the second component at some (x, y, k) is changed from FALSE in A_1 to TRUE in A_2. This relaxes the requirement that the transition be impossible, without insisting that it be possible.

Forming an abstraction as in (i) will be valuable for us because the result may be a model with far fewer states, so that model-checking is much easier. Forming abstractions as in (ii) and (iii) is also valuable, as we may not need to use all the information available, and it will be safe to use 'default' values of FALSE in the first component, and TRUE in the second. So, when we form an abstraction (a_1, a_2) of a model m, we only set $a_1(U, V, k)$ equal to TRUE if for all states x in U there is a y in V such that $m(x, y, k)$, and we expect this to be relevant to what we want to prove. Similarly, we only set $a_2(U, V, k)$ equal to FALSE if for all states x in U there is no y in V such that $m(x, y, k)$, and we expect this to be relevant. This will reduce the amount of work that needs to be done to prove that one model is an abstraction of another, in the following formal sense. We give the definition of a predicate 'abstraction', where abstraction $m \; a \; g$ means that the abstract model a is an abstraction of the abstract model m using the function g from states of m to states of a, so that the states of a are the sets of states of m on which g has the same value.

```
fun abstraction (m1,m2) (a1,a2) g =
   forall u,v,k.
      a1(u,v,k)
      ->> forall x.
            g x == u ->> exists y. g y == v /\ m1(x,y,k)
   /\
   forall x,y,k.
      m2(x,y,k) ->> a2((g u),(g v),k);
```

Thus a is an abstraction of m if every 'must' transition in a corresponds to a 'must' transition in m, and every 'may' transition in m corresponds to a 'may' transition in a.

There is a close relationship between abstraction as we have defined it and the relation of refinement between specifications as defined by Larsen and Thomsen. Essentially, our abstract transitions systems with the abstraction relation are a special case of modal transition systems: if a model a is an abstraction of a model m in our sense, then the specifications associated with m are refinements of those associated with a. It would be more satisfactory theoretically to allow for a more general relation of abstraction, in line with the refinement relation (which is a generalization of bisimulation). In particular, there could be abstractions not arising from a function between the states of two models. But we want the proof that one model is an abstraction of another to be easy, as it will have to be carried out interactively in LAMBDA, and so we have not gone beyond our present simple notion of abstraction for transition systems.

2.3 Relationships between Models

The main result is that if a formula is satisfied at some state u in some abstraction a of a model m, then it is satisfied at all states of m that map to u. As a formal theorem in LAMBDA:

Theorem 1

```
|- forall m,a,g,x,u,f,v.
   (abstraction m a g) /\ (g x == u) /\ (abssat a u f v)
   ->> abssat m x f (lift g v)
```

Here lift $g \; v$ is the valuation that maps any variable Z to the inverse image under g of $v(Z)$.

The proof is by induction on the structure of the formula f, retaining quantification over x, u, and v. We just consider the cases of fixed-point formulae (all the others are simpler).

Suppose that f is Nu (c, h), and let x be a state of m, and $u = g(x)$ a state of a. Assume that f is satisfied at u in a, for the valuation v. Then there is some subset E of the states of a such that $x \in E$ and h is satisfied at all y in E, for the valuation $v' = v[E/c]$. Let $D = g^{-1}(E)$. By the inductive hypothesis, h is satisfied at all z in D, for the valuation lift(g, v'), which is (lift$(g, v))[D/c]$. Since x is in D, this means that Nu (c, h) is satisfied at x in m, for the valuation lift(g, v). This completes the proof of this inductive case.

There does not seem to be a straightforward proof along these lines for the case where f is Mu (c, h). We give an alternative proof using transfinite induction, for an alternative definition of the satisfaction relation for 'mu' formulae. For ordinals α, we define a satisfaction relation sat$^{\alpha}$ by:

sat^0 m x (Mu (c, f)) v = FALSE
sat$^{\alpha+1}$ m x (Mu (c, f)) v = sat m x f $v[Q/c]$,
where $Q = \{y \mid$ sat$^{\alpha}$ m y (Mu (c, f)) $v\}$

and for limit ordinals β,
sat$^{\beta}$ m x (Mu (c, f)) v = for some $\alpha < \beta$, sat$^{\alpha}$ m x (Mu (c, f)) v

Then sat m x (Mu (c, f)) v = for some ordinal α, sat$^{\alpha}$ m x (Mu(c, f)) v. The definition for abssat is similar.

In proving the inductive case where f is Mu (c, h), we first prove by transfinite induction on α that

$$P(\alpha) : \text{abssat}^{\alpha} \ a \ u \ (\text{Mu} \ (c, h)) \ v \Rightarrow \text{abssat}^{\alpha} \ m \ x \ (\text{Mu} \ (c, h)) \ (\text{lift} \ g \ v)$$

The case $\alpha = 0$ is trivial, and the case of a limit ordinal is also easy. We prove that that $P(\alpha + 1)$ holds, assuming that $P(\alpha)$ holds, and also assuming, using the structural inductive hypothesis, that, for any v, h is satisfied at x for the valuation (lift g v) if it is satisfied at u for the valuation v.

Suppose that abssat$^{\alpha+1}$ a u (Mu (c, h)) v. By definition, sat a u h $v[Q/c]$, where $Q = \{y \mid$ sat$^{\alpha}$ a y (Mu (c, h)) $v\}$. Therefore sat m x h lift$(g$ $v[Q/c])$, that is, sat m x h lift$(g$ $v)[g^{-1}Q/c]$. We need to prove that sat$^{\alpha+1}$ m x (Mu (c, h)) (lift g v), in other words that sat m x h (lift g $v)[R/c]$, where $R = \{y \mid$ sat$^{\alpha}$ m y (Mu (c, h)) $v\}$. This differs from the formula we have obtained only in that $g^{-1}Q$ is replaced by R. Now it is immediate from $P(\alpha)$ that $g^{-1}Q \subseteq R$. Since all the formulae are positive, it can be proved that if a satisfaction relation holds for some valuation v, it still holds if $v(c)$ is increased for some variable c. Thus we do indeed have the result we require, and can conclude that $P(\alpha)$ holds for all α.

It is now easy to complete the proof of the inductive case, for if Mu (c, h) holds at u for v, then abssat$^{\alpha}$ a u (Mu (c, h)) v for some α, and for that α, abssat$^{\alpha}$ m x (Mu (c, h)) (lift g v), so that Mu (c, h) holds at x for (lift g v).

The proof of all the inductive cases except for this last one has been carried out in LAMBDA. Further, the following result has been fully proved in LAMBDA.

Theorem 2

```
|- forall m,x,f,v.
       abssat (makeabstract m) x f v ->> sat m x f v
```

This states that a formula holds at some state in a concrete model if it holds at that state in the abstract version of that model formed by duplicating the set of transitions. Together with Theorem 1, it enables one to reduce the problem of proving that a formula holds in some concrete model M to that of proving that the formula holds in some abstract model A that is an abstraction of the abstract version of M. The proof of Theorem 2 is very straightforward, by induction on the structure of f.

If a formula is not satisfied at some state u in an abstract model, no conclusions about more concrete models can be inferred, and in particular it does not follow that the formula is not satisfied at states x in the more concrete model that map to u. This is one reason why we have not included negation in our version of the modal mu-calculus: Theorem 1 would not remain valid if there were a constructor Neg such that (Neg f) was satisfied precisely when f was not satisfied. It would be possible to define a satisfaction relation for negations of formulae, but in order to understand the meaning of negated formulae for abstract models, it would probably be easiest to convert them explicitly into positive formulae by moving the negation inwards.

If we want to prove assertions of the form 'f_1 is satisfied at x_1 and f_2 is satisfied at x_2', for some concrete model M, then it will be sufficient to prove that f_1 satisfied at u_1 and f_2 is satisfied at u_2 at appropriate states u_1 and u_2 in some abstraction A of M; and similarly if 'and' is replaced by 'or' in the assertion. But the position is different for assertions such as 'f_1 is satisfied at x_1 implies that f_2 is satisfied at x_2', which implicitly involve negation. Such assertions do arise in proofs of inductive steps, and it will be useful to have a theorem which allows us to convert some such assertions into single assertions of the form 'f is satisfied at x', which can then be checked in an abstraction of the model.

A simple version of such a theorem is the following. Let M be a (concrete) model, and x and y states in M. Let f be a formula such that Z is the only free variable in f. Let v be a valuation such that $v(Z)$ is the singleton set $\{x\}$. Then the assertion that f holds at y for the valuation v is equivalent to the following: for all formulae p and valuations w, if p holds at x for v, then $f[p/Z]$ holds at y for w, where $f[p/Z]$ is the formula obtained by substituting p for Z in f.

We have formalized this theorem in LAMBDA, for the sake of simplicity at present only considering formulae f that have no variables at all other than the free variable Z (and so no fixed-point operators). We define a datatype of contexts:

```
datatype 'b context = Hole |
                      CTt| CFf |
                      CBoth of 'b context * 'b context |
                      CEither of 'b context * 'b context |
                      CBox of ('b -> om) * 'b context |
                      CDiamond of ('b -> om) * 'b context;
```

and a function that substitutes a formula p for the hole in a context, defined as indicated here for a few patterns:

```
fun
placeincontext Hole p = p |
placeincontext CTt p = Tt |
placeincontext (CBoth (g,h)) p
               = Both ((placeincontext g p) (placeincontext h p)) |
placeincontext (CDiamond (q,h)) p
               = Diamond (q, (placeincontext h p)) |
....
```

Then:

Theorem 3

```
|- forall m,f,x,y.
   (forall p,v.
      sat m x p v ->> sat m y (placeincontext f p) v)
   ==
   sat m y (placeincontext f (Var Z) [(Z, char x)]
```

(The valuation is implemented as a list of pairs, and char x is the characteristic function of the set $\{x\}$.)

3 An Example

We introduced abstract transition systems to allow for an approach to verification that uses information about a model in one of only two ways:

(i) To prove that some other model is an abstraction of that model;

(ii) To verify some satisfaction relation using a specialized package.

Thus any assertions about infinite models, even proofs of simple inductive steps, must be converted to assertions about abstractions of the models. It is not immediately obvious how this can be done, even for a simple model such as the set of non-negative integers, with transitions from n to $n-1$ for each positive integer n. For this example, we will demonstrate a suitable abstraction that enables us to prove an assertion about termination. The approach would not be recommended for such a simple case, but it does illustrate the techniques that can be used, and show how small is the amount of information that needs to be retained in particular abstract models.

Consider the labelled transition system M with the set N of natural numbers as its states, and the singleton set $\{L\}$ as its labels. Its transitions are the triples (x, y, L) for which $x = y + 1$. Thus, from any state, after some finite number of transitions with label L, the system has counted down to zero, and there can be no further transitions. This is the assertion that we have formalized and proved in LAMBDA, using Theorems 1, 2, and 3.

We define **trueatL** to be TRUE at the single element L of the type of labels, and we define a function **stopsformula** on the natural numbers by

```
fun stopsformula 0 = Ff |
    stopsformula 1'n = Box (trueatL, (stopsformula n));
```

($1'$ is the successor function.) Thus, **stopsformula** n would be written as $[\{L\}]^n\text{ff}$ in a more common notation. We call the model **countdown**. We set the goal

```
|- forall n. exists t. sat countdown n (stopsformula t)  □
```

We will need to create abstract versions of **countdown**; we propose the following abstract model, parameterized by a natural number n. There are four states, IN, LAST, THIS, and OUT. (For each n, we plan to map n in **countdown** to THIS, $n+1$ to LAST, all numbers greater than $n+1$ to IN, and all other numbers to OUT.) We will not be interested in transitions from IN and OUT, so we include all triples (IN,y, L) and (OUT,y, L) as 'may' transitions, and none of these as 'must' transitions. There are both 'may' and 'must' transitions $(\text{LAST},\text{THIS},L)$, and no other transitions from LAST. The transitions from THIS to OUT are the only ones that depend on the value of n: there are both 'may' and 'must' transitions if $n > 0$, and no transitions if $n = 0$. There are no other transitions from THIS. So the abstract model could be viewed as

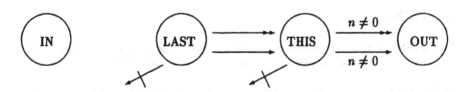

Here the single-headed arrows indicate 'may' transitions and the double-headed arrows 'must' transitions. The crossed arrows mean that there are no 'may' transitions except where specifically indicated.

We have indicated how **countdown** is to be mapped to the abstract model for each value of n. We can define **absfn** n as the parameterized function that effects the abstraction, and we name the parameterized abstract models **fourstates** n. We can now prove the crucial lemma:

Lemma 1

> `|- forall n. abstraction (makeabstract countdown) (fourstates n)`
> `(absfn n)`

The proof is painless, as the abstract models are so simple. It is also useful to record the following fact as an easy lemma:

Lemma 2

> `|- forall n. absfn n n == THIS`

We now proceed towards proving our goal. The first thing is to apply induction: we consider just the proof of the inductive step (the case $n = 0$ is simpler). After a little rearrangement we have the subgoal

> `|- sat countdown n' (stopsformula t') []`
> `->> sat stopsformula (1'n') (stopsformula t1) []`

Here n' and t' are rigid variables; $t1$ is a flexible variable, which we can instantiate as we choose. Since the number of successive transitions that can be made from $(1'n')$ is one more than the number than can be made from n, it is sensible to set $t1 = 1't'$. We can then rewrite using the definition of **stopsformula** to obtain the subgoal

> `|- sat countdown n' (stopsformula t') []`
> `->> sat countdown (1'n') (Box (trueatL, stopsformula t')) []`

At this stage we use Theorem 3. We can introduce the conclusion of Theorem 3 as a hypothesis, and instantiate f in the theorem as **CBox (trueatL, Hole)**. As things stand, this has to be done more or less explicitly, but it would be possible to write tactics to make the instantiation, together with rewriting using the definition of **placeincontext**. It is now straightforward to reduce the goal to

> `|- sat countdown (1'n') (Box (trueatL, Var Z)) [(Z, char n')]`

Using Theorem 2, and then Theorem 1 together with Lemmas 1 and 2, this is reduced to

> `|- abssat (fourstates n') BEFORE (Box (trueatL, Var Z))`
> `                                        [(Z, char THIS)]`

It would be no trouble to prove this using the definition of the abstract models, but we consider that our task is finished at this stage. We just have to check some formula in a finite model, and we are assuming that we have some automated tool to do this. The fact that the model is parameterized is not a problem: in general, we would expect to be able to split into a small number of cases depending on the value of n.

4 Conclusions

We have formalized a simple notion of abstraction for transition systems, and indicated how it can be used to reduce model-checking tasks to ones that can be performed by fully automated tools. Our abstraction technique is quite general, and could be applied to more complex systems than the very simple one we have examined. Whether it can substantially reduce the amount of work that has to be done within LAMBDA for more complex systems remains to be seen. For such systems, it is likely that one will want to use a mixture of techniques. Our demonstration that our technique is viable in principle indicates that it may have a place in a mixed approach to verification tasks, and encourages one to think that it is worth devoting serious effort to building systems that support such an approach.

References

1. J.C. Bradfield. A proof assistant for symbolic model checking. LFCS Report Series ECS-LFCS-92-199, Laboratory for the Foundations of Computer Science, University of Edinburgh, March 1992.
2. J.C. Bradfield and Colin Stirling. Local model checking for infinite state spaces. *Theoretical Computer Science*, 96:157–174, 1992.
3. Randal E. Bryant. Graph-based algorithms for boolean function manipulation. *IEEE Transactions on Computers*, C-35(8):677–691, August 1986.
4. E.M. Clarke, E. Emerson, and A.P. Sistla. Automatic verification of finite state concurrent systems using temporal logic specifications. *ACM Transactions on Programming Languages and Systems*, 8(2):244–263, April 1986.
5. Rance Cleaveland, Joachim Parrow, and Bernhard Steffen. The Concurrency Workbench. In J. Sifakis, editor, *Automatic Verification Methods for Finite State Systems*, pages 24–37. Springer-Verlag, 1989. Lecture Notes in Computer Science 407.
6. E.M.Clarke, O.Grumberg, and D.E.Long. Model checking and abstraction. In *Proceedings of the 19th Annual ACM Symposium on Principles of Programming Languages*, 1992.
7. Mick Francis, Simon Finn, Ellie Mayger, and Roger B. Hughes. *Reference Manual for the Lambda System*. Abstract Hardware Limited, Version 4.2.1 edition, 1992.
8. Jeffrey J. Joyce and Carl-Johan H. Seger. Linking BDD-based symbolic evaluation to interactive theorem-proving. In *Proceedings of the 30th Design Automation Conference*. To appear.
9. D. Kozen. Results on the propositional mu-calculus. *Theoretical Computer Science*, 27:333–354, 1983.
10. Kim G. Larsen and Bent Thomsen. A modal process logic. In *Proceedings of the Third Annual Symposium on Logic in Computer Science*, pages 203–210, 1988.
11. Monica Nesi. Formalising a modal logic for CCS in the HOL theorem prover. In Luc Claesen and Michael Gordon, editors, *Higher Order Logic Theorem Proving and its Applications, Leuven, 1992*, pages 279–294. North-Holland, 1993.
12. Carl-Johan Seger and Jeffrey J. Joyce. A mathematically precise two-level formal verification methodology. Report 92-34, Department of Computer Science, University of British Columbia, December 1992.

Report on the UCD Microcoded Viper Verification Project*

Tej Arora, Tony Leung, Karl Levitt, Thomas Schubert**, and Phillip Windley

Systems Verification Laboratory
Department of Computer Science
University of California, Davis

Abstract. The formal verification of a microprocessor involves demonstrating that a *specification* of the microprocessor is satisfied by its *implementation*. The incomplete proof of Viper (mostly because it became too time consuming) is of particular interest. Our view of the incomplete proof is that the jump in abstraction between the electronic block model and the specification is too great. By introducing intermediate levels between the two extreme models, the overall proof becomes one of establishing more, but simpler, proofs.

We present our effort to apply Windley's generic interpreter model to a microcoded version of Viper. We redesigned Viper as a hierarchy of five interpreters, each of which is an instance of the generic interpreter. The top level specifies the Viper instruction set and the lowest level specifies an abstraction of a conventional electronic block model. The design and verification was carried out in approximately one person-year.

1 Introduction

Computers are being used with increasing frequency in areas where the correct implementation of the computer hardware is critical. These include safety-critical applications, security–critical applications, as well as mass produced consumer goods. In these and other applications, it is vital that the computer system be correct. Although the verification of large programs is beyond the capability of current verification technology, the verification of commercial microprocessors should be realistic. Our justifications for being optimistic about microprocessor verification are as follows:

- The specification for a microprocessor is not difficult to produce, largely expressing the functional behavior of each instruction.
- The implementation for many microprocessors is conceptually straightforward, largely involving iterative structures (such as registers) and control logic to resolve the many different cases. The algorithms represented by the

---

* This work was sponsored under Boeing Contract NAS1-18586, Task Assignment No. 3, with NASA-Langley Research Center.
** Corresponding author may be reached at: Department of Computer Science, Portland State University, P.O. Box 751, Portland, OR 97006, USA

implementation, even for arithmetic, are usually extremely simple compared with those associated with programs.

However, the detail involved in microprocessor proofs rapidly becomes staggering. This was the experience of Avra Cohn in attempting to verify Viper.

1.1 Viper

Viper was designed by RSRE [5] in the mid-1980's. Not intended by its designers to push the envelope of microprocessor design, Viper was designed to be simple and verifiable. Of interest to us here, are the attempts to verify Viper, in particular [3]. The top–level specification defines the **NEXT** state as a function of the current state and the current instruction. The elements of the state are main memory, five registers, and a few status bits—abstracting away a large fraction of the implementation state. The implementation is described in terms of logical blocks (ALU, registers, flip-flops, multiplexors, etc.). Both the specification and the electronic block model were provided to Cohn by RSRE. The proof was to demonstrate that the electronic block model implies the specification.

Cohn's work represents a significant contribution, having formalized the electronic block model in HOL and having developed a methodology and many lemmas that could be used to carry out the proof. However, the proof was not completed. As it progressed, it became clear that approximately 1 person-week was required to prove the implementation of each of the 128 cases in the specification. The difficulty was due to a number of factors, including:

1. RSRE's specification is extremely unstructured; essentially it is almost totally non-orthogonal. Although not conceptually difficult, the specification is quite more unstructured than what is normally expected of the instruction set architecture for a computer with the instruction set power of Viper.
2. Although not particularly complicated as compared with state-of-the-art microprocessors, the implementation is still quite long. If this were a program being verified, by all measures it would be of nontrivial length.
3. The jump in abstraction between the specification and the electronic block model is too large to be carried out in one step.
4. There was insufficient support in HOL for the kinds of low level reasoning associated with words, bit strings, etc.

Item (3) is of particular concern. We conjectured that through intermediate abstractions, the proof effort required for Viper could be simplified to the point where it would be realistic. It is still necessary to verify the lowest level of abstraction, defined in seven pages of HOL logic, ultimately with respect to the highest level of abstraction, occupying three pages of specification representing 128 cases. However, if the next to the lowest level of abstraction has fewer cases, the lowest level will be easier to verify. Similarly, if the next-to-highest level of abstraction is shorter, it will be relatively easy to verify with respect to the specification. The handcrafting of levels of abstractions is what is needed to simplify the verification of complex systems. In creating these abstractions, there will be tradeoffs among the number of cases, the size of the abstraction's specifications, and the jump in data abstraction between adjacent abstractions.

The specification of the electronic block model of our Viper machine is simpler than that of Cohn's, with respect to omitted details not pertinent to our proof. For example, we do not specify in detail the logic of the ALU; instead it is declared to perform one of 32 unspecified functions. This incompleteness, of course, appears at all levels, including the top level. As noted by Brock and Hunt [2] with respect to a similar, but less glaring weakness in the RSRE specifications of Viper, the top-level specification does not permit proofs of programs that depend on the semantics of these operations to be carried out. However, the incompleteness in the electronic block model is not relevant to the main purpose of our verification effort: to verify that the sequence of actions at the electronic block model assure (among many other things) that the correct ALU control lines are asserted with respect to the instructions under execution.

Viper has many more features that make it suitable for use in safety-critical applications, but are not modeled at the top-level. These include input signals for resetting the machine, single-stepping, forcing the machine into an error state, and extending read/write cycles. Output signals are also provided to indicate the state of the STOP and B flags, and whether the machine is currently fetching or executing an instruction. Viper also incorporates a time-out facility in its interaction with the memory. Because these features are inconsequential to the top-level specification, they can safely be ignored in the block level specification. However, for the purpose of verification with respect to the top-level instructions, certain assumptions about the behavior of these signals must be made. For example, the reset signal is assumed to be false throughout the execution of an instruction and the STOP flag is assumed to be false at the beginning of an instruction. In addition, a simple memory model in which memory responds in a fixed and known number of cycles is being assumed, although the design of Viper supports more complex memory protocols.

1.2 Hierarchical Decomposition

Viewing a complex program as a hierarchy of abstractions is a well-known approach to simplifying the verification of such a system. Hierarchical decomposition can lead to significant reductions in the amount of effort used to structure and complete a correctness proof.

Verification requires at least two formal descriptions of the computer system: one behavioral, $\mathbf{B}$, and one structural, $\mathbf{S}$. Verification consists of showing through formal proof techniques that

$$\mathbf{S} \Rightarrow \mathbf{B}.$$

One need not be limited, of course, to one level of abstraction. Supposing that $\mathbf{B}_1$ through $\mathbf{B}_n$ represent increasingly abstract specifications of the system's behavior, one could verify its correctness by proving:

$$\mathbf{S} \Rightarrow \mathbf{B}_1 \Rightarrow \ldots \Rightarrow \mathbf{B}_n.$$

In order to use abstraction in the design and verification of microprocessors, Windley formalized the concept of interpreters [9]. Figure 1 shows how this principle can be applied to the specification of a microprogrammed microprocessor. At the bottom of the hierarchy is the usual structural specification of the electronic block model.

This specification describes the computer's implementation—for our purpose, the connections among its various components. At the top is the behavioral specification corresponding to the programmer's model of the microprocessor. In between these are two additional abstraction levels: one for the microcode interpreter and one specifying the phase (or subcycle) behavior. Our Viper design has two macro levels: the topmost is the RSRE specification and the next lower specifies an orthogonal instruction set containing 20 instructions.

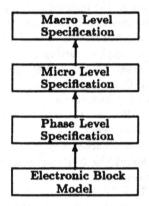

Fig. 1. A microprocessor specification can be decomposed hierarchically.

Generic Interpreters With one exception, each of the levels in the specification hierarchy shown in Figure 1 has the same structure. The bottom level specification is a structural description, but the other specifications all share a common structure. Each of the abstract behavioral descriptions can be specified using an *interpreter model*. In addition, our level that corresponds to the RSRE instruction set does not fit exactly our interpreter model.

Since each of the behavioral descriptions in the specification hierarchy are similar, we utilize a general model of an interpreter and use this model in our specification rather than treating each level in the hierarchy separately.

A generic interpreter specification consists of a number of parts: abstract state, instructions, selectors for instructions, mapping to next lower state, description of implementation, etc. To verify the instantiation of a generic interpreter involves the verification of *obligations*, the most difficult of which is that the each instruction is correctly implemented.

1.3 What we have accomplished vis-a-vis Viper

Our goal was to show that through the use of the generic interpreter methodology, a microprocessor as complex as Viper could be verified. Since Viper was
not designed as a hierarchy of interpreters, the RSRE Viper design could not
be verified using this methodology. Hence, we designed a microprocessor that
would realize the Viper instruction set as specified by RSRE. The design is in
terms of the five levels of abstraction, as follows:

1. Viper instruction level: the RSRE specification with a few minor simplifications. This is what the assembly-language programmer sees. In the RSRE
   specification, all functions (except a few arithmetic functions) are defined. In
   our specification some functions (such as the comparison of two words) are
   uninterpreted. The exact meaning of functions used to define the instructions is not relevant to a proof that shows that the appropriate ALU signals
   are asserted for each instruction, and that operands are fetched from and
   stored to the specified locations.
2. Macro Level: the high-level Viper specification as an interpreter; it consists
   of 20 instructions (the top level of figure 1). This level, as opposed to the
   RSRE specification, represents the Viper instruction set in terms of com-
   paratively few instructions with orthogonal fields. It is emphasized that this
   level is equivalent in power to the RSRE specification. It was necessary to
   demonstrate that this level realizes the RSRE specification at level (1).
3. Micro Level: this level provides approximately 128 microinstructions. Each
   macro instruction is implemented as a linear (loop-free) sequence of a subset
   of the microinstructions.
4. Phase Level: this level implements each micro-instruction in a sequence of 3
   phases.
5. Electronic Block Model level: the control structure and datapaths that implement each of the phases.

Our experience has convinced us that the generic methodology has simplified
the proof effort by half, as compared with Cohn's experience. Furthermore, the
use of hierarchical abstractions has permitted us to divide up the proof.

As Cohn has noted it is important to emphasize what it means to verify a
microprocessor—and what has not been verified. Our proof demonstrates that
our Electronic Block Model implements the RSRE instruction set. It is important
to note that the ALU is a component of the Electronic Block Model. But having
just specifications for the ALU without implementation, means that we are not
verifying that the ALU when stimulated with signals that are assumed to cause
it to add two numbers, actually does carry out the add operation. Of course, we
could carry out the verification down to the gate-level—and verify the ALU, de-
coders, flip-flops, registers—and the other components taken as primitives of the
Electronic Block Model. Such proofs are within current verification capabilities
and in fact have been performed routinely by many verification teams.

To summarize, our verification shows the following: for each instruction of the
RSRE specifications, the Electronic Block Model causes the proper sequencing

of actions to take place, the operands are fetched from the right place (registers or memory), the results are stored in the right place, and the right signals are asserted on the primitive functional units (such as the ALU). Since there are many ways the Electronic Block Model could sequence activities (most of them incorrect) what is verified is far from trivial.

1.4 Related Microprocessor Verification Efforts

There have been numerous efforts to verify microprocessors. Many of these have used the same implicit behavioral model. In general, the model uses a state transition system to describe the microprocessor. The microprocessor specification has four important parts:

1. A representation of the state, S. This representation varies depending on the verification system being used.
2. A set of state transition functions, J, denoting the behavior of the individual instructions of the microprocessor. Each of these functions takes the state defined in step (a) as an argument and returns the state updated in some meaningful way.
3. A selection function, N, that selects a function from the set J according to the current state.
4. A predicate, I, relating the state at time $t+1$ to the state at time t by means of J and N.

In some cases, the individual state transition functions, J, and the selection function, N, are combined to form one large state transition function. Also, a functional specification would use a function for part (4) instead of a predicate. The specifications, however, are largely the same.

After the microprocessor has been specified, we can verify that a machine description, M, implements it by showing

$$\forall s \in S \; M(s) \Rightarrow I(s).$$

That is, I has the same effect on the state, s, that M does. This theorem is typically shown by case analysis on the instructions in J by establishing the following lemma:

$$\forall j \in J \; M(s) \Rightarrow (\forall t : \text{time } C(j, s, t) \Rightarrow s(t + n_j) = j(s(t)))$$

where C is a predicate expressing the conditions for instruction j's selection, $s(t)$ is the state at time t, and n_j is the number of cycles that it takes to execute j. This lemma says that if an instruction j is selected, then applying j to the current state yields the state that results by letting the implementing interpreter M run for n_j cycles. We call this lemma the instruction correctness lemma.

Table 1 summarizes the designs of the four other verified microprocessors. The table, like all such tabulations, cannot hope to capture all of the important characteristics of the microprocessors, but the data presented does provide some basis for judging relative complexities.

Table 1. Comparison of verified microprocessors

	Tamarack	FM8501	Viper	SECD
User Registers	2	8	4	4
Instructions	8	26	20	21
Microcoded	yes	yes	no	yes
Microstore size	32 words	16 words	N/A	512 words
Interrupts	yes	no	no	no
Memory Model	async	async	sync	sync
Word Width	16-bit	16-bit	32-bit	32-bit
Memory Size	8K	64K	1M	16K

2 The Five-level Structure of our Viper Implementation

The following sections describe the architecture of each of the hierarchical levels and summarize the proof strategy used to verify Viper. The hierarchical decomposition approach uses five levels described in section 1.3.

2.1 Viper Instruction Level

Viper's high-level architecture consists of three general purpose 32-bit registers (A, X and Y), a 20-bit program counter (P), and a single-bit boolean register (B) that holds the results of comparison instructions. There is also a STOP flag that is not accessible to a programmer, but indicates an error condition in the machine. Any illegal operation, arithmetic overflow or computation of an illegal address causes the STOP flag to be set.

The address space is divided into a memory space and a peripheral space each addressed by 20 bits. A memory address is 20 bits, but the memory itself has 32-bit words. Only the least significant 20 bits of the program counter are meaningful and loading a '1' into any of the top 12 bits will cause the machine to halt (viz., the STOP flag becomes true).

An instruction word is 32 bits long and consists of an operation code in the most significant 12 bits and a 20-bit address. The address field is also used as an offset or constant by some instructions. The opcode is further subdivided as shown in figure 2. The opcode subfields are not orthogonal and are interdependent in an intricate way.

2.2 The Macro Level

Although the 12 opcode bits allow 4096 possible instructions, many of the combinations have redundant subfields, or represent impossible conditions, so that there are only 256 unique possibilities. These possibilities can be supported by 20 distinct instructions; falling into six categories: shifts, comparisons, arithmetic and logical operations, procedure calls, memory read/writes and input/output instructions (see tables 2 and 3).

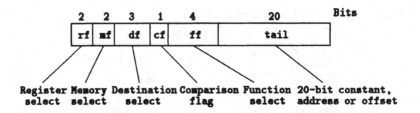

Fig. 2. Viper Instruction Format

For example, the SHLS instruction is one of 20 instructions in our macro level. If the stop field is set, there is no state change. The new value for the program counter is computed by adding 1 to the current value. If the address is invalid, the stop field is set. Otherwise, the register to be shifted is determined, and the shift performed. Finally, the shifted result is written to the appropriate register and the overflow bit is set if appropriate.

To verify that the macro-level realizes the Viper instruction level, it is necessary to map each of the 20 macroinstructions to the 12 opcode bits of the Viper level. A decoder function is introduced that maps the 12 opcode bits into a 5 bit instruction field (for 20 instructions) and (nearly) orthogonal fields corresponding to source register select (2 bits), memory mode select (2 bits) and destination register select (2 bits). The comparison flag and function select fields of the Viper instruction level are not needed at the macro level.

2.3 Micro Level

Our proof of Viper is based on a micro-coded design in order to be able specify Viper as a hierarchy of interpreters using the paradigm described in [9]. As a result, we are able to take advantage of the proof simplification afforded by this method. Each macro level instruction is implemented by a series of microinstructions. For example, the microinstruction trace for the SHLS instruction is illustrated in table 4.

The microprogram that implements the SHLS instruction uses 10 of the approximately 128 microinstructions supported by the micro level. Many instructions use the same microinstructions, e.g. for fetching instructions, incrementing the program counter, etc. The microinstruction AXY_WRITE assures that the destination register is one of a, x, y. For this instruction, the destination cannot be the program counter. The microinstruction SHLS_n1 performs the actual shift and the write to the destination register.

2.4 Phase Level

The phase level, although it is the lowest level interpreter in the hierarchy, is more properly considered to be equivalent to the EBM level, rather than an abstraction of it. In particular, the phase and EBM levels share the same state

Table 2. Viper macroinstructions

Mnemonic	Operands	Effect
NOOP	dreg, sreg	No operation
SHRS	dreg, sreg	dreg := sreg shifted right (copy sign bit)
SHRB	dreg, sreg	dreg:= sreg shifted right through B
SHLS	dreg, sreg	dreg := sreg shifted left; STOP := overflow
SHLB	dreg, sreg	dreg := sreg shifted left through B
COMPARE	ff, sreg, m	compare sreg and m, depending on ff
ADDB	dreg, sreg, m	dreg := sreg + m; B := carry
ADDS	dreg, sreg, m	dreg := sreg + m; STOP := overflow
SUBB	dreg, sreg, m	dreg := sreg - m; B := borrow
SUBS	dreg, sreg, m	dreg := sreg - m; STOP := overflow
NEG	dreg, m	dreg := -m
ANDM	dreg, sreg, m	dreg := sreg AND m
NOR	dreg, sreg, m	dreg := sreg NOR m
XOR	dreg, sreg, m	dreg := sreg XOR m
ANDMBAR	dreg, sreg, m	dreg := sreg AND m-complement
CALL	m	Y := P; P := m
WRITEMEM	sreg, addr	mem[addr] := sreg
READMEM	dreg, mem	dreg := m (from memory space)
WRITEIO	sreg, addr	io[addr] := sreg
READIO	dreg, mem	dreg := m (from io space)

Table 3. Decoding operand fields

sreg = source register (one of A, X, Y, P)
dreg = destination register (one of A, X, Y, P)
STOP = flag which indicates machine has stopped
B = flag set by comparison operators and if overflow occurs

m=		addr =	
tail	if mf=0	tail	if mf=1
(tail)	if mf=1	tail+X	if mf=2
(tail+X)	if mf=2	tail+Y	if mf=3
(tail+Y)	if mf=3		

and clock. Each phase in the system clock is associated with an instruction in the phase-level interpreter. The inputs to the phase-level interpreter consist of a bit-translation of the microinstructions defined for the micro level. In this way, the phase-level interpreter implements the micro-level interpreter.

Each microinstruction requires the execution of three phase cycles. The specification for the phase level has a separate definition for each of the phase cycles. The events that occur during each phase are described in section 2.5.

The phase level state consists of a list of general purpose registers (including a, x, y, p and others), registers to hold temporary results, the current instruc-

Table 4. Microinstruction sequence for SHLS

Cycle	uCode	uLoc	Comment
t	fetch_u1	0	fetch macro instruction
$t+1$	fetch_u2	1	increment pc
$t+2$	fetch_u3	2	invalid address (> 20 bits)?
$t+3$	fetch_u4	3	ir ← macro instruction
$t+4$	jmp_reqm	4	require memory?
$t+5$	jmp_opc	5	jump to noop+instruction number
$t+6$	AXY_WRITE	10	destination must be register A, X or Y
$t+7$	SHLS_u1	11	shls operation
$t+8$	NO_OVL	12	result must not overflow
$t+9$	NOOP	13	jump to fetch next macro instruction

tion, data in and data out to memory (or I/O), the memory, b and **stop** bits, the memory address register and a result register for the ALU, the microprogram counter, the microinstruction register, the micro-ROM contents, 2 latches, and phase bits (to indicate the current and next phases). If the **stop** bit is set there is no state change, except to indicate there is no next phase. Otherwise, the contents of the micro-ROM, as defined by the microprogram, are fetched and control proceeds to phase 2. The other phases are similar, but much more complex, due to the complexity of the steps performed.

2.5 Electronic Block Level

The Electronic Block Model of Viper used in the proof differs from the original RSRE design in several ways. In addition to being microcoded, the external interface design does not include certain input and output signals that have no effect with regard to the top-level specification. These signals were also ignored in Cohn's proof effort [4].

The Data Path The data path (see figure 3) consists of the registers at the phase level in addition to a few others that are used as internal scratchpad registers: an instruction register (**INS**), a temporary register used in operand computation (**M**), a register holding the numerical constant '1' (**ONE**). Register values may be feed into a 32-bit **ALU** via one of two internal buses (the **m** or **r** buses). Registers can be loaded with either the ALU result or the word fetched from memory (**DIN**). The least-significant 20 bits of P and **INS** can also be output to the **MAR** input bus. The overflow and result of an operation are fed into both the register block and the micro-sequencing logic unit, which sets the STOP flag when an invalid result is generated in some contexts.

To communicate with memory, there is a 20-bit memory address register **MAR**, and two 32-bit data registers **DIN** and **DOUT**. The **MAR** can be loaded in parallel with an ALU operation. The **MAR** and **DIN** registers are loaded only if the **r** signal is set, and **DOUT** is loaded only if the **w** signal is set.

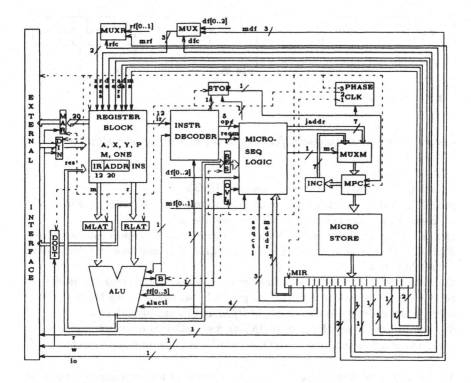

Fig. 3. UCD Viper EBM

The instruction decoder unit takes in 12 bits of opcode from the I**N**S register and the B flag, and sets the STOP flag if the opcode is illegal. Otherwise, it generates a condensed opcode. It also generates a signal, r**e**q**m**, that denotes whether or not the instruction requires computation of an operand. This information is used by the microcode for branching purposes. The STOP flag is set by both the instruction decoder and the micro-sequencing logic units. The machine may halt for two reasons: illegal instruction format (four static error cases) and illegal operations during instruction execution (six dynamic error cases).

The Control Unit

Microinstruction Format: A microinstruction is 31 bits long. Its format is as shown in figure 4. The interpretation of the microinstruction fields is given below.

maddr: address in the microcode, 7 bits.
seqctl: 3 control lines for the micro sequencing logic. The interpretation of this
    field determines the address of the next microinstruction.
aluctl: 4 control lines for the ALU.

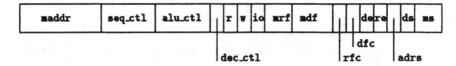

Fig. 4. Microinstruction Format

dec_ctl: control line to disable/enable the instruction decoder stop output.
r: read signal.
w: write signal.
io: read/write from io (if true) or memory (if false).
mdf: destination select for alu result (for intermediate operations); 3 lines.
mrf: source register select (for intermediate operations).
rfc: MUXR control line to decide which of rf/mrf is used to select source register.
dfc: MUXD control line to decide which of df/mdf is used to select destination of alu result.
de: data enable, to enable data from memory to be written into reg block.
re: res enable, to enable the ALU output to be written into reg block.
adrs: address select, to choose one of P/ADDR to put out address.
ds: data select, to select one of M/INS as destination of data from mem/io.
ms: m select, to select one of M/ONE/ADDR to come out on the m bus.

Microinstruction Specification: The state relevant to a microinstruction is that of the micro level: the list of general purpose registers, the temporary (m), instruction, data input and data output registers, the memory, the overflow and stop bits, the memory address register, the (ALU) result register, the microprogram counter, and the reset bit. The RSF field determines the source field—the register whose contents are to be shifted. Assuming the stop bit is not set, the register determined by the DSF field receives the shifted contents of the source register, and the microprogram counter is incremented. All other state variables are unaffected.

Microinstruction Timing: Each microcycle is composed of three phase cycles, and the net effect of a microinstruction is an accumulation of effects of the three phases in sequence. Briefly, the events during each of the phases are as follows:

1. Load the next microinstruction to be executed into the microinstruction register MIR.
2. Gate the register values into MLATCH and RLATCH. Load MAR with P or ADDR if r (read signal) is true. Load DOUT if w (write signal) is true. Set the STOP flag if either of the two stop conditions is true.
3. Load DIN with the value from memory if the read signal is true. Load the ALU result of data from memory into the register block. Load MPC with the address of the next microinstruction. Load RES and OVL with the ALU result and ALU overflow, respectively.

3 Proof Methodology

The basis of this verification is the use of an abstract representation of functions and a generic model for interpreters. These two methodologies provide a way to separate critical control aspects from unnecessary details of concrete data operations.

Even when using the interpreter model to organize the proof effort, the verification of the RSRE Viper micro-processor still involved a large number of cases to be verified, each of them very large. As explained previously, we have solved this problem by designing the architecture of the processor as five levels.

The interpreter model is used in all the proof levels. For example, at one level we consider the instantiation of the interpreter where the instruction list consists of the macro-instructions and the implementation is given by the micro-code. At another level, there is another instantiation of the generic processor, this time with the instruction set containing the micro-instructions and the implementation consisting of the phase level description of the architecture.

The details may be found in [1]. Each of the proofs consists of specifying the instruction set and the implementation, proving all the numerous lemmas—one for each instruction—which constitute the proof obligations, and then instantiating the proofs of correspondence for that level.

4 Conclusions

This task was initiated because previous attempts to verify the design of the Viper microprocessor using mechanical theorem provers were not completed. Since Cohn's incomplete verification effort was published in its entirety, we had the opportunity to attempt to determine why it was so difficult to complete. One reason is the large jump in abstraction between the instruction specification and the implementation. The second reason is the complexity of the specification itself. Many machines have clearly identified instructions with orthogonal fields to define addressing modes, register selection, etc. This is not the case for Viper. Thus, although the instruction architecture is not complex, 256 unique cases must be separately considered in verifying the implementation.

Based on the success Windley achieved using a hierarchical methodology to verify a simpler microprocessor (AVM-1), we decided to apply the methodology to Viper. Windley's methodology depends on viewing the design of a microprocessor as a hierarchy of interpreters, the topmost providing the abstraction of the instructions accessible to the assembly language programmer and the lowest the implementation that is to be verified. To address the issue of the complexity of the specification, we introduced a level below the Viper instruction level which provides the functionality of Viper but in terms of 20 orthogonal instructions.

Our verification demonstrates the following: corresponding to a Viper object program instruction occupying the 12-bit opcode field, the logic of the electronic block model is such that the correct ALU function will be invoked, the arguments (if any) will be drawn from the correct register and main memory locations, the results (if any) will be stored in the correct register (and flag bit) or main memory location, and the program counter will be correctly updated (incremented by one or set to the correct jump address). Since our design is microcoded, the proof entails (among many other things) showing that the microprogram corresponding to each instruction is correct. What the verification does not guarantee is important to disclose:

- Our specification of the electronic block model does not capture the semantics of the functions, such as add, shift-left, xor, etc. Hence, it is not possible to use our specifications to reason about the computations of assembly language programs. We decided not to provide such a specification as our main goal was to verify the control logic of the microprocessor.
- Viper has external control lines, such as a reset button. The RSRE specification does not consider these lines, nor do we.
- We have assumed that the main memory responds essentially instantaneously to read or write requests. Viper can support an asynchronous interaction between the processing unit and main memory. Techniques are known for modeling such an interaction, but we did not use them here.
- Main memory is assumed to be a black box. It is certainly feasible to consider a less abstract model of memory. Again, verification is not the best approach to reason about the details of a memory system.

References

1. Tej Arora, Mark Heckman, Sara Kalvala, Tony Leung, Karl Levitt, Tom Schubert, and Phillip Windley. Verification of a microcode implementation of the VIPER microprocessor. Technical report, University of California at Davis, September 1991.
2. Bishop Brock and Warren Hunt. Report on the formal specification and partial verification of the VIPER microprocessor. Contractor Report 187540, NASA Langley Research Center, 1991.
3. A. Cohn. A proof of correctness of the viper microprocessor: the first level. *VLSI Specification, Verification, and synthesis, G. Birtwhistle and P. S ubrahmanyam, eds.*, pages 27–71, 1988.
4. A. Cohn. A proof of correctness of the viper microprocessor: the second level. *University of Cambridge computer Laboratory Technical Report*, 1989.
5. W. J. Cullyer. Implementing safety critical systems: The VIPER microprocessor. In G. Birtwhistle and P.A Subrahmanyam, editors, *VLSI Specification, Verification, and Synthesis*, pages 1–25. Kluwer Academic Press, 1988.
6. Brian Graham and Graham Birtwhisle. Formalising the design of an SECD chip. In M. Leeser and G. Brown, editors, *Workshop on Hardware Specification, Verification, and Synthesis: Mathematical Aspects*, Lecture Notes in Computer Science. Springer-Verlag, 1989.
7. W. A. Hunt. A verified microprocessor. Technical Report 47, The University of Texas at Austin, Dec. 1985.
8. Jeffrey J. Joyce. *Multi-Level Verification of Microprocessor-Based Systems*. PhD thesis, Cambridge University, December 1989.
9. P. J. Windley. The formal verification of generic interpreters. *Ph.D Thesis*, 1990.

Verification of the Tamarack-3 Microprocessor in A Hybrid Verification Environment*

Zheng Zhu, Jeff Joyce, Carl Seger
Department of Computer Science
The University of British Columbia
Vancouver, B.C. Canada V6T 1Z2

Abstract. HOL-Voss is a hybrid verification system which combines symbolic simulation and model-checking with the HOL system. The purpose of HOL-Voss is to provide an environment for verification which requires less general theorem-proving expertise and to explore the efficient and automated symbolic trajectory evaluation. To verify Tamarack-3 in HOL-Voss , we need to translate a behavioral description of Tamarack-3 in HOL to a more informative switch-level description. Maintaining consistency between different levels of description was one the major focuses in the exercise. Therefore, providing a systematic approach to translations of specifications is an important goal of our research. In this report, we discuss three aspects in the translation: Implementation of Tamarack-3 instructions by sequences of microinstructions; integrating circuit implementation parameters; and factorization of the internal memory description to make it external.

1 An Introduction

Due to fast growing complexities of hardware designs, it has become increasingly difficult to maintain consistency between levels of abstraction during a design process. Consequently, there has been a growing interest in developing formal methods to verify hardware designs. Up to date, most verification approaches are suitable to certain levels of abstraction and become inadequate to others. The HOL-Voss system [1, 2], which links the HOL theorem-proving system to a symbolic simulator (Voss), combines the strengths of theorem-proving approach and the symbolic trajectory evaluation to offer a promise in bridging the gap between symbolic simulation and general reasoning methodologies such as abstraction and induction. In this paper, we report a project of verifying the Tamarack-3 microprocessor [3] by the HOL-Voss system. Our primary motivation of the project is to demonstrate the applicability of the HOL-Voss system

* This research was supported, in part, by operating grants OGPO 109688 and OGPO 046196 from the Natural Sciences Research Council of Canada, fellowships from the Province of British Columbia Advanced Systems Institute, and by research contract 92-DJ-295 from the Semiconductor Research Corporation.

to hardware verification, and to expose areas where further developments are needed in order to make the HOL-Voss system a more realistic hardware verification environment.

A major focus of this project is to obtain a more detailed instruction level specification for the Voss system which must be consistent with the behavioral description in HOL. In this paper, we discuss the translation from a register transfer level description of Tamarack-3 , written in HOL, to a switch level description, in a language called HCL. The rest of this paper is organized as follows: Section 2 briefly introduces the HOL-Voss system and the HCL specification language. Section 3 describes the Tamarack-3 microprocessor architecture and its behavioral specification. Section 4 discusses the translation of the specification (obtained in Section 3) to an HCL description. Finally, we briefly discuss a general translation framework and future works to enchance the HOL-Voss system.

2 The HOL-Voss System

The HOL-Voss system is a hybrid verification tool based on an interface between the HOL system and the Voss system. This interface is more than an *ad hoc* translation of outputs from one verification tool into inputs for another. A considerable amount of the development effort has focused on the establishment of a "mathematical interface" between symbolic trajectory simulation and interactive theorem-proving as a sound foundation for the development of a tool interface. The cornerstone of the system is the definitions of several new predicates in the HOL system which establishes a mathematical link between the specification language of the HOL system and that of the Voss system. This includes the formal definition of Voss verification. In an HOL jargon, the establishment of this link can be described as a "semantic embedding" of Voss within the higher-order logic. The establishment of this mathematical link causes the specification language of Voss to become a subset of the language of the HOL system.

A simple switch level circuit specification language, named HCL, is developed as a user interface to HOL-Voss. This language consists of a number of constructs for specifying waveforms – that is, for specifying temporal relationship between values of circuit nodes (vectors of circuit nodes). For example, HCL includes a operators is (isv) for expressing the instantaneous constraints that a particular node (vector of nodes) is equal to some Boolean value (or vector of Boolean values). Another HCL operator, during, is used to express the temporal constraint that an instantaneous constraint holds during some specified temporal interval. The following program of HCL specifies the constraint that the node denoted by phi has the value 0 from time 0 until time 100 and then has the value 1 until time 200. The program also specifies the constraint that a node vector of 16 bits, named as Na, is equal to a little-endian, unsigned binary representation of natural number a from time 80 to time 200.

```
(((phi is F) during (0, 100)) and
 ((phi is T) during (100, 200)) and
 ((Na isv (sized 16 (num2bv a))) during (80, 200))
)
```

The HOL-Voss system can be used to verify assertions of the form

(Antecedent, Consequent)

where Antecedent and Consequent are HCL formulae as mentioned above. The assertion means that for every possible sequence of circuit states, if the sequence satisfies the formula Antecedent, then the same sequence must also satisfy the formula Consequent.

3   The Specification of Tamarack-3

From the programmer's perspective, the Tamarack-3 microprocessor [3] contains an accumulator (acc); a program counter (pc); a return address register (rtn) for storing the current content of pc when an interrupt is granted; an interrupt acknowledgement register iack. iack having value T indicates that there is an interrupt being serviced thus prohibiting other interrupt requests; an ALU logic; and a tester which tests whether the content of acc is zero. Fig 1 is a schematic of Tamarack-3 .

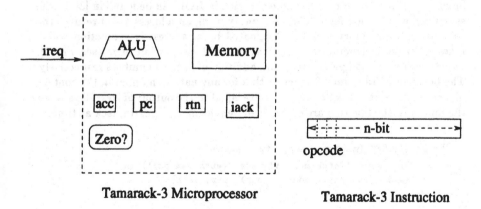

Tamarack-3 Microprocessor   Tamarack-3 Instruction

Fig. 1. Schematic of Tamarack-3 and Instruction Format

The microprocessor has 8 instructions, whose meanings are briefly described in the following table:

Instruction	Opcode	Meaning
JZR	000	Jump if acc is zero
JMP	001	Unconditional jump
ADD	010	Add content the of memory cell to that of acc
SUB	011	Subtract content the memory cell from that of acc
LDA	100	Load the content of the memory cell to acc
STA	101	Store the content of acc to the memory location
RFI	110	Return from interrupt: rtn ⇒ pc
NOP	111	No operation

The microprocessor has a 1-bit interrupt request input ireq. At the beginning of each instruction cycle, if ireq is high and the register iack is low (no interrupt is being processed), then the microprocess sets iack to high, saves the current content of pc to the register rtn and changes the content of pc to 0. Tamarack-3 has only one interrupt service routine which starts from the memory location 0. The last instruction of the interrupt service routine is RFI which restores the content of pc from rtn. If there is not any interrupt request, the microprocessor fetches an instruction from memory and executes it.

Memory is an integrated part of the microprocessor's behavioral specification. Access and update to the memory are represented by functions fetch and store. At the behavioral level, memory is treated a (large) register and access and update to the memory are treated the same as those to registers.

The behavioral specification of Tamarack-3 is given in terms of a "next state" function which specifies the change of each registers of the microprocessor. What follows shows a portion of the specification. Due to the space limit, we omit some auxiliary definitions. A complete specification in HOL can be found in [3]. In the specification, function fetch takes a memory mem, an address a and returns the content of the memory location designated by a. address is a function which takes an n-bit instruction and returns the address portion of it (see Fig 1). Functions inc and add perform $+1$ and addition of natural numbers respectively. The behavior of Tamarack-3 specifies that for any natural number n, the content of mem, acc, pc, rtn, iack at time $n+1$ are equal to the output of NextState when the inputs to the function are contents of ireq, mem, acc, pc, rtn, iack at time n.

$\vdash_{def}$ ADD_SEM (mem, pc, acc, rtn, iack)=
    let opr = (fetch mem (address (fetch mem pc))) in
    (mem, (inc pc), (add acc opr), rtn, iack)

$\vdots$

$\vdash_{def}$ TamBeh (ireq, mem, pc, acc, rtn, iack) =
    ∀ n:time.
    (mem(n+1), pc(n+1), acc(n+1), rtn(n+1), iack(n+1)) =

```
let opcval = OpcVal (read(mem, pc)) in
((ireq ∧ iack) => IRQ_SEM (mem, pc, acc, rtn, iack) |
(opcval = 0) => JZR_SEM (mem, pc, acc, rtn, iack) |
(opcval = 1) => JMP_SEM (mem, pc, acc, rtn, iack) |
(opcval = 2) => ADD_SEM (mem, pc, acc, rtn, iack) |
(opcval = 3) => SUB_SEM (mem, pc, acc, rtn, iack) |
(opcval = 4) => LDA_SEM (mem, pc, acc, rtn, iack) |
(opcval = 5) => STA_SEM (mem, pc, acc, rtn, iack) |
(opcval = 6) => RFI_SEM (mem, pc, acc, rtn, iack) |
(opcval = 7) => NOP_SEM (mem, pc, acc, rtn, iack))
```

4 Translation of Tamarack-3 Specifications

In this project, we manually translated the HOL specification to HCL, according to the microprogramming level description and the implementation parameters. There are two steps involved in the translation. The first step is to translate the behavioral description (in HOL) to a microprogramming level description. In Tamarack-3 's microprogramming level description, each of Tamarack-3 instructions is implemented by a sequence of microinstructions (Fig 2). There are 16 microinstructions in total. A 4-bit microprogram counter mpc is used to indicate, at any microinstruction cycle, which microinstruction is being executed. In Figure 2, each circle in the flowchart represents a microinstruction and the number in the circle is the value of mpc. The meaning of each microinstruction can be found in [3]. The second step is to integrate the parameters of implementation into the microprogramming level description obtained from the first step. These parameters include, for example, clocking scheme, latch setup time, hold time etc. In the rest of this section, we give a brief discussion of these transformations.

4.1 Behavioral Description to Microprogramming Level Description

The translation from the behavioral description to the microprogramming level description is guided by the following two principles:

1. Every Tamarack-3 instruction is implemented by a sequence of microinstruction. The content of every register at "the end of the sequence" has to be consistent to what specified in the behavioral description.
2. Since every Tamarack-3 instruction is implemented by a sequence of microinstruction, we need to introduce an *invariant* which signifies the beginning of each sequence. According to the microinstruction level control flowchart (Fig 2), mpc = 0 indicates that the microprogram control is at the start of a new sequence of microinstructions. In other words, at the end of each sequence, mpc will become 0 in order to start the execution of another Tamarack-3 instruction from the next microprogramming cycle.

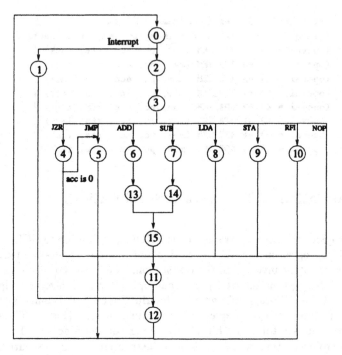

Fig. 2. The Tamarack-3 Microprogram Control Flowchart

As an example, we consider how to generate a microprogramming level description of the instruction ADD. In the behavioral specification of Tamarack-3, ADD_SEM specifies that an execution of ADD instruction results in unchanged mem, rtn and iack. It updates pc to (inc pc) and acc to the value represented by the expression

(add acc (fetch mem (address (fetch mem (address pc)))))

According to Fig 2, ADD is implemented by a sequence of 8 microinstructions:

$$0, 2, 3, 6, 13, 15, 11, 12$$

Therefore, at the end of 8th clock cycle, memory mem and registers pc, acc, rtn, iack should have the the same content as they do at the beginning of the instruction cycle, *pc* should contain the increment of the value it had at the 0th clock cycle, etc. We use a predicate

(Val-Is reg value n)

to specify "the register reg's content is value at the n-th clock cycle". The exact meaning of Val-Is becomes clear when we consider the clocking scheme of the implementation. For the moment, we are content with this informal definition. The ADD instruction requires 8 clock cycles to execute, therefore,

$$\text{pc} = (\text{inc pc})$$

is translated to the following pair of formulae:

$$\text{(Val-Is pc pcvalue 0)} \quad \text{and} \quad \text{(Val-Is pc (inc pcvalue) 8)}$$

where pcvalue is a variable introduced to represent the content of pc at the beginning of the instruction cycle.

Assume memvalue, pcvalue, accvalue, rtnvalue, and iackvalue are the respective values of mem, pc, acc, rtn and iack before the execution of the microinstruction sequence (when mpc=0). That is, the antecedent is:

(Val-Is mem memvalue 0) $\wedge$
(Val-Is pc pcvalue 0) $\wedge$
(Val-Is acc accvalue 0) $\wedge$
(Val-Is rtn rtnvalue 0) $\wedge$
(Val-Is iack iackvalue 0)

Then the execution of ADD can be specified by the following HCL style formula:

(Val-Is mem memvalue 8) $\wedge$
(Val-Is pc (inc pcvalue) 8) $\wedge$
(Val-Is acc (add accvalue (fetch memvalue (address (fetch memvalue (address pcvalue)))))
8) $\wedge$
(Val-Is rtn rtnvalue 8) $\wedge$
(Val-Is iack iackvalue 8)

Finally, the microprogramming level description should also specify that mpc=0 at the 8-th clock cycle: (Val-Is mpc 0 8).

In general, the translation proceeds as follows: An HOL statement

$$(reg_1, \cdots, reg_k) = (F_1(reg_1, \cdots, reg_k), \cdots, F_k(reg_1, \cdots, reg_k)) \tag{1}$$

is translated to a pair of formulae:

$$\text{(Val-is } reg_1 \text{ regvalue}_1 \text{ 0)} \wedge$$
$$\vdots \tag{2}$$
$$\text{(Val-is } reg_k \text{ regvalue}_k \text{ 0)}$$

$$\text{(Val-is } reg_1 \text{ } F_k(\text{regvalue}_1, \text{regvalue}_2, \cdots, \text{regvalue}_k) \text{ n)} \wedge$$
$$\vdots \tag{3}$$
$$\text{(Val-is } reg_k \text{ } F_k(\text{regvalue}_1, \text{regvalue}_2, \cdots, \text{regvalue}_k) \text{ n)}$$

where $regvalue_1, \cdots, regvalue_k$ are unused variables and n is the number of clock cycles needed to execute the instruction. (2) and (3) serve as the antecedent and consequent respectively. Informally, this pair means that for every possible trajectory of nodes $reg_1, \cdots, reg_k$, if the trajectory satisfies formula (2), then it also satisfies the formula (3).

4.2 Microprogramming Level Description to the HCL Specification

Although the microprogramming level description is more informative than its behavioral counterpart, it still lacks the following important information of an implementation to be verified by Voss:

1. Latch parameters: These parameters include setup, data hold and latch-delay of latches used in an implementation. They are given in terms of *number of simulation steps*. A simulation step is the time unit used in the circuit model as a unit delay. These latch parameters serve as a conservative estimation of signal delays during latch operations.
2. Clocking scheme. The implementation we verified uses a two-phase clocking scheme. A clock cycle is 1,000 simulation steps and the value-one duration of each phase is 250 simulation steps. Fig 3 shows the timing diagram of this clocking scheme.

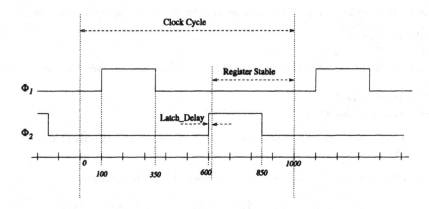

Fig. 3. The Clocking Scheme

3. Circuit Initialization. The circuit we verified requires two clock cycles to become stable after it is powered on. This means that the circuit presents the behavior of Tamarack-3 from the third chock cycle after it is powered on. This changes the control flowchart in Fig 2 to the one in Fig 4.

In our exercise, these parameters are given as HCL definitions. For example, the following HCL definitions specify setup, hold, and latch_delay as 20, 20 and 10 simulation steps respectively:

$\vdash_{def}$ SETUP = 20
$\vdash_{def}$ HOLD = 20
$\vdash_{def}$ LATCH_DELAY = 10

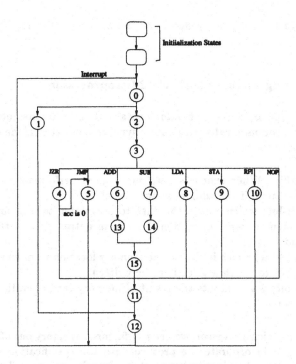

Fig. 4. The Tamarack-3 Control Flowchart with Initialization

In the following definitions, parameters of a two-phase clocking scheme is defined. PHI1R i, PHI1F i specify the rise and fall of the first phase of the ith clock cycle. PHI2R i, PHI2F i specify the rise and fall of the second phase of the ith clock cycle:

$$\vdash_{def} \quad \text{CYCLE k} = (k+1) \times 1000$$
$$\vdash_{def} \quad \text{PHI1R i} = (\text{CYCLE i}) + 100$$
$$\vdash_{def} \quad \text{PHI1F i} = (\text{CYCLE i}) + 350$$
$$\vdash_{def} \quad \text{PHI2R i} = (\text{CYCLE i}) + 600$$
$$\vdash_{def} \quad \text{PHI2F i} = (\text{CYCLE i}) + 850$$

The adoption of the two-phase clocking scheme in the implementation allows us to give a precise definition of Val-Is used in the microprogramming level description. In this clocking scheme, register values become stable from 10 simulation steps (LATCH_DELAY) after the rise of the second phase to the end of the clock cycle [4], as indicated in Fig 3. Therefore, Val-Is can be formally defined as:

$$\vdash_{def} \text{ Val-Is reg dv n} =$$
$$((\text{reg isv (bvNOT dv)}) \text{ during } (((\text{PHI2R (n-1)}) + \text{LATCH\_DELAY}), (\text{CYCLE n})))$$

(bvNOT dv) in the definition reflects the fact that all registers output negated values of their contents.

4.3 Separating Memory from the Microprocessor

Although the Tamarack-3 specification contains an internal memory, the actual microprocessor communicates with a fully synchronous external memory module via the following ports:

- wmem (1-bit). When wmem is low, the memory performs a read operation (or no-operation, if the address is not available);
- dataout (n-bit) which carries the data to be written to a memory location during a memory write operation. This is an output port of the microprocessor; and
- addr (($n-3$)-bit) which indicates the memory location from which a value is to be read or to which a value is to be written.
- datain (n-bit) which inputs the result of memory read operation to the microprocessor.

To verify such a microprocessor, we need to factorize memory out of the original specification and incorporate the protocol into the specification. In order to do this, we first need a description of the external memory and the protocol (method of synchronization) it uses. We used an HCL style specification of a memory. It consists of an antecedent (mem-ante) and a consequent (mem-cons). The antecedent stipulates the conditions for memory-read operation ($x_w = 0$) and memory write operation ($x_w = 1$) respectively. Mem-cons gives a description of the outputs of the memory under the conditions specified in the antecedent. In both antecedent and consequent, n is a variable, referring to the n-th clock cycle. Functions used in the specification, such as CYCLE, are as defined in the previous subsection. Part of mem-ante and mem-cons which specify the memory read operation are given below:

mem-ante:
(when $(x_w = 0)$
        (wmem isv x_w during ((CYCLE n-1), (CYCLE n))) $\wedge$
        (mem isv x_m during ((PHI2F n-1)+LATCH_DELAY, (CYCLE n-1))) $\wedge$
        (addr isv x_a during ((PHI2F n-1)+LATCH_DELAY, (PHI2F n))))

mem-cons:
(when $(x_w = 0)$
        (mem isv x_m during ((PHI2F n)+LATCH_DELAY, (CYCLE n))) $\wedge$
        (datain isv (fetch x_m x_a)
                during (((PHI1R n)-SETUP), ((PHI1F n)+HOLD)))))

During an execution of ADD, there are two memory-read operations:

- fetch of the instruction; and
- fetch of an operand

By the microprogramming level control flowchart, the fetch of an instruction happens in the first clock cycle (mpc = 2), and the fetch of an operand happens in the fourth clock cycle (mpc = 13). According to the memory specification, the microprocessor should put the current content of pc in the first clock cycle to the address line addr, and put the address portion of the instruction fetched in the first clock cycle on the address line in the fourth clock cycle, while keeping wmem low throughout the entire microinstruction sequence to prevent the memory from being updated during an ADD operation. After these are done, datain line will carry

$$(\text{fetch memvalue pcvalue})$$

during the period

$$[(\text{PHI1R 1})\text{-SETUP}, (\text{PHI1F 1})\text{+HOLD}]$$

and

$$(\text{fetch memvalue (address (fetch memvalue pcvalue))})$$

during the period

$$[(\text{PHI1R 4})\text{-SETUP}, (\text{PHI1F 4})\text{+HOLD}]$$

respectively while the memory content stays unchanged throughout the execution of the ADD operation. For this purpose, we define an HOL predicate

$$(\text{addr-is address n})$$

to specify "addr carries address during the n-th clock cycle", according to the memory specification given earlier:

$\vdash_{def}$ addr-is address n =
$\qquad$ ((addr isv address) during ((PHI1R n)-SETUP, (PHI1F n)+HOLD))

and a predicate

$$(\text{datain-is value n})$$

to define "datain carries value during the n-th clock cycle":

$\vdash_{def}$ datain-is value n =
$\qquad$ ((datain isv value) during (((PHI1R n)-SETUP), ((PHI1F n)+HOLD)))

If we assume (datain-is inst 1) and (datain-is operand 4), $i.e.$ (datain-is inst 1)∧(datain-is operand 4) is a part of antecedent, then the microprogramming level description given in Section 4.1 can be modified to:

$\vdash_{def}$ (Val-Is mpc 0 8) ∧

```
(Val-Is pc (inc pcvalue) 8) ∧
(Val-Is acc operand 8) ∧
(Val-Is rtn rtnvalue 8) ∧
(Val-Is iack iackvalue 8) ∧
((wmem is F) during (0, 8000)) ∧
(addr-is pcvalue 1) ∧
(addr-is (address inst) 4)
```

Finally, we summarize conditions under which the above specification of ADD holds true. These conditions are also part of the antecedent:

1. During the 0-th clock cycle, acc, pc, rtn, iack have the values accvalue, pcvalue, rtnvalue, iackvalue respectively.
2. During the first clock cycle, datain carries the value inst, and during the third clock cycle, datain carries the value operand.

Formally, these conditions (the antecedent for ADD operation) can be written as:

$$\vdash_{def}$$
```
(Val-Is acc accvalue 0) ∧
(Val-Is pc pcvalue 0) ∧
(Val-Is rtn rtnvalue 0) ∧
(Val-Is iack iackvalue 0) ∧
(datain-is inst 1) ∧
(datain-is operand 4)
```

5 Conclusions and Future Works

The HOL-Voss system rigorously links a general theorem-proving system with a symbolic simulator to provide an environment for hardware design verification. In the project reported in this paper, we verified a CMOS netlist implementation of Tamarack-3 against its detailed instruction level specification. This is the first experiment of verifying a mid-size circuit ($\geq 2,000$ transistors) by the HOL-Voss system. The experiment also provided good insight to the future improvement of the HOL-Voss system to make it applicable to large circuit verification. The verification of an 8-bit implementation on an SPARCstation ELC with 16MB memory takes about 6 minutes to complete. Such an implementation has $2,376$ transistors and roughly $2,000$ nodes.

The key issue in this exercise is to obtain a correct HCL level specification of Tamarack-3 from a high-level HOL specification. There are three major steps in the process of obtaining an HCL specification:

– Translating an instruction level (behavioral level) description to a clock cycle level description. This translation is based on the microprogramming level specification of Tamarack-3 .

- Integrating implementation parameters such as clocking scheme, latch-delay, setup time of latches. These parameters allows us to define the precise meaning of predicates such as "the register acc has the value x after ADD operation".
- Factorizing memory from the microprocessor specification according to a memory specification. It is this step that allows us to actually verify an implementation which separates the memory from the microprocessor.

Our experience from the exercise suggests that the future works of enchancing the HOL-Voss system should be in the following directions:

- Further development of HOL-Voss infrastructure to increase the usability of the HOL-Voss system. Our goal is to reduce interactive theorem-proving expertise required in achieving non-trivial verification results.
- Provide assistance in translating general HOL specifications to HCL specifications. We are exploring the idea of providing HOL-Voss users mechanism to introduce information needed for such translations, and to complete the translation automatically, according to user-provided information. In a hardware verification context, a behavioral HOL specification may not contain enough information of hardware implementation needed to run symbolic simulation. There are at least two aspects in this area of development:
  1. Adopting a restricted style of behavioral specifications (e.g. generic interpreter style [5].) Such a restricted style of specification should be general enough to cover a large number of circuits. In the mean time, it should provide a good starting point for necessary transformations.
  2. Providing a suitable mechanism of inputing implementation parameters. Such a mechanism should provide users flexibility to specify circuits which may use unusual design methods and styles. On the other hand, it should be able to fill in default parameters if these parameters are not available. In our exercise, the definition of the predicate val-is can have a default definition as given in this paper, which is decided by the clocking scheme used in the implementation, unless the designer explicitly states otherwise.
- Composing verification results from symbolic simulation. There are two approaches to achieve the goal of verifying large systems. The first one is to decompose HOL specifications during their translation to HCL ones. The correctness of implementation is then an consequence of the correctness of decomposition and establishment of some verification conditions which can be proved by either HOL or Voss. The second approach is to compose verification results of components of a large system at HCL level. This approach also requires proving necessary verification conditions, either inside or outside of HOL.

An approach we are exploring is to design a simple language to allow users to input those parameters and provide a description of decomposition of a higher level description in the HCL language. The exercise reported here provides a good experience and insight in designing such a translation system to enchance the usability of the HOL-Voss system.

References

1. JOYCE, J., AND SEGER, C.-J. The HOL-Voss system: Model-checking inside a general-purpose theorem-prover. In *Proceedings of HUG'93* (1993).
2. JOYCE, J., AND SEGER, C. Linking bdd-based symbolic evaluation to interactive theorem-proving. Tech. Rep. Technical Report 93-18, Department of Computer Science, The University of British Columbia, Vancouver, British Columbia, Canada, 1993.
3. JOYCE, J. *Multi-level Verification of Microprocessor-Based Systems*. PhD thesis, Computer Laboratory, Cambridge University, 1989.
4. MEAD, C., AND CONWAY, L. *Introduction to VLSI Systems*. Addison-Wesley, 1980.
5. WINDLEY, P. *The Formal Verification of Generic Interpreters*. PhD thesis, University of California at Davis, 1990.

Abstraction Techniques for Modeling Real-World Interface Chips

David A. Fura[1,3], Phillip J. Windley[2], and Arun K. Somani[3,4]

[1] Flight Critical Information Processing, Boeing Defense & Space Group, Seattle, WA, 98124, USA
[2] Laboratory for Applied Logic, Computer Science Department, Brigham Young University, Provo, UT, 84602, USA
[3] Electrical Engineering Department, University of Washington, Seattle, WA, 98195, USA
[4] Computer Science & Engineering Department, University of Washington, Seattle, WA, 98195, USA

Abstract. We describe techniques for specifying the requirements of real-world interface chips, using as a test target a processor interface unit (PIU). The PIU modeling problem is explained, and current interpreter specification practices are shown to be inadequate for the PIU requirements. General modeling techniques for handling difficult aspects of the PIU are explained, and a new approach that implements these techniques is presented and applied to the PIU requirements specification.

1 Introduction

This paper describes part of our on-going work to formally specify and verify application-specific interface chips. Our initial target is a processor interface unit (PIU) for a commercially-developed fault-tolerant computer system. The Boeing Fault-Tolerant Embedded Processor (FTEP) is designed for applications in avionics and space requiring extremely high levels of mission reliability, extended maintenance-free operation, or both. Since the need for high-quality design assurance in these systems is an undisputed fact, further development and application of formal methods is vital as these systems see increasing use in modern society.

We describe part of our early progress in transferring formal methods from academic settings to real-world hardware applications. As the test case for our initial attempt at this, the PIU has turned out to be a good choice in that it successfully exploits recent academic research, and, at the same time, has helped to focus new research towards the problems affecting real-world hardware modeling.

Current hardware modeling practices fail to address some special problems presented by interface chips. One distinction between interface-chip modeling and most of the earlier work is that this prior work dealt with *standalone* systems, whereas interface chips are *embedded* subsystems. For example, 'microprocessor' verifications to date have not been of microprocessors, per se, but instead complete microcomputer systems — microprocessor plus memory (e.g., [Hun86][Joy90][Win90]). These systems were modeled as output-free interpreters. (An interpreter is state machine whose behavior is partitioned into a set of instructions.) Because of the PIU's role as an interface subsystem however, its output behavior is a prominent part of its overall behavior and cannot be so easily disregarded.

1.1 Contributions of this Paper

In this paper we introduce a new approach to hardware interpreter modeling and demonstrate its application to the specification of an interface chip. To our knowledge, our work on the PIU requirements spec-

This research was sponsored by NASA-Langley Research Center under contract NAS1-18586, Tasks 9 and 10. The NASA technical monitor was Sally Johnson.
Windley was at the University of Idaho when this work was done.

ification (and verification) represents the first successful application of an interpreter-based approach to *transaction-level* modeling.

Transaction Packets. We introduce transaction packets as the basic data type for transaction-level inputs and outputs; comparable to the records of the programming language Pascal, they aggregate the information sent or received during an abstract operation. In modeling bus transactions, for example, a packet contains the expected address and data-block fields; in addition, packet *opcodes* are introduced to abstract concrete-level control behavior such as bus arbitration, handshaking, and tri-state buffer enabling. In grouping sequentially-arriving and -departing concrete signals into a single entity, transaction packets are fundamental to achieving interpreter-based transaction modeling.

Interval Abstraction. Current interpreter abstraction techniques relate the state, input, and (sometimes) output of a system at the boundaries of abstract operations. Our approach extends these methods by relating *intermediate* concrete values to the abstract level. This capability is necessary to implement the abstract transaction-level packets above; in addition, it provides a high-quality solution to the problem of modeling shared state (e.g., [Sch92]).

Hierarchical Pre-Post Logic. To address the issues presented by transaction-level modeling we modify an existing interpreter model [Win90], augmenting it with a new technique for representing temporal and data abstraction. The new model employs Hoare-like preconditions and postconditions, in conjunction with execution predicates, to accommodate a wide range of behaviors. *Abstraction predicates* support the modeling of complicated abstraction relationships and locate them more properly within the system specification hierarchy, rather than within the interpreter correctness statement as in current practice.

State-Machine Modeling of Transactions. Modeling approaches based on state machines are notoriously bad at representing a distributed collection of components because of the state-space explosion that can occur. In this paper we demonstrate two techniques: (a) transaction-level component composition and (b) behavioral decomposition, that help to circumvent this problem.

1.2 Related Work

To date, hardware modeling work has generally targeted low levels of abstraction and/or standalone state-transition systems. Much of this is based on techniques described in [Mel90]. Although not sufficient by itself for our particular modeling problem, this work forms the basis for the techniques described here.

Process Algebras. In [Sch92] Schubert describes a modeling approach based on the process algebra CCS [Mil89]. This work had similar objectives to ours in its targeting of abstract-level hardware specification and composition. Our work differs from Schubert's in that we integrate the abstract level into a standard hardware specification hierarchy, where it can be verified with respect to its implementation. Schubert's work focused on proving that systems consisting of interconnected CCS processes worked correctly, rather than proving that the CCS specifications were themselves implemented correctly. Our work addresses both.

'Light-weight' Specification. In [Bai92] Bainbridge, Camilleri, and Fleming describe what they term a 'light-weight' modeling approach for an industrial-designed processor-memory interface. Although our specification target is similar to this, our work is different in that we model a much more complete set of behavior and employ an interpreter-based approach throughout. The 'light-weight' approach, facing severe time constraints, was limited to representing two key properties: *liveness* and *correct event ordering*, using a more ad hoc modeling approach.

State Transition Assertions. In [Gor92] Gordon describes a specification approach for real-time software based on state transition assertions (STAs). Our work is similar to STAs in that we also describe behavior using predicates defined over *sequences* of system inputs and outputs. However, our approach differs by expressing abstract-level behavior at a single level of temporal abstraction, rather than the two levels of the STA approach. Abstraction predicates help to purge implementation detail from abstract-level specifications, that would otherwise become quite complicated in a hierarchy of hardware interpreters.

The rest of this paper is divided into four sections. Section 2 overviews the PIU and explains the problems it poses for current modeling practice. Section 3 explains general modeling approaches to handle the problems brought out in Section 2. Section 4 introduces our hierarchical pre-post modeling approach and describes how it can be used to implement the techniques presented in Section 3. Section 5 provides a concluding discussion.

2 The PIU Modeling Problem

The PIU is an application-specific integrated circuit (ASIC) providing the 'glue' functions necessary to interconnect a commercial microprocessor, memory chips, and system bus. Its role as an interface chip makes it quite different from the microprocessors generally targeted by the formal methods community, and it presents some interesting modeling problems.

The PIU provides memory-interface, bus-interface, and additional support services within the Processor-Memory Module (PMM) of the FTEP system. The PIU's position within the PMM structure is shown in Figure 1. A PMM, itself a single block within an FTEP Core, interconnects three internal PMM subsystems: the local processors, the local memory, and the Core Bus (C-Bus).

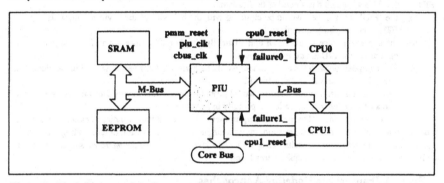

Figure 1: Block Diagram of the Processor-Memory Module (PMM).

We have developed a PIU specification hierarchy extending from the top-most transaction level down to the component models of the silicon compiler used by the design team. Working bottom-up, we developed a multi-level *design specification* using the Generic Interpreter Theory described in [Win90]. In contrast, the top-level *requirements specification* turned out to be difficult for a number of reasons.

Multiple-Process Problem. PIU requirements modeling is complicated by the large number of independent tasks involved:

    (a) the PIU handles memory accesses initiated by the local processor;
    (b) the PIU handles memory accesses sourced by the C-Bus;
    (c) the PIU provides timekeeping and interrupt support for the local processor; and
    (d) the PIU performs PMM initialization upon system reset.

All of these activities proceed in parallel during system operation. Using a standard interpreter modeling approach we might be tempted to group these activities into a single machine description. However, this would result in a virtually incomprehensible description of PIU requirements.

Shared-State Problem. The shared-state problem is one that has been described in earlier work (e.g., [Win90][Sch92]) and can arise in situations where two or more independently-modeled processes have access to a common memory resource. The FTEP PMM includes two such resources: the PMM local memory and the PIU register file.

The problem can be easily understood from the point of view of the local-CPU process. For example, a CPU data load assembly-language instruction is normally modeled similar to the following:

CPU_Reg [Rd] (t + 1) = LMem [Adr] (t).

This states that the new value for a destination register within the CPU is equal to the current value of a targeted memory location.

However, if the C-Bus is accessing local memory during the time a CPU memory-read request arrives at the PIU, then the CPU request must wait. If during this time the C-Bus modifies the value at the location to be read by the CPU, then the behavior described by the above relation cannot be proven to hold—the value read into the destination register (CPU_Reg(Rd)(t+1)) can be different from the memory value at the time of the read request (LMem[Adr](t)).

Many-to-Many Problem. The PIU handles bus transactions sourced by both the local processor and the C-Bus. For either of these sources a single transaction can involve the transfer of a block of data containing as many as four words. Such transfers are implemented as a *sequence* of data movements over a fixed set of signal wires.

The approach favored in prior microprocessor specifications avoids the need to model sequential outputs by integrating the interfacing environment into the specification itself. For the PIU, this approach would incorporate the CPU interface, the local-memory interface, and the C-Bus interface into a single 'PIU' model. This approach has a number of disadvantages:

(a) the 'PIU' specification would be cluttered with the behavior of the environment, which, in the FTEP system, can be complex;

(b) the C-Bus interface would be an unrealistically simple model since not all future system applications for the PIU are currently known;

(c) PIU composition with the environment would be performed at a very low level in the hierarchy, where its verification is most difficult;

(d) verifying this difficult composition would probably be necessary every time the PIU is designed into a new system configuration, especially in light of (b).

To use an interpreter model to represent requirements, we are left with a choice of either representing behavior at a level of abstraction corresponding to a single-word data transfer, or else finding a new data representation. The first choice results in a specification level that we call the *microtransaction* level that we believe is too low to act as a requirements level.

3 General Transaction Modeling Approaches

State machines are a natural choice to base a transaction-level specification model. Among their advantages they are executable, composable, and widely understood within both the verification and design communities. In this section we describe some basic approaches to support state-machine transaction modeling in the face of the problems explained in the last section.

3.1 Behavioral Decomposition

Standard specification methods describe behavior using the next-state and output functions of a single machine. Behavioral decomposition is achieved by introducing an instruction decoding function that serves to define an instruction set for the system being modeled. A state machine defined this way is called an *interpreter* (e.g., [Win90]).

As explained in the last section, a single interpreter model for the entire PIU is a poor choice for representing PIU requirements. A better approach is to further decompose PIU behavior by defining an interpreter for each *class* of behavior. This not only avoids a multiplicative growth in instruction set size, but also serves to restrict the scope of each instruction to its individual class.

Figure 2 shows a portion of the specification hierarchy developed for the PIU. Residing at the top of the figure are the four behavior classes (or processes) listed in Section 2. The P process defines memory transactions initiated by the local processor. The processes C, R, and S represent the C-Bus-initiated transactions, register timers and interrupts, and startup behavior, respectively.

The boxes at the very bottom of the figure represent (most of) the PIU implementation in terms of the five major blocks (called *ports*) of the PIU. Each of these port models implements one or more *transaction-*

level models, as indicated by the directed lines. When appropriately composed, these models implement the four processes. For example, the darkened boxes in the figure indicate those models participating in the *P* process specification.

We have found the breakdown of PIU behavior into these four classes to significantly simplify the transaction-level models. This is extremely important in making feasible the composition of transaction-level components and subsystems into larger systems.

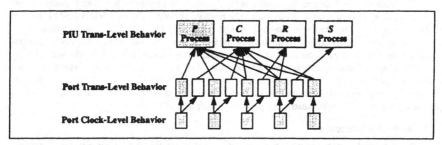

Figure 2: Approximate Implementation Relationships among PIU Specification Models.

3.2 Transaction Packets

Transaction packets are the data type for transaction-level inputs and outputs. By aggregating all of the information sent or received during a transaction into a single entity, they permit interpreter-based specification of transactions.

Although all four PIU processes have their own set of packet definitions, for brevity we demonstrate the idea using those of the *P* process only. Figure 3 shows the *P*-process packets being exchanged with the local processor (on the right), the local memory (on the left), the C-Bus (on the bottom), and the Fault-Tolerant Clock Unit (at the top).

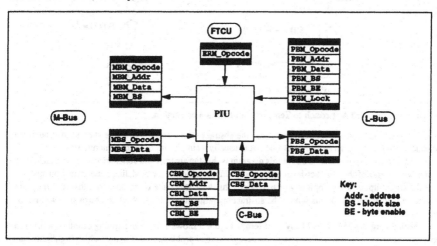

Figure 3: Packet I/O Perspective of the PIU P Process.

As seen in the figure, most of the packet fields have a close correspondence to similarly-named counterparts in the data sheets of the two microprocessors being targeted by the FTEP system—the Intel 80960MC [Int89] and the MIPS R3000 [Kan87]. Similar fields are found in other microprocessors as well.

The address and data fields contain the information suggested by their names. The 4-word data field aggregates the (up to) four words within a block access, and demonstrates the ability of packets to solve the many-to-many problem.

The block-size field defines the number of data words to be read or written. The byte-enable field defines which bytes within the four words are to be replaced on writes. The lock field is used by the Intel 80960 to specify whether the current transaction is part of an atomic read-modify-write operation.

The opcode fields are somewhat different in that they have no direct counterparts in a typical microprocessor data sheet. Instead these fields abstract low-level communication and control behavior, including bus arbitration, handshaking, and tri-state buffer enabling. Opcodes that are not 'illegal' represent scenarios in which the sender is correctly implementing its portion of the relevant protocol. The important role that these fields play in defining interpreter execution is explained in Section 4.1. The fields are themselves explained in more detail in the abstraction predicate descriptions of Section 4.2.

3.3 Interval Abstraction

The traditional approach to interpreter abstraction is shown in Figure 4. In this diagram, an abstract machine, represented by a next-state function NS and state S, is implemented by a concrete machine with a next state function NS' and state S'. Each unit of coarse-grained abstract time t corresponds to multiple units of fine-grained concrete time t'. Temporal abstraction relates the two time sequences. This is implemented by a predicate defined over the concrete state (and perhaps inputs not shown here) that defines the time boundaries of the abstract operations. A typical example of such a predicate is one that returns true whenever a microcode-level program counter reaches the address zero, designating the completion of an assembly-language operation of a microprocessor.

Figure 4: Standard Approach to Temporal and Data Abstraction.

The data abstraction function Abs relates the abstract state S and the concrete state S'. Although the abstract state function can be quite complex, it is usually a simple subset of the concrete state.

The important point to note about this diagram is that the abstract and concrete states are related only at the *boundaries* of the abstract-level operations. This is sufficient for modeling state-transition systems that, lacking outputs, are completely characterized this way. It is quite clear, however, that if outputs are produced at intermediate points within the abstract operation then this approach to abstraction will not be adequate.

Addressing the Many-to-Many Problem. Figure 5 shows a flexible mapping of intra-transaction concrete signals that permits a solution to the many-to-many problem. In this figure, which highlights *output* rather than next-state behavior, the Addr and Data fields of a packet are shown being related to the (80960) concrete signal L_ad; Addr is the value at concrete time tp' and Data[0] is the value at time t_d0', which might represent the time that a control signal (such as L_ready_ of the 80960) becomes active. This flexible mapping of intermediate concrete signal values is the key to embedding transaction packets within a traditional specification hierarchy.

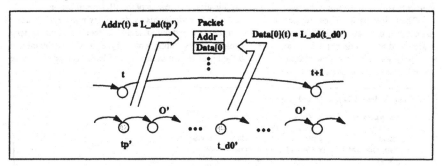

Figure 5: Interval Abstraction to Address the Many-to-Many Problem.

Addressing the Shared-State Problem. Figure 6 shows an approach to the shared-state problem that also exploits interval abstraction. In this case, one concrete PIU state variable (P_Rqt') defined at the beginning of the transaction (concrete time tp' here) is related to its associated abstract variable (P_Rqt). Another concrete state variable, PIU_Reg', representing the PIU register file (for example), is related to its associated abstract state variable PIU_Reg at an *intermediate* point of time ti'.

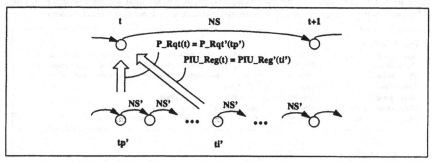

Figure 6: Interval Abstraction to Address the Shared-State Problem.

This abstraction can effectively address the shared-state problem if ti' represents the time that the local CPU actually accesses the PIU register file. In this case the data load instruction specification shown in Section 2 can be verified. Again, the key is the association of the abstract PIU register state at time t with the concrete state at an intermediate concrete time ti', the point at which the local CPU actually owns the register file.

4 Hierarchical Pre-Post Logic

At the start of this project we intended to use the generic interpreter model (GIM) [Win90], which has already proven itself on numerous verifications, including several microprocessors and support chips. While we were able to use the GIM to model the lower levels of the PIU design, the abstraction requirements of the transaction level ultimately exceeded its current capabilities.

Our long-term goal is to create a new generic theory for transaction-level modeling. As a first step towards this, we have developed a very general modeling and verification approach for immediate application to the PIU. We plan to use the experience gained here to support future generic theory development.

In this section we describe a hierarchical pre-post logic (HPL) modeling approach that we used to specify the requirements and design of the PIU. Subsection 4.1 provides a general description of the approach, while Subsection 4.2 demonstrates its application to portions of the PIU specification.

The following variables are used throughout the HOL code in this section. They represent, in order, instructions, time, state signals, environment (input) signals, and output signals. When a variable or type is decorated with a prime it denotes a *concrete*-level, or implementing, entity. Type variables containing a * are polymorphic and denote abstract types. All theorems shown in this section have been proven using the HOL theorem proving system [Gor93].[1]

Common Variables and Their Types

Instructions:	k :*instr, k' :*instr'
Time:	t :time, t' :time'
State:	s :time→*state, s' :time'→*state'
Environment:	e :time→*env, e' :time'→*env'
Output:	p :time→*out, p' :time'→*out'

4.1 General Description

In this section we describe four major aspects of our modeling approach: (a) the interpreter model, (b) abstraction between levels, (c) interpreter liveness, and (d) the definition of interpreter correctness. For simplicity, we use generic operators to denote several key functions whose meanings are explained below. Briefly, the first three operators are the execution predicate, the precondition, and the postcondition. They are followed by the event predicate and the state, environment, and output abstraction functions.

Generic Operators

Execution Predicate:	EXEC rep	:*instr→(time→*state)→(time→*env)→(time→*out)→time→bool
Precondition:	PREC rep	:*instr→(time→*state)→(time→*env)→(time→*out)→time→bool
Postcondition:	POSTC rep	:*instr→(time→*state)→(time→*env)→(time→*out)→time→bool
Event Predicate:	G rep	:time'→bool
State Abstraction:	SAbs rep	:(time'→*state')→time'→*state
Environment Abstraction:	EAbs rep	:(time'→*state')→(time'→*env')→time'→*env
Output Abstraction:	PAbs rep	:(time'→*state')→(time'→*out')→time'→*out

---

Interpreter Model

Our interpreter model is based on existing approaches, but has been structured to handle the needs of the transaction level. An interpreter is defined in terms of an execution predicate (**EXEC rep**) and a postcondition (**POSTC rep**) as shown in the following definition. Interpreter correctness is stated informally as "whenever an instruction is executed its postcondition is satisfied." This represents the standard behavior used in other models. Where ours differs is: (a) in the use of an explicit instruction variable, k, to facilitate verification case splits (similar to the GIM) and (b) in the separation of instruction decoding and execution into their own predicates. As seen below, these predicates play distinct roles in interpreter modeling.

Interpeter Definition:
⊢ INTRP rep s e p = ∀ k t. EXEC rep k s e p t ⊃ POSTC rep k s e p t

---

1. In addition to the normal logical connectives ∀, ∃, ⊃, ∧, ∨, and ¬, the HOL code in this paper uses T and F to represent true and false; the form "e1 ⇒ e2 | e3" to represent "if e1 then e2 else e3;" the form "λ v. u" to represent a function such that (λ v. u) w = u w (juxtaposition represents function application); and the form "ε v . P" to represent an arbitrary value of v's type satisfying predicate P.

While this definition can faithfully model interpreter behavior, we found it impractical for direct use in transaction-level implementation proofs. In these proofs, correct operation requires that certain state variables contain specified initial values at the beginning of a transaction. A convenient way to handle this requirement is to postulate these values using a precondition (PREC rep) and then prove the simpler theorem shown next.[2]

Preconditioned Interpreter Definition:

⊢ INTRP_PREC rep s e p =
    ∀ k t . EXEC rep k s e p t ∧ PREC rep k s e p t ⊃ POSTC rep k s e p t

---

The need for transaction preconditioning results in an additional theory obligation; i.e., that the precondition is initially established at time 0, and that it is propagated at each successor time. The following definition of 'precondition satisfaction' describes this property.

Precondition Satisfaction:

⊢ PREC_SAT rep s e p =
    (∀ k. EXEC rep k s e p 0 ⊃ PREC rep k s e p 0) ∧
    (∀ k k1 t. POSTC rep k s e p t ∧ EXEC rep k1 s e p (SUC t) ⊃ PREC rep k1 s e p (SUC t))

While the need for the above is expected, the following proof obligation did surprise us initially. It states that when an instruction is executed at some (nonzero) time then there must have been an instruction executed at the previous time.

Instruction Sequence Liveness:

⊢ SEQ_LIVE rep s e p = ∀ k t. EXEC rep k s e p (SUC t) ⊃ (∃ k1 . EXEC rep k1 s e p t)

'Instruction sequence liveness' is an issue because we are verifying an instruction set rather than a program. In a program we know that a prior instruction is executed by virtue of its position within the code. Instruction set verification does not permit this solution, instead we must explicitly prove it.

In summary, interpreter correctness can be verified by satisfying the following three proof obligations. The justification theorem states that when they are satisfied, then the interpreter is correct.

Modified Proof Obligations:

∀ rep s e p. INTRP_PREC rep s e p
∀ rep s e p. PREC_SAT rep s e p
∀ rep s e p. SEQ_LIVE rep s e p

Justification Theorem:

⊢ ∀ rep s e p.
    INTRP_PREC rep s e p ⊃
    PREC_SAT rep s e p ⊃
    SEQ_LIVE rep s e p ⊃
    INTRP rep s e p

Interpreter Abstraction

Our approach to interpreter abstraction uses many of the same widely-used abstraction techniques described in [Mel90]. However, we differ from existing practice in two fundamental ways: (a) we use more-powerful data abstraction functions to implement interval abstraction and (b) we implement our abstraction within 'abstraction predicates' rather than embedding it within the interpreter correctness statement.

Fundamental to existing abstraction approaches is the notion of an 'event predicate,' G rep, shown among the following HOL expressions. This predicate, when true, marks an 'instruction boundary event.'

---

2. Another approach to handle this would be to prove a theorem that directly establishes the necessary conditions at *all* transaction beginnings. This approach also requires the two theorems that follow.

The 'instruction boundary predicate' **Istimeof (G rep) t t'** relates the abstract and concrete times through the event **G rep**. The predicate is read "**G rep** is true for the t'th time at concrete time t'." The 'temporal abstraction function,' **t_abs**, maps abstract time to concrete time using the instruction boundary predicate. For each abstract time t, it returns a concrete time, t', such that **G rep** is true for the t'th time at t'.

Event Predicate: (G rep) :time'→bool
Instruction Boundary Event: G rep t' = T
Instruction Boundary Predicate: Istimeof (G rep) t t'
Temporal Abstraction Function: t_abs :time→time' = λ t . ε t' . Istimeof (G rep) t t'

The following HOL statements show the differences between our HPL abstraction predicates and current practice. State abstraction, for example, is currently implemented by mapping the concrete state (s' t') to the abstract state (s t) via the function s_abs (not SAbs rep) shown below. Note that this t' is the concrete time associated with abstract time t (i.e., t' = t_abs t). The environment and output abstractions are similar.

Current Abstraction Relationships:

   s t = (s_abs o s' o t_abs) t
   e t = (e_abs o e' o t_abs) t
   p t = (p_abs o p' o t_abs) t

where s_abs :*state'→*state
etc.

HPL Abstraction Predicate:

⊢ INTRP_ABS rep s o p s' o' p' =
   ∀ t . let t' = t_abs t in
    ((s t = SAbs rep s' t') ∧
    (e t = EAbs rep s' e' t') ∧
    (p t = PAbs rep s' p' t'))

HPL abstraction uses the same temporal operator but strengthens the other functions to support interval abstraction. For example, the state abstraction is implemented using **SAbs rep** above, which operates on the concrete state *signal*, s' (type :time'→*state'), rather than the state *value*, s' t' (type :*state'), as discussed above. This gives the abstraction function freedom to map multiple (temporal) instances of a concrete signal to the abstract level (as in Figure 5), and allows concrete times other than t' to be mapped from (as in Figures 5 and 6 with tp' corresponding to t' here).

HPL abstraction is defined within an 'abstraction predicate,' in contrast to current practice where it is defined within the interpreter correctness statement (see below). In our view, this more closely matches an intuitive understanding of abstraction, as well as cleaning up the correctness statement.

Interpreter Liveness

Interpreter correctness proofs require the predicate **Istimeof (G rep) t t'** to be true for the abstract instruction executed at time t, otherwise the concrete and abstract variables cannot be related. Existing approaches do this by either assuming or proving a theorem similar to the 'universal liveness' statement shown here.

Universal Liveness:

⊢ INTRP_LIVE rep s e p e' =
   ∀ t . ∃ t' . Istimeof (G rep) t t'

Conditional Liveness:

⊢ COND_INTRP_LIVE rep s e p e' =
   ∀ k t . EXEC rep k s e p t
    ⊃ ∃ t' . Istimeof (G rep) t t'

A universal liveness proof can usually be obtained when the event predicate **G rep** is a function of the interpreter state s'. However, for the PIU **G rep** is a function of the concrete input e' and universal liveness cannot be proven. It is not desirable to assume it either since, for example, in the S process **G rep** is true only when an external reset is received—this is not an infinitely-occurring event.

To handle finite-event predicates for the PIU, we use the 'conditional liveness' definition shown above. This states that if an instruction is executed at abstract time t then the event predicate is true for some concrete time t'. We have used this definition to prove implementation correctness for parts of the PIU. In these proofs we have assumed interpreter liveness, but we are currently studying ways to prove it.

Interpreter Correctness

Current interpreter verifications normally use a correctness statement similar to the one shown next, on the left. This statement says: "if we assume the given implementation and interpreter liveness, then the instance of the specification interpreter 'defined by the abstraction' is correct."

Current Interpreter Correctness Definition	*HPL Interpreter Correctness Definition*
∀ rep s' e' p' .	∀ rep s e p s' e' p' .
INTRP_imp rep s' e' p' ∧	INTRP_imp rep s' e' p' ∧
INTRP_LIVE rep s'	COND_INTRP_LIVE rep s e p e' ∧
⊃ INTRP_spec rep (s_abs o s' o t_abs)	INTRP_ABS rep s e p s' e' p'
(e_abs o e' o t_abs)	⊃ INTRP_spec rep s e p
(p_abs o p' o t_abs)	

The HPL interpreter correctness definition is shown on the right-hand side. This statement says: "if we assume the given implementation and interpreter liveness, and if the abstraction is defined according to the abstraction predicate there, then the specification interpreter is correct." In our view, this statement is the more intuitively sound of the two: (a) the theorem consequence contains the specification interpreter, rather than a particular instance of the interpreter, and (b) abstraction is clearly (and correctly) treated as an *asserted* entity by its very position within the correctness statement.

The practical differences between the two definitions may be small however. By modifying the abstraction functions in the current approach we have demonstrated that interval abstraction can be implemented there. Hierarchical (vertical) composition seems to be easier with the new approach however, while differences in component (horizontal) composition are currently being evaluated.

4.2 Application to the PIU P Process

The PIU contains five major blocks: the processor port, memory port, register port, Core Bus port, and startup controller. The processor port (or P-Port) interfaces the local CPU to the PIU's internal bus (I-Bus). In this section, we show how HPL can be applied to the P-Port abstraction definition and transaction-level specification. Because of space limitations, we show only enough detail to give the reader a taste of the modeling style. The P-Port precondition is also omitted.

The P-Port transaction-level behavior is partitioned into two instructions, PT_Write and PT_Read, implementing the behavior implied by their names. We use the same variables here as before to represent instructions (k), state (s), inputs (e), and outputs (p). Several accessor functions, which operate on s, e, and p, are used to denote individual values of the given data structure; they should be evident from the context.

P-Port Abstraction

The P-Port is an interesting modeling test case, in part, because it has two different temporal bases, denoted by the temporal variables tp' and ti' below. The variable tp' defines the clock-level time that a transaction request is received from the L-Bus, while ti' marks the time such a request is relayed onto the I-Bus. Since an I-Bus request is delayed when the C-Port occupies the bus, the relationship tp' ≤ ti' holds. The signals ale_sig_pb e' and ale_sig_ib p' are functions of the clock-level signals defining the conditions under which the two requests occur; they are concrete instantiations of the generic operator G rep used earlier.

Having two temporal bases permits a solution to the shared-state problem, since a single abstract time t now corresponds to the clock-level times that transactions are transmitted over *both* the L-Bus and the I-Bus. When the P-Port transmits over the I-Bus, the C-Port is unable to interfere with its memory access, thus a memory access at abstract time t can be verified according to its standard definition.

The address relationships shown below are among the simplest of all the P-Port variables. The functions ASel and BSel select the phase-A and phase-B values of a clock-level signal, respectively; SUBARRAY f (m,n) returns the elements m downto n of an array f; and wordnVAL maps a 4-valued logic array into a boolean array. The interesting aspect to these variables is their use of the different temporal bases: tp' and ti'.

P-Port Abstraction Relationships:

Two temporal streams
tp' = εt'. lstimeof (ale_sig_pb e') t t'
ti' = εt'. lstimeof (ale_sig_ib p') t t'

Reset input abstraction
Rst_Opcode_InE (e t) =
   (∀ t'. ¬BSel(RstE (e' t')))
      ⇒ RM_NoReset | RM_Illegal

I-Bus slave input abstraction
let valid_ack =
   (∃ u'. STABLE_TRUE_THEN_FALSE (belg l_srdy_E e') (ti'+1, u')) ∧
   (∀ u'. rdy_sig_ib e' p' u' ⊃ (∃ v'. STABLE_TRUE_THEN_FALSE (belg l_srdy_E e') (u'+1, v'))) in
IB_Opcode_InE (e t) = valid_ack ⇒ IBS_Ready | IBS_Illegal

L-Bus address input abstraction
PB_Addr_InE (e t) =
   SUBARRAY (ASel(L_ad_InE (e' tp')) (25,2)

I-Bus address output abstraction
IB_Addr_outO (p t) =
   SUBARRAY (wordnVAL (BSel(l_ad_outO (p' ti'))) (23,0)

The 'reset' input (Rst_Opcode_In) abstraction was implied earlier in the paper. The arrival of an opcode of RM_NoReset is seen to be equivalent to a universally-F clock-level Rst signal, which is sampled during clock phase B.

The 'I-Bus slave opcode' (IB_Opcode_In) defines the correct behavior of the slave's l_srdy_ clock-level handshake signal. As can be inferred from this definition, to transmit a valid opcode the slave must send an active-F l_srdy_ (during phase B) sometime after the I-Bus transaction start (at ti'). In addition, if the P-Port sends an inactive-T l_last_ when the slave is sending the l_srdy_ (i.e., rdy_sig_ib e' p' u' is true) then the slave must send another active-F l_srdy_ sometime in the future. As explained below, a correctly functioning I-Bus slave is critical to the P-Port's ability to successfully transmit a majority of its I-Bus output.

Execution Predicate

The predicate PT_EXEC shown below defines the conditions under which each instruction k is executed. There are three parts to this predicate. The first part specifies a 'reset opcode' of RM_No_Reset, which, as explained above, defines the clock-level Rst input to be inactive during the transaction. The second part specifies an 'I-Bus slave arbitration opcode' of IBAS_Ready, which requires the C-Port's clock-level bus arbitration behavior to be correct. This includes, for example, the fact that the C-Port will relinquish the I-Bus sometime after the P-Port requests it.

The third part of the predicate specifies an appropriate 'L-Bus opcode.' All of these opcodes correspond to correct L-Bus handshaking and tri-state buffer control by the local CPU; they are distinguished

---

P-Port Execution Predicate:

⊢ PT_EXEC k s e p t =
   (Rst_Opcode_InE (e t) = RM_NoReset)   ∧
   (IBAS_Opcode_InE (e t) = IBAS_Ready)   ∧
   (((k = PT_Write) ⇒
      ((PB_Opcode_InE (e t) = PBM_WrLM) ∨
       (PB_Opcode_InE (e t) = PBM_WrPIU) ∨
       (PB_Opcode_InE (e t) = PBM_WrCB))
   % (k = PT_Read) % |
      ((PB_Opcode_InE (e t) = PBM_RdLM) ∨
       (PB_Opcode_InE (e t) = PBM_RdPIU) ∨
       (PB_Opcode_InE (e t) = PBM_RdCB)))

P-Port Postcondition:

⊢ PT_POSTC k s e p t =
   ¬(PT_fsm_stateS (s (t+1)) = PD) ∧
   ¬PT_RqtS (s (t+1)) ∧
   (IB_Opcode_outO (p t) = PB_Opcode_InE (e t)) ∧
   (IB_Addr_outO (p t) = PB_Addr_InE (e t)) ∧
   ((IB_Opcode_InE (e t) = IBS_Ready) ⊃
      (((k = PT_Write) ⊃
         (IB_Data_outO (p t) = PB_Data_InE (e t)) ∧
         (IB_BS_outO (p t) = PB_BS_InE (e t)) ∧
         (IB_BE_outO (p t) = PB_BE_InE (e t)))) ∧
   (IBA_Opcode_outO (p t) = IBAM_Ready) ∧
   ((IB_Opcode_InE (e t) = IBS_Ready) ⊃
      ((PB_Opcode_outO (p t) = PBS_Ready) ∧
      ((k = PT_Read) ⊃
         (PB_Data_outO (p t) = IB_Data_InE (e t)))))

by the transmitted address and read/write information in the expected way. For example, the three opcodes for the instruction **PT_Write** imply a value of logic-**T** for the clock-level signal **L_wr** of the Intel 80960.

Postcondition

The postcondition defines the expected behavior of the P-Port. The first two lines define the required next-state values for the state variables **PT_fsm_state** and **PT_Rqt**; the rest of the predicate specifies the outputs for the I Bus, the I-Bus arbitration, and the L-Bus, respectively.

At the transaction level the P-Port's I-Bus outputs nearly mirror the L-Bus inputs. The first two values listed, **IB_Opcode_out** and **IB_Addr_out**, are the same as the received values. The other three are preconditioned, however, on the receipt of an 'I-Bus slave opcode' of **IBS_Ready**, which, as explained above, defines the correct behavior of the slave's **L_srdy_** signal. It is, in fact, not possible to prove that the data, block size, and byte enables are transmitted correctly unless the slave cooperates. For example, the slave can cause the P-Port to hang indefinitely by refusing to transmit an active **L_srdy_**.

An interesting aspect to this is that, while the slave can interfere with three of the P-Port's outputs, it cannot prevent the correct transmission of the opcode and address. In other words, the P-Port doesn't require anything of the slave to transmit these outputs. This is important because the slave ports cannot themselves transmit a valid slave opcode unless they *first* receive a valid opcode from the P-Port. This explains the positioning of the slave opcode within the postcondition here, rather than in the execution predicate. If it were in the execution predicate then neither the P-Port nor the slave could be verified because neither of the execution predicates could be proven to hold. This is the same type of potential deadlock scenario that was described in [Bai92].

The transmitted I-Bus data is preconditioned by the instruction itself (read or write) in the expected way. The output opcodes for the I-Bus arbitration (**IBA_Opcode_out**) and the L-Bus protocol (**PB_Opcode_out**) follow the same pattern as the others, with the suffix '**_Ready**' indicating adherence to the appropriate standard. A valid I-Bus slave opcode is needed to ensure correct transmission onto the processor's L-Bus.

Transaction-level modeling has proven to be very successful at suppressing implementation detail from the top-level specification. The primary reason for this is the isolation of the complex abstraction definition within the abstraction predicate. The top-level specification is lifted through this temporal-logic description to achieve a simple description in the form of an interpreter. The abstraction predicate can always be consulted when one *is* interested in studying the relationships there.

5 Conclusions and Future Work

We have successfully specified (and verified) significant portions of a commercial processor interface unit (PIU) using a new interpreter modeling approach. Previous models were seen to be useful for the lower levels of the PIU specification hierarchy but were unable to handle the top-most level. In general, the current theorem-proving research emphasis towards microprocessors seems to have left the application-specific integrated circuit (ASIC) problem relatively unexplored. This is extremely unfortunate since ASICs are now being designed into safety-critical systems; and, because the user base for a given ASIC is so much smaller than for a mature commercial microprocessor, it can be argued that a design flaw in an ASIC is more likely to see its way into a deployed critical system.

Current interpreter modeling techniques fall short mainly in their handling of the interpreter interface with the environment. The current emphasis on models with purely state-to-state behavior is impractical for ASICs, and for highly-distributed avionics systems in general. Greater attention to input and output modeling is required to support the necessary level of abstraction for both subsystem modeling and subsystem composition.

We have developed a practical approach for interpreter-based modeling of distributed hardware. By partitioning independent behavior among separate interpreters, we can avoid a multiplicative growth in model size as models are composed with one another. Abstract, transaction-level modeling ensures that the component models are themselves simple.

One key to achieving abstract-level composition is a carefully-defined data type for inputs and outputs. Transaction packets, which aggregate several concrete signals into single entities, are implemented using an enhanced abstraction mechanism that permits intermediate concrete values to be mapped to the transaction level. This interval abstraction also supports a solution to the problem of modeling shared state.

The following table illustrates the utility of these approaches for the actual *P* Process specification of the PIU [Fur93]. For example, in each port the (page count) complexity of the transaction-level model is roughly half that of its clock-level model. In fact, the difference would have been even greater had we defined the transaction level using the concise P-Port definition style shown in the last section. Beyond this benefit, due to abstraction, the behavioral decomposition associated with the *P*-Process partition leads to a very small increase in complexity for the model of the PIU compared to the individual *port* models.

Table 1: Approximate *P*-Process Behavioral Specification Sizes.

PIU Subsystem	Clock-Level	Transaction-Level
P-Port	3.5 pages	2.5 pages
M-Port	5.0 pages	3.0 pages
R-Port	8.0 pages	5.0 pages
C-Port	15.0 pages	3.0 pages
PIU	N.A.	6.0 pages

Aside from finishing the specification and verification of the PIU, much work remains to be done to increase the efficiency of these types of modeling tasks. We are currently investigating a generic interpreter theory that allows interdependent temporal and data abstraction, and we believe that this will provide the basis for a practical transaction modeling (and verification) tool. We intend to exploit our lessons learned from this task in targeting additional interface chips within the FTEP system.

6 References

[Bai92] S. Bainbridge, A. Camilleri, and R. Fleming, "Theorem Proving as an Industrial Tool for System Level Design," in V. Stavridou, T.F. Melham, R.T. Boute (eds.), *Theorem Provers in Circuit Design*, Elsevier Science Publishers, 1992.

[Fur93] D.A. Fura, P.J. Windley, and G.C. Cohen, "Towards the Formal Specification of the Requirements and Design of a Processor Interface Unit," *NASA Contractor Report 4521*, 1993.

[Gor92] M.J.C. Gordon, "Verifying Real-Time Programs: A Case Study," in Hoare and Gordon (eds.), *Mechanized Reasoning and Hardware Design*, Prentice Hall, 1992.

[Gor93] M.J.C. Gordon and T.F. Melham, *Introduction to HOL: A Theorem Proving Environment for Higher Order Logic*, Cambridge University Press, 1993.

[Hun86] W.A. Hunt Jr., *FM8501: A Verified Microprocessor*, Ph.D. thesis and Institute for Computing Science Technical Report 47, University of Texas at Austin, February 1986.

[Int89] Intel Corporation, *80960MC Hardware Designer's Reference Manual*, 1989.

[Joy90] J.J. Joyce, *Multi-Level Verification of Microprocessor-Based Systems*, Ph.D. thesis and Technical Report No. 195, Computer Laboratory, University of Cambridge, May 1990.

[Kan87] G. Kane, *MIPS R2000 RISC Architecture*, Prentice Hall, 1987.

[Mel90] T.F. Melham, *Formalizing Abstraction Mechanisms for Hardware Verification in Higher Order Logic*, Ph.D. thesis and Technical Report 201, University of Cambridge, August 1990.

[Mil89] R. Milner, *Communication and Concurrency*, Prentice Hall, 1989.

[Sch92] E.T. Schubert, "Verification of Composed Hardware Systems Using CCS," in M. Archer, J. Joyce, K. Levitt, and P. Windley (eds.), *Proceedings of the 1991 International Workshop on the HOL Theorem Proving System and its Applications*, IEEE Computer Society Press, 1992.

[Win90] P.J. Windley, *The Formal Verification of Generic Interpreters*, Ph.D. thesis and Research Report CSE-90-22, Division of Computer Science, University of California, Davis, July 1990.

Implementing a Methodology for Formally Verifying RISC Processors in HOL[*]

Sofiène Tahar[1] and Ramayya Kumar[2]

[1] University of Karlsruhe, Institute of Computer Design and Fault Tolerance
(Prof. D. Schmid), P.O. Box 6980, 76128 Karlsruhe, Germany
e-mail:tahar@ira.uka.de

[2] Forschungszentrum Informatik, Department of Automation in Circuit
Design,Haid-und-Neu Straße 10-14, 76131 Karlsruhe, Germany
e-mail:kumar@fzi.de

Abstract. In this paper a methodology for verifying RISC cores is presented. This methodology is based on a hierarchical model of interpreters. This model allows us to define formal specifications at each level of abstraction and successively prove the correctness between the neighbouring abstraction levels, so that the overall specification is correct with respect to its hardware implementation. The correctness proofs have been split into two steps so that the parallelism in the execution due to the pipelining of instructions, is accounted for. The first step shows that the instructions are correctly processed by the pipeline and the second step shows that the semantic of each instruction is correct. We have implemented the specification of the entire model and performed parts of the proofs in HOL.

1 Introduction

Completely automating the verification of general complex systems is practically impossible. Hence appropriate heuristics for specific classes of circuits such as finite state machines, arithmetic circuits, microprocessors, signal processors, etc., are needed for aiding the verification process.

Microprocessors build a particular class of well structured hierarchical circuits, that are used in a wide range of applications. Many microprocessor verification efforts have been done in the recent past; some using the HOL system [Cohn88,Joyc89,Wind90] and others based on functional calculi [Hunt87,SrBi90]. Most of them handle only one specific microprogrammed processor. The objective of our endeavour is the development of a generic methodology (comparable to that of Windley [Wind90]) for the formal verification of a large number of realistic RISC processor cores. Furthermore, we plan to integrate this methodology into a general verification framework [KuSK93]. In our previous work, we have developed and formalized a general hierarchical model for RISC cores and sketched the formal verification aspects using this model [TaKu93a,TaKu93b]. The aim of this paper is to show the practicability of our methodology by implementing it in HOL.

---

[*] This work has been partly financed by a german national grant, project Automated System Design, SFB No.358.

The organization of this paper is as follows. Section 2 presents briefly the hierarchical verification model. Section 3 describes the different abstractions used by the model with a focus on temporal abstraction. Section 4 gives a formalization of the abstract specifications and implementation in HOL. Section 5 discusses the verification steps and strategies adopted and includes HOL implementations of some proofs. Section 6 finally concludes the paper. Further, the presented methodology and implementation are exercised throughout this paper by means of a RISC example: DLX[1] [HePa90].

2 The Hierarchical Verification Model

In some related works, it has been shown that the verification of microprogrammed processors can be simplified by the use of a hierarchical model, called *interpreter model* [Joyc89,Wind90]. This model is based on abstraction levels which are introduced between the top-level specification and the circuit implementation EBM (Electronic Block Model). Structuring the verification of RISC processors using this interpreter model is not possible, since RISCs function differently. RISC instructions are executed in a number of physical steps, called pipeline stages (e.g. IF, ID, EX, WB) [HePa90], whose duration corresponds to one machine clock period. We define a *stage instruction* as the set of transfers, which occur during the corresponding pipeline segment. Each stage operation is partitioned into a number of sub-operations corresponding to the phases of a multiple phase non-overlapping clock. We define a *phase instruction* of a specific stage as the set of sub-transfers that occur during that clock phase. Usually, the RISC phase instructions are stage dependent and the stage instructions differ from one instruction to another (table 1). These dependencies lead to a big number of verification steps between the EBM and the phase level. In order to circumvent this problem, we exploit the notion of instruction classes[2] and introduce the *class level* as the top level of our interpreter model. The stage and phase instructions are then parameterized in accordance with the class abstraction.

The overall verification model is given in Fig. , where the arrow between the levels means that the lower level is an abstraction of the next upper one. Using this hierarchical model, we are able to successively prove the correctness of all phase, stage and class instructions. The correctness of the architectural instructions can then be shown by a simple instantiation of the proofs for each particular instruction.

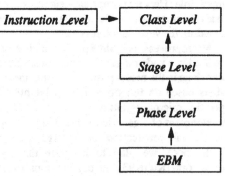

Fig. 1. RISC Interpreter Model

---

[1] DLX is an hypothetical RISC which includes the most common features of many existing RISCs (Intel i860, Motorola M88000, SPARC, MIPS R3000).

[2] An instruction class intuitively corresponds to the set of instructions with similar semantics, e.g. ALU, FLP, LOAD, CONTROL.

Table 1. DLX pipeline structure[3]

	ALU	LOAD	STORE	CONTROL
IF	IR ← MEM[PC] PC ← PC+4	IR ← MEM[PC] PC ← PC+4	IR ← MEM[PC] PC ← PC+4	IR ← MEM[PC] PC ← PC+4
ID	A $\leftarrow_2$ RF[rs1] B $\leftarrow_2$ RF[rs2] IR1 ← IR	A $\leftarrow_2$ RF[rs1] B $\leftarrow_2$ RF[rs2] IR1 ← IR	A $\leftarrow_2$ RF[rs1] B $\leftarrow_2$ RF[rs2] IR1 ← IR	BTA $\leftarrow_1$ f (PC) PC $\leftarrow_2$ BTA
EX	ALUout ← A op B or ALUout ← A op (IR1)	DMAR ← A+(IR1)	DMAR ← A+(IR1) SMDR ← B	
MEM	ALUout1 ← ALUout	LMDR ← MEM[DMAR]	MEM[DMAR] ← SMDR	
WB	RF[rd] $\leftarrow_1$ ALUout1	RF[rd] $\leftarrow_1$ LMDR		

3 Abstraction Types

Within this verification model, we have different abstraction levels which have to be related to each other. It is important to consider a correct abstraction mechanism for the formal specification as well as for the formal verification. There are four kinds of abstractions [Melh88]: structural abstraction, behavioural abstraction, data abstraction and temporal abstraction. Structural and behavioural abstractions are a natural consequence of the hierarchical model. In the following we will briefly handle data abstraction and then we will focus on the concept of temporal abstraction.

3.1 Data Abstraction

Throughout our approach, we will let the specifications be based on the data types that the microprocessor is to manipulate, i.e. bit-vectors. The problem with using such concrete data representations is the presence of the implementation details of the data types. In [Joyc89,Wind90] uninterpreted data types were used. Here we use data types of a bit-vector theory. Since bit-vectors are naturally used in the description of the instruction set, we do not make any data abstractions and use bit-vectors through all abstraction levels. Thus, we save a lot of mapping functions between the concrete data type and an abstract one, e.g. natural numbers.

3.2 Temporal Abstraction

Temporal abstraction relates the different time granularities which occur in the formal specifications at various levels of abstractions [Melh88]. The class and the instruction level use the same time granularity, which corresponds to instruction cycles. The stage level granularity is that of clock cycles and the phase level granularity corresponds to the duration of single phases of the clock (Fig. 2).

---

[3] In table 1, "←" represents those stage transfers that are not broken down into phase transfers. "$\leftarrow_1$", "$\leftarrow_2$" represent the transfers which take place in phases 1 and 2, respectively.

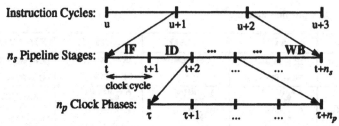

Fig. 2. Temporal abstraction

An ADD-instruction at the instruction level can be specified as follows:

$ADD\_SPEC (PC, I\text{-}MEM, RF, D\text{-}MEM):=$

$\forall u: Inst\_cycle.\ RF(u+1)[rd(u)] = RF(u)[rs1(u)] + RF(u)[rs2(u)]$ [4]   (1)

Referring to table 1, the ADD-instruction, using the more concrete time granularity, is:

$ADD\_IMP (PC, I\text{-}MEM, RF, D\text{-}MEM):=$

$\forall t: Clk\_cycle.\ RF (t+5)[rd(t)] = RF(t+1)[rs1(t)] + RF(t+1)[rs2(t)]$ (2)

A mapping function that relates abstract time scales in (1) to concrete ones in (2) is not linear since a unit of time at the abstract time scale does not necessarily correspond to one discrete time point of the next concrete time scale. For example, the same time point u, used in (1) for computing the addresses and reading the register file RF, have to be related to t and $t+1$, corresponding to the IF and ID-stage, respectively.

In a pipelined execution, state changes at the abstract level can take place at some time between the two discrete end-points of its time interval, i.e. at some time between u and $u+1$ depending upon the implementation. A mapping time abstraction function should therefore take some implementation contexts into consideration while converting the abstract time to a more concrete one. A context parameter could be given as a tuple involving a read/write information and a pipeline stage identifier. For example, let f be the temporal abstraction function reading during the ID-stage yields -"$f ([read, ID], u) = t+1$"- and writing during the WB-stage yields -"$f ([write, WB], u+1) = t+5$"-.

Semantically, the affected state values in (1) have to be interpreted as follows: $X(u)$ and $X(u+1)$ correspond to the read and written values of X in the open intervals [u;u+1[and]u;u+1], respectively. Syntactically, these values of the specification must be supplemented with an abstraction function f in the verification goal, i.e. $(X\ f)\ (u)$ instead of $X(u)$, where " " is the composition suffix. (see Sect. 5.2)

At a lower level, i.e. between the stage and the phase levels, we have a similar temporal relationship since state transitions can occur at time points within the clock cycle interval at some specific phases. In a similar manner, a time abstraction function has to be used in the appropriate verification goal. An implementation of a general time

---

[4] RF is the register file and rd, rs1, rs2 are the destination and source addresses of some registers. These addresses correspond to fields of the actual instruction word, which is addressed by the program counter PC.

abstraction function for both abstraction levels can be defined in HOL as follows:

```
val Time_abs = new_definition ("Time_abs",
    --`Time_abs n C u =
        let RW = (EL 0 C) in
        let ORD = (EL 1 C) in
            n*(u-RW) + ORD + RW `--);
```

The parameter *n* corresponds to the total number of implemented pipeline stages or clock phases. Assuming that stage and phase identifiers are ordered in some manner, the context parameter *C* is a tuple comprising of a read/write information and the ordinal value of the corresponding stage or phase identifier, e.g. *ORD (EX) = 2*. This definition of the time abstraction function assumes that written values are considered at the end of a discrete time interval while read values are those at the beginning.

This abstraction function has the advantage, that it allows specifications to be abstract and implementation independent. The various instantiations, i.e. *n* and *C*, are given later when making the appropriate verification goal (see Sect. 5.2).

4 Formal Specification

4.1 Instruction and Class Level

Referring to the instruction set manual of a specific processor, the semantic of an instruction is given as a state transition occurring in an instruction cycle which implicitly involves time. For example the semantic of an ADD-instruction is defined as:

$$ADD := RF[rd] \leftarrow RF[rs1] + RF[rs2]$$

A Class instruction abstracts the semantic of a group of architectural instructions. The semantic of the ALU-class instruction, for example, can therefore be given as follows, where *op* abstracts all ALU-operations:

$$ALU := RF[rd] \leftarrow RF[rs1] \, op \, RF[rs2]$$

Let *u* be a unit of time for an instruction cycle. This ALU-instruction can be described formally in HOL by means of a predicate involving time as follows, where the function *BITS* extracts the bit-vector portion specified by the first two parameters:

```
val ALU_SPEC = new_definition ("ALU_SPEC",
    --`ALU_SPEC (PC,I_MEM,RF,D_MEM) op =
        let rs1 = BITS(25,21,(I_MEM PC)) in
        let rs2 = BITS(20,16,(I_MEM PC)) in
        let rd  = BITS(15,11,(I_MEM PC)) in
        !u. RF (u+1) (rd u) = op (RF u (rs1 u))
                                 (RF u (rs2 u))`--);
```

A first description of the implementation of the ALU-class, without structural abstraction and in terms of clock cycle units *t* can be given as follows:

```
val ALU_IMP = new_definition("ALU_IMP",
    --`ALU_IMP (PC,I_MEM,RF,D_MEM) op =
        let rs1 = BITS(25,21,(I_MEM PC)) in
        let rs2 = BITS(20,16,(I_MEM PC)) in
        let rd  = BITS(15,11,(I_MEM PC)) in
        !t. RF (t+5) (rd t) = op (RF (t+1) (rs1 t))
                                 (RF (t+1) (rs2 t))`--);
```

4.2 Stage Level

A stage instruction is defined as a set of elementary state transitions, that implement the corresponding semantic. Formally, a stage instruction is specified as a predicate on the visible states at this level. It is a conjunction of simple transfers that can be directly read off from the pipeline architecture (table 1) and encoded formally. For example, the ID-stage instruction common to the ID-row and the class columns ALU, LOAD and STORE is specified in HOL with the following predicate:

```
val ID_SPEC = new_definition ("ID_SPEC",
    --`ID_SPEC (A,B,RF,IR,IR1) =
        let rs1 = BITS(25,21,IR) in
        let rs2 = BITS(20,16,IR) in
        ! t. (A (t+1) = RF t (rs1 t)) /\
             (B (t+1) = RF t (rs2 t)) /\
             (IR1 (t+1) = IR t) `--);
```

An implementation of this stage instruction in terms of clock phases can be specified as follows:

```
val ID_IMP = new_definition ("ID_IMP",
    --`ID_IMP (A,B,RF,IR,IR1) =
        let rs1 = BITS(25,21,IR) in
        let rs2 = BITS(20,16,IR) in
        ! tau. (A (tau+2) = RF (tau+1) (rs1 tau)) /\
               (B (tau+2) = RF (tau+1) (rs2 tau)) /\
               (IR1 (tau+2) = IR tau) `--);
```

4.3 Phase Level

Phase instructions are also specified as conjunctions of elementary state transitions. These transfers are also read off directly from the pipeline structure (table 1) and encoded formally. State values of phase transitions that are not explicitly marked in the pipeline architecture, e.g. between IR and IR1-register in the ID-stage, are left unchanged until the last phase. From the ID-row and the class columns ALU, LOAD and STORE, the two phase instructions of the ID-stage are specified in HOL as follows:

```
val ID_Ph1 = new_definition ("ID_Ph1",
    --`ID_Ph1 (A,B,RF,IR,IR1) tau =
            IR (tau+1) = IR tau`--);

val ID_Ph2 = new_definition ("ID_Ph2",
    --`ID_Ph2 (A,B,RF,IR,IR1) tau =
        let rs1 = BITS(25,21,IR) in
        let rs2 = BITS(20,16,IR) in
        (A (tau+1) = RF tau (rs1 tau)) /\
        (B (tau+1) = RF tau (rs2 tau)) /\
        (IR1 (tau+1) = IR tau) `--);
```

4.4 Electronic Block Model

While the abstract levels of our interpreter model are behavioural descriptions, the EBM describes the structure of the hardware at the RT-level. However, the visible states and the temporal refinement used are the same as those of the phase level. At the top most level, the EBM is composed of the RISC processor and the interfaced instruction and data memory (cache). The processor consists conventionally of a datapath and a control unit, which are themselves compositions of simpler blocks (Fig. 3).

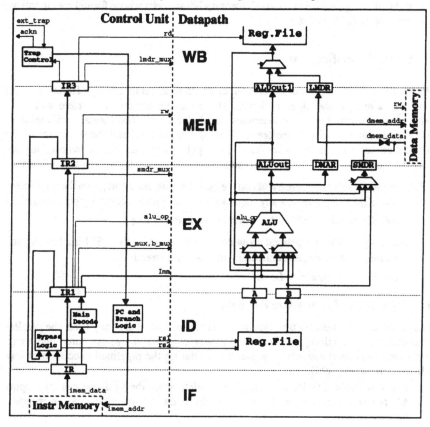

Fig. 3. Electronic Block Model of DLX (simplified)

Formally, the EBM is specified as a complex hierarchy of predicates, which are composed using conjunctions. The top level implementation of the EBM looks formally as follows, where the processor predicate is expanded into datapath and control unit:

```
EBM (PC,I_MEM,RF,D_MEM,A,B,ALUout,ALUout1,DMAR,SMDR,LMDR,
     IR,IR1,IR2,IR3,BTA,IAR,ext_trap,ackn,clk1,clk2) =
       ? rs1,rs2,rd,Imm,a_mux,b_mux,alu_op,smdr_mux,lmdr_mux,
         imem_addr,imem_data,dmem_addr,dmem_data,rw.
     (DataPath RF,A,B,ALUout,ALUout1,DMAR,SMDR,LMDR,
               dmem_addr,dmem_data,rs1,rs2,rd,Imm,a_mux,b_mux,
               alu_op,smdr_mux,lmdr_mux,clk1,clk2) /\
       (Control_Unit PC,IR,IR1,IR2,IR3,BTA,IAR,rw,ext_trap,ackn,
                     imem_addr,imem_data,rs1,rs2,rd,Imm,a_mux,
                     b_mux,alu_op,smdr_mux,lmdr_mux, clk1,clk2) /\
        (Instr_Memory I_MEM,imem_addr,imem_data,clk2) /\
         (Data_Memory D_MEM,dmem_addr,dmem_data,rw,clk2)
```

Our approach is embedded within the MEPHISTO verification framework and the formal description of the circuit is obtained automatically from an EDIF output of a schematic representation, entered within a CAD tool [KuSK93]. The sub-blocks of the hierarchical design are broken into elementary library cells whose formal specifications are contained in a HOL-library.

5 Formal Verification

Potentially, a RISC processor executes n_s instructions in parallel, in n_s different pipeline stages (where n_s is the pipeline depth). These instructions could interfere with each other so that semantical inconsistencies could also occur. The formal verification of RISC processors consists therefore in showing that on one hand the semantics of the instruction set and on the other hand the pipelined sequencing of instructions are correctly implemented. Thus we split the correctness proof into two steps as follows:

1. given some software constraints on the actual architecture and given the implementation EBM, we prove that any sequence of instructions is correctly pipelined, i.e.:

$$SW\_Constraints, EBM \vdash Correct\_Instr\_Pipelining \qquad (3)$$

2. assuming a correct pipelined sequencing, we prove that the EBM implements the semantic of each single architectural instruction correctly, i.e.:

$$EBM \vdash Instruction Level \qquad (4)$$

5.1 Correctness of Instruction Pipelining

The proof of correct instruction pipelining means to show that at any time, the parallel execution of instructions does not lead to any conflicts. There are three classes of conflicts, also called hazards, that can appear during the pipelined execution of any RISC machine [HePa90]:

- *structural hazards* arise from resource conflicts when the hardware cannot support all possible combinations of instructions during simultaneous overlapped execution.
- *data hazards* arise when an instruction depends on the results of a previous instruction, i.e. data dependencies exist between them.
- *control hazards* arise from the pipelining of branches and other instructions that change the PC, i.e. interruption of the linear instruction flow.

The correctness of the pipelined execution of instructions is the direct consequence of the absence of all these hazards. We specify each hazard class formally as a predicate and define *Hazard_Pred* as their conjunction, i.e.:

$$Hazard\_Pred = Struct\_Pred \wedge Data\_Pred \wedge Control\_Pred$$

Thus, the correctness statement (3) remains in showing that the predicate *Hazard_Pred* is never true and the corresponding proof can be split as follows:

$$SW\_Constraints, EBM \vdash \neg Hazard\_Pred \qquad (5)$$

$$\neg Hazard\_Pred \vdash Correct\_Instr\_Pipelining \qquad (6)$$

The software constraints in (5) are those conditions which are to be met for designing the software so as to avoid conflicts, e.g. the number of delay slots to be introduced between the instructions while using a software scheduling technique. Further the EBM is assumed to include some conflict resolution hardware. The correctness of (6) includes the proof that all possible sequence combinations of n_s instructions are executed correctly.

In the following, we will sketch how hazards are specified and handled. The proofs of these steps are subject of ongoing work and are not reported here.

5.1.1 Handling Single Hazards

The different hazard predicates mentioned are closely related to our hierarchical interpreter model, since the data and control hazards are related only to the visible states at the instruction (class) level and the structural hazards are related to the stage and phase levels in a similar manner. For example, structural hazards at the stage level can be defined as follows:

$Stage\_Struct\_Pred := \exists\ r: Resource.\ \exists\ S_i, S_j: Stage\_Instruction.$

$(\ S\_type(S_i)\quad S\_type(S_j)\ )\ )\ \wedge S\_used\_Pred(S_i, r) \wedge S\_used\_Pred(S_j, r)$

where $S\_type(S_i)$ is a function which gives the logical type of the stage instruction S_i, e.g. ID or EX, and $S\_used\_Pred(S_i, r)$ means that the resource r, visible at the stage level, is used by the instruction S_i.

The formal specification of these hazards can be used for different purposes depending on the kind of verification applied:

- *on the fly verification*: the hazard predicates help the designer in synthesizing the hardware and software constraints needed for conflict resolution at a given step of the design process. For example, when a designer finds out that the predicate *Stage_Struct_Pred* is true for a given resource, that is used simultaneously by the class instructions i and j at stages S_n and S_m (Fig. 4), respectively, then either a software or a hardware constraint is extracted.

- *post verification*: the hazard predicates help the designer in validating some existing software or hardware constraints for conflict resolution.

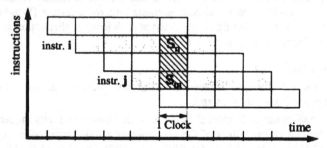

Fig. 4. Structural pipeline hazard

5.1.2 Handling Hazard Combinations

The proof of the predicate *Correct_Instr_Pipelining* ensures that all possible combinations of instructions that occur in the n_s pipeline stages are executed correctly. Exploiting the notion of class instructions, as described in Sect. 2, the case analysis explosion is avoided by considering the combinations of few classes.

In step (5), we have shown that structural, data and control hazards do not occur between any two instructions. In proving step (6), we have to show that for all possible permutations of the class instructions, the combinations of the instructions executing in the different pipeline stages do not lead to conflicts. This is done by splitting the proof into three steps, each corresponding to one kind of hazard. Each of these steps can then be proved by instantiating the theorems obtained in step (5), since they show the absence of conflicts between any two instructions in the pipeline.

5.2 Correctness of Single Instructions

Assuming the non existence of conflicts when instructions are executed in parallel within the pipeline, the correctness of each architectural instruction has to be shown. Using the notion of instruction classes and the hierarchical verification model, as described in Sect. 2, we first hierarchically prove the correctness of the class level, i.e.:

$$EBM \Rightarrow Phase\ Level \Rightarrow Stage\ Level \Rightarrow Class\ Level. \qquad (8)$$

and then through instantiation, we show the correctness of the instruction level i.e.:

$$EBM \Rightarrow Phase\ Level \Rightarrow Stage\ Level \Rightarrow Instruction\ Level. \qquad (9)$$

According to the abstract levels, each of these steps is broken into the following steps:

$$Stage\ Level \Rightarrow Class\ Level, \qquad (10)$$

$$Phase\ Level \Rightarrow Stage\ Level\ and \qquad (11)$$

$$EBM \Rightarrow Phase\ Level\ . \qquad (12)$$

Due to this structuring of the verification task, the verification goals at different levels are simple and show some similarities. The proofs are manageable easier and a lot of general proof strategies could be developed. The goal settings and proofs are done automatically using generic functions and tactics. Further, each abstraction level can be verified independently, so that a designer is able to successively refine and verify the design.

In many aspects, this part of the verification process is similar to that used by Windley for the verification of microprogrammed microprocessors [Wind90]. In the following, we present the verification process at each level of abstraction and we then show how the instantiations are handled.

5.2.1 Stage-Class Level

Since the time granularities between the class and stage levels are different, the correctness proof is done in two steps:

Step I: We show that individual class implementations are correctly implemented by the cumulative effect of the corresponding stage instructions, i.e. for the ALU-class: $IF\_SPEC \wedge ID\_SPEC \wedge ... \wedge WB\_SPEC \Rightarrow ALU\_IMP$

Step II: This class implementation must now imply the abstract class specification, i.e.: $ALU\_IMP \Rightarrow ALU\_SPEC$

The specifications in step I are at the same time granularity (that of clock cycles). Hence no temporal abstraction is needed in the verification goal. Using the formal descriptions of the class implementation and the stage specification, as described in sections 4.1 and 4.2, respectively, the verification goal and proof of this step for the ALU-class can be described in HOL as follows, where CLASS_IMP_TAC is a simple general tactic based mainly on several rewrites:

```
set_goal([],(--`!op.
                 IF_SPEC (PC,I_MEM,IR) /\
                 ID_SPEC (A,B,RF,IR,IR1) /\
                 EX_SPEC (A,B,ALUout,IR1,IR2) op /\
                 MEM_SPEC(ALUout1,ALUout,IR2,IR3) /\
                 WB_SPEC (RF,ALUout1,IR3)
                    ==> ALU_IMP (PC,I_MEM,RF,D_MEM) op`--));
e(CLASS_IMP_TAC(ALU_IMP,IF_SPEC,ID_SPEC,MEM_SPEC,EX_SPEC,WB_SPEC));
save_top_thm("ALU_IMP_CORRECT");
```

The specifications in step II use different time granularities. Therefore the verification goal should include a temporal abstraction function and context parameters (as discussed in Sect. 3.2). The temporal abstraction function is instantiated with the corresponding pipeline depth n_s and is applied to the different context parameters. These implementation dependent context parameters are introduced for each state component in the specification formula. These parameters are existentially quantified and have to be instantiated appropriately later during the proof. For the proof, we use a general common tactic COMN_SPEC_TAC with the following parameters: the pipeline depth n_s, the pipeline stage identifiers, the corresponding class implementation and a list of the needed context parameters. An implementation in HOL of the COMN_SPEC_TAC tactic is given in following:

```
fun COMN_SPEC_TAC (depth, segments_list, imp, context_list) =
    MAP_EVERY EXISTS_TAC context_list THEN
    REPEAT (CONV_TAC let_CONV) THEN
    REWRITE_TAC [imp] THEN
    REPEAT STRIP_TAC THEN
    MAP_EVERY ASM_REWRITE_TAC
             [segments_list, [depth,R,W],
              [Time_abs,o_THM,EL,HD,TL,num_CONV(--`1`--)]] THEN
    BETA_TAC THEN
    CONV_TAC(DEPTH_CONV let_CONV) THEN
    ASM_REWRITE_TAC [ADD1,ADD_CLAUSES,ADD_SUB,SUB_0,
                     LEFT_ADD_DISTRIB,MULT_CLAUSES];
```

Using the specifications of the class level, as described in Sect. 4.1, and the time abstraction function, as described in Sect. 3.2, the verification goal and proof of step II for the ALU-class look formally in HOL as follows:

```
set_goal([],(--`
    !op.
       ?c1 c2 c3.
         let f = Time_abs ns in
           let rs1 = BITS(25,21,(I_MEM PC)) in
           let rs2 = BITS(20,16,(I_MEM PC)) in
           let rd  = BITS(15,11,(I_MEM PC)) in
               ALU_IMP (PC,I_MEM,RF,D_MEM) op
                  ==> !u. (RF o f c1) (u+1) ((rd o f c3) u) =
                          op ((RF o f c2) u ((rs1 o f c3) u))
                             ((RF o f c2) u ((rs2 o f c3) u))`--));
e (GEN_TAC THEN
   COMN_SPEC_TAC (ns, [IF,ID,EX,MEM,WB], ALU_IMP,
                  [(--`[W;WB]`--),(--`[R;ID]`--),(--`[R;IF]`--)])));
save_top_thm("ALU_SPEC_CORRECT");
```

The context parameters used in the proof are directly derived from the implemented pipeline architecture (table 1), e.g. $c2 = [R;ID]$ corresponds to a *read*-operation during the *ID-stage*.

5.2.2 Phase-Stage Level

The correctness proof between the phase and the stage level is done in a manner similar to the previous section, by the two steps as follows:

Step I: proof that each stage instruction implementation is implied by the conjunction of the corresponding phase instructions, e.g. for the ID-stage:

$$ID\_Ph1 \wedge ID\_Ph2 \Rightarrow ID\_IMP$$

Step II: proof that the stage implementations imply the corresponding abstract stage specifications, e.g.: $ID\_IMP \Rightarrow ID\_SPEC$

Using the specifications of the stage and phase levels, as described in sections 4.2 and 4.3, respectively, the verification goal and proof of step I for the ID-stage look formally in HOL as follows, where STAGES_IMP_TAC is a simple general tactic based mainly on several rewrites:

```
set_goal([], (--`(!tau.ID_Ph1 (A,B,RF,IR,IR1) tau /\
                       ID_Ph2 (A,B,RF,IR,IR1)(tau+1) )
                     ==> ID_IMP (A,B,RF,IR,IR1)`--));
e (STAGES_IMP_TAC (ID_IMP,ID_Ph1,ID_Ph2));
save_top_thm("ID_IMP_CORRECT");
```

Similar to the class level, the verification goal for step II involves temporal abstraction. The time abstraction function is instantiated with the corresponding number of clock phases n_p and context variables are included for each state component. The tactic COMN_SPEC_TAC is again used for the proof with the appropriate parameters: the number of clock phases n_p, the clock phase identifiers, the corresponding stage implementation and a list of the needed context parameters. These context parameters are directly derived from the implemented pipeline architecture (table 1).

5.2.3 EBM-Phase Level

The phase level lies directly above the EBM. This step of the verification is different from the previous proofs since the EBM is a structural specification, while the phase level is a behavioural one. However, due to the advantage of having used the hierarchical interpreter model, we only have to show the correctness of simple transfers. Although the specification of EBM is quite complex, a large amount of automation has been achieved in the domain of verification, e.g. MEPHISTO [KuSK93]. The goal to be proved is successively broken down into a number of smaller subgoals which can then be solved more or less automatically by the MEPHISTO verification framework.

We have implemented a general tactic which proves all phase instructions automatically. This tactic, PHASES_IMP_TAC, is based on the tactics available in MEPHISTO. Using the specifications of the phase level and EBM, as described in sections 4.3 and 4.4, respectively, the verification goal and proof of the ID-phase2, for example, are given in HOL as follows:

```
set_goal([--` !t. (clk1 t = F) /\ (clk2 t = T)`--],
  (--`EBM (PC,I_MEM,RF,D_MEM,A,B,ALUout,ALUout1,DMAR,SMDR,
          LMDR,IR,IR1,IR2,IR3,BTA,IAR,ext_trap,ackn,clk1,clk2)
      ==> (!tau. ID_Ph2 (A,B,RF,IR,IR1) tau) `--) );
e PHASES_IMP_TAC;
save_top_thm("ID_PHASE2_CORRECT");
```

An implementation of PHASES_IMP_TAC is given in following, where EXPAND_DEFS_TAC, FLAT_HW_TAC and COMB_ELIM_TAC extracts all definitions of the current theory and its parent theories and expands the given goal, flattens a hierarchical description and eliminates combinatorial line variables, respectively:

```
val PHASES_IMP_TAC = EXPAND_DEFS_TAC THEN
                     FLAT_HW_TAC THEN
                     COMB_ELIM_TAC THEN
                     ASM_REWRITE_TAC[] THEN
                     REPEAT STRIP_TAC THEN
                     ASM_REWRITE_TAC[];
```

5.2.4 Instantiations

In this last part of the verification task, we deal with the correctness proofs of each instruction of the given architecture. This is obtained by simply instantiating the proofs at the class, stage and phase levels.

Since the class level is an abstraction of the instruction level (see Sect. 2), we first show the equivalence of particular instruction specifications to the instantiated related class specifications, e.g. for the ADD-Instruction we obtain the following theorem:

```
|- ADD_IMP (PC,I_MEM,RF,D_MEM) = ALU_IMP (PC,I_MEM,RF,D_MEM) add
```

Further, all correctness proofs at the class level have to be instantiated for the particular instructions of the instruction level. For example, we instantiate the proven theorem ALU_IMP_CORRECT of the ALU-class (cf. Sect. 5.2.1) with the operator constant *add* and obtain the following theorem for the ADD-instruction:

```
val ADD_IMP_CORRECT = SPEC (--`add`--) ALU_IMP_CORRECT;
val ADD_IMP_CORRECT =
 |- IF_SPEC (PC,I_MEM,IR) /\
    ID_SPEC (A,B,RF,IR,IR1) /\
    EX_SPEC (A,B,ALUout,IR1,IR2) add /\
    MEM_SPEC(ALUout1,ALUout,IR2,IR3) /\
    WB_SPEC (RF,ALUout1,IR3)
      ==> ALU_IMP (PC,I_MEM,RF,D_MEM) add : thm
```

From these two theorems, the correctness of the ADD-instruction is easily shown.

At the stage and phase levels, only the correctness of those instructions which include a class specific parameter has to be shown. The related theorems are obtained in a similar way as for the class level by simple instantiations.

6 Conclusions and Future Work

In this paper, we have presented a partly implemented methodology for the verification of RISC cores in HOL. This methodology is based on a hierarchical interpreter model which closely reflects the design hierarchy used for designing RISCs. Due to the parallelism in the execution of instructions resulting from the pipelined architecture of RISCs, a meticulous temporal abstraction has been developed and implemented in HOL. Further, the verification process could be ensured by splitting the proof goal into two parts, namely the correctness of the pipelined execution of various instructions and the correct implementation of the semantic of each single instruction by the hardware. The ease of formalizing the specifications at each level of abstraction and the similarity of the proofs between the levels have been illustrated by some simple examples implemented in HOL. The methodology is generic, in that it provides a pattern to follow when verifying RISC cores. The specification and verification templates give which definitions must be specified and which goals must be proved to verify the machine.

The implementations have been done in HOL90 within the MEPHISTO verification framework. We have implemented different specifications and proof strategies at each level of abstraction. Proof strategies for the multiple conflict cases are at present under development. We have also implemented the hardware for the DLX in a commercial VLSI design environment and are using it as one of the benchmarks for our methodology.

References

[Cohn88]   Cohn, A.: A Proof of the Viper Microprocessor: The First Level; In: VLSI Specification, Verification and Synthesis, Eds. G. Birtwistle and P.A. Subrahmanyam, Kluwer, 1988.

[HePa90]   Hennessy, J., Patterson, D.: Computer Architecture A Quantitative Approach; Morgan Kaufmann Publishers, Inc. San Mateo, California, 1990.

[Hunt87]   Hunt, W.: The Mechanical Verification of a Microprocessor Design; In: From HDL Description to Guaranteed Correct Circuit Designs, Ed. D. Borrione, North-Holland, 1987.

[Joyc89]   Joyce, J.: Multi-Level Verification of Microprocessor-Based Systems; PhD thesis, Cambridge University, December 1989.

[KuSK93] Kumar, R., Schneider, K., Kropf, Th.: Structuring and Automating Hardware Proofs in a Higher-Order Theorem-Proving Environment; Journal of Formal Methods in System Design, Vol. 2, pp. 165-230, 1993.

[Melh88]   Melham, Th.:Abstraction Mechanisms for Hardware Verification; In: VLSI Specification, Verification and Synthesis, Eds. G. Birtwistle and P. A. Subrahmanyam, Kluwer, 1988.

[SrBi90]   Srivas, M., Bickford, M.: Verification of a Pipelined Microprocessor Using Clio; In: Hardware Specification, Verification and Synthesis: Mathematical Aspects, Eds. M. Leeser and G. Brown, Springer, 1990.

[TaKu93a] Tahar, S., Kumar, R.: A Formalization of a Hierarchical Model for RISC Processors; to appear in Proc. of Euro-ARCH'93, Munich, Germany, Springer Verlag, 1993.

[TaKu93b] Tahar, S., Kumar, R.: Towards a Methodology for the Formal Hierarchical Verification of RISC Processors; to appear in Proc. of the 1993 International Conference on Computer Design, Cambridge, Massachusetts, IEEE, 1993.

[Wind90]   Windley, P.: The Formal Verification of Generic Interpreters; PhD thesis, University of California, Davis, Division of Computer Science, July 1990.

Domain Theory in HOL

Sten Agerholm

Computer Science Department, Aarhus University, DK-8000 Aarhus C, Denmark

Abstract. In this paper we present a formalization of domain theory in HOL. The notions of complete partial order, continuous function and inclusive predicate are introduced as semantic constants in HOL and fixed point induction is a derived theorem, just as we can derive other techniques for recursion. We provide tools which prove certain terms are cpos, continuous functions or inclusive predicates, automatically.

1 Introduction

Domain theory is the study of complete partial orders (cpos) and continuous functions — the mathematical foundation of denotational semantics. It provides the concepts and techniques useful to reason about formal semantics, for instance the semantics of nontermination and recursive definitions (least fixed points). Its applicability is documented in a good many text books, e.g. [Wi93, Sc86, St77].

Scott's Logic of Computable Functions was given a semantics using domain theory such that types in the logic were cpos and functions between types were continuous functions between cpos. Scott's logic has been implemented in the theorem prover LCF [Pa87, GMW79]. LCF provides a first order logic and fixed point induction to prove recursive (continuous) functions satisfy inclusive predicates (called chain-complete or admissible predicates in [Pa87, GMW79]). Inclusiveness is a semantic notion and one of the disadvantages of LCF is that the test for inclusiveness in LCF is a syntactic one which is not complete. Thus, there are inclusive predicates which cannot be used for induction in LCF. Another problem in LCF is that fixed point induction is the only way to reason about recursive definitions (structural induction is derived from fixed point induction) but using other techniques (see e.g. [Wi93]) or reasoning directly about fixed points allow us to prove more theorems than with just fixed point induction. Finally, we can mention that LCF has very few built-in theories, tools and libraries which makes it tedious to use the system.

In this paper we present a formalization of domain theory in HOL that does not suffer from such problems. The notions of complete partial order, continuous function and inclusive predicate are introduced as semantic constants in HOL and one can therefore prove terms satisfy the semantic conditions in an ad hoc way in the HOL system. But we can also build proof tools, i.e. ML programs, which prove certain terms satisfy the cpo, continuity and inclusiveness conditions, automatically, based on the syntax of these terms. For the syntax a number of constructions on cpos and continuous functions have been introduced. Another point is that in our formalization fixed point induction is a derived theorem (requiring functions are continuous), just as we can derive other techniques for recursion and reason about fixed points directly. Finally, we are able to exploit the many built-in theories and libraries of HOL in applications of the formalization because the set part of cpos are subsets of HOL types. For instance, the set can be the booleans or the natural numbers.

A limitation of our approach is that the inverse limit construction cannot be directly formalized, and therefore recursive domain equations cannot be solved at the moment. If we switch to other models as information systems [Wi93] or $P\omega$ [Sc76, Ba84] we would be able to construct solutions to equations but we might not be able to use HOL types as directly as we do now (see below).

Other formalizations of domain theory exists. Petersen [Pe93] has formalized the $P\omega$ model (see the HOL contrib library) such that all recursive domain equations can be solved. However, domains live in $P\omega$ only and it is still not clear how to lift HOL types and functions to $P\omega$ and back. Therefore very few of HOL's facilities can be exploited directly. Camilleri mechanized a theory of cpos and fixed points (see the HOL contrib library) which he used to define recursive operators in CSP trace theory [Ca90]. However, he did not consider constructions on cpos and continuous functions but proved continuity in an ad hoc way in the HOL system. A major problem with his approach is that it does not allow the continuous function space construction which is fundamental to our work. The problem is that the set of continuous functions between two cpos is a dependent subset of all HOL functions. We have solved this problem using dependent subtypes, also called term parameterized types. Though these are not provided by the HOL logic itself they can be simulated by predicates denoting subsets of types (the same approach is used in [JM93]).

The contribution of this work is a formalization of domain theoretical concepts in HOL and a number of proof tools to make the formalization more useful. The formalization is based on the book by Winskel [Wi93]. An interface and various definition tools are implemented on top of the basic tools for cpos, continuity and inclusiveness. In this way we obtain an integrated system where cpo and continuity facts are proved behind the scenes automatically (though the system is still a prototype and rather simple).

The goal is eventually to provide a system for reasoning about functional programs in HOL based on domain theory and its techniques. We hope that such a system can benefit from being embedded in as powerful and flexible a system as the HOL system (compared to a direct implementation of domain theory as in LCF), for instance by ensuring the HOL logic and tools are enherited and by allowing domain and set theoretical reasoning to be mixed.

The formalization is described in section 2, basic proof tools in section 3 and the interface in section 4. In section 5 a few examples and derived definition tools are provided to illustrate the use of the system.

2 Domain Theory

We give a brief overview of our formalization of domain theory in HOL.

2.1 Representation

A complete partial order (cpo) is a pair consisting of a set and a binary relation such that a number of conditions are satisfied. In literature a cpo is usually thought of as a set which has an associated ordering relation [Wi93, Sc86, Pa87]. Thus, it is common to say that something is an element of a cpo when one should really say it is an element of the underlying set, and so on. We can provide much the same confusion in HOL by introducing the following constants

```
"rel:cpo -> (* -> * -> bool)"
"ins:* -> cpo -> bool"
```

where we use cpo to abbreviate the type of cpos ":(* -> bool)#(* -> * -> bool)" (this is not a valid HOL abbreviation). Terms of this type, or instantiations of this type, are called cpo pairs in this paper. Applying rel to a cpo (really a cpo pair) yields the ordering relation of the cpo and using the infix ins we can express whether some term is an element of a cpo. The constant rel is defined by

```
|- rel = SND
```

and ins is defined by

```
|- !a A. a ins A = a IN (FST A)
```

where IN is used to express that some term is an element of a set (see the pred_sets library version 2.01).

2.2 Basic Definitions

Assume A is a cpo pair. Then A is called a partial order (po) if the relation is reflexive, transitive and antisymmetric on all elements of the underlying set

```
|- !A. po A = refl A /\ trans A /\ antisym A
```

Reflexivity is formalized by

```
|- !A. refl A = (!x. x ins A ==> rel A x x)
```

and the other properties are introduced in a similar way.

The notions of upper bound and least upper bound (lub) are introduced by the following defining theorems

```
|- !a B A.
    a is_ub (B,A) =
    a ins A /\ po A /\ B subset A /\
    (!b. b IN B ==> rel A b a)
|- !a B A.
    a is_lub (B,A) = a is_least ({b | b is_ub (B,A)},rel A)
```

where `subset` and `is_least` are defined as follows

```
|- !B A. B subset A = B SUBSET (FST A)
|- !a A.
     a is_least A = a ins A /\ (!b. b ins A ==> rel A a b)
```

We use the sets library of HOL which provides the set notation and the constant `SUBSET` . The definition

```
|- !B A. lub(B,A) = (@a. a is_lub (B,A))
```

gives an expression for a least upper bound if a lub exists. Then it is unique.

A partial order `D` is called a complete partial order when all chains in `D` have a least upper bound in `D`

```
|- !X. cset X = {X n | 0 <= n}
|- !X A.
     chain(X,A) =
     (!n. (X n) ins A) /\ (!n. rel A(X n)(X(n + 1)))
|- !D.
     cpo D =
     po D /\ (!X. chain(X,D) ==> (?d. d is_lub (cset X,D)))
```

A pointed cpo is a cpo with a bottom element (i.e. an element which is smaller than all other elements w.r.t. the ordering relation)

```
|- !E. pcpo E = cpo E /\ (?bt. bt is_least E)
```

Not all cpos have a bottom but it is unique if it exists. The definition

```
|- !E. bottom E = (@bt. bt is_least E)
```

gives an expression for the bottom (if it exists).

We shall consider functions on cpos and in particular continuous functions which are monotonic functions that preserve lubs of chains

```
|- !f D1 D2.
     order_pres f(D1,D2) =
     (!d.
       d ins D1 ==>
       (!d'.
         d' ins D1 ==> rel D1 d d' ==> rel D2(f d)(f d')))
|- !f D.
     determined f D = (!x. ~x ins D ==> (f x = ARB))
|- !f D1 D2.
     map f(D1,D2) =
     (!d. d ins D1 ==> (f d) ins D2) /\ determined f D1
|- !f D1 D2.
     mono f(D1,D2) =
     cpo D1 /\ cpo D2 /\ map f(D1,D2) /\ order_pres f(D1,D2)
```

```
|- !f D1 D2.
   cont f(D1,D2) =
   mono f(D1,D2) /\
   (!X.
      chain(X,D1) ==>
      (f(lub(cset X,D1)) = lub(cset(\n. f(X n)),D2)))
```

A HOL function is said to be determined by its action on elements of a cpo D when it is equal to a fixed arbitrary value ARB (predefined constant in HOL) on elements outside D . This notion is needed because a continuous function is in fact a representative for an equivalence class of HOL functions. The equivalence classes are induced by a function equality which works on subsets of types (continuous functions are specified on subsets of types only) unlike the HOL function equality which works on all elements of a type (extensional equality). Since it is very cumbersome to work with equivalence classes of functions we pick instead a certain fixed representative in each equivalence class by requiring functions are determined.

In order to ensure functions are determined by their actions we write functions using a dependent lambda abstraction which is introduced as follows

```
|- !D f. lambda D f = (\x. (x ins D => f x | ARB))
```

This is used to define all function constructors and they therefore become a kind of dependent constructors which depend on the domains on which they work.

2.3 Constructions

In this section we introduce a number of constructions on cpos and continuous functions. We do not present the theorems stating the actual cpo and continuity facts though they have been proved in HOL. An I (abbreviating internal) is used as the last letter of all names of function constructors in order to distinguish names of the internal and external (interface level) syntax (see section 4).

Discrete

```
|- !Z.
   discrete Z =
   Z,(\d1 d2. d1 IN Z /\ d2 IN Z /\ (d1 = d2))
```

The discrete construction which associates the discrete ordering with a set is useful for making HOL sets into cpos.

There are no function constructions associated with the discrete cpo construction. However, all determined functions from a discrete cpo to some cpo are continuous. This fact and instances of this fact are useful for extending HOL functions to continuous functions.

Continuous Function Space

```
|- !D1 D2.
   cf(D1,D2) =
   {f | cont f(D1,D2)},
   (\f g.
     cont f(D1,D2) /\ cont g(D1,D2) /\
     (!d. d ins D1 ==> rel D2(f d)(g d)))
```

Thus this construction is equipped with the pointwise ordering on functions.

Two well-known constructions are associated with the continuous function space, namely application and currying

```
|- !D1 D2.
   ApplyI(D1,D2) = lambda(prod(cf(D1,D2),D1))(\(f,x). f x)
|- !D1 D2 D3.
   CurryI(D1,D2,D3) =
   lambda
   (cf(prod(D1,D2),D3))
   (\g. lambda D1(\x. lambda D2(\y. g(x,y))))
```

The constant **prod** is the cpo product construction and it is defined below. It is not necessary to use **ApplyI** to apply a function to an argument. HOL application by juxtaposition can also be used.

Product

```
|- !D1 D2.
   prod(D1,D2) =
   {(d1,d2) | d1 ins D1 /\ d2 ins D2},
   (\x y. rel D1(FST x)(FST y) /\ rel D2(SND x)(SND y))
```

Thus the ordering relation is defined componentwise. The product of for instance three cpos is written as "prod(D1,prod(D2,D3))".

Associated with the product construction there are four function constructions. Of course we have the two projection functions

```
|- !D1 D2. Proj1I(D1,D2) = lambda(prod(D1,D2))FST
|- !D1 D2. Proj2I(D1,D2) = lambda(prod(D1,D2))SND
```

Besides, the tupling of elements can be extended to tupling of functions and we have the function product construction

```
|- !D1 D2 D3.
   TuplingI(D1,D2,D3) =
   lambda
   (prod(cf(D1,D2),cf(D1,D3)))
   (\(f,g). lambda D1(\x. (f x,g x)))
|- !D1 D2 D3 D4.
```

```
ProdI(D1,D2,D3,D4) =
lambda
(prod(cf(D1,D3),cf(D2,D4)))
(\(f,g). lambda(prod(D1,D2))(\(x,y). (f x,g y)))
```

Lifting

```
|- !D.
   lift D =
   {Bt} UNION {Lft d | d ins D},
   (\x y.
     (x = Bt) \/
     (?d d'. (x = Lft d) /\ (y = Lft d') /\ rel D d d'))
```

Using the lifting construction we can add a bottom element **Bt** to a cpo. The HOL constant **Bt** is part of the abstract datatype **(*)lty = Bt | Lft *** where **Lft** abbreviates 'Lift'.

There are two function constructions associated with lifting

```
|- !D. LiftI D = lambda D Lft
|- !D1 D2.
   ExtI(D1,D2) =
   lambda
   (cf(D1,D2))
   (\f.
     lambda
     (lift D1)(\x. ((x = Bt) => bottom D2 | f(unlift x))))
```

where the codomain of the function argument of **ExtI** must be a cpo with bottom. The constant **unlift** is defined by |- !d. unlift(Lft d) = d.

Sum The sum construction is similar to the product construction but it is based on the HOL sum type. The injection functions "**InlI D**" and "**InrI D**" are continuous and so is the function sum construction "**SumI(D1,D2,D3)**" which applies one of two function arguments to an element depending on which component of a sum the element belongs to.

Identity and Composition Identity and composition are not associated with any cpo constructions. The two function constructions are introduced as follows

```
|- !D. IdI D = lambda D I
|- !D1 D2 D3.
   CompI(D1,D2,D3) =
   lambda
   (prod(cf(D2,D3),cf(D1,D2)))(\(f,g). lambda D1(f o g))
```

The constants **I** and **o** are the built-in identity and composition operators, respectively.

Fixed Point Operator Finally, we introduce the least fixed point construction which is used to give semantics to recursion. It is defined as follows

```
|- (!D. pow D 0 = lambda(cf(D,D))(\f. bottom D)) /\
   (!D n.
     pow D(SUC n) =
     lambda
     (cf(D,D))
     (\f. ApplyI(D,D)(f,ApplyI(cf(D,D),D)(pow D n,f))))
|- !D. FixI D = lub(cset(pow D),cf(cf(D,D),D))
```

where **pow** is used to name the usual infinite number of continuous functions forming a chain of which the fixed point is the least upper bound.

2.4 Fixed Point Induction

Based on the definition of fixed points introduced above we can derive the following fixed point induction theorem

```
|- !E.
   pcpo E ==>
   (!f.
    f ins (cf(E,E)) ==>
    (!P.
      inclusive(P,E) /\ (bottom E) IN P /\
      (!x. x IN P ==> (f x) IN P) ==>
      (FixI E f) IN P))
```

A subset P of a cpo admits induction when it is inclusive

```
|- !P D.
   inclusive(P,D) =
   P subset D /\
   (!X.
     chain(X,D) /\ (!n. (X n) IN P) ==>
     (lub(cset X,D)) IN P)
```

Inclusiveness states that all chains in P must have a least upper bound in P. From this the fixed point induction theorem follows easily. There is no difference between sets and predicates in HOL except perhaps that the former usually are written using the set notation provided by the set library. We can therefore use 'inclusive subset' and 'inclusive predicate' interchangeable.

Since inclusive predicates must be subsets of cpos it is convenient to introduce the following constant

```
|- !D P. mk_pred(D,P) = {x | x ins D /\ x IN P}
```

for writing inclusive predicates.

3 Basic Proof Tools

Based on a number of theorems about the cpo and continuous function constructions introduced above we can implement proof tools to prove certain terms are cpos and continuous functions. These are ML programs called the cpo and continuity provers, respectively. We also provide a tool called the inclusive prover to prove certain predicates are inclusive.

The cpo prover proves that terms built using the cpo constructors are cpos, for instance

|- cpo(cf(prod(discrete X,discrete Y),lift(discrete Z)))

Actually, this term is also a pointed cpo and there is a tool to prove this fact too.

The continuity prover proves that functions built using the function constructors are continuous. It is a recursive program which derives continuity from the continuity of subterms. Therefore it is allowed to write functions which include terms different from the function constructions as long as the continuity prover is provided a number of theorems which say which cpos the terms belong to. A similar facility is provided for the cpo prover. We will not describe the syntax of continuous functions more precisely here. It will appear in the examples of the following two sections.

We can prove predicates are inclusive automatically if they satisfy certain syntactic conditions. For this the inclusive prover has been implemented. The inclusive prover can prove predicates of the form "mk_pred(D,\x.e[x])" are inclusive predicates on the cpo D if all chains in D are finite, i.e. chains are constant from a certain point, or if e has the following form

- v , any term not containing x .
- "rel E f1[x] f2[x]", f1[x] and f2[x] continuous in x .
- "f1[x] = f2[x]", f1[x] and f2[x] continuous in x .
- "~(f[x] = Bt)", f[x] continuous in x .
- "~(rel E f[x] v)", v in E and f[x] continuous in x .
- "e1[x] /\ e2[x]", e1[x] and e2[x] inclusive in x .
- "e1[x] \/ e2[x]", e1[x] and e2[x] inclusive in x .
- "e1[x] ==> e2[x]", ~e1[x] and e2[x] inclusive in x .
- "!a. a IN A ==> e[a,x]", e[a,x] inclusive in x for all a in A .

When we say something is continuous in a variable we mean the lambda abstracted term over the variable is continuous. Similarly when we say something is inclusive we mean the mk_pred term is inclusive. The syntactic checks performed by the inclusive prover appear to be very similar to those preformed by LCF (see [Pa87], page 199-200).

4 An Interface

Since function constructors are parameterized by cpos it is difficult to read and write functions. Let us consider the following example

```
"lambda
 (cf(Nat,lift Nat))
 (\f.
   ExtI
   (Nat,prod(lift Nat,lift Nat))
   (TuplingI(Nat,lift Nat,lift Nat)(f,f)))"
```

where Nat is the discrete cpo constructed from the set of all elements of type
":num" (see section 5). Writing such a function we must calculate very carefully
the cpo parameters and reading the function is difficult because of the cpo parameters on ExtI and TuplingI. Note that these are not necessary for our
understanding of what the function does. The interface provides the following
syntax

```
"\f :: Dom(cf(Nat,lift Nat)). Ext(Tupling(f,f))"
```

Parsing translates this term to the term above and pretty-printing does the
opposite translation. The constant Dom picks the set part of a cpo.

The current prototype version of the parser does not infer the domain of
the variable of a lambda abstraction. It calculates the domain of a term from
its subterms and therefore it needs the domains of variables explicitly. But it
is usually no problem to decide what the domain of a variable must be writing
an abstraction. For the same reason domains of constants different from the
constructors presented above must be made available to the interface (see section
5 for examples on how this can be done).

Just as there are new constants Ext and Tupling for the interface level
syntax of ExtI and TuplingI there are new constants for each of the other
constructors (except lambda). The difference between the interface constants
and the constructors is that the constants do not take cpo parameters. The
interface constants are not introduced by definition, only by type, since we do
not want them to abbreviate anything, merely to ensure interface level terms
type in a certain way (if they were variables it might be necessary sometimes to
give types to terms explicitly).

5 Examples

The interface and various definition and other tools constitute an integrated system where cpo, continuity and inclusiveness facts are proved behind the scenes.
Thus the user can use the HOL system in much the same way as usual without
worrying too much about these notions. Below we consider a few examples on
the use of the system.

5.1 Booleans and Natural Numbers

First we introduce the cpo of booleans and and the cpo of natural numbers using
the discrete construction and the universal set

```
#let Bool_DEF, Bool_CPO = new_cpo_definition 'Bool_DEF' 'Bool_CPO'
#    "Bool = discrete(UNIV:bool->bool)";;
Bool_DEF = |- Bool = Bool
Bool_CPO = |- cpo Bool
```

```
#let Nat_DEF, Nat_CPO = new_cpo_definition 'Nat_DEF' 'Nat_CPO'
#    "Nat = discrete(UNIV:num->bool)";;
Nat_DEF = |- Nat = Nat
Nat_CPO = |- cpo Nat
```

Here, the program new_cpo_definition is used to define two new constants
Bool and Nat and to prove they are cpos using the cpo prover. The new cpos
are added to the system which means that their definitions are unfolded/folded
by the parser/pretty-printer. This is the reason why cpo definitions are pretty-
printed as something of the form "x = x". The definitions of other things (e.g.
continuous functions) are not unfolded/folded by the system. Instead an 'ins
fact', saying which cpos these belong to, is added to the system (see below).

A strict conditional associated with the cpo of booleans can be introduced
in the following way

```
#new_insfact(IDUP "COND ins (cf(Bool,cf(Nat,cf(Nat,Nat))))");;
|- COND ins (cf(Bool,cf(Nat,cf(Nat,Nat))))
```

```
#let Cond_DEF, Cond_CF = new_constant_definition
#    'Cond_DEF''Cond_CF'
#    "Cond =
#    Ext
#    (\x::Dom Bool.
#       Ext
#       (\y::Dom Nat.
#          Ext(\z::Dom Nat. Lift(x => y | z)))))";;
Cond_DEF =
|- Cond =
   Ext
   (\x :: Dom Bool.
     Ext(\y :: Dom Nat. Ext(\z :: Dom Nat. Lift(x => y | z))))
Cond_CF =
|- Cond ins (cf(lift Bool,cf(lift Nat,cf(lift Nat,lift Nat))))
```

The strict conditional Cond is defined using the built-in conditional COND
which is parsed and pretty-printed as _=>_|_ (see the HOL Description Man-
ual). The result of _=>_|_ is lifted using Lift because the codomain of the
function argument of Ext must be a cpo with bottom. In order for the parser
and the continuity prover to know which cpo COND belongs to we first prove
this fact and add it to the system. The program IDUP (Ins Discrete Universal
Prover) proves that a term belongs to a discrete universal cpo (any term does
that as long as its type is right). It can be used because the continuous function

space of discrete universal cpos is a discrete universal cpo. The strict conditional Cond is introduced using new_constant_definition which derives continuity from subterms of functions using the continuity prover. This definition program makes the continuity theorems it proves available to the system such that a new constant can be used to define other continuous functions.

Three operations associated with the cpo of natural numbers are introduced: test for zero, predecessor and addition. The test for zero is defined as follows

```
#let iszero_DEF, iszero_CF = new_discrete_univ_definition
#    'iszero_DEF''iszero_CF'
#    ("iszero = (\n::Dom Nat. n=0)", "Bool");;
iszero_DEF |- iszero = (\n :: Dom Nat. n = 0)
iszero_CF |- iszero ins (cf(Nat,Bool))
```

```
#let Iszero_DEF, Iszero_CF = new_constant_definition
#    'Iszero_DEF''Iszero_CF'
#    "Iszero = Ext(\n::Dom Nat. Lift(iszero n))";;
Iszero_DEF |- Iszero = Ext(\n :: Dom Nat. Lift(iszero n))
Iszero_CF |- Iszero ins (cf(lift Nat,lift Bool))
```

The program new_discrete_univ_definition can be used to define a function from a discrete cpo to a (lifted) discrete cpo when both are constructed from a universal set. It proves the function is in the continuous function space (all functions between such cpos are) and add this fact to the system. Predecessor and addition are introduced in the same way as test for zero and the conditional respectively

```
|- pred = (\n :: Dom Nat. ((n = 0) => Bt | Lft(PRE n)))
|- pred ins (cf(Nat,lift Nat))
|- Pred = Ext pred
|- Pred ins (cf(lift Nat,lift Nat))
```

```
|- $+ ins (cf(Nat,cf(Nat,Nat)))
|- Add = Ext(\n :: Dom Nat. Ext(\m :: Dom Nat. Lift(n + m)))
|- Add ins (cf(lift Nat,cf(lift Nat,lift Nat)))
```

Note that the predecessor of zero is undefined (i.e. equals Bt) whereas the built-in predecessor PRE of zero is zero—all HOL functions must be total.

The following reduction theorems show how the strict versions of these functions work on their 'input'

```
|- (!e1 e2. Cond Bt e1 e2 = Bt) /\
   (!b e2. Cond b Bt e2 = Bt) /\
   (!b e1. Cond b e1 Bt = Bt) /\
   (!e1 n. Cond(Lft T)e1(Lft n) = e1) /\
   (!n e2. Cond(Lft F)(Lft n)e2 = e2)

|- (Iszero Bt = Bt) /\
```

```
      (Iszero(Lft 0) = Lft T) /\
      (!n. Iszero(Lft(n + 1)) = Lft F)

  |- (Pred Bt = Bt) /\
     (Pred(Lft 0) = Bt) /\
     (!n. Pred(Lft(n + 1)) = Lft n)

  |- (!n. Add Bt n = Bt) /\
     (!n. Add n Bt = Bt) /\
     (!n m. Add(Lft n)(Lft m) = Lft(n + m))
```

These could have been the definitions in a functional programming language. Here, however, they have been proved from the denotational semantics of the functions. The theorems give us more confidence in believing that the functions are defined in the right way. The proofs of these do not use induction but simply reduction by definition and sometimes a case split on whether or not an argument is bottom. In comparision, an induction on the natural numbers (with bottom) is used in LCF (see [Pa87], page 267) to show that addition terminates on non-bottom input. Paulson needs an induction because his definition is primitive recursive. Our addition is just defined in terms of the built-in HOL addition.

5.2   Using Fixed Point Induction

As an example of the use of fixed point induction we will show that the following (double recursive) function

```
GG_DEF =
|- GG =
   Fix
   (\g :: Dom(cf(lift Nat,lift Nat)).
     \n :: Dom(lift Nat).
      Cond(Iszero n)(Lift 1)(Pred(Add(g(Pred n),g(Pred n)))))
GG_CF = |- GG ins (cf(lift Nat,lift Nat))
```

always returns one when it terminates

```
#set_goal([],
#"!n. n ins (lift Nat) ==> ~(GG n = Bt) ==> (GG n = Lft 1)");;
"!n. n ins (lift Nat) ==> ~(GG n = Bt) ==> (GG n = Lft 1)"
```

This example is taken from [Sc86] (see section 6.6.3 and proposition 6.27).

Proving inclusiveness and continuity behind the scenes, a fixed point induction tactic FPI_TAC based on the fixed point induction theorem and the proof tools for cpos, continuity and inclusiveness reduces the goal to two subgoals

```
#expand(PORT[GG_DEF] THEN FPI_TAC(rand(concl(GG_DEF))));;
OK..
2 subgoals
```

```
"!x.
  x ins (cf(lift Nat,lift Nat)) ==>
  (!n. n ins (lift Nat) ==> ~(x n = Bt) ==> (x n = Lft 1)) ==>
  (!n.
    n ins (lift Nat) ==>
    ~(ggfun x n = Bt) ==>
    (ggfun x n = Lft 1))"
```

```
"!n.
  n ins (lift Nat) ==>
  ~(bottom(cf(lift Nat,lift Nat))n = Bt) ==>
  (bottom(cf(lift Nat,lift Nat))n = Lft 1)"
```

where PORT is PURE_ONCE_REWRITE_TAC and we have written the GG functional as ggfun (i.e. it abbreviates the lambda abstraction argument of Fix above). Knowing the theorem

```
      |- bottom(cf(lift Nat,lift Nat)) =
         (\x :: Dom(lift Nat). Bt)
```

the first subgoal (written below the second one) can be finished off since the second antecedent reduces to false.

The second subgoal requires a bit more work. First we strip the goal apart, beta convert ggfun applied to its two arguments (obtaining a conditional) and do a case split on n (the argument of ggfun)

```
2 subgoals
"~(Cond
    (Iszero(Lft n'))
    (Lft 1)
    (Pred(Add(x(Pred(Lft n')))(x(Pred(Lft n')))))) =
    Bt) ==>
  (Cond
   (Iszero(Lft n'))
   (Lft 1)
   (Pred(Add(x(Pred(Lft n')))(x(Pred(Lft n')))))) =
   Lft 1)"
      [ "x ins (cf(lift Nat,lift Nat))" ]
      [ "!n. n ins (lift Nat) ==> ~(x n = Bt) ==> (x n = Lft 1)" ]
      [ "(Lft n') ins (lift Nat)" ]

"~(Cond
    (Iszero Bt)(Lft 1)(Pred(Add(x(Pred Bt))(x(Pred Bt)))) = Bt) ==>
  (Cond
   (Iszero Bt)(Lft 1)(Pred(Add(x(Pred Bt))(x(Pred Bt)))) = Lft 1)"
      [ "x ins (cf(lift Nat,lift Nat))" ]
      [ "!n. n ins (lift Nat) ==> ~(x n = Bt) ==> (x n = Lft 1)" ]
      [ "Bt ins (lift Nat)" ]
```

The first of these follows from the reduction theorems above because Iszero and Cond are strict functions. Again the antecedent becomes false.

The proof continues with one case split after the other and simplification using the reduction theorems above. The second subgoal is split into two subgoals based on whether n' is zero or not. If it is zero and the third component of the conditional is bottom then we are finished because the conditional is strict (and the antecedent is false). If it is zero and the third component is not bottom we are finished because the consequent is true. Otherwise n' is equal to m+1 for some m . If x applied to m lifted is bottom then the result follows as before because our functions are strict. Else it is not bottom and therefore by assumption equal to 1 lifted. Now, the predecessor of one plus one is one which concludes the proof.

Acknowledgements

This work was supported by the DART project funded by the Danish Research Council. Thanks to Flemming Andersen, Richard Boulton, Esben Dalsgård, Tom Melham, Kim Dam Petersen, Laurent Thery and Glynn Winskel for discussions concerning this work. Special thanks to the reviewer who did an excellent job.

References

[Ba84] H.P. Barendregt, *The Lambda Calculus: Its Syntax and Semantics.* North Holland, 1984.

[Ca90] A.J. Camilleri, 'Mechanizing CSP Trace Theory in Higher Order Logic'. *IEEE Transactions on Software Engineering*, Vol. 16, No. 9, September 1990.

[GMW79] M.J.C. Gordon, R. Milner and C.P. Wadsworth, *Edinburgh LCF: A Mechanised Logic of Computation.* Springer-Verlag, LNCS 78, 1979.

[JM93] B. Jacobs and T. Melham, 'Translating Dependent Type Theory into Higher Order Logic'. In the *Proceedings of the International Conference on Typed Lambda Calculi and Applications*, Utrecht, 16–18 March 1993. Springer-Verlag, LNCS 664, 1993.

[Pa87] L.C. Paulson, *Logic and Computation.* Cambridge Tracts in Theoretical Computing 2, Cambridge University Press, 1987.

[Pe93] K.D. Petersen, 'Graph Model of LAMBDA in Higher Order Logic'. In the *Proceedings of the 1993 International Meeting on Higher Order Logic Theorem Proving and Its Applications*, Vancouver Canada, 11–13 August 1993 (Springer-Verlag, LNCS Series).

[Sc86] D.A. Schmidt, *Denotational Semantics.* Allyn and Bacon, 1986.

[Sc76] D. Scott, 'Data Types as Lattices'. *SIAM J. Comput.*, Vol. 5, No. 3, September 1976.

[St77] J.E. Stoy, *Denotational Semantics: The Scott-Strachey Approach to Programming Language Theory.* The MIT Press, 1977.

[Wi93] G. Winskel, *The Formal Semantics of Programming Languages.* The MIT Press, 1993.

Predicates, Temporal Logic, and Simulations

Ching-Tsun Chou

(chou@cs.ucla.edu)

Computer Science Department, University of California at Los Angeles
Los Angeles, CA 90024, U.S.A.

Abstract. We present an *integrated* theory of predicates, temporal logic, and simulations, developed using the mechanical theorem prover HOL. Both the predicate-based formulation of temporal logic and the formal theory of simulations have been investigated before in the context of HOL. What is new here is the study of the interaction between the two. In particular, we develop a formal theory of simulations in which certain kinds of temporal properties of the simulating program can be translated into those of the simulated program. Thus, if the simulated program is hard to reason about but the simulating program is easy to reason about, then our theory enables us to perform as much reasoning as possible about the "easy" program and then translate the results back to the "hard" program. The translatable temporal properties include many useful liveness properties.

1 Introduction

Temporal logic and simulation are two important tools for reasoning about concurrent programs. Although both have been investigated before in the context of HOL (see Section 9 for details), there has been hardly any work done on the interaction between the two. In particular, we are interested in the following question. Suppose that $Prog_L$ is a "concrete" program that is to be verified and $Prog_R$ is an "abstract" view of $Prog_L$ that is easier to reason about. (It should be stressed, however, that the theory developed in this paper does *not* depend on these programs being "concrete" or "abstract" in any sense. That's why we have chosen to use the subscripts $_L$ and $_R$, which simply mean "left" and "right".[1]) Is it possible to reason about the "abstract" program $Prog_R$ and then "translate" the results back to the "concrete" program $Prog_L$? The main contribution of this paper consists in a precise formulation of sufficient conditions under which this question can be answered in the affirmative.

The main ingredient of the sufficient conditions is, of course, the notion of simulations. Roughly speaking, $Prog_L$ is simulated by $Prog_R$ via the joint invariant Inv iff (if and only if) the following statements are true:[2]

---

[1] There is, however, an unintended pun: $Prog_R$ can often be thought of as the "right" way to view $Prog_L$ in actual applications of our theory.

[2] The actual definitions of simulations used in this paper are slightly more complicated; see Sections 6 and 7 for details.

1. For every initial state of $Prog_L$, there exists an initial state of $Prog_R$ such that the two initial states jointly satisfy Inv.
2. For every joint state that satisfies Inv and every transition of $Prog_L$ out of that joint state, there exists a transition of $Prog_R$ out of the same joint state such that the two destination states jointly satisfy Inv again.

Then an inductive argument shows that for every (infinite) behavior of $Prog_L$, there exists a behavior of $Prog_R$ such that the invariant Inv holds throughout the joint behavior obtained by splicing together those two behaviors. Therefore, if a temporal property Q_R is satisfied by every behavior of $Prog_R$, then it is at least plausible that Q_R can be "translated", via the joint invariant Inv, into another temporal property Q_L that is satisfied by every behavior of $Prog_L$. As a typical example, assume that Q_R is of the form:[3]

$$\Box(\Diamond C_R) \tag{1}$$

where $\Box$ means "always in the future", $\Diamond$ means "sometime in the future", and C_R is a predicate on the states of $Prog_R$. Then Q_R can be translated into Q_L of the form:

$$\Box(\Diamond C_L) \tag{2}$$

where C_L is a predicate on the states of $Prog_L$ such that:

$$C_L(s_L) \;=\; \exists s_R.\, C_R(s_R) \wedge Inv(s_L, s_R)$$

Note that (1) and (2) have the same form. In other words, the translation happens at the state, instead of the behavior, level. This is also true for other forms of translatable temporal properties.

The translatable temporal properties are of the following four forms:

$$\Box(C_R), \quad \Diamond(C_R), \quad \Box(\Diamond(C_R)), \quad \Diamond(\Box(C_R))$$

Since the following equations hold for any temporal property Q:

$$\begin{aligned}
\Box(\Box(Q)) &= \Box(Q) \\
\Diamond(\Diamond(Q)) &= \Diamond(Q) \\
\Diamond(\Box(\Diamond(Q))) &= \Box(\Diamond(Q)) \\
\Box(\Diamond(\Box(Q))) &= \Diamond(\Box(Q))
\end{aligned}$$

any temporal property built up from a single predicate on states and arbitrary numbers of $\Box$ and $\Diamond$ is translatable.

The theory presented in this paper is formalized in a higher-order logic, which is called the HOL logic because it is essentially the logic mechanized by the HOL system [5]. The HOL logic is as powerful as Zermelo set theory and is sufficient for expressing most ordinary mathematical theories as definitional

---

[3] In fact, C_R must be "coerced" into a temporal property before $\Box$ and $\Diamond$ can be applied, but we shall ignore this technicality for now; see Section 8 for details.

extensions. In particular, we can *embed* many other logics in the HOL logic by formalizing their semantics in the HOL logic. The semantic embedding approach has two advantages. Firstly, *security*: the embedded logic becomes a definitional extension of the HOL logic and hence no inconsistency is introduced. Instead of being postulated, all inference rules[4] of the embedded logic are proved as theorems of the HOL logic. Secondly, *flexibility*: reasoning in the embedded logic and reasoning in the HOL logic can be mixed in arbitrary ways. We no longer need to worry about the completeness of inference rules of the embedded logic, for we can always introduce new rules by proving them in the HOL logic.

A *predicate* P is a function whose type is of the form $\tau \rightarrow$ bool, where τ is called the *domain* of P. Predicates are important because most semantic embeddings depend on them. The basic scheme is as follows:

1. Propositions in the embedded logic are represented by predicates over suitable domains in the HOL logic.
2. The validity assertion and the logical operators are *lifted* from booleans to predicates over the same domain to furnish the logical infrastructure.
3. Lifted logics on different domains are interrelated via the image and inverse image operators.
4. Extra-logical operators, such as temporal modalities, are represented by exploiting the structures of the domains of predicates.

The temporal logic used in this paper—a simplified version of the temporal logic of actions (TLA) [6]—is embedded in the HOL logic using exactly this scheme. TLA is a linear-time, future-tense temporal logic in which the basic assertions include not only predicates on states (called *conditions*) but also predicates on transitions (called *actions*). We shall use TLA to express both semantics and temporal properties of concurrent programs.

Finally, some words that put this work in perspective are in order. This work is part of the author's ongoing research into machine-assisted verification of distributed algorithms, such as the well-known distributed minimum spanning tree algorithm of Gallager, Humblet, and Spira (GHS) [4]. This research was motivated by two beliefs. The first belief is that the only way to gain a high degree of confidence in the correctness of a concurrent program is to construct a formal proof. This belief is based both on the fact that most concurrent programs are so nondeterministic as to render testing useless, and on the author's own experience [1] of discovering in the program given in [4] a bug that, despite the degree to which the GHS algorithm had been scrutinized by numerous researchers, did not seem to be noticed before. The second belief is that formal correctness proofs of concurrent programs should be constructed and checked with the help of mechanical theorem provers. As a rule, formal proofs are long and tedious. It is unrealistic to expect that a human prover maintain the necessary logical rigor throughout a proof. Besides, the human prover approach is a terrible waste of human ingenuity, which machines lack. It is more profitable to let machines do the tedious work of maintaining logical rigor and humans do the creative work

---

[4] Axioms are regarded as a special kind of inference rules.

of directing mechanical provers. To be sure, the state of the art of mechanical theorem proving is still quite primitive. But the author believes that at present the human/machine prover approach is already superior to the purely human prover approach.

The rest of this paper is organized as follows. Section 2 discusses the pertinent features of the HOL logic and some notational conventions, Section 3 the lifted logic of predicates, Section 4 the semantic embedding of TLA, Section 5 state transition systems (which are used to represent concurrent programs), Section 6 simulations between state transition systems without fairness, Section 7 simulations between state transition systems with fairness, Section 8 how to use simulation results to translate temporal properties, and Section 9 related work. Due to space limitations, proofs are sketchy, if given at all.

Acknowledgements. The author is grateful to Professors Eli Gafni and David Martin for their guidance and to Vernon Austel, Peter Homeier, Brad Pierce, and Ray Toal for their friendly criticism.

2 Higher-Order Logic

The logic used in this paper is the higher-order logic supported by the HOL system [5], with the minor addition of sequence types as discussed in Section 2.1. For the purpose of understanding this paper, those who are not familiar with higher-order logic may think of it as a generalization of first-order logic in which variables can range over not only individuals but also functions on individuals, functions on functions on individuals, and so on. To avoid such troubles as Russell's Paradox, every term in higher-order logic has a *type* and term formation must obey certain typing rules. In particular, if term t denotes a function and term u an element, then the term tu (called a *combination*) that denotes the value of t at u can be formed only when t's type is of the form $\alpha \rightarrow \beta$ and u's type is α. In addition to $\rightarrow$, other (possibly nullary) type constructors that we shall use are: bool, the type of booleans, num, the type of natural numbers, and $\times$, the cartesian product.

Formulas are simply terms of type bool. The meanings of the following logical operators should be clear: $\neg, \wedge, \vee, \Rightarrow, \forall, \exists$. Later we shall use the **bold** versions of these symbols for logical operators on predicates (see Section 3). Free variables in a theorem are implicitly universally quantified.

2.1 Sequences

For any type τ, let τ^k be the type of sequences of length k over τ, where k is always one of 1, 2, and ω in this paper. This kind of sequence types is not available in HOL, but we use it nonetheless in order to have a uniform notation. We can easily implement τ^1, τ^2, and τ^ω using τ, $\tau \times \tau$, and num $\rightarrow \tau$ respectively.

A sequence consisting of elements $x_0, x_1, \ldots$ is denoted by $\langle\!\langle x_0, x_1, \ldots \rangle\!\rangle$. The componentwise pairing and projection operators on sequences of the same length (which can be 1 or 2) are defined by:

$$\langle\!\langle x_0, x_1, \ldots \rangle\!\rangle \bullet \langle\!\langle y_0, y_1, \ldots \rangle\!\rangle \quad \triangleq \quad \langle\!\langle (x_0, y_0), (x_1, y_1), \ldots \rangle\!\rangle$$

$$\mathsf{p}_L \langle\!\langle (x_0, y_0), (x_1, y_1), \ldots \rangle\!\rangle \quad \triangleq \quad \langle\!\langle x_0, x_1, \ldots \rangle\!\rangle$$

$$\mathsf{p}_R \langle\!\langle (x_0, y_0), (x_1, y_1), \ldots \rangle\!\rangle \quad \triangleq \quad \langle\!\langle y_0, y_1, \ldots \rangle\!\rangle$$

where $\triangleq$ means equality by definition.

2.2 Notational Conventions

Type variables are represented by Greek letters, term variables by identifiers in *Italic* font, (type or term) constants by identifiers in sans serif font or by special symbols. Identifiers may have subscripts and superscripts.

Combinations associate to the left, so $tuvw$ means $((tu)v)w$. Infix operators associate to the right, so $t \wedge u \wedge v \wedge w$ means $t \wedge (u \wedge (v \wedge w))$. Some special symbols used in this paper are listed below in increasing binding power from left to right:

$$\triangleq \quad = \quad \begin{matrix} \forall \\ \exists \end{matrix} \quad \Rightarrow \quad \begin{matrix} \wedge \\ \vee \end{matrix} \quad \neg \quad \models \quad \begin{matrix} \forall \\ \exists \end{matrix} \quad \Rightarrow \quad \begin{matrix} \wedge \\ \vee \end{matrix} \quad \neg$$

All other combinations bind more tightly than these symbols.

3 Lifted Logic of Predicates

Let $P_1, \ldots, P_n$, and Q be predicates over the same domain τ. The *validity* of Q under the assumptions $P_1, \ldots, P_n$, denoted $P_1, \ldots, P_n \models Q$, is defined by:

$$P_1, \ldots, P_n \models Q \quad \triangleq \quad \forall x.\, P_1(x) \wedge \ldots \wedge P_n(x) \Rightarrow Q(x)$$

The left side of $\models$ is a list of predicates and the definition above is ultimately based on primitive recursion on lists.

Logical operators on predicates are defined by pointwise lifting of the corresponding operators on booleans:

$$\mathsf{T}(x) \quad \triangleq \quad \mathsf{T}$$

$$\mathsf{F}(x) \quad \triangleq \quad \mathsf{F}$$

$$(\neg P)(x) \quad \triangleq \quad \neg P(x)$$

$$(P \wedge Q)(x) \quad \triangleq \quad P(x) \wedge Q(x)$$

$$(P \vee Q)(x) \quad \triangleq \quad P(x) \vee Q(x)$$

$$(P \Rightarrow Q)(x) \quad \triangleq \quad P(x) \Rightarrow Q(x)$$

$$(\forall R)(x) \quad \triangleq \quad \forall i.\, R(i)(x)$$

$$(\exists R)(x) \quad \triangleq \quad \exists i.\, R(i)(x)$$

where $x : \tau$, $P : \tau \to$ bool, $Q : \tau \to$ bool, and $R : \iota \to \tau \to$ bool. The last two definitions essentially say the following: if $P[i]$ is a predicate that may contain variable $i : \iota$ free, then the following *families* of equations hold:

$$(\forall i. P[i])(x) \;=\; \forall i. P[i](x)$$
$$(\exists i. P[i])(x) \;=\; \exists i. P[i](x)$$

Note that the variable i is bound on the right sides by $\forall$ and $\exists$.

The lifted logical operators and validity behave exactly like their unlifted counterparts. For instance, the following are the lifted versions of the elimination and introduction rules for implication:

$$(\models P \Rightarrow Q) \;\Rightarrow\; (\models P \Rightarrow \models Q)$$
$$(P \models Q) \;\Rightarrow\; (\models P \Rightarrow Q)$$

Since HOL itself is based on a sequent calculus, the sequent formulation of lifted validity (*i.e.*, the presence of assumptions on the left side of $\models$) facilitates theorem proving in lifted logics by allowing the lifting of tactics and tacticals in HOL. For details, see [2].

3.1 Images and Inverse Images

Let $P : \sigma \to$ bool, $Q : \tau \to$ bool, and $f : \sigma \to \tau$. The *image* of P under f, $\triangleright f(P)$, is defined by:

$$\triangleright f(P)(y) \;\triangleq\; \exists x. P(x) \wedge (y = f(x))$$

The *inverse image* of Q under f, $\triangleleft f(Q)$, is defined by:

$$\triangleleft f(Q)(x) \;\triangleq\; Q(f(x))$$

In this paper it will always be the case that f is some kind of "projection" so that $\triangleright f(P)$ can be viewed as "existential quantification" over the components discarded by the projection f and $\triangleleft f(Q)$ the "coercion" of a predicate on τ into one on σ.

3.2 Important Theorems

Predicates over the same domain are partially ordered by lifted implication: P is *stronger* than Q (or, Q is *weaker* than P) iff $\models P \Rightarrow Q$ holds.

Theorem 1. [*Monotonicity*]

$$\models P \Rightarrow Q \;\Rightarrow\; \models \triangleright f(P) \Rightarrow \triangleright f(Q)$$
$$\models P \Rightarrow Q \;\Rightarrow\; \models \triangleleft f(P) \Rightarrow \triangleleft f(Q)$$

Theorem 2. [*Absorption*]

$$\triangleright f(P) \wedge Q \;=\; \triangleright f(P \wedge \triangleleft f(Q))$$

Theorem 3. [*Distributivity*]

$$
\begin{aligned}
\blacktriangleleft f(\mathbf{T}) &= \mathbf{T} \\
\blacktriangleleft f(\mathbf{F}) &= \mathbf{F} \\
\blacktriangleleft f(\neg P) &= \neg \blacktriangleleft f(P) \\
\blacktriangleleft f(P \wedge Q) &= \blacktriangleleft f(P) \wedge \blacktriangleleft f(Q) \\
\blacktriangleleft f(P \vee Q) &= \blacktriangleleft f(P) \vee \blacktriangleleft f(Q) \\
\blacktriangleleft f(P \Rightarrow Q) &= \blacktriangleleft f(P) \Rightarrow \blacktriangleleft f(Q) \\
\blacktriangleleft f(\forall i.\, P[i]) &= \forall i.\, \blacktriangleleft f(P[i]) \qquad (3) \\
\blacktriangleleft f(\exists i.\, P[i]) &= \exists i.\, \blacktriangleleft f(P[i]) \qquad (4)
\end{aligned}
$$

Note that (3) and (4) are families of theorems.

4 Temporal Logic of Actions

In TLA there are three kinds of objects over which predicates are needed: *states*, *transitions*, and *behaviors*. A state can be anything but usually is a tuple of values of program variables, a transition is a pair of states representing a step of program execution, and a behavior is an infinite sequence of states representing a complete history of program execution. Predicates on states, transitions, and behaviors are called respectively *conditions*, *actions*, and *temporal properties*. To summarize, if the type of states is σ, then:

Object	Type	Predicate	Type
State	σ^1	Condition	$\sigma^1 \to \mathbf{bool}$
Transition	σ^2	Action	$\sigma^2 \to \mathbf{bool}$
Behavior	σ^ω	Temporal property	$\sigma^\omega \to \mathbf{bool}$

In the rest of this paper, we shall use s with subscripts to range over states, C, I conditions, A, B, N actions, and P, Q, R temporal properties.

4.1 Projections and Coercions

Several *projections* map between states, transitions, and behaviors:

$$
\begin{aligned}
\mathsf{ts}_0 \langle\!\langle s_0, s_1 \rangle\!\rangle &\triangleq \langle\!\langle s_0 \rangle\!\rangle \\
\mathsf{ts}_1 \langle\!\langle s_0, s_1 \rangle\!\rangle &\triangleq \langle\!\langle s_1 \rangle\!\rangle \\
\mathsf{bs} \langle\!\langle s_0, s_1, \ldots \rangle\!\rangle &\triangleq \langle\!\langle s_0 \rangle\!\rangle \\
\mathsf{bt} \langle\!\langle s_0, s_1, \ldots \rangle\!\rangle &\triangleq \langle\!\langle s_0, s_1 \rangle\!\rangle
\end{aligned}
$$

These projections are used in combination with the inverse image operator to *coerce* a condition into an action, an action into a temporal property, and so on:

Theorem 4.

$$\text{⊲ts}_0(C)\langle\!\langle s_0, s_1 \rangle\!\rangle \quad = \quad C\langle\!\langle s_0 \rangle\!\rangle$$
$$\text{⊲ts}_1(C)\langle\!\langle s_0, s_1 \rangle\!\rangle \quad = \quad C\langle\!\langle s_1 \rangle\!\rangle$$
$$\text{⊲bs}\,(C)\langle\!\langle s_0, s_1, \ldots \rangle\!\rangle \quad = \quad C\langle\!\langle s_0 \rangle\!\rangle$$
$$\text{⊲bt}\,(A)\langle\!\langle s_0, s_1, \ldots \rangle\!\rangle \quad = \quad A\langle\!\langle s_0, s_1 \rangle\!\rangle$$

4.2 Enabled, May, and Must

An action A is *enabled* at a state s_0 iff there exists an A-step from s_0:

$$\text{En}(A)\langle\!\langle s_0 \rangle\!\rangle \quad \triangleq \quad \exists s_1.\,A\langle\!\langle s_0, s_1 \rangle\!\rangle$$

A *may-A* step is either an A-step or a *stuttering* step:

$$[A]\langle\!\langle s_0, s_1 \rangle\!\rangle \quad \triangleq \quad A\langle\!\langle s_0, s_1 \rangle\!\rangle \vee (s_1 = s_0)$$

A *must-A* step is a non-stuttering A-step:

$$\langle A \rangle\langle\!\langle s_0, s_1 \rangle\!\rangle \quad \triangleq \quad A\langle\!\langle s_0, s_1 \rangle\!\rangle \wedge (s_1 \neq s_0)$$

4.3 Temporal Operators

The two basic temporal operators are $\Box$ and $\Diamond$, which means respectively "always in the future" and "sometime in the future" ("the future" includes "now"):

$$\Box P\langle\!\langle s_0, s_1, \ldots \rangle\!\rangle \quad \triangleq \quad \forall n.\,P\langle\!\langle s_n, s_{n+1}, \ldots \rangle\!\rangle$$
$$\Diamond P\langle\!\langle s_0, s_1, \ldots \rangle\!\rangle \quad \triangleq \quad \exists n.\,P\langle\!\langle s_n, s_{n+1}, \ldots \rangle\!\rangle$$

A is *weakly fair* iff A is infinitely often fired or infinitely often disabled:

$$\text{WF}(A) \quad \triangleq \quad \Box(\Diamond(\text{⊲bt}(A))) \vee \Box(\Diamond(\text{⊲bs}(\neg\text{En}(A))))$$

A is *strongly fair* iff A is infinitely often fired or almost always disabled:

$$\text{SF}(A) \quad \triangleq \quad \Box(\Diamond(\text{⊲bt}(A))) \vee \Diamond(\Box(\text{⊲bs}(\neg\text{En}(A))))$$

4.4 Important Theorems

Theorem 5. *[Monotonicity]*

$$[N], \text{⊲ts}_0(I), \text{⊲ts}_1(I) \models A \Rightarrow B \,\wedge\, I \models \text{En}(B) \Rightarrow \text{En}(A)$$
$$\Rightarrow \quad \Box(\text{⊲bt}[N]), \Box(\text{⊲bs}(I)) \models \text{WF}(A) \Rightarrow \text{WF}(B)$$

This monotonicity property holds for SF as well.

Theorem 6. [*Strengthening*]

$$\Box(◁\mathsf{bs}(I)) \;\models\; \Box(◁\mathsf{bs}(C)) \Rightarrow \Box(◁\mathsf{bs}(C \wedge I))$$
$$\Box(◁\mathsf{bs}(I)) \;\models\; \Diamond(◁\mathsf{bs}(C)) \Rightarrow \Diamond(◁\mathsf{bs}(C \wedge I))$$
$$\Box(◁\mathsf{bs}(I)) \;\models\; \Box(\Diamond(◁\mathsf{bs}(C))) \Rightarrow \Box(\Diamond(◁\mathsf{bs}(C \wedge I)))$$
$$\Box(◁\mathsf{bs}(I)) \;\models\; \Diamond(\Box(◁\mathsf{bs}(C))) \Rightarrow \Diamond(\Box(◁\mathsf{bs}(C \wedge I)))$$

Theorem 7. [*Distributivity*]

$$◁\mathsf{p}_L(◁\mathsf{ts}_0(C)) \;=\; ◁\mathsf{ts}_0(◁\mathsf{p}_L(C))$$
$$◁\mathsf{p}_L(◁\mathsf{ts}_1(C)) \;=\; ◁\mathsf{ts}_1(◁\mathsf{p}_L(C))$$
$$◁\mathsf{p}_L(◁\mathsf{bs}\,(C)) \;=\; ◁\mathsf{bs}\,(◁\mathsf{p}_L(C))$$
$$◁\mathsf{p}_L(◁\mathsf{bt}\,(A)) \;=\; ◁\mathsf{bt}\,(◁\mathsf{p}_L(A))$$

$$◁\mathsf{p}_L(\mathsf{En}(A)) \;=\; \mathsf{En}(◁\mathsf{p}_L(A))$$

$$◁\mathsf{p}_L(\Box P) \;=\; \Box(◁\mathsf{p}_L(P))$$
$$◁\mathsf{p}_L(\Diamond P) \;=\; \Diamond(◁\mathsf{p}_L(P))$$

$$◁\mathsf{p}_L(\mathsf{WF}(A)) \;=\; \mathsf{WF}(◁\mathsf{p}_L(A))$$
$$◁\mathsf{p}_L(\,\mathsf{SF}(A)) \;=\; \mathsf{SF}(◁\mathsf{p}_L(A))$$

All these equations hold for $◁\mathsf{p}_R$ as well.

Theorem 8. [*Distributivity*]

$$▷\mathsf{p}_L(\Box(◁\mathsf{bs}(C))) \;=\; \Box(◁\mathsf{bs}(▷\mathsf{p}_L(C)))$$
$$▷\mathsf{p}_L(\Diamond(◁\mathsf{bs}(C))) \;=\; \Diamond(◁\mathsf{bs}(▷\mathsf{p}_L(C)))$$

$$▷\mathsf{p}_L(\Box(\Diamond(◁\mathsf{bs}(C)))) \;=\; \Box(\Diamond(◁\mathsf{bs}(▷\mathsf{p}_L(C))))$$
$$▷\mathsf{p}_L(\Diamond(\Box(◁\mathsf{bs}(C)))) \;=\; \Diamond(\Box(◁\mathsf{bs}(▷\mathsf{p}_L(C))))$$

All these equations hold for $▷\mathsf{p}_R$ as well. Note that the distributivity laws for $▷\mathsf{p}_L$ and $▷\mathsf{p}_R$ are weaker than those for $◁\mathsf{p}_L$ and $◁\mathsf{p}_R$ in Theorem 7.

5 State Transition Systems

A *state transition system* (STS) is a pair consisting of a condition I and an action N over the same type σ of states:

$$(\, I : \sigma^1 \to \mathsf{bool}\,, \; N : \sigma^2 \to \mathsf{bool}\,)$$

The STS (I, N) represents a program whose initial states are characterized by I and whose possible transitions by N. Consequently the possible behaviors of the program is characterized by:

$$[\![(I, N)]\!] \;\triangleq\; ◁\mathsf{bs}(I) \wedge \Box(◁\mathsf{bt}[N])$$

In fact, we shall identify (I, N) with the program and view $[\![(I, N)]\!]$ as its semantics, hence the $[\![\cdot]\!]$ notation.

Note that stuttering steps are always allowed by an STS, so nothing is guaranteed to happen in its behaviors. To ensure something will eventually happen, explicit *fairness assumptions* must be added. A *state transition system with fairness* (STS-F) is a quadruple of the form:

$$(I : \sigma^1 \to \mathsf{bool}, \; N : \sigma^2 \to \mathsf{bool}, \; A : \alpha \to \sigma^2 \to \mathsf{bool}, \; B : \beta \to \sigma^2 \to \mathsf{bool} \,)$$

Its semantics is:

$$[\![(I, N, A, B)]\!] \;\triangleq\; [\![(I, N)]\!] \wedge (\,\forall i.\, \mathsf{WF}\langle A(i)\rangle\,) \wedge (\,\forall j.\, \mathsf{SF}\langle B(j)\rangle\,)$$

The idea is that A is an indexed family of actions each of which must be weakly fair in every behavior of (I, N, A, B); a similar remark applies to B.

Since weak fairness and strong fairness can be treated in the same manner, we shall drop the strong fairness part and focus on *state transition systems with weak fairness* (STS-WF's) in the remainder of this paper, where an STS-WF is a triple of the form:

$$(I : \sigma^1 \to \mathsf{bool}, \; N : \sigma^2 \to \mathsf{bool}, \; A : \alpha \to \sigma^2 \to \mathsf{bool} \,)$$

and its semantics is:

$$[\![(I, N, A)]\!] \;\triangleq\; [\![(I, N)]\!] \wedge (\,\forall i.\, \mathsf{WF}\langle A(i)\rangle\,)$$

6 Simulations between STS's

Consider the following two STS's, which are called, for lack of better names, the *left STS* (subscript $_L$) and the *right STS* (subscript $_R$):

$$(I_L : \sigma_L^1 \to \mathsf{bool}, \; N_L : \sigma_L^2 \to \mathsf{bool} \,)$$
$$(I_R : \sigma_R^1 \to \mathsf{bool}, \; N_R : \sigma_R^2 \to \mathsf{bool} \,)$$

It is important to note that the two STS's may have entirely different types of states. We say that the left STS is *simulated* by the right STS via the joint action $Par : (\sigma_L \times \sigma_R)^2 \to \mathsf{bool}$ and the joint invariant $Inv : (\sigma_L \times \sigma_R)^1 \to \mathsf{bool}$ iff the following two statements are true:

1. For any initial state $\langle\!\langle s_L \rangle\!\rangle$ of the left STS, there exists a corresponding initial state $\langle\!\langle s_R \rangle\!\rangle$ of the right STS such that the joint state $\langle\!\langle s_L \rangle\!\rangle \bullet \langle\!\langle s_R \rangle\!\rangle$ satisfies Inv.

2. For any joint state $\langle\!\langle s_L \rangle\!\rangle \bullet \langle\!\langle s_R \rangle\!\rangle$ satisfying Inv and any transition $\langle\!\langle s_L, s_L' \rangle\!\rangle$ of the left STS, there exists a corresponding transition $\langle\!\langle s_R, s_R' \rangle\!\rangle$ of the right STS such that the joint transition $\langle\!\langle s_L, s_L' \rangle\!\rangle \bullet \langle\!\langle s_R, s_R' \rangle\!\rangle$ satisfies Par and the joint state $\langle\!\langle s_L' \rangle\!\rangle \bullet \langle\!\langle s_R' \rangle\!\rangle$ satisfies Inv.

Formally,

$$Sim(I_L, N_L)(I_R, N_R)(Par)(Inv) \triangleq$$
$$(\forall s_L . I_L \langle\!\langle s_L \rangle\!\rangle \Rightarrow$$
$$\exists s_R . I_R \langle\!\langle s_R \rangle\!\rangle \wedge Inv(\langle\!\langle s_L \rangle\!\rangle \bullet \langle\!\langle s_R \rangle\!\rangle)) \tag{5}$$
$$\wedge \quad (\forall s_L \, s_R . Inv(\langle\!\langle s_L \rangle\!\rangle \bullet \langle\!\langle s_R \rangle\!\rangle) \Rightarrow$$
$$\forall s_L' . N_L \langle\!\langle s_L, s_L' \rangle\!\rangle \Rightarrow$$
$$\exists s_R' . N_R \langle\!\langle s_R, s_R' \rangle\!\rangle \wedge Par(\langle\!\langle s_L, s_L' \rangle\!\rangle \bullet \langle\!\langle s_R, s_R' \rangle\!\rangle)$$
$$\wedge Inv(\langle\!\langle s_L' \rangle\!\rangle \bullet \langle\!\langle s_R' \rangle\!\rangle)) \tag{6}$$

Theorem 9. [*Simulation Theorem for STS's*]

$$Sim(Prog_L)(Prog_R)(Par)(Inv) \Rightarrow$$
$$\models [\, Prog_L \,] \Rightarrow \triangleright p_L (\, \triangleleft p_R [\, Prog_R \,] \wedge \Box(\triangleleft bt[Par]) \wedge \Box(\triangleleft bs(Inv)))$$

Proof: Let $Prog_L = (I_L, N_L)$ and $Prog_R = (I_R, N_R)$. Then the conclusion of the theorem is equivalent to the following statement:

$$\forall \langle\!\langle s_L^0, s_L^1, \ldots \rangle\!\rangle . \ I_L \langle\!\langle s_L^0 \rangle\!\rangle \wedge (\forall n . [N_L] \langle\!\langle s_L^n, s_L^{n+1} \rangle\!\rangle) \Rightarrow$$
$$\exists \langle\!\langle s_R^0, s_R^1, \ldots \rangle\!\rangle . \ I_R \langle\!\langle s_R^0 \rangle\!\rangle \wedge (\forall n . [N_R] \langle\!\langle s_R^n, s_R^{n+1} \rangle\!\rangle) \wedge$$
$$(\forall n . [Par](\langle\!\langle s_L^n, s_L^{n+1} \rangle\!\rangle \bullet \langle\!\langle s_R^n, s_R^{n+1} \rangle\!\rangle)) \wedge (\forall n . Inv(\langle\!\langle s_L^n \rangle\!\rangle \bullet \langle\!\langle s_R^n \rangle\!\rangle))$$

Intuitively, we construct the right behavior $\langle\!\langle s_R^0, s_R^1, \ldots \rangle\!\rangle$ from the left behavior $\langle\!\langle s_L^0, s_L^1, \ldots \rangle\!\rangle$ in a state-by-state fashion:

1. s_R^0 exists due to (5) and the existence of s_L^0;
2. s_R^{n+1} exists due to (6) and the existence of s_R^n, s_L^n, and s_L^{n+1}, for each n.

Formally, we use the selection operator ε and primitive recursion on n to define the right behavior in terms of the left behavior and prove the desired result by induction on n. Note that skolemization does *not* work here because of the clause about *Par* in (6). ∎

7 Simulations between STS-WF's

Consider the following two STS-WF's:

$$(\ I_L : \sigma_L^1 \rightarrow \text{bool}, \ N_L : \sigma_L^2 \rightarrow \text{bool}, \ A_L : \alpha \rightarrow \sigma_L^2 \rightarrow \text{bool}\ )$$
$$(\ I_R : \sigma_R^1 \rightarrow \text{bool}, \ N_R : \sigma_R^2 \rightarrow \text{bool}, \ A_R : \alpha \rightarrow \sigma_R^2 \rightarrow \text{bool}\ )$$

Note that the A_L and A_R must be indexed by the same type. We say that the left STS-WF is *simulated* by the right STS-WF via the joint action *Par* : $(\sigma_L \times \sigma_R)^2 \rightarrow \text{bool}$ and the joint invariant *Inv* : $(\sigma_L \times \sigma_R)^1 \rightarrow \text{bool}$ iff the following three statements are true:

1. The left STS (I_L, N_L) is simulated by the right STS (I_R, N_R) via the joint action *Par* and the joint invariant *Inv*.

2. For each $i : \alpha$ and under the assumptions $[Par]$, $\triangleleft ts_0(Inv)$, and $\triangleleft ts_1(Inv)$, $\langle A_L(i) \rangle$ is stronger than $\langle A_R(i) \rangle$ when both are viewed as actions on the joint states.

3. For each $i : \alpha$ and under the assumption Inv, $En\langle A_L(i) \rangle$ is weaker than $En\langle A_R(i) \rangle$ when both are viewed as conditions on the joint states.

Formally,

$$SimWF(I_L, N_L, A_L)(I_R, N_R, A_R)(Par)(Inv) \triangleq$$

$$Sim(I_L, N_L)(I_R, N_R)(Par)(Inv) \tag{7}$$

$$\wedge \ (\forall i. \ [Par], \triangleleft ts_0(Inv), \triangleleft ts_1(Inv) \models \triangleleft p_L \langle A_L(i) \rangle \Rightarrow \triangleleft p_R \langle A_R(i) \rangle) \tag{8}$$

$$\wedge \ (\forall i. \ Inv \models \triangleleft p_R(En\langle A_R(i) \rangle) \Rightarrow \triangleleft p_L(En\langle A_L(i) \rangle)) \tag{9}$$

Theorem 10. [*Simulation Theorem for STS-WF's*]

$$SimWF(Prog_L)(Prog_R)(Par)(Inv) \Rightarrow$$

$$\models [\![Prog_L]\!] \Rightarrow \triangleright p_L(\triangleleft p_R [\![Prog_R]\!] \wedge \Box(\triangleleft bt[Par]) \wedge \Box(\triangleleft bs(Inv)))$$

Proof: Let $Prog_L = (I_L, N_L, A_L)$ and $Prog_R = (I_R, N_R, A_R)$. It follows from (7) and Theorem 9 that:

$$\models [\![Prog'_L]\!] \Rightarrow \triangleright p_L(\triangleleft p_R [\![Prog'_R]\!] \wedge \Box(\triangleleft bt[Par]) \wedge \Box(\triangleleft bs(Inv))) \tag{10}$$

where $Prog'_L = (I_L, N_L)$ and $Prog'_R = (I_R, N_R)$. Strengthening both sides of $\Rightarrow$ with $(\forall i. WF\langle A_L(i) \rangle)$ in (10), it then follows from Theorems 2, 3, and 7 that:

$$\models [\![Prog'_L]\!] \wedge (\forall i. WF\langle A_L(i) \rangle) \Rightarrow$$

$$\triangleright p_L \left(\begin{array}{c} \triangleleft p_R [\![Prog'_R]\!] \wedge (\forall i. WF(\triangleleft p_L \langle A_L(i) \rangle)) \\ \wedge \Box(\triangleleft bt[Par]) \wedge \Box(\triangleleft bs(Inv)) \end{array} \right) \tag{11}$$

By (8), (9), and Theorems 5 and 7, the argument of $\triangleright p_L$ in (11) is stronger than $(\forall i. WF(\triangleleft p_R \langle A_R(i) \rangle))$. This implies the desired result by Theorem 1 and the definition of $[\![\cdot]\!]$ for STS-WF's. ∎

8 How to Use Simulations

The conclusion of Theorem 10 is of the form:

$$\models [\![Prog_L]\!] \Rightarrow \triangleright p_L(\triangleleft p_R [\![Prog_R]\!] \wedge \Box(\triangleleft bt[Par]) \wedge \Box(\triangleleft bs(Inv))) \tag{12}$$

where $Prog_L$ and $Prog_R$ are both STS-WF's. How does a result like (12) help us? The answer is that it enables us to *translate* certain kinds of temporal properties of $Prog_R$ into those of $Prog_L$, in the following manner.

A temporal property Q_R of $Prog_R$ is satisfied by each behavior of $Prog_R$, *i.e.*,

$$\models [\![Prog_R]\!] \Rightarrow Q_R \tag{13}$$

Our translation technique applies to Q_R's of the following forms:

$$\Box(\textrm{◁bs}(C_R)) , \quad \Diamond(\textrm{◁bs}(C_R)), \quad \Box(\Diamond(\textrm{◁bs}(C_R))), \quad \Diamond(\Box(\textrm{◁bs}(C_R)))$$

where C_R is a *condition* on the right states. We shall illustrate the translation method on the third form, so let us assume that:

$$Q_R = \Box(\Diamond(\textrm{◁bs}(C_R))) \tag{14}$$

The other three cases are all similar. It follows from assumptions (12), (13), (14), and Theorem 1 that:

$$\models [\![Prog_L]\!] \Rightarrow {\triangleright}\mathsf{p}_L(\ {\triangleleft}\mathsf{p}_R(\Box(\Diamond(\textrm{◁bs}(C_R)))) \wedge \Box(\textrm{◁bs}(Inv))\ ) \tag{15}$$

Pushing ${\triangleleft}\mathsf{p}_R$ all the way in using Theorem 7 and then invoking Theorems 1 and 6, we get:

$$\models [\![Prog_L]\!] \Rightarrow {\triangleright}\mathsf{p}_L(\Box(\Diamond(\textrm{◁bs}({\triangleleft}\mathsf{p}_R(C_R) \wedge Inv)))) \tag{16}$$

Finally, Theorem 8 implies that:

$$\models [\![Prog_L]\!] \Rightarrow \Box(\Diamond(\textrm{◁bs}({\triangleright}\mathsf{p}_L({\triangleleft}\mathsf{p}_R(C_R) \wedge Inv)))) \tag{17}$$

The upshot of (12)–(17) is that a temporal property of $Prog_R$ of the form:

$$\Box(\Diamond(\textrm{◁bs}(C_R)))$$

is translated into a temporal property of $Prog_L$ of the form:

$$\Box(\Diamond(\textrm{◁bs}({\triangleright}\mathsf{p}_L({\triangleleft}\mathsf{p}_R(C_R) \wedge Inv))))$$

Note that:

$$\triangleright\mathsf{p}_L(\triangleleft\mathsf{p}_R(C_R) \wedge Inv)\langle\!\langle s_L \rangle\!\rangle \ = \ \exists s_R.\ C_R\langle\!\langle s_R \rangle\!\rangle \wedge Inv(\langle\!\langle s_L \rangle\!\rangle \bullet \langle\!\langle s_R \rangle\!\rangle)$$

So, while the temporal part $\Box(\Diamond(\textrm{◁bs}(\cdot)))$ is unchanged, a condition C_R on the right states is translated into a condition on the left states by relating a left state s_L with a right state s_R that satisfies C_R using the joint invariant Inv and then existentially "quantifying away" the right state s_R. To summarize:

Theorem 11.

$$SimWF(Prog_L)(Prog_R)(Par)(Inv) \Rightarrow$$
$$(\models [\![Prog_R]\!] \Rightarrow \Box(\textrm{◁bs}(C_R)) \Rightarrow$$
$$\models [\![Prog_L]\!] \Rightarrow \Box(\textrm{◁bs}({\triangleright}\mathsf{p}_L({\triangleleft}\mathsf{p}_R(C_R) \wedge Inv)))\ )$$
$$\wedge\ (\models [\![Prog_R]\!] \Rightarrow \Diamond(\textrm{◁bs}(C_R)) \Rightarrow$$
$$\models [\![Prog_L]\!] \Rightarrow \Diamond(\textrm{◁bs}({\triangleright}\mathsf{p}_L({\triangleleft}\mathsf{p}_R(C_R) \wedge Inv)))\ )$$
$$\wedge\ (\models [\![Prog_R]\!] \Rightarrow \Box(\Diamond(\textrm{◁bs}(C_R))) \Rightarrow$$
$$\models [\![Prog_L]\!] \Rightarrow \Box(\Diamond(\textrm{◁bs}({\triangleright}\mathsf{p}_L({\triangleleft}\mathsf{p}_R(C_R) \wedge Inv))))\ )$$
$$\wedge\ (\models [\![Prog_R]\!] \Rightarrow \Diamond(\Box(\textrm{◁bs}(C_R))) \Rightarrow$$
$$\models [\![Prog_L]\!] \Rightarrow \Diamond(\Box(\textrm{◁bs}({\triangleright}\mathsf{p}_L({\triangleleft}\mathsf{p}_R(C_R) \wedge Inv))))\ )$$

9 Related Work

HOL, both the logic and the system, was described in detail by Gordon and Melham in [5]. TLA was proposed by Lamport in [6].

Much of the material of Sections 3 and 4 appeared in an earlier paper of the author's [2]. However, all theorems in Sections 3.2 and 4.4 except Theorem 3 are new. Also new is the use of the image operator.

The semantic embedding of TLA in HOL was also studied by von Wright *et al.* [9, 10]; see [2] for a comparison between von Wright's approach and ours.

Loewenstein [7] developed in HOL a formal theory of simulations between infinite automata. Lynch and Vaandrager [8] surveyed a variety of simulation techniques for I/O-automata. Their work differs from ours in two major ways. Firstly, their notions of simulations use either external state components [7] or external events [8] to relate automata, while ours is based on joint invariants and joint actions which do not distinguish between external and internal state components or use events at all. Secondly, they do not use temporal logic or treat general liveness properties. The integration of simulation techniques and temporal logic developed in this paper seems to be new. On the other hand, both [7] and [8] treat more general kinds of simulations than we do here.

References

1. Ching-Tsun Chou, "A Bug in the Distributed Minimum Spanning Tree Algorithm of Gallager, Humblet and Spira", unpublished note, 1988.
2. Ching-Tsun Chou, "A Sequent Formulation of a Logic of Predicates in HOL", pp. 71–80 of [3].
3. Luc J.M. Claesen and Michael J.C. Gordon (eds.), *Higher Order Logic Theorem Proving and Its Applications*, North-Holland, 1993.
4. R.G. Gallager, P.A. Humblet and P.M. Spira, "A Distributed Algorithm for Minimum-Weight Spanning Trees", in *ACM Trans. on Programming Languages and Systems*, Vol. 5, No. 1, pp. 66–77, Jan. 1983.
5. Michael J.C. Gordon and Tom F. Melham (eds.), *Introduction to HOL: A Theorem-Proving Environment for Higher-Order Logic*, Cambridge University Press, 1993.
6. Leslie Lamport, "The Temporal Logic of Actions", DEC SRC Research Report 79, Dec. 1991.
7. Paul Loewenstein, "A Formal Theory of Simulations between Infinite Automata", pp. 227–246 of [3].
8. Nancy A. Lynch and Frits W. Vaandrager, "Forward and Backward Simulations, Part I: Untimed Systems", CWI Report CS-R9313, 1993.
9. J. von Wright, "Mechanising the Temporal Logic of Actions in HOL", in *Proc. of the HOL Tutorial and Workshop*, 1991.
10. J. von Wright and T. Långbacka, "Using a Theorem Prover for Reasoning about Concurrent Algorithms", in *Proc. of Workshop on Computer-Aided Verification*, 1992.

Formalization of Variables Access Constraints to Support Compositionality of Liveness Properties

ISWB Prasetya

Rijksuniversiteit Utrecht, Vakgroep Informatica
Postbus 80.089, 3508 TB Utrecht, Nederland
Email: wishnu@cs.ruu.nl

Abstract. Because reasoning about programs' liveness behavior is difficult people become interested in the potential of theorem provers to aid verification. In extending a theorem prover with a lifeness logic it would be nice if compositionality is also supported since it is a property of a great practical interest: it allows modularity in design. However, a straightforward extension that only embodies the essence of the logic will fail to do so. In implementing such an extension we should therefore be aware of the technical details required for compositionality. In particular, compositionality of progress under parallel composition depends on the concept of variable accessibility. Therefore, this concept has to be explicitly present in the extension. This paper is about the formalization of access constraints to support compositionality.

Category: Research Report

1 Introduction

Reasoning about liveness behavior of a program is inherently difficult. A verification tool may therefore be of a great assistance. To use a theorem prover such as HOL as a general environment for mechanical verification, programming logics are formalized into it. For example, formalization of Interval Temporal Logic into HOL was done by Roger Hale [5], Temporal Logic of Actions by J von Wright and T Långbacka [9], UNITY by Flemming Andersen [1] and also by Wishnu Prasetya [7].

Liveness properties can be distinguished into properties telling what a program is not allowed to do, these are the so-called *safety properties*, and properties telling what the program is guaranteed to do, the so-called *progress properties*. It is of a great practical interest that the formalization supports the compositionality of liveness under various program compositions. We will concentrate on the parallel composition.

The compositionality of progress under parallel composition is a difficult issue. If a progress in a component program is achieved by relying on certain behavior of shared variables, then parallel composition will likely destroy this progress. One might indeed say that a progress in a component program P can be preserved by parallel composition by requiring the other component program

Q to respect the behavior assumed by P. However, during the design process one does not always want to fix beforehand what precisely the required behavior should be. This may be something one wishes to postpone until a later design stage.

Under certain circumstances, however, progress is compositional. The condition is given by Ambuj K Singh [8], and seems to be a technically acceptable. It turns out however that the theorem makes a silent but crucial reference to some properties of shared variables and local predicates. Indeed, we usually take them for granted. But these properties too, need to be formalized. Their formalization, in turn, requires that the formalization of the programming logic does not only capture the essence of the logic itself, but also expresses about read and write variables, and defines which variables a behavior expression is allowed to refer to. This article discusses a formalization of UNITY that also supports compositionality of progress as given by Singh.

UNITY is a programming logic invented by Chandy and Misra [3]. The logic is designed to reason about liveness behavior of programs, especially in parallel environment. The logic is not as powerful as the linear temporal logic but it is becoming more popular because of its simplicity. The theorem prover we use is HOL, developed by Mike Gordon. HOL supports reasoning in *higher order logic* (hence the name HOL), that is, a version of predicate calculus in which variables can range over functions and predicates. The logic is also *typed*. See [4] for an introduction to HOL.

2 Notation

The notation used is a mix of HOL notation and notation common to computer science community. This is to make the formulas more readable, especially for those who are new to HOL, and at the same time to show how formulas are represented in HOL.

Function application is denoted with a small dot: f applied to x is written as $f.x$. Some functions are treated as unary operator and an application of such a function is thus written without dot like NOT p and UNITY Pr. Function abstraction is denoted by λ. For type $\mathcal{T}$, $x : \mathcal{T}$ or $x \in \mathcal{T}$ means x is an object of type $\mathcal{T}$.

Predicate calculus operators are denoted as common by $\neg, \wedge, \vee, \Rightarrow, \forall$ and $\exists$. These operators *bind* most weakly compared with other operators.

Universal quantification over the whole domain is written $\forall i . P.i$. Universal quantification over a restricted part of the domain characterized by predicate V is written $\forall i :: V . P.i$. Notice that the *big dot* marks the end of domain specification while *small dot* denotes function application. The same notation applies for other quantified operators.

A predicate is a function from a given domain to **bool**. *Predicate operators* are defined as the point-wise lift of the predicate calculus operators:

Definition 1. Lifted Operators
Let $p, q \in \mathcal{D} \to$ **bool** be predicates.

$$\begin{aligned}
\text{NOT } p &= (\lambda s.\ \neg p.s) \\
p \text{ AND } q &= (\lambda s.\ p.s \wedge q.s) \\
p \text{ OR } q &= (\lambda s.\ p.s \vee q.s) \\
p \text{ IMP } q &= (\lambda s.\ p.s \Rightarrow q.s)
\end{aligned}$$

We use the terms 'set' and 'predicate' interchangely as they are isomorphic. In some places set notations $\cup$ and $\cap$ are used instead of **OR** and **AND** .

For predicate p, the *everywhere* operator is denoted by $[p]$. It means that in every point s in the domain, $p.s$ holds.

3 A Brief Review on UNITY

UNITY views a program as a collection of atomic actions. There is no ordering imposed on the execution of the actions, so the actions can be executed either sequentially, or in full parallel, or in anyway in between. There is only a single restriction, namely that the implementation should guarantee fairness.

More formally, a UNITY program is a pair (P, In) where P is a non-empty set of always enabled actions, that is, actions that do not abort and always terminate, and In is a predicate describing the initial condition.

Here, an action is a relation on the state space. For a given action A and states s, t, $A.s.t$ means that t is a possible final state if A is executed in state s. In the sequel the following notation is used: **State** denotes the entire state space, **Action** denotes the collection of all actions, **Pred** denotes the collection of all predicates over states.

Definition 2. Always Enabled Statement
For all $A \in$ **Action**:

$$\text{ALWAYS ENABLED } A = \forall s.\ \exists t.\ A.s.t$$

Definition 3. UNITY Program
For all $P \in$ **Action** $\to$ **bool** and $In \in$ **Pred**:

$$\text{UNITY } (P, In) = (\exists A.\ P.A) \wedge (\forall A :: P.\ \text{ALWAYS ENABLED } A)$$

In the sequel **Uprog** abbreviates (**Action** $\to$ **bool**) $\times$ **Pred**. The predicate **UNITY** characterizes which objects of **Uprog** are UNITY programs.

To denote the components of a pair in **Uprog**, the following destructors are used. For all $(P, In) \in$ **Uprog** : **PROG**.$(P, In) = P$ and **INIT**.$(P, In) = In$.

The UNITY's model of execution is as follows. A program execution starts from any state satisfying the initial condition and goes on forever; in each step of execution some action is selected nondeterministically and executed. The selection process is required to be (weakly) fair: each action should be selected infinitely many times. In UNITY's view, there is no such thing as termination although it can always be simulated by fixed points, that is, states that are invariant under the execution any actions in the program.

UNITY is however intended to reason about liveness instead of stationary properties such as termination. To do this UNITY logic provides three primitive operators, namely UNLESS , ENSURES , and $\mapsto$ (leads-to).

Definition 4. UNLESS

For all $p, q \in$ Pred and $Pr \in$ Uprog:

p UNLESS q in $Pr = \forall A :: $ PROG.$Pr.$ $\{p$ AND (NOT $q)\}$ A $\{p$ OR $q\}$

where $\{p\}$ A $\{q\}$ denotes a Hoare Triple and is formally defined as follows. For all $A \in$ Action:

$\{p\}$ A $\{q\} = \forall s, t. (p.s \wedge A.s.t) \Rightarrow q.t$

Informally, p UNLESS q means that once p holds during the execution of a program, it will remain to hold at least until q becomes true. So, UNLESS expresses safety.

UNLESS is monotonic in its second argument, and universally disjunctive in its first argument.

Theorem 5. Monotonicity of UNLESS

$\forall Pr :: $ UNITY. $\forall p, q, r.$
  p UNLESS q in Pr $\wedge$ $[q$ IMP $r]$ $\Rightarrow$ p UNLESS r in Pr

---

Theorem 6. General Disjunction Law

$\forall Pr :: $ UNITY. $\forall W, P, Q.$
  $(\forall i :: W. P.i$ UNLESS q in $Pr)$ $\Rightarrow$ $(\bigvee i :: W. P.i)$ UNLESS q in Pr

where $\bigvee$ denotes the universally quantified OR .

Another important property is the following.

Theorem 7. Conjunction Law

$\forall Pr :: \text{UNITY} \cdot \forall p, q, r, s \cdot$
  p **UNLESS** q in $Pr \ \wedge \ r$ **UNLESS** s in Pr
  $\Rightarrow$
  $(p$ **AND** $r)$ **UNLESS** $((p$ **AND** $s)$ **OR** $(r$ **AND** $q)$ **OR** $(q$ **AND** $s))$ in Pr

---

ENSURES expresses progress: p **ENSURES** q implies that every time p holds during the execution, eventually q will hold. More specifically, progress is achieved by holding p stable as long as q is not yet true, thus p **UNLESS** q, and by requiring the existence of an action that will indeed establish q when executed in a state satisfying p. Fairness guarantees that this particular action will eventually be executed and thereby establishing q.

Definition 8. ENSURES
For all $Pr \in \text{Uprog}$ and $p, q \in \text{Pred}$:

$$p \text{ ENSURES } q \text{ in } Pr \ = \ \begin{cases} \text{UNITY } Pr \ \wedge \ p \text{ UNLESS } q \text{ in } Pr \ \wedge \\[2mm] (\exists A :: \text{PROG}.Pr \cdot \ \{p \text{ AND (NOT } q)\} \ A \ \{q\}) \end{cases}$$

Like **UNLESS**, **ENSURES** is monotonic in its second argument.

Theorem 9. Monotonicity of ENSURES

$\forall Pr :: \text{UNITY} \cdot \forall p, q, r \cdot$
  p **ENSURES** q in $Pr \wedge [p \text{ IMP } r] \ \Rightarrow \ p$ **ENSURES** r in Pr

---

ENSURES, however, does not cover all progress. Progress is a transitive and disjunctive property whereas **ENSURES** is not. So what is required is a transitive and disjunctive closure of **ENSURES**. Let the relation **LeadsToRel** characterize all such closures. Leads-to, denoted by $\mapsto$, is defined as the least **LeadsToRel** closure.

Definition 10. LeadsToRel
For all $Pr \in \text{Uprog}$ and $R \in \text{Pred} \rightarrow \text{Pred} \rightarrow \text{Uprog} \rightarrow \text{Bool}$:

$\text{LeadsToRel}.R.Pr \ =$
$$\forall p, q, r, W \cdot \begin{cases} (p \text{ ENSURES } q \text{ in } Pr \ \Rightarrow \ R.p.q.Pr) \ \wedge \\ (R.p.r.Pr \wedge R.r.q.Pr \ \Rightarrow \ R.p.q.Pr) \ \wedge \\ ((\forall p' :: W \cdot R.p'.q.Pr) \ \Rightarrow \ R.(\bigvee p' :: W \cdot p').q.Pr) \end{cases}$$

> **Definition 11. Leads-To**
> For all $Pr \in$ Uprog and $p, q \in$ Pred:
>
> $$p \mapsto q \text{ in } Pr = (\forall R. \text{ LeadsToRel}.R.Pr \Rightarrow R.p.q.Pr)$$

Since $\mapsto$ is a proper closure of **ENSURES** it covers more progress than **ENSURES**. It does not however coincides with the operational view of progress as shown by Jan Pachl in [6]: progress expressible by $\mapsto$ is valid in the operational semantic, but not the other way around. Nevertheless, $\mapsto$ still covers enough interesting progress properties. It is also characteristic that in a sufficiently powerful logic system, provability and validity do not coincide.

Progress induction ($\mapsto$ induction [3]) plays an important role in reasoning about progress. It states that if a relation X satisfies:

1. it includes **ENSURES**, that is, p **ENSURES** $q \Rightarrow X.p.q$
2. it is transitive, that is, $X.p.q \wedge X.q.r \Rightarrow X.p.r$
3. it is universally disjunctive in its first argument, that is, $(\forall i :: W . X.(P.i).q) \Rightarrow X.(\bigvee i :: W . P.i).q$, for all index domain W

then

$$\forall a, b. \ a \mapsto b \Rightarrow X.a.b$$

holds. This is an immediate consequence of the fact that $\mapsto$ is the least transitive and disjunctive closure of **ENSURES**. In fact, the induction principle above actually states that any relation X that satisfies **LeadsToRel** includes $\mapsto$, which is evident since $\mapsto$ is the least relation satisfying **LeadsToRel**. In fact, every least closure induces an induction principle [2].

4 Parallel Composition

Parallel Composition and Superposition are the only compositions of practical interest in UNITY. The latter will not be discussed in this article. Since UNITY does not impose any ordering on the execution actions, there is no sequential composition in UNITY.

It is of a great practical importance to have theorems about the compositionality of UNITY operators with respect to parallel composition. Compositionality is what allows us to conclude the validity of a property in a *composite* program from the properties of its component. During the design process this means we can implement a specification by splitting it into specifications of component programs. Thus, modularity.

Because UNITY does not impose any ordering in which the actions are to be executed, parallel composition can be modeled simply by the union of the component programs. More specifically:

Definition 12. Parallel Composition
For all $Pr, Qr \in$ Uprog:

$Pr \| Qr$ = (PROG.$Pr \cup$ PROG.Qr , INIT.Pr AND INIT.Qr)

Parallel composition is commutative and is closed within the space of UNITY programs.

Compositionality of UNLESS is straightforward. If we want p UNLESS q to hold in $Pr \| Qr$, by definition this requires that the Hoare Triple

$\{p$ AND (NOT $q)\}\; A\; \{p$ OR $q\}$

is satisfied by every action in PROG.($Pr \| Qr$). But PROG.($Pr \| Qr$) = PROG.$Pr \cup$ PROG.Qr, so with other words, p UNLESS q should hold in each component program.

Theorem 13. Compositionality of UNLESS

$\forall p, q, J, Pr, Qr$.
    p UNLESS q in Pr $\wedge$ p UNLESS q in Qr
    $\Rightarrow$
    p UNLESS q in $(Pr \| Qr)$

Compositionality of ENSURES is also straight forward. Since by definition p ENSURES q holds if and only if p UNLESS q and there exists an action A that will indeed establish q, it follows that its validity in $Pr \| Qr$ requires the validity of UNLESS part (which we already know how to achieve from Theorem 13) and the existence of that particular action A, either in Pr or Qr.

Theorem 14. Compositionality of ENSURES

$\forall Pr :: $ UNITY. $\forall p, q, J, Qr$.
    p UNLESS q in Pr $\wedge$ p ENSURES q in Qr
    $\Rightarrow$
    p ENSURES q in $(Pr \| Qr)$

As mentioned before, ENSURES is too restricted to cover all interesting progress properties. Besides, by requiring the existence of a specific action ENSURES is too close to implementation. During the design process it is certainly convenient to be able to talk about progress without having to know the actions that establish it. Whether the progress is achieved by a single action, or by mutual effect of several actions, it is a design decision one generally postpones as far as possible. Unfortunately no elegant compositionality statement is known for $\mapsto$, which is actually not so much a surprise in a shared variables environment. Theorem for compositionality of $\mapsto$ is given Ambuj K Singh in [8]. The formalization of this theorem in HOL will be discussed in the next section.

5 Compositionality of $\mapsto$

One may indeed argue that for the sake of verification it is not necessary to formalize the compositionality of $\mapsto$. During the design process one works top-down, starting from a general specification down to the program. Progress will be likely specified in terms of $\mapsto$, and then refined by a set of ENSURES properties. Verification does not need to follow the same path. It may even be more economic to do it bottom-up. In this case, one only needs compositionality of ENSURES as progress by $\mapsto$ can always be reconstructed from the underlying progress by ENSURES . If the verification fails, however, one may have difficulty in fixing the design steps as the verification goes a direction in opposite to the design steps. So, there is a practical reason for formalizing the compositionality of $\mapsto$.

Consider the composition $Pr\|Qr$. Consider the progress $p \mapsto q$ in Pr. Very often such a progress will also depend on the behavior of the shared variables of Pr and Qr. Unfortunately, Qr may destroy this progress by writing to the shared variables, thus tampering with the assumed behaviors of the shared variables. So, Qr is required to respect this assumed behavior. The problem is that the exact behavior of the shared variables required for the progress may not be known until some later design stage. However, if it is known that whenever Qr modifies the shared variables it also raises a flag b, then one may still conclude that p leads either to q, or Qr has written to the shared variables in which case the flag b will be raised. Formally, it is given in the following theorem by Ambuj K Singh [8].

Theorem 15. Compositionality of $\mapsto$ [Singh]
Let V denote the vector $v_1, v_2, \ldots$ of *all* variables read by Pr and written by Qr.

$$\frac{p \mapsto q \text{ in } Pr \qquad \forall C \cdot (V = C) \text{ UNLESS } b \text{ in } Qr}{p \mapsto (q \text{ OR } b) \text{ in } (Pr\|Qr)}$$

Unfortunately the theorem is not derivable with the previously chosen definition of a UNITY program since it knows nothing about read or written variables. So first of all, these notions have to be integrated into the formal definition.

Secondly, the theorem silently assumes that p and q in the condition $p \mapsto q$ in Pr are predicates on the state space of Pr. This seems quite natural, since we are talking about the behavior of Pr anyway. This suggests that the type of permissible $\mapsto$ expressions of a program is constrained by the space of read variables that belongs to the program. As HOL does not support dependent types we have to encode this with with predicates.

We consider a behavior expression to be defined on the universal state space over *all available* variable names. Consequently, our definition of UNITY operators (Section 3) allows expressions like $z > 0 \mapsto x > 0$ in Pr even though Pr does not have anything to do with z. However, if we allow this kind of expressions then Singh's theorem is not provable. To see this is not very difficult. Consider the following two programs, Pr and Qr:

Name: Pr	Name: Qr
Read Vars = Write Vars : x	Read Vars = Write Vars : b
Init: TT	Init: TT
$x := x + 1$	$b := 0$

Notice that Pr and Qr do not share variables, so the second condition in the Singh's theorem is automatically satisfied. The theorem then states that by choosing $b = \text{FF}$ any progress $p \mapsto q$ in Pr also holds in $Pr \| Qr$. However, $(x = 0 \text{ AND } b = 1) \mapsto (x = 1 \text{ AND } b = 1)$ holds in Pr, but definitely not in $Pr \| Qr$.

This shows that in formalizing UNITY we also need to formalize the taken for granted assumption that a progress by $\mapsto$ is only considered relevant when it does a statement exclusively about the state of variables of the program in question.

5.1 Extending the Definition of UNITY Program

A UNITY program is now extended to be a tuple (P, In, R, W) where P is a non-empty set of always enabled actions, In is a predicate defining the initial condition, R is the set of read variables, and W is the set of write variables. In addition, $W \subseteq R$ and we explicitly require that the *Writability Constraint* is satisfied, that is, the program does not modify any variable not in W. The analogous Readability Constraint can be required, but this will require the formalization of the programming language, and this is beyond the scope of this article. So, the predicate UNITY is now extended as below. From now on Var denotes the set of all available variable names and VARS denote Var $\to$ bool or $\mathcal{P}(\text{Var})$.

Definition 16. UNITY Program
For all $Pr \in$ Action $\to$ bool, $In \in$ Pred, and $R, W \in$ VARS:

$$\text{UNITY } (P, In, R, W) =$$
$$\begin{cases} (\exists A \cdot P.A) \ \wedge \\ (\forall A :: P \cdot \text{ ALWAYS ENABLED } .A) \ \wedge \\ (\forall A :: P \cdot (\forall s, t, z \cdot A.s.t. \wedge \neg W.z \Rightarrow (s.z = t.z))) \ \wedge \\ (\forall x \cdot W.x \Rightarrow R.x) \end{cases}$$

The definition of Uprog is updated accordingly and to denote new components of a tuple $Pr \in$ Uprog the following destructors are added. For all $P \in$ ACTIONS, $In \in$ Pred: READ.$(P, In, R, W) = R$ and WRITE.$(P, In, R, W) = W$.

The definition of parallel composition is extended accordingly.

Definition 17. Parallel Composition
For all $Pr, Qr \in$ Uprog:

$$Pr|Qr = (\text{PROG}.Pr \cup \text{PROG}.Qr \ , \ \text{INIT}.Pr \text{ AND INIT}.Qr \ ,$$
$$\text{READ}.Pr \cup \text{READ}.Qr \ , \ \text{WRITE}.Pr \cup \text{WRITE}.Qr)$$

We want that progress $p \mapsto q$ in Pr is only valid if both p and q are indeed predicates on the state space of a program Pr, or as we will call it: they should be *local predicates* of Pr. A predicate p is local to Pr iff p allows any variable not in READ.Pr to have any value or no value at all. With other words, p only carries information about the read variables of Pr, and nothing else. To formalize this the relation LocalPred is introduced.

Definition 18. Local Predicate
For all $Pr \in$ Uprog and $p \in$ Pred:

$$p \text{ LocalPred in } Pr \ = \ (\forall s,t \text{ . } (\forall x :: \text{READ}.Pr \text{ . } s.x = t.x) \Rightarrow p.s = p.t)$$

All lifted predicate calculus operators (NOT , AND , IMP ,...) preserve local predicates. TT and FF are always local predicates. Also, a local predicate of a component program is also a local predicate of the composite program.

The definition of LeadsToRel is now extended by integrating the notion of local predicate.

Definition 19. LeadsToRel
For all $Pr \in$ Uprog and $R \in$ Pred $\rightarrow$ Pred $\rightarrow$ Uprog $\rightarrow$ Bool:

$$\text{LeadsToRel}.R.Pr \ = \ \forall p, q, r, W \text{ . }$$
$$\begin{cases} (p \text{ ENSURES } q \text{ in } Pr \ \wedge \ p \text{ LocalPred in } Pr \ \wedge \ q \text{ LocalPred in } Pr \\ \quad \Rightarrow \ R.p.q.Pr) \ \wedge \\ (R.p.r.Pr \wedge R.r.q.Pr \ \Rightarrow \ R.p.q.Pr) \ \wedge \\ (\neg(W = \text{FF}) \wedge (\forall p' :: W \text{ . } R.p'.q.Pr) \ \Rightarrow \ R.(\bigvee p' :: W \text{ . } p').q.Pr) \end{cases}$$

The definition of $\mapsto$ remains the same, namely that it is the least relation satisfying LeadsToRel. The difference is that now $p \mapsto q$ in Pr can only hold if both p and q only depends on the variables of Pr. Or even stronger: the progress $p \mapsto q$ in Pr can only be valid if it does not at any point depend on the behavior of any variable not in READ.Pr.

Let p be a local predicate in Pr and X be the vector of *all* shared variables of Pr and Qr. A natural and expected (and usually taken for granted) property of local predicates is: the only way Qr can disturb p is by modifying the shared variables X. This property is necessary to prove compositionality of $\mapsto$, so it needs to be formalized too.

So far we have used the term *shared variables* without mentioning what it exactly means. A possibility is to define it as the intersection of the read variables of the component programs. However, here we will formally define shared variables between Pr and Qr, denoted by $DVa.Pr.Qr$, as the variables read by Pr and written by Qr (so DVa is not symmetric). Singh's Theorem becomes more general with this definition of shared variables.

Definition 20. Shared Variables
For all $Pr, Qr \in$ Uprog:

$$DVa.Pr.Qr = \text{READ}.Pr \cap \text{WRITE}.Qr$$

Let $V = DVa.Pr.Qr$. The sub-expression $V = C$ in $(V = C)$ UNLESS b in the condition of Singh's theorem (Theorem 15) is an abuse of notation because $E1 = E2$ has the type **bool** while **UNLESS** expects a predicate. What is actually meant is a predicate characterizing those states where the value of each variable $v \in V$ is $C.v$. To formalize this, angled brackets lifting is introduced.

Definition 21. Angled Brackets Lifting
Let $\mathcal{D}$ be the space of values of variables. For all $V \in$ VARS, $\supset: \mathcal{D} \to \mathcal{D} \to$ Bool, and $C \in$ Var $\to \mathcal{D}$:

$$\langle V \supset C \rangle = (\lambda s : \text{State.} \ (\forall x :: V. \ s.x \supset C.x))$$

So, the condition in Singh's theorem can now be formally written as

$$\forall C. \ \langle DVa.Pr.Qr = C \rangle \text{ UNLESS } b \text{ in } Qr$$

The previously mentioned property of local predicates is formalized in the theorem below.

Theorem 22. Local Predicate Safety

$$\forall Pr, Qr :: \text{UNITY.} \ \forall p.$$
$$p \ \text{LocalPred in } Pr$$
$$\Rightarrow$$
$$(\forall C. \ (p \text{ AND } \langle DVa.Pr.Qr = C \rangle) \text{ UNLESS}$$
$$(\text{NOT } \langle DVa.Pr.Qr = C \rangle) \text{ in } Qr)$$

Proof:
The proof can be sketched as follows. Let $\mathcal{D}$ denote the value space of the variables. For any $s \in$ Var $\to \mathcal{D}$ and V a set of variables, let $s \uparrow V$ denote the projection of s on V. Let V stand for $DVa.Pr.Qr$, that is, the shared variables of Pr and Qr.

$(p$ AND $\langle V = C \rangle)$ UNLESS (NOT $\langle V = C \rangle$) in Qr

$=$ { Definition of UNLESS and predicate calculus }

$\forall A :: \text{PROG}.Qr.$ $\{p$ AND $\langle V = C \rangle\}$ A $\{p$ OR (NOT $\langle V = C \rangle)\}$

$=$ { Definition Hoare Triple, angled brackets, and projection }

$\forall A :: \text{PROG}.Qr.$ $\forall s, t : \text{State}.$
$A.s.t \wedge p.s \wedge (t \uparrow V = C \uparrow V) \Rightarrow p.t \vee (s \uparrow V \neq C \uparrow V)$

$=$ { predicate calculus }

$\forall A :: \text{PROG}.Qr.$ $\forall s, t : \text{State}.$ $A.s.t \wedge p.s \wedge (s \uparrow V = C \uparrow V = t \uparrow V) \Rightarrow p.t$

$\Leftarrow$ { p LocalPred in Pr }

$\forall A :: \text{PROG}.Qr.$ $\forall s, t : \text{State}.$
$A.s.t \wedge p.s \wedge (s \uparrow V = t \uparrow V) \Rightarrow p.s \wedge (s \uparrow \text{READ}.Pr = t \uparrow \text{READ}.Pr)$

$\Leftarrow$ { Qr is a UNITY program, hence it satisfies the Writability Constraint. Let U denote $\text{READ}.Pr - \text{WRITE}.Qr$ }

$\forall A :: \text{PROG}.Qr.$ $\forall s, t : \text{State}.$
$(s \uparrow U = t \uparrow U) \wedge (s \uparrow V = t \uparrow V) \Rightarrow (s \uparrow \text{READ}.Pr = t \uparrow \text{READ}.Pr)$

$=$ { properties of projection }

$\forall A :: \text{PROG}.Qr.$ $\forall s, t : \text{State}.$
$(s \uparrow (U \cup V) = t \uparrow (U \cup V)) \Rightarrow (s \uparrow \text{READ}.Pr = t \uparrow \text{READ}.Pr)$

$=$ { by definition of U and V we have $U \cup V = \text{READ}.Pr$ }

true

$\square$

Notice that the fact that a UNITY program satisfies the Writability Constraint is crucial to the proof. This shows the practical merit of the complication we have added in Definition 16.

The other direction of Theorem 22 does not necessarily hold. This is not so difficult to show. For example if p mentions only variables which are neither in Pr or in Qr, then p is not a local predicate of Pr while it is stable in Qr (that is, p UNLESS false in Qr).

The Singh's theorem of compositionality is formalized as follows.

Theorem 23. Compositionality of $\mapsto$

$\forall Pr, Qr :: \text{UNITY}.$ $\forall p, q, b.$
  b LocalPred in $(Pr\|Qr)$ $\wedge$ $p \mapsto q$ in Pr $\wedge$
  $(\forall C.$ $\langle DVa.Pr.Qr = C \rangle$ UNLESS b in $Qr)$
  $\Rightarrow$
  $p \mapsto (q$ OR $b)$ in $(Pr\|Qr)$

The theorem can be proved by using $\mapsto$ induction mentioned in the end of Section 3. Below the proof will be sketched for the ENSURES case to show the essential part where the property of local predicates comes into play.

Theorem 24. Compositionality of $\mapsto$, The ENSURES Case

$\forall Pr, Qr :: \text{UNITY}. \;\; \forall p, q, b.$

    p ENSURES q in Pr $\land$ p LocalPred in Pr $\land$

    $(\forall C. \; \langle DVa.Pr.Qr = C\rangle$ UNLESS b in $Qr)$

    $\Rightarrow$

    p ENSURES $(q$ OR $b)$ in $(Pr\|Qr)$

Proof:

p ENSURES q in Pr holds by assumption. ENSURES is right monotonic (Theorem 9), hence

    (1) p ENSURES $(q$ OR $b)$ in Pr

holds too. Then we derive:

    p LocalPred in Pr

$\Rightarrow$ { Theorem 22 }

    $(p$ AND $\langle DVa.Pr.Qr = C\rangle)$ UNLESS (NOT $\langle DVa.Pr.Qr = C\rangle)$ in Qr

$\Rightarrow$ { $\langle DVa.Pr.Qr = C\rangle$ UNLESS b in Qr holds as assumption, Theorem 7 of UNLESS conjunction }

    $(p$ AND $\langle DVa.Pr.Qr = C\rangle)$ UNLESS b in Qr

$\Rightarrow$ { Theorem 6 of UNLESS disjunction }

    p UNLESS b in Qr

$\Rightarrow$ { Theorem 5: UNLESS is right monotonic }

    p UNLESS $(q$ OR $b)$

Hence (2) p UNLESS $(q$ OR $b)$ in Qr

By compositionality of ENSURES (Theorem 14), it follows from (1) and (2) that p ENSURES $(q$ OR $b)$ holds in $Pr\|Qr$. $\square$

Note the crucial role of the condition p LocalPred in Pr in above proof. Compositionality of $\mapsto$ can be motivated in much the same way. Recall that by the new definition of $\mapsto$ progress $p \mapsto q$ can only be valid in Pr if the progress does not at any point depend on the behavior of local variables of Qr. So, by Theorem 22 Qr can only disturb this progress by modifying the shared variables $DVa.Pr.Qr$. However, the assumption also says that every time Qr modifies $DVa.Pr.Qr$ it will also raise the flag b. So, either the progress $p \mapsto q$ is maintained in $Pr\|Qr$, or Qr disturbs this progress —in which case b is raised. Hence $p \mapsto (q$ OR $b)$ in $(Pr\|Qr)$.

This concludes the discussion about formalizing compositionality of progress by $\mapsto$.

6 Conclusion

Usually, when people start to formalize a programming logic, they start from the theoretical point of view rather than practical. The resulting formalization tends to be more straightforward, carrying only the essence of the logic. However, there are aspects that may seem uninteresting from the theoretical point of view, but actually have a great impact on the practical use of logic.

Compositionality of progress by $\mapsto$ is a topic that is of great practical interest in designing parallel programs. The formalization of compositionality requires us to be explicit in dealing with variable accessibility. So far, the formalization of variable accessibility has been avoided as it seems only to obscure the novelty of the programming logic used. In this paper we have summerized the ingredients that have to be explicitly present in the formalization of a programming logic in order to formally derive compositionality of liveness behavior. At least, if one is aware of them then one can consciously choose, depending on his purpose, whether to make his formalization powerful enough to have compositionality, or to keep his formalization less powerful but simple.

Since the problem of compositionality described in this paper is not typical of UNITY, but appears in any programming logic which deals with liveness properties of programs that have shared variables, our results are also applicable to such a logic.

7 Acknowledgement

I thank Johan Jeuring for his correction to my English.

References

1. Flemming Andersen. *A Theorem Prover for UNITY in Higher Order Logic.* PhD thesis, Technical University of Denmark, 1992.
2. Juanito Camilleri and Tom Melham. Reasoning with inductively defined relations in the hol theorem prove. Technical Report 265, University of Cambridge, 1992.
3. K.M. Chandy and J. Misra. *Parallel Program Design – A Foundation.* Addison-Wesley Publishing Company, Inc., 1988.
4. Mike JC Gordon and Tom F Melham. *Introduction to HOL.* Cambridge University Press, 1993.
5. Roger W.S. Hale. *Programming in Temporal Logic.* PhD thesis, University of Cambridge, 1988.
6. J. Pachl. Three definitions of *leads-to* for UNITY. *Notes on UNITY*, 23-90, December 1990.
7. ISWB Prasetya. *Documentation of HOL-Library UNITY.* University of Utrecht, 1993. Will appear as a technical report.
8. A.K. Singh. Leads-to and program union. *Notes on UNITY*, 06-89, 1989.
9. J. von Wright and T. Långbacka. Using a theorem prover for reasoning about concurent algorithms. In *Proc. 4th Workshop on Computer-Aided Verification*, Montreal, Canada, June 1992. Springer-Verlag.

The Semantics of Statecharts in HOL

Nancy Day and Jeffrey J. Joyce

Integrated Systems Design Laboratory, Department of Computer Science,
University of British Columbia, Vancouver, B.C., V6T 1Z2, Canada

Abstract. Statecharts are used to produce operational specifications in the CASE tool STATEMATE. This tool provides some analysis capabilities such as reachability of states, but formal methods offer the potential of linking more powerful requirements analysis with CASE tools. To provide this link, it is necessary to have a rigorous semantics for the specification notation. In this paper we present an operational semantics for statecharts in quantifier free higher order logic, embedded in the theorem prover HOL.

1 Introduction

Statecharts are an extended finite state machine, graphical formalism for real-time systems which alleviates many of the problems such as state explosion, encountered with other state machine notations [4]. They are the notation used to give operational specifications in the commercial CASE tool STATEMATE. This tool provides some analysis capabilities such as reachability of states, but formal methods offer the potential of linking more powerful requirements analysis with CASE tools.

To use these methods, it is necessary to have a rigorous semantics for the notation. Previous work has given semantic interpretations for statecharts[2][7][8], but these did not completely agree with our intuitive idea of their behaviour. In this paper we present an operational semantics for statecharts in quantifier free higher order logic, embedded in the theorem prover HOL. The type-checking facilities of HOL and the expressiveness of higher order logic were very useful in writing the semantics.

The first section informally describes the operation of statecharts and the remaining parts of the paper formalize these ideas, pointing out situations where it is not obvious what the behaviour of the statechart will be. Particular attention is given to issues such as what it means to take a step, race conditions, and multiple actions associated with one transition. The semantics are given by a next configuration predicate which holds true if one complete system configuration is a successor to another.

These semantics have been used to create a model checker for statecharts in the hybrid verification tool HOL-VOSS[9].

2   An Introduction to Statecharts

There is a great deal of interest from both academia and industry in the state-charts formalism. It is an extended state transition notation for expressing the concurrent operation of real-time systems. It is often described as:

state-diagrams + depth + orthogonality + broadcast-communication[2]

In statecharts, the diagrammatic layout of the notation has meaning beyond just the labels on states and transitions. A hierarchy of states is portrayed in a style similar to set inclusion in Venn diagrams to reduce the complexity of the model and therefore make it more readable. In light of this, we rely on the graphical notation to introduce statecharts through an example. The reader is referred to Harel[3] for an explanation of the origins of statecharts as a type of higraph that combines the elements of graphs and Venn diagrams.

The STATEMATE manual describes a traffic light system controlling a two-way intersection which is a simple but effective example of the expressiveness of statecharts[5]. The statechart for this controller is given in Fig.1 and will be referred to throughout this section.

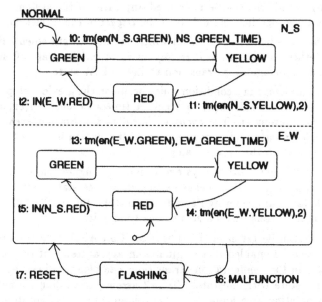

Fig. 1. Traffic light statechart

A statechart models the system as being in a number of *states*, depicted by rounded boxes, which describe its operation. For example, the state labelled NORMAL, at the top of the page, represents the normal operation of the lights in both directions. The dashed line through its middle splits it into two substates, north-south(N_S) and east-west(E_W), which operate concurrently (orthogonality), making NORMAL an AND-state. N_S is an OR-state since it has substates, labelled red, yellow, and green, and the model can only be in one of them at any time (exclusive OR). The representation of these substates within the larger rounded box creates a hierarchy of states (depth). Within this hierarchy, the state NORMAL is an *ancestor* of N_S and E_W. Similarly, N_S.GREEN is a *descendant* of N_S. When a state is not decomposed into AND- or OR-states, it is called a basic state.

At a given moment, the *configuration* of the model includes the states the system is currently in and the values for all data-items. The current set of states alone is called the *state configuration*. A set of basic states is a *legal state configuration* if it satisfies the constraints of the hierarchy. A discrete notion of time is used where the system moves between configurations as a result of stimulus generated both from within the system and externally.

States are connected by *transitions* with labels of the form *event [condition] / action*. The tags t0, t1, etc. are used for reference only. If the system is currently in the source state of a certain transition, labelled e[c]/a, and the event e occurs when the condition c is satisfied then the transition is *enabled*. *Broadcast communication* is used which means that all events and the values of any data-items can be referenced anywhere in the system. The event and condition are together referred to as the *trigger* of the transition.

A *condition* is a boolean expression which can include statements like $IN(x)$ to check whether the system is currently in state x. These are often used to synchronize components as in transition t5 in the E_W state.

An *event* is a change in a condition which occurs in the previous step. Entering a state x causes the event $en(x)$ to occur. A timeout event, $tm(ev, x)$, occurs x steps after the event ev. The event ev is called the *timeout event* and x is the *timeout step number*. The configuration of the system includes the relevant events which occur in the previous step.

Enabled transitions move the system from one configuration to another. Following, or *taking* a transition means exiting its source state, carrying out the actions on its label, and entering its destination state. Informally, following a set of these transitions constitutes a *step* or one time unit.

Transitions can be taken in the substates of an AND-state simultaneously. A transition can be enabled if it originates in any ancestor of the current set of basic states in the configuration. Transitions can also terminate at the outer boundary of a state with substates. *Default arrows*, given diagrammatically as open circles pointing at a state, lead the system into a configuration of basic states. For example, when transition t7 is followed, it terminates at the state NORMAL which is made up of parallel components. The default arrows for each of its substates point at E_W.GREEN and N_S.RED.

If a transition is followed, the action part of the label is carried out and the system moves into the destination state. Actions include generating events or modifying values of variables through assignment statements.

Statecharts often include elements like history states, connectors for compound transitions, static reactions, and transitions with multiple source and destination states. For simplicity, these are not considered here but they are discussed in [1].

3 Ambiguities in Statecharts

The notation described above may seem very straightforward, however, statecharts can be created where their intended meaning is not so obvious.

3.1 What is a Step?

There is no inherent model of timing associated with statecharts other than the movement between states by following transitions. Following a set of transitions and carrying out their actions is considered one time unit or *step*. There are different interpretations of what constitutes this set of transitions.

Briefly, the factors to consider are:

- How are conflicts among enabled transitions resolved ? (i.e. when they can not all be taken)
- Can events generated by the actions of transitions followed in this step trigger transitions which are also followed in this step? In Fig.2a, if the system starts in the states **A** and **C**, and transition **t0** is followed generating the event **f**, is **t1** then enabled and followed in the same step?
- are transitions only from the current set of source states considered or can we move through multiple states in a path in one step? From Fig.2b, we can see that this could lead to infinite loops within a step[5].

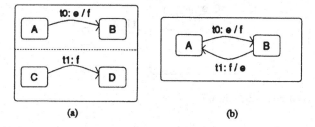

(a)   (b)

Fig. 2. What is a step?

3.2 Multiple Actions on a Transition

A transition may have multiple actions. If more than one of these actions modify the same variable, what will the value of the variable be at the end of the transition? For example, the action $/x := 1; x := x + 2$, with 0 as the value of x in the current configuration, has the following possible interpretations:

1. The actions are taken sequentially so that after following the transition x has the value 3.
2. The actions are taken relative to the beginning of the step but their effects are evaluated sequentially, therefore the second action takes precedence and the result is that x becomes 2.
3. The actions are evaluated relative to the beginning of the step, and they are not assumed to happen in any particular order, however, the actions are atomic and do not conflict. With this interpretation, the result is that x could be 1 or 2 after the transition is taken.

3.3 Race Conditions

A race condition occurs if transitions followed simultaneously in parallel components modify the same variable in a step. This is similar to the situation described in the previous section, but has the added possible interpretation that the actions could conflict with each other (i.e. they are not atomic), and the value for x would then be indeterminate.

3.4 Non-determinism

Statecharts have a hierarchy of states and transitions can originate from states at any level in the hierarchy. If multiple transitions are enabled from states which are descendants or ancestors of each other in the hierarchy, as in Fig.3 where **A** is the parent of **B**, which transition is taken or should both be followed?

Transitions may also originate at exactly the same state and if both are enabled either could be followed.

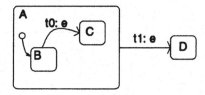

Fig. 3. Structural non-determinism

3.5 Timeouts

The statechart for the traffic light in Fig.1 uses several timeouts to trigger different transitions, such as **t0** or **t1** . When should the system begin to consider the event upon which the timeout is based? Is it the last time the timeout event occurs? Or must the timeout event occur after we have entered the source state and then the system waits the appropriate number of steps before following the transition?

When the timeout step number is symbolic, there is the further question of when to evaluate it. Is it evaluated when the system arrives in the transition's source state (i.e. the first time the transition could be enabled)? Or can the value change between steps? An example of a situation where this might occur in the traffic light is if the **NS_GREEN_TIME** is affected by a pedestrian button which indicates someone wants to cross the street.

3.6 Transitions Among AND Component States

The orthogonal components of AND-states operate concurrently, so it is difficult to see the need for transitions which go between them. However, it is possible, depending on the definition of a legal statechart. In Fig.4a, we can see that following transition t0 leads into the state C, but the system must remain in some state of X at all times. At this point, should it follow the default transition of X into A to reach a legal state configuration?

The situation could occur where two transitions cross AND-state boundaries at the same time. In Fig.4b, if t0 and t3 are followed at the same time, the system will arrive in states B and C, which is a legal state configuration.

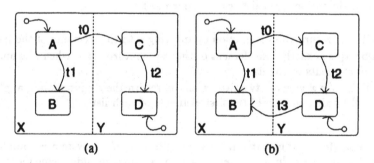

(a)   (b)

Fig. 4. Transitions among AND component states

4 Textual Representation of Statecharts

The graphical notation described in the previous section can be directly translated to a textual representation. A translation program extracts information about the statechart from STATEMATE and outputs ML code to create a HOL definition for the statechart. This definition can be given directly as an argument (referred to as the variable *sc*) to the semantic functions. The semantics use functions which extract information from the HOL definition.

5   A Semantics for Statecharts

Meaning is given to the syntax describing the statechart by translating it into a relation over the current configuration and next configuration. Other semantics for statecharts, given previously by Harel[2], Pnueli[8], and Leveson[7], differ from the semantics used here both in content and form. This paper presents only how we resolved the major difficulties in writing the semantics. The full semantics and a comparison to other approaches can be found in [1].

Our semantic functions were written with the intention of using them with a model checker, where they are executed in a Binary Decision Diagram(BDD) package which understands higher order logic with quantification only over Boolean variables. Limited quantification is not a major restriction since we are dealing with a finite domain. However, it is natural to express conditions over all transitions or all variables. In order to create the abstraction of existential and universal quantification over these sets, the semantics use definitions like EVERY, which takes the conjunction of applying a predicate p, to all elements of a list x:

$$\mathsf{EVERY}\, p\, x = (x = [\,]) \Rightarrow \mathsf{T}\, |\, p\,(\mathsf{HD}\, x) \wedge \mathsf{EVERY}\, p\,(\mathsf{TL}\, x)$$

Analogous definitions over lists can be given for:

EXISTS p x: return the disjunction of applying p to the elements of the list x
X_EXISTS p x: exactly one element of the list x returns true when p is applied to all elements in the list
PAIR_EVERY p x y: given two lists, x and y, return the conjunction of applying p to the pair made up of the first elements of each list

Modelling the configuration. The configuration of the system is completely represented by the values of a set of variables which include elements for the basic states, data-items, and events.

We can describe a configuration as a function mapping variables to values:

$$\mathrm{Config} \equiv \mathrm{Variable} \rightarrow \mathrm{Value}$$

The variable cf will be used in functions to represent a configuration.

The meaning of expressions can be given compositionally using functions which take a configuration as an argument and return the result of evaluating the expression in that configuration. For example, the meaning of a variable v is given by the function: $\mathsf{SemVAR}\, v = \lambda cf.\, cf\, v$. This style of function has been used previously to give the semantics of a small imperative language where all of its elements are compositional[6]. Expressions are evaluated relative to the current configuration and actions assign the values of these expressions to the variables in the next configuration.

Hierarchy of states. Each basic state is represented by one Boolean variable which indicates whether or not the system is currently in that state. Higher level states (stn) are given meaning through the values of the basic states:[1]

$$\textsf{INSTATE } sc \; cf \; stn = ((\textsf{TYP } sc \; stn = \textsf{B}) \wedge \textsf{SemVAR } stn \; cf) \vee$$
$$((\textsf{TYP } sc \; stn = \textsf{A}) \wedge \textsf{EVERY}(\textsf{INSTATE } sc \; cf)(\textsf{SUBSTATES } sc \; stn)) \vee$$
$$((\textsf{TYP } sc \; stn = \textsf{O}) \wedge \textsf{EXISTS}(\textsf{INSTATE } sc \; cf)(\textsf{SUBSTATES } sc \; stn))$$

The difficulty in giving the semantics is in expressing both which transitions can be taken and what the result is of following these transitions. The remaining semantic functions can be grouped into three areas:

- determining if events occur in a step (EVENT_COND)
- conditions on the set of transitions which can be taken, including hierarchy, priority and triggers (TRANS_COND)
- conditions on the variables in the next configuration modified in this step (VAR_COND)

The validity of these semantics, both in our interpretation of the operation of statecharts, and in the correctness of expressing this interpretation in higher order logic, has been checked using the theorem prover HOL by reducing the semantic functions to Boolean expressions over the variables for particular problems. Through this process, errors were discovered and fixed, and we have increased confidence in the result.

5.1 Events

Events are interpreted as Boolean expressions which depend on whether changes in values occur between steps. For events other than timeouts, this is relative to the previous step, but for timeouts, it can involve checking several time units earlier. In order to minimize the number of values used in the overall expression, we must determine the truth value of an event relative to the current configuration only. A counter for each relevant event is created which gets reset to zero when the event occurs and otherwise is incremented in each step. For a timeout, we test if the counter is equal to the timeout step number. This allows us to determine the truth value of events relative to the current configuration only.

Since each counter has a maximum value, we have to ensure that it does not falsely indicate that the event occurs when it overflows. This is done by incrementing it only up to its maximum value. This maximum value can never be used to indicate an event occurring, therefore a counter must be larger than its associated timeout step number.

To determine if a timeout event occurs, the function SemTM checks if the counter(c) for the event equals the expressions (e) for the timeout step number:

$$\textsf{SemTM}(c, e) = \lambda cf. \; \neg \textsf{MAXVALUE}(\textsf{SemVAR } c \; cf) \wedge \textsf{SemEQUAL}(c, e) \, cf$$

---

[1] States can have type A (AND), O(OR), or B(basic).

The timeout step number is evaluated in the current configuration resolving the questions raised in Sect.3.5.

The counter is reset to zero in the next configuration if the event occurs in this step. For example, to determine if the system enters a state (stn) in this step, we use the function:

SemEN $sc\ stn = \lambda(cf, cf').\ \neg$INSTATE $sc\ cf\ stn \wedge$ INSTATE $sc\ cf'\ stn$

The next configuration relation must include a condition which updates the events for each transition:

EVENT_COND $sc\ transset\ cf\ cf' =$ (1)
EVERY $(\lambda tlabel.$ UpdateEvent (EVENT (TRANS $sc\ tlabel$)) $(cf, cf'))\ transset$

where $transset$ is the set of numeric transition labels.

5.2 Transition Condition

A *step* means following a set of transitions which satisfy a number of conditions. Each transition is represented by a Boolean flag indicating if the transition is taken in this step. Because statecharts can describe non-deterministic operation, there will be several different sets which are eligible. If a vector of transition flags satisfies the transition condition then it represents a legal set. We resolve the issues raised in Sects.3.1 and 3.4 by giving the conditions which this set must satisfy:

1. A transition is enabled if the system is in its source state and its trigger is true. Any transitions which are followed must be enabled.
2. If two or more transitions are enabled and have the same source state, only one will be taken but it is indeterminate as to which will be chosen.
3. Within an OR-state, only one transition can be followed.
4. Transitions may be followed within each of the substates of an AND-state.
5. If a transition from a parent state is enabled, it has precedence over one from a descendant.
6. If one or more transitions are enabled then some set of transitions will be followed.

The conditions on transitions relative to the hierarchy of states is determined by their source state.

We make the assumption that transitions do not go between components of an AND-state (Sect.3.6) and at the present time, we will also assume that destination states for transitions which are chosen do not conflict.[2] This second assumption should be relaxed in the final version of the semantics. With these assumptions, the above conditions ensure that more than one chosen transition does not modify the same basic state. Provided that the system is currently

---

[2] Destination states for chosen transitions will never conflict if transitions only go between substates of the same parent state.

in a legal state configuration, the next configuration will be legal if the set of transitions taken satisfy these conditions. For our purposes, a step does not include transitions triggered by events occurring in this step and therefore only transitions out of the set of states at the beginning of the step are considered (Sect.3.1).

To satisfy the first three conditions, we can consider the transitions among the substates of a given OR state within the hierarchy. At this one level, we can take exactly one of the transitions providing it is enabled:

$$ONE\_LEVEL\ cf\ sc\ index\ bvtrs = X\_EXISTS\ (\lambda y.\ y)\ bvtrs\ \wedge$$
$$PAIR\_EVERY\ (\lambda(flag, i).\ flag \implies TRIGGER\ sc\ i\ cf)\ bvtrs\ index$$

where $index$ is the list of transition labels for this level and $bvtrs$ is the list of associated flags. The fourth condition is given by saying that the function ONE_LEVEL must be true in all components of an AND-state.

We now have to consider multiple levels in the hierarchy and the priority among these levels. The priority of transitions is given by checking if there is any way to satisfy the function ONE_LEVEL for a given level by existentially quantifying over its transition flags. Only if there is no way to satisfy this function, do we consider transitions originating at lower levels. The function EXISTSN creates a number of existentially quantified Boolean variables in a bit vector ($bvtrans$) which is given as a parameter to the second argument of EXISTSN[9]. In pseudo-code, the transition condition can be given as:

$$TRANS\_COND\ cf\ sc\ bvtrans\ stn = \qquad\qquad (2)$$
$$(TYP\ sc\ stn = B) \Rightarrow T\ |$$
$$(TYP\ sc\ stn = A) \Rightarrow$$
$$\qquad EVERY\ (TRANS\_COND\ cf\ sc\ bvtrans)\ (SUBSTATES\ sc\ stn)\ |$$

let $here$ = set of transitions at this level in

let $prioritytest$ = EXISTSN (LENGTH $here$) (ONE_LEVEL cf sc $here$) in

$prioritytest \Rightarrow$ (ONE_LEVEL cf sc $here$ (transition flags for this level) $\wedge$

$\qquad\qquad$ (all flags for lower levels set to false)) |

$\qquad\qquad$ ((all flags for this level set to false) $\wedge$

$\qquad$ EVERY (TRANS_COND cf sc $bvtrans$) (SUBSTATES sc stn))

5.3 Variable Condition

Given the set of transitions which can be taken, we now have to determine the effects of these transitions on the whole system. The function RESULT returns the set of modifications for exiting the source state, carrying out the actions, and entering the destination state.

Results of a transition. The transitions are labelled with actions which modify the data items in the system. These actions can all be defined in terms of an assignment statement. The semantic function for an assignment statement returns a pair, (v, e), which indicates that the expression e, evaluated in the current configuration, should be assigned to the variable v in the next configuration.

Executing a transition modifies the configuration not only by the actions but also by leaving the source state and entering the destination state. The variables for the basic states of the source state must be set to false and the ones for the destination should be be set to true by following default entrances. If the destination modifications overlap with the ones for the source (for example if a transition loops), then the changes for the destination have precedence.

Combining results of executing transitions. The values of the variables in the next configuration must satisfy the following three properties:

1. If a given transition is taken, at the end of the step the system will be in a configuration which includes the destination state of the transition and all its actions will be carried out except where conflicts occur among the actions of all transitions.
2. If more than one modification is made to the same variable (i.e. a conflict occurs) then exactly one of these modifications will be true in the next configuration.
3. If a variable is not modified by any transition in a step, then it retains its previous value.

Modifications from all transitions are considered together, whether they came from the same or different transitions. The variable condition resolves conflicts among assignments (CH) as discussed in Sects.3.2 and 3.3 and ensures the last property for variables which are not changed (UNCH):

$$\text{VAR\_COND } sc \ cf \ cf' \ transset \ bvtrans \ varlist = \qquad (3)$$
$$\text{EVERY}(\lambda v. \ \text{UNCH } sc \ v \ cf \ cf' \ transset \ bvtrans \ \lor$$
$$\text{CH } sc \ v \ cf \ cf' \ transset \ bvtrans) \ varlist$$

where sc is the textual representation for the statechart, $varlist$ is the set of variables, cf is the current configuration, cf' is the next configuration, $transset$ is the set of labels for the transitions, and $bvtrans$ is a bit vector containing the flags for the transitions.

Classification of variables. Only variables under this system's control should necessarily keep their previous value if they are not modified, i.e. internal variables. External data-items and events may not retain their previous values between steps. The variables for the basic states are all internal, but the classification of events and data items as external or internal must be given. We assume that $varlist$ includes only the internal variables.

Unchanged variables. The function UNCH uses the set of modifications returned by RESULT and the transition flags to determine if a variable, v, has not been changed in a step, and therefore should keep its previous value[3]:

$$\text{UNCH } sc \; v \; cf \; cf' \; transset \; bvtrans =$$
$$\text{EVERY} \, (\lambda tlabel. \; \neg \text{EL } tlabel \, bvtrans \; \vee$$
$$\neg \text{MEMBER} \, v \, (\text{CHANGEDVAR} (\text{RESULT } sc \; tlabel))) \, transset \; \wedge$$
$$\text{EQUAL} \, (\text{SemVAR } v \; cf)(\text{SemVAR } v \; cf')$$

Resolving conflicts. When more than one assignment is made to the same variable, the actions are treated atomically and exactly one of the possible modifications occurs. The function ACT looks at the list of modifications (*modlist*) and forms the disjunction of all possible modifications to a variable (v), evaluating the expressions relative to the current configuration:

$$\text{ACT} \, v \; modlist \; cf \; cf' =$$
$$\text{EXISTS} \, (\lambda asn. \, (\text{FST } asn = v) \wedge \text{EQUAL} \, (cf' \; v) \, ((\text{SND } asn) \, cf)) \; modlist$$

Applying the ACT function to the results of each transition and then taking the disjunction of these clauses for all chosen transitions produces the effect of taking the disjunction of all possible modifications to a variable (v) in a step:

$$\text{CH} \, sc \; v \; cf \; cf' \; transset \; bvtrans =$$
$$\text{EXISTS} \, (\lambda tlabel. \, \text{EL } tlabel \, bvtrans \wedge \text{ACT} \, v \, (\text{RESULT } sc \; tlabel) \, cf \; cf') \, transset$$

5.4 Next Configuration Relation

The next configuration predicate combines all of the restrictions given above to produce a relation between the previous configuration (cf) and the next configuration (cf') for a particular statechart (sc). The set of internal variables is given as a parameter (*varlist*). The variable *root* is the ancestor of all states.

$$\text{NC} \, sc \; varlist \; cf \; cf' =$$
$$\text{let } root = \text{ROOT} \, sc \text{ and } transset = \text{GET\_TRANS\_LABELS} \, sc \text{ in}$$
$$\text{let } transnum = \text{SUC} \, (\text{MAX\_TRANS} \, transset) \text{ in}$$
$$\text{EXISTSN } transnum \, (\lambda bvtrans.$$

$$\text{EVENT\_COND} \, sc \; transset \; cf \; cf' \; \wedge \qquad\qquad (1)$$
$$\text{TRANS\_COND} \, cf \; sc \; bvtrans \; root \; \wedge \qquad\qquad (2)$$
$$\text{VAR\_COND} \, sc \; cf \; cf' \; transset \; bvtrans \; varlist) \qquad (3)$$

---

[3] EL n l returns the nth element of l.

6 Embedding the Semantics in HOL

Most of the preceding definitions can be input directly into HOL, but in a few cases recursive definitions over the hierarchy of the statechart are used. Since HOL does not provide general recursion, these need to be phrased in terms of primitive recursion. We do this by giving the length of the list of states in the statechart as an extra argument to recurse over. This is an upper bound on the recursion since the statechart hierarchy would have to be a degenerate tree to reach this bound. Definitions like **INSTATE** become:

$$(\text{INSTATE } 0 \; sc \; cf \; stn = \text{F}) \wedge$$
$$(\text{INSTATE}\,(\text{SUC}\,n)\; sc \; cf \; stn =$$
$$((\text{TYP}\,sc\;stn = \text{B}) \wedge (\text{SemVAR}\,stn\;cf)) \vee$$
$$((\text{TYP}\,sc\;stn = \text{A}) \wedge \text{EVERY}\,(\text{INSTATE}\,n\;sc\;cf)\,(\text{SUBSTATES}\,sc\;stn)) \vee$$
$$((\text{TYP}\,sc\;stn = \text{O}) \wedge \text{EXISTS}\,(\text{INSTATE}\,n\;sc\;cf)\,(\text{SUBSTATES}\,sc\;stn)))$$

7 Model Checking

The next configuration relation forms the basis for a model checking function written in HOL, and executed in VOSS using the efficient representation of its BDD package. This model checker tests boolean expressions relative to the current configuration. These may be invariants to prove safety properties, functional or timing requirements.

The model checking function quantifies over the bits used to represent the configuration and iteratively checks that the expression holds through a limited number of steps. By starting in any possible system configuration, we can show that the property holds for all time. A full description of the model checker and examples of its use can be found in [1].

The end result is that an operational specification can be created in the CASE tool STATEMATE and then analyzed using a BDD-based model checker. We expect it is possible to make the semantic functions more efficient to speed up this analysis.

8 Conclusion

This paper presents a high-level view of an operational semantics for a working subset of statecharts in quantifier free higher order logic as a next configuration relation. It is important to note that since HOL only deals with total functions, every statechart has an interpretation.

The difficulty in giving these semantics is that the components of statecharts are not completely compositional. The meaning of expressions and actions are expressed simply by examining their parts. Timeout events use counters so they can be evaluated relative the current configuration only. But the overall conditions on transitions and values of variables in the next configuration have to consider all parts of the statechart. Our definition of a *step* is simpler than that used by other versions of the semantics but is easier and clearer to express. It implements non-determinism among transitions on the same level and priority for transitions from states related in the hierarchy. Race conditions and multiple conflicting actions on a transition are resolved by considering all actions together when assigning values to the variables in the next configuration. Interpretations for enabled transitions with conflicting destination states and those which go between orthogonal components have not yet been included.

These semantics form the basis of a model checker for statecharts, but they could also be used to examine their properties. Since they are presented within the framework of HOL, it is open for others to use this theorem-prover to examine these semantics.

9 Acknowledgements

This work was completed while the first author was funded by a Canadian Natural Science and Engineering Research Council Post Graduate Scholarship. We are indebted to the Integrated Systems Design Laboratory for interesting discussions where many of these ideas originated and the graduate students at UBC for comments on drafts of this paper.

References

1. Nancy Day. A model checker for statecharts. Master's thesis, University of British Columbia, 1993. In preparation.
2. D. Harel, A. Pnueli, J.P. Schmidt, and R. Sherman. On the formal semantics of statecharts. In *Proceedings of the 2nd IEEE Symposium on Logic in Computer Science*, pages 54–64, Ithaca, New York, June 1987.
3. David Harel. On visual formalisms. *Communications of the ACM*, 31(5):514–530, May 1988.
4. David Harel, H. Lachover, et al. Statemate: A working environment for the development of complex reactive systems. *IEEE Transactions on Software Engineering*, 16(4):403–414, April 1990.
5. i-Logix Inc., Burlington, MA. *Statemate 4.0 Analyzer User and Reference Manual*, April 1991.
6. Jeffrey J. Joyce. Totally verified systems: Linking verified software to verified hardware. Technical Report No. 178, University of Cambridge Computer Laboratory, September 1989.
7. Nancy G. Leveson, Mats P.E. Heimdahl, Holly Hildreth, and Jon D. Reese. Requirements specification for process-control systems. Technical Report 92-106, University of California, Irvine, Information and Computer Science, 1992.
8. A. Pnueli and M. Shalev. What's in a step: On the semantics of statecharts.
9. Carl-Johan H. Seger and Jeffrey J. Joyce. A mathematically precise two-level formal hardware verification methodology. Technical Report 92-34, University of British Columbia, Department of Computer Science, December 1992.

Value-Passing CCS in HOL *

Monica Nesi

Computer Laboratory, University of Cambridge
New Museums Site, Pembroke Street
Cambridge CB2 3QG, England

mn@cl.cam.ac.uk

Abstract. Value-passing CCS is a process algebra in which actions consist of sending and receiving values through communication ports, and the transmitted data can be tested using a conditional construct. In a previous work, a verification environment for pure CCS agents (no value-passing, only synchronization) has been mechanized in the HOL system. This paper presents an embedding of value-passing CCS in HOL that translates value-passing agents into pure ones. Once the translation has been defined, the HOL verification environment for pure CCS is used to derive a proof environment for the value-passing calculus.

1 Introduction

Value-passing CCS has been introduced in [13, 6]. In [13] Milner presents this calculus by defining its signature and giving the semantics to the operators by adopting the standard approach of translating value-passing agents into the pure calculus. The inference rules of the operational semantics, the definitions of the behavioural semantics as well as their laws, can then be derived from those of the pure calculus due to the translation. Hennessy [6] instead does not resort to any translation and defines the operational semantics for the value-passing calculus by directly following Plotkin's SOS approach. Several proof systems are then proposed [8] which provide powerful methods for reasoning about value-passing agents, and the verification tool VPAM is based on such proof systems [9].

In a previous work [14], a formal theory for pure CCS has been embedded in the HOL theorem prover [4]. The resulting formalization is the basis for verification by mechanized formal proofs, such as proofs of correctness by mathematical induction for parameterized specifications and verification of modal properties [14, 15].

Our aim is to extend the verification environment developed for pure CCS to deal with value-passing agents. In this paper we discuss how value-passing CCS can be formalized in HOL by exploiting the environment for pure CCS. The translation from the value-passing calculus into its pure version is defined in the HOL system. This formalization illustrates the use of polymorphism, and how

---
* The research described in this paper has been partially supported by Girton College, Cambridge.

indexed summations of agents can be mechanized in HOL by means of function types in recursive type definitions. Then, several properties about the transition relation which defines the operational semantics of value-passing agents, and the algebraic laws for some behavioural semantics are derived. These theorems are proved using the translation and the corresponding properties and laws for pure CCS. Reasoning about value-passing agents can thus be performed by applying the derived theorems without resorting to their pure versions.

2 Value-Passing CCS

Given a domain V of values, the syntax of value-passing CCS expressions, ranged over by E, is as follows:

$$E ::= X \mid \sum_{i \in I} E_i \mid a(x).E \mid \bar{a}(e).E \mid \tau.E \mid E \setminus L$$
$$\mid E[f] \mid E \mid E \mid \mathbf{rec}\, X.\, E \mid \mathbf{if}\ b\ \mathbf{then}\ E$$

where

- X denotes an agent variable, which is bound in a recursive agent $\mathbf{rec}\, X.\, E$[1].
- I is an indexing set.
- The label a ranges over the set A of ports at which data can be input, and x denotes a variable in which value constants $v \in V$ can be received. For any input port a there is a corresponding output port $\bar{a}$, from which value expressions e can be sent. The set $A \cup \overline{A}$ of labels of ports is ranged over by l, and L ranges over subsets of labels.
- The complement operation has the property that $\bar{\bar{l}} = l$. This operation is essential for defining communication between agents. The result of a communication is the action τ, the so-called silent or invisible action. This action together with input actions $a(v)$ $(v \in V)$ and output actions $\bar{a}(e)$ gives the set of basic actions, ranged over by u, that an agent can perform.
- A relabelling f is a function from labels to labels with the property that $f(\bar{l}) = \overline{f(l)}$. A relabelling f is then extended to actions by defining $f(\tau) = \tau$.
- b ranges over boolean expressions.

The meaning of the operators is the following.

- The agent $\sum_{i \in I} E_i$ is the (possibly infinite) summation of all expressions E_i as i ranges over the indexing set I. This agent behaves like one of its summands $E_j, j \in I$. Special cases of the summation operator are the inactive agent **nil** which cannot perform any action (when I is the empty set), and the binary summation $E_1 + E_2$ (when $I = \{1, 2\}$).

---

[1] In this syntax the *rec*-notation is used for the definition of recursion, thus differing from the notations in [1, 13], where recursion is encoded as a set of recursive equations. See Section 6 for a brief comment about this.

- The agent $a(x).E$ can perform the action $a(v)$ of receiving any value $v \in V$ at the input port a, and then behaves like $E\{v/x\}$, which is the agent expression obtained by substituting the value v for all free occurrences of the variable x in the expression E.
- The agent $\overline{a}(e).E$ can output the expression e by performing the action $\overline{a}(e)$ and then behaves like E.
- The agent $\tau.E$ can perform the silent action τ and then behaves like E.
- The agent $E \setminus L$ behaves like E but cannot perform an action u if either the input or the output port at which u is performed is in L.
- The actions of $E[f]$ are renamings of those of the agent E via the relabelling function f.
- The agent $E_1 \mid E_2$ can perform the actions of E_1 and E_2 in parallel; moreover, the agents E_1 and E_2 can synchronize through the action τ and communicate data whenever they are able to perform complementary actions.
- The expression **rec** $X.\ E$ denotes a recursive agent.
- The conditional expression **if** b **then** E behaves like E if the boolean expression b is true, and cannot perform any action otherwise.

3 Translating Value-Passing CCS into Pure CCS in HOL

The operational semantics for the operators of the value-passing calculus can be given in two different ways. The first one is *by translation* into the pure version of the calculus for which the operational semantics has already been given [13]. The second one is to follow Plotkin's SOS approach and inductively define the transition relation $E \xrightarrow{u} E'$ over value-passing expressions by a set of inference rules [6]. In this paper we consider the approach based on the translation.

In [14] the formal theory for the observational semantics and for a modal logic over pure CCS has been embedded in the HOL logic [4]. Many properties and laws which characterize the operational and observational semantics have been formally proved. The proof environment contains a collection of conversions and tactics, which allows one to prove the correctness of specifications using powerful techniques, like mathematical induction [14], within various logics, like modal logic [15].

Given the HOL formalization for pure CCS, the translation from the value-passing syntax into the pure one allows us to derive the properties and the laws for the various semantics of the value-passing calculus without having to redo all the proofs. These proofs can be simply carried out by translating the value-passing expressions into their pure versions, and then using the corresponding results already proved for pure CCS. Once these basic properties and laws have been derived, reasoning about the value-passing calculus will be performed without translating the expressions into their pure versions (which would contain possibly infinite indexed summations).

Not only is the translation worth mechanizing, but it is also interesting from the point of view of its HOL formalization, since it raises the problem of defining an operator of indexed summation over recursively defined agents. As described in [13], the translation of an input prefixed agent $a(x).E$, denoted by $\widehat{a(x)}.E$, gives rise to an indexed summation of prefixed agents, $\sum_{v \in V} a_v.E\{v/x\}^2$.

In the following sections, we discuss the issues and problems arising when defining the translation of value-passing CCS into the pure version in HOL, based on the formalization of pure CCS developed in [14].

3.1 Translating the Actions

When translating an input prefixed agent $a(x).E$ into an indexed summation of prefixed agents $\sum_{v \in V} a_v.E\{\widehat{v/x}\}$, any input action $a(v)$ is translated into an action a_v of the pure CCS syntax for each $v \in V$.

A label is simply the name of a port at which communication is performed, and a port can be either an input or an output port. Labels can be restricted and renamed via the restriction and relabelling operators respectively, while values (which appear in actions) are neither restricted nor renamed.

In the HOL formalization of pure CCS [14] strings of characters are used to encode labels. The syntax of labels and actions is defined by two concrete data types using the derived principle for (recursive) type definitions [10] as follows:

$$label = \text{name } string \mid \text{coname } string$$

$$action = \text{tau} \mid \text{label } label$$

where name, coname, tau and label are distinct constructors. These definitions imply that labels are assumed to be strings with information about the status of the ports, i.e. given a string s, name s denotes the input port s, while coname s stands for the output port $\bar{s}$. The HOL derived rule for (recursive) type definition [10] automatically proves the following theorems of higher order logic, which characterize the types *label* and *action* in a complete and abstract way:

$$\vdash \forall f0\ f1.\ \exists!\ fn.\ (\forall s.\ fn\,(\text{name } s) = f0\ s)\ \wedge\ (\forall s.\ fn\,(\text{coname } s) = f1\ s)$$

$$\vdash \forall e\ f.\ \exists!\ fn.\ (fn\,\text{tau} = e)\ \wedge\ (\forall l.\ fn\,(\text{label } l) = f\ l)$$

These theorems assert the admissibility of defining functions over the types *label* and *action* by primitive recursion. Structural induction theorems for both types are provided as well.

Value-passing labels and actions could be formalized in HOL in a similar way:

$$labelv = \text{in } string \mid \text{out } string$$

$$actionv = \text{tauv} \mid \text{labelv } labelv\ \beta$$

where in, out, tauv and labelv are distinct constructors, and β is a type variable which represents the domain V of values.

---

[2] The translation is not the first time when an operator of indexed summation appears in the CCS syntax. The presentation of both pure and value-passing syntaxes given in [13] includes an indexed summation operator, and the inactive agent and the binary summation are derived operators.

When translating a value-passing action $a(x)$, represented as labelv(in a) x in HOL, into a pure action a_v, the formalization of a_v will be an action label (name s) for some string s which is the result of a "combination" of the port a and the value $v : \beta$ which replaces $x : \beta$ when executing the action $a(v)$. This "combination" can be easily obtained if we define a *polymorphic* version of the syntax for pure CCS, in which labels can be of any type α. The polymorphic version $(\alpha)label$ and $(\alpha)action$ for the syntactic types *label* and *action* is therefore the following:

$$label = \text{name } \alpha \mid \text{coname } \alpha$$

$$action = \text{tau} \mid \text{label } (\alpha)label$$

The type variable α is appropriately instantiated depending on the particular application. For example, the action label (name '*bye*') is of type $(string)action$, and label (coname 7) is an action of type $(num)action$.

The operations of complement and relabelling of labels are defined on the polymorphic type $(\alpha)label$, and then extended to the type $(\alpha)action$, similarly to their definitions for the string-based version of pure CCS [14].

When moving from the formalization of pure CCS in which labels are strings, to the more general one in which labels are represented by a type variable α, no relevant changes occur in the HOL code. All the proofs of properties and laws for the CCS operators go through unchanged, apart from adding some type information. The only interesting (and expected) change occurs in the definition of the conversions for applying the laws for the restriction, relabelling and parallel operators. In order to apply these laws, it is necessary to decide whether two labels are equal or not, or whether a label is in a given set of labels. In the polymorphic formalization, these conversions have a procedure as a parameter to decide equality over the type which instantiates the type variable α. In the string-based formalization such a decision procedure was defined in a straightforward way using the built-in conversion string_EQ_CONV for equality of strings.

The polymorphic versions $(\alpha)labelv$ and $(\alpha, \beta)actionv$ of the types for the value-passing labels and actions are defined as follows:

$$labelv = \text{in } \alpha \mid \text{out } \alpha$$

$$actionv = \text{tauv} \mid \text{labelv } (\alpha)labelv \ \beta$$

where α is a type variable for the labels of the ports, and the type variable β represents the domain V of values. For example, labelv (in 'a') $(x : num)$ is an input action of type $(string, num)actionv$, the labels of ports and the values under consideration being strings and integers respectively, and labelv (out T) '*hello*' is an action of type $(bool, string)actionv$.

The presence of a type variable for the ports in the polymorphic formalization of pure CCS allows us to translate the value-passing actions into the pure ones, by simply mapping each input action $a(x)$ into a pair (a, v) for each $v \in V$ which is received in x. If *port* and *value* are the types that instantiate the type variables α and β respectively, and $a(x)$ is of type $(port, value)actionv$, then $a(x)$ is translated into the pure action (a, v) of type $(port\#value)action$, where the compound type *port#value* instantiates the type variable in $(\alpha)action$. For example, the value-passing action labelv(in 'a') $(x : num)$ is translated into the pure action label(name('a', 5)) : $(string\#num)action$, when the value $5 : num$ is received in the variable x. Similarly, any output action $\bar{a}(e) : (port, value)actionv$ is translated into a pure action $\overline{(a, e)} : (port\#value)action$.

The operations of complement and relabelling are defined on the polymorphic type $(\alpha)labelv$ similarly to their definitions in the string-based formalization of pure CCS, and then extended to the type $(\alpha, \beta)actionv$ in a straightforward way. The same holds of the polymorphic type $(\alpha, \beta)relabellingv$ for relabelling functions: it is defined similarly to its definition in [14] as the set of functions of type $(\alpha, \beta)actionv \rightarrow (\alpha, \beta)actionv$, such that relabelling respects complements and τ is renamed as τ.

3.2 An Operator of Indexed Summation

As mentioned above, an agent $a(x).E$ is translated into a summation of agents $\sum_{v \in V} a_v.E\widehat{\{v/x\}}$, which may be infinite, depending on the value domain V. In [14] the syntax for pure CCS with the inactive agent and the binary summation operator has been mechanized in HOL. In what follows, we briefly discuss the HOL formalization of the syntax for pure CCS which includes an operator of indexed summation.

A summation operator $\sum_{i \in I} E_i$ can be formalized by means of a function f, defined on a set I of indexes, which returns a CCS term E_i for each $i \in I$. In order to define the recursive type of pure CCS expressions in HOL, the package for recursive type definition [10] could be used. As already seen in the previous section when formalizing labels and actions, the input to this definition mechanism is an informal specification of the syntax of the type operators, written in terms of existing types and recursive calls to the type being defined. The specification for the type $(\sigma, \alpha)CCS$ of pure CCS expressions, where the type variables σ and α represent the domains of indexes and ports respectively, is the following:

$$
\begin{aligned}
CCS = \ &\mathsf{var}\ string\ | \\
&\mathsf{summ}\ (\sigma \rightarrow CCS)\ (\sigma)set\ | \\
&\mathsf{prefix}\ (\alpha)action\ CCS\ | \\
&\mathsf{restr}\ CCS\ ((\alpha)label)set\ | \\
&\mathsf{relab}\ CCS\ (\alpha)relabelling\ | \\
&\mathsf{par}\ CCS\ CCS\ | \\
&\mathsf{rec}\ string\ CCS
\end{aligned}
$$

where var, summ, prefix, restr, relab, par and rec are distinct constructors. In this specification the type constructor summ has an argument of type $\sigma \rightarrow CCS$, where CCS is the type being defined. This kind of type specification cannot be handled by the current implementation of the type definition package, but as discussed in [5], the package can be extended to deal with specifications of the form

$$
ty = \ldots\ |\ C_i \ldots (\sigma \rightarrow ty) \ldots\ |\ \ldots
$$

provided that the type ty does not occur in σ. Recently David Shepherd has done some work on extending the type definition package in such a way that specifications like the one above can be dealt with, and a theorem of higher order logic which characterizes the recursive type can be derived automatically [16]. His idea is to represent the recursive type by a pair of types, where $\sigma \to CCS$ elements are encoded as a set of pairs which define the function. The inductive type definition package [2] is then used to introduce a relation which will play the role of the uniquely defined function in a recursive type definition. Hence, this relation is shown to induce such a unique function (i.e. each element has exactly one target in the relation).

By following these steps starting from the above type definition for pure CCS, the theorem of higher order logic for the type $(\sigma, \alpha)CCS$ has been derived:

$$\vdash \forall f_1 \ f_2 \ f_3 \ f_4 \ f_5 \ f_6 \ f_7.$$
$$\exists! \ fn.$$
$$(\forall s1. \ fn(\mathsf{var} \ s1) = f_1 \ s1) \land$$
$$(\forall f1 \ s2. \ fn(\mathsf{summ} \ f1 \ s2) = f_2 \ (fn \circ f1) \ f1 \ s2) \land$$
$$(\forall A1 \ C2. \ fn(\mathsf{prefix} \ A1 \ C2) = f_3 \ (fn \ C2) \ A1 \ C2) \land$$
$$(\forall C1 \ s2. \ fn(\mathsf{restr} \ C1 \ s2) = f_4 \ (fn \ C1) \ C1 \ s2) \land$$
$$(\forall C1 \ R2. \ fn(\mathsf{relab} \ C1 \ R2) = f_5 \ (fn \ C1) \ C1 \ R2) \land$$
$$(\forall C1 \ C2. \ fn(\mathsf{par} \ C1 \ C2) = f_6 \ (fn \ C1) \ (fn \ C2) \ C1 \ C2) \land$$
$$(\forall s1 \ C2. \ fn(\mathsf{rec} \ s1 \ C2) = f_7 \ (fn \ C2) \ s1 \ C2)$$

where o denotes composition of functions. Given this theorem, functions can be defined over the type $(\sigma, \alpha)CCS$ by primitive recursion. Moreover, functions which allow one to prove that the above type constructors are distinct and one-to-one, plus theorems for structural induction and case analysis, are provided similarly to the ones in Tom Melham's type definition package.

The inactive agent and the binary summation operator can then be defined as instantiations of the indexed summation **summ** $f \ I$. The inactive agent is formalized in HOL as a summation over an empty indexing set

$$\mathsf{nil} = \mathsf{summ} \ \mathsf{ARBccs} \ \{\}$$

where ARBccs is an arbitrary function. The operator '+' can be defined as a summation over an indexing set of two distinct elements as follows:

$$\forall E \ E' : (\sigma, \alpha)CCS.$$
$$\mathsf{sum} \ E \ E' =$$
$$\mathsf{let} \ x = \varepsilon x : \sigma.\mathsf{T} \ \mathsf{in}$$
$$\mathsf{let} \ y = \varepsilon y : \sigma. \sim (x = y) \ \mathsf{in}$$
$$\mathsf{summ} \ (\lambda i. \ (i = x) \to E \mid E') \ \{z \mid (z = x) \lor (z = y)\}$$

Properties like distinctness can be easily proved for these operators starting from their definitions and the corresponding properties for the basic operators.

The syntax for value-passing CCS expressions can be defined in HOL similarly to that for pure CCS. In this definition, besides the clause for indexed summation, a function type is also used for the mechanization of the input prefix operator.

In fact, an input prefixed agent $a(x).E$ is formalized in HOL by means of the operator In which takes as arguments a port $a : \alpha$ and a function $f : \beta \to CCSv$. This function will be given as an abstraction $\lambda x. E[x]$ which binds the variable $x : \beta$ in the expression $E[x]$.

The specification for the type $(\sigma, \beta, \alpha)CCSv$, where the type variables σ, β and α represent the domains of indexes, values and ports respectively, is thus the following:

$$
\begin{aligned}
CCSv = \; &\text{Var } string \mid \\
&\text{Summ } (\sigma \to CCSv)\,(\sigma)set \mid \\
&\text{In } \alpha\,(\beta \to CCSv) \mid \\
&\text{Out } \alpha\,\beta\,CCSv \mid \\
&\text{Tau } CCSv \mid \\
&\text{Restr } CCSv\,((\alpha)labelv)set \mid \\
&\text{Relab } CCSv\,(\alpha, \beta)relabellingv \mid \\
&\text{Par } CCSv\,CCSv \mid \\
&\text{Rec } string\,CCSv \mid \\
&\text{Cond } bool\,CCSv
\end{aligned}
$$

where Var, Summ, In, Out, Tau, Restr, Relab, Par, Rec and Cond are distinct constructors. The following theorem of higher order logic characterizes the type $(\sigma, \beta, \alpha)CCSv$:

$$
\begin{aligned}
\vdash \; &\forall f_1\ f_2\ f_3\ f_4\ f_5\ f_6\ f_7\ f_8\ f_9\ f_{10}. \\
&\exists!\ fn. \\
&(\forall s1.\ fn(\text{Var } s1)\ =\ f_1\ s1)\ \wedge \\
&(\forall f1\ s2.\ fn(\text{Summ } f1\ s2)\ =\ f_2\ (fn \circ f1)\ f1\ s2)\ \wedge \\
&(\forall x1\ f2.\ fn(\text{In } x1\ f2)\ =\ f_3\ (fn \circ f2)\ x1\ f2)\ \wedge \\
&(\forall x1\ x2\ C3.\ fn(\text{Out } x1\ x2\ C3)\ =\ f_4\ (fn\ C3)\ x1\ x2\ C3)\ \wedge \\
&(\forall C1.\ fn(\text{Tau } C1)\ =\ f_5\ (fn\ C1)\ C1)\ \wedge \\
&(\forall C1\ s2.\ fn(\text{Restr } C1\ s2)\ =\ f_6\ (fn\ C1)\ C1\ s2)\ \wedge \\
&(\forall C1\ R2.\ fn(\text{Relab } C1\ R2)\ =\ f_7\ (fn\ C1)\ C1\ R2)\ \wedge \\
&(\forall C1\ C2.\ fn(\text{Par } C1\ C2)\ =\ f_8\ (fn\ C1)\ (fn\ C2)\ C1\ C2)\ \wedge \\
&(\forall s1\ C2.\ fn(\text{Rec } s1\ C2)\ =\ f_9\ (fn\ C2)\ s1\ C2)\ \wedge \\
&(\forall b1\ C2.\ fn(\text{Cond } b1\ C2)\ =\ f_{10}\ (fn\ C2)\ b1\ C2)
\end{aligned}
$$

This theorem is the basis for reasoning about value-passing CCS expressions in HOL. The formalization of the inactive agent nil and of the binary summation '+' over the value-passing calculus, represented in HOL by the operators Nil and Sum respectively, is similar to the one for the operators nil and sum in pure CCS.

3.3 The Translation for Value-Passing CCS

The translation from value-passing CCS into pure CCS is given recursively on the structure of expressions. The definition of the translation appears not to be primitive recursive, as in the clause for the input prefix constructor, $\widehat{a(x).E} = \sum_{v \in V} a_v.\widehat{E\{v/x\}}$, the recursive occurrence of the translation function is not applied to the subterm E but to the expression $E\{v/x\}$. However, the formalization of the input prefix operator through λ-abstraction and composition of functions makes it possible to define the translation in HOL as a primitive recursive function over the type $(\sigma, \beta, \alpha)CCSv$. This is achieved by composing the translation function with the function argument of the input prefix operator and using the extended version of the function new_recursive_definition [16].

Given an expression $E : (\sigma, \beta, \alpha)CCSv$ and a set V of values, the function CCSv_To_CCS : $(\sigma, \beta, \alpha)CCSv \rightarrow (\beta)set \rightarrow (\sigma + \beta, \alpha\#\beta)CCS$ translates E into a pure expression whose domains of indexes and ports are $\sigma + \beta$ and $\alpha\#\beta$ respectively. The disjoint sum instantiates the type variable for the indexing domain in $(\sigma+\beta, \alpha\#\beta)CCS$ because, when translating an input prefixed agent of type $(\sigma, \beta, \alpha)CCSv$, the result is a summation over the value domain $V : (\beta)set$.

$\vdash (\forall X\, V.\ \text{CCSv\_To\_CCS}\, (\text{Var}\, X)\, V = \text{var}\, X) \land$
 $\ (\forall f\, I\, V.$
 $\quad \text{CCSv\_To\_CCS}\, (\text{Summ}\, f\, I)\, V =$
 $\quad ((\sim (I = \{\})) \supset$
 $\quad \text{summ}\, (\lambda i.\ (\text{CCSv\_To\_CCS} \circ f)\, (\text{Outl}\, i)\, V)\, \{\text{Inl}\, j \mid j \in I\}\ \mid\ \text{nil})) \land$
 $\ (\forall p\, f\, V.$
 $\quad \text{CCSv\_To\_CCS}\, (\text{In}\, p\, f)\, V =$
 $\quad \text{summ}$
 $\quad (\lambda v.\ \text{prefix}\, (\text{label}\, (\text{name}\, (p, \text{Outr}\, v)))\, ((\text{CCSv\_To\_CCS} \circ f)\, (\text{Outr}\, v)\, V))$
 $\quad \{\text{Inr}\, v' \mid v' \in V\}) \land$
 $\ (\forall p\, e\, E\, V.$
 $\quad \text{CCSv\_To\_CCS}\, (\text{Out}\, p\, e\, E)\, V =$
 $\quad \text{prefix}\, (\text{label}\, (\text{coname}\, (p, e)))\, (\text{CCSv\_To\_CCS}\, E\, V)) \land$
 $\ (\forall E\, V.\ \text{CCSv\_To\_CCS}\, (\text{Tau}\, E)\, V = \text{prefix}\, \text{tau}\, (\text{CCSv\_To\_CCS}\, E\, V)) \land$
 $\ (\forall E\, L\, V.$
 $\quad \text{CCSv\_To\_CCS}\, (\text{Restr}\, E\, L)\, V =$
 $\quad \text{restr}$
 $\quad (\text{CCSv\_To\_CCS}\, E\, V)$
 $\quad (\{\text{name}\, (p, v) \mid (\text{in}\, p) \in L \land v \in V\} \cup$
 $\quad \{\text{coname}\, (p, v) \mid (\text{out}\, p) \in L \land v \in V\})) \land$
 $\ (\forall E\, rf\, V.$
 $\quad \text{CCSv\_To\_CCS}\, (\text{Relab}\, E\, rf)\, V =$
 $\quad \text{relab}$
 $\quad (\text{CCSv\_To\_CCS}\, E\, V)$
 $\quad (\text{ABS\_Relabelling}\, (\text{Relab\_TR}\, (\text{REP\_Relabellingv}\, rf)))) \land$
 $\ (\forall E\, E'\, V.$
 $\quad \text{CCSv\_To\_CCS}\, (\text{Par}\, E\, E')\, V =$
 $\quad \text{par}\, (\text{CCSv\_To\_CCS}\, E\, V)(\text{CCSv\_To\_CCS}\, E'\, V)) \land$
 $\ (\forall X\, E\, V.\ \text{CCSv\_To\_CCS}\, (\text{Rec}\, X\, E)\, V = \text{rec}\, X\, (\text{CCSv\_To\_CCS}\, E\, V)) \land$
 $\ (\forall b\, E\, V.\ \text{CCSv\_To\_CCS}\, (\text{Cond}\, b\, E)\, V = (b \supset \text{CCSv\_To\_CCS}\, E\, V\ \mid\ \text{nil}))$

The functions Outl (Outr) and Inl (Inr) are the left (right) projection and the left (right) injection functions for the disjoint sum operation on types. The function Relab_TR translates a relabelling that renames value-passing actions into a relabelling that renames pure actions. ABS_Relabelling and REP_Relabellingv are the abstraction and representation functions of the types $(\alpha)relabelling$ and $(\alpha, \beta)relabellingv$ respectively.

The translation of the derived operators Nil and Sum into the corresponding pure CCS operators can be derived from their definitions by rewriting with CCSv_To_CCS.

The formalization of an input prefixed agent $a(x).E$ through a λ-abstraction that binds the variable x in the expression E, allows one to use the β-reduction in the HOL logic to mechanize the substitution of value constants v for variables x in value-passing expressions, like $E\{v/x\}$, in boolean expressions b and in value expressions e. Let E be the agent expression

$$E \equiv in(x). (\text{if } x < 5 \text{ then } \overline{out}(x+1).\text{nil})$$

where the labels of ports are strings, the value domain V is the set of natural numbers, and the indexing domain is represented by any type σ. The agent expression E is represented in HOL as the term $t : (\sigma, num, string)CCSv$

$$\text{In 'in' } (\lambda x.\, \text{Cond } (x < 5)\, (\text{Out 'out' } (x+1)\, \text{Nil}))$$

By applying the translation function CCSv_To_CCS to the term t and the universe of natural numbers, Univ : $(num)set$, and by rewriting with the definition of composition of functions and applying β-reduction, we get the following theorem:

$\vdash$ CCSv_To_CCS (In 'in' ($\lambda x.$ Cond $(x < 5)$ (Out 'out' $(x+1)$ Nil))) Univ =
summ
($\lambda v.$
  prefix
  (label (name ('in', Outr v)))
  ((Outr v) $< 5 \supset$ prefix (label (coname ('out', (Outr v) $+ 1$))) nil | nil)
{Inr v' | $v' \in$ Univ}

whose right-hand side is the translation of t of type $(\sigma+num, string\#num)CCS$. This term is similar to the schema of expressions given in [13], where the bound variable x has been replaced by the value Outr v both in the boolean expression $x < 5$ and in the value expression $x + 1$.

When reasoning about value-passing agents, we would like to be able to work on the value-passing expressions without resorting to their translation into pure CCS. In the next section, we outline the development of a proof environment for the value-passing calculus based on the translation and the HOL formalization of pure CCS. Once this environment has been derived, verification will be performed on the value-passing agents without translating them into pure expressions.

4 Deriving Properties and Laws for Value-Passing CCS

The behaviour of any value-passing CCS expression E can be explored by translating E into its pure CCS version E' and then using the inference rules of the operational semantics for pure CCS. Similarly, the equivalence of any two value-passing agents $E1$ and $E2$, with respect to a given behavioural semantics, can be verified by checking the behavioural equivalence between the two expressions $E1'$ and $E2'$ resulting from the translation into pure CCS. In what follows, we show how some of the properties and the laws for the observational equivalence over value-passing CCS can be derived. The same method applies to any other behavioural semantics as well.

Behavioural equivalences over value-passing expressions can be directly defined in terms of the corresponding equivalences over pure CCS agents, without having to resort to the notion of bisimulation. The relation

$$\text{Obs\_Equiv\_v} : (\sigma, \beta, \alpha)CCSv \rightarrow (\sigma, \beta, \alpha)CCSv \rightarrow bool$$

denotes the observational equivalence in HOL over value-passing expressions, and is defined in terms of the HOL relation **Obs_Equiv** for the observational equivalence over pure CCS as follows:

$\forall E\, E' : (\sigma, \beta, \alpha)CCSv.$
Obs_Equiv_v $E\, E' = (\forall V.\ \text{Obs\_Equiv}\,(\text{CCSv\_To\_CCS}\ E\ V)\ (\text{CCSv\_To\_CCS}\ E'\ V))$

Properties of the observational equivalence over value-passing expressions can be easily derived from the corresponding properties over pure expressions. For example, the fact that **Obs_Equiv_v** is an equivalence relation is given by the following theorems which assert that **Obs_Equiv_v** is a reflexive, symmetric and transitive relation:

OBS_EQUIV_REFLv: $\vdash \forall E : (\sigma, \beta, \alpha)CCSv.\ \text{Obs\_Equiv\_v}\ E\ E$

OBS_EQUIV_SYMv: $\vdash \forall E\, E' : (\sigma, \beta, \alpha)CCSv.$
                    $\text{Obs\_Equiv\_v}\ E\ E' \supset \text{Obs\_Equiv\_v}\ E'\ E$

OBS_EQUIV_TRANSv: $\vdash \forall E\, E'\, E'' : (\sigma, \beta, \alpha)CCSv.$
                    $\text{Obs\_Equiv\_v}\ E\ E' \land \text{Obs\_Equiv\_v}\ E'\ E'' \supset$
                    $\text{Obs\_Equiv\_v}\ E\ E''$

These and other properties about the observational equivalence over value-passing CCS are trivially derived. The standard proof consists of rewriting the property to be proved with the definition of **Obs_Equiv_v** and of applying the corresponding property over pure CCS expressions.

The same reasoning applies to the proofs of the algebraic laws of the behavioural semantics. For example, one of the τ-laws for the observational equivalence, $E + \tau.\, E =_{obseq} \tau.\, E$, is formalized in HOL by the following theorem:

TAU2_EQUIVv: $\vdash \forall E : (\sigma, \beta, \alpha)CCSv.\ \text{Obs\_Equiv\_v}\,(\text{Sum}\ E\ (\text{Tau}\ E))\ (\text{Tau}\ E)$

The standard proof of the algebraic laws for a given behavioural equivalence Beh over the value-passing calculus, simply consists of rewriting the goal with the definition of Beh and of the translation CCSv_To_CCS, and then applying the corresponding law for Beh over pure CCS. In our formalization of value-passing CCS, the algebraic laws for strong equivalence, observational equivalence and observational congruence have been derived from the corresponding ones over pure CCS.

5 Related Work

There exist several tools for verifying properties of concurrent specifications [18, 19, 20]. These systems mainly deal with pure CCS, and most of them are based on a finite state machine representation of agents, which works fine when only finite state specifications are considered. As soon as we address communication of values, a finite state machine representation still works only if the value domains under consideration are finite, as value-passing agents can be translated into finite state pure agents.

In [1] the syntax and the operational semantics of a language for value-passing CCS is defined independently of pure CCS. This language is then implemented via a translation into pure CCS, such that the result of the translation can be verified in a finite state machine environment like the Concurrency Workbench [3].

In [6] Hennessy proposes a proof system which provides powerful methods for reasoning about value-passing agents. The main idea is to separate reasoning about the data from reasoning about the process behaviour as much as possible. Entirely separate proof systems are provided for reasoning on the data, and are used by the main proof system which works on the process behaviour. In a value-passing framework the behaviour of an agent depends on the data it receives, and therefore the structure of a particular proof will depend on properties of the data expressions involved.

The verification tool VPAM [9] for value-passing CCS is based on a proof system which deals with data and boolean expressions *symbolically* [7]. This means that, when value-passing agents are analyzed, boolean and value expressions are not evaluated, and input variables are not instantiated. In this way, reasoning about data is separated from reasoning about agents, and is performed by extracting "proof obligations" which can be verified by another theorem prover later or on-line with the main proof about the process behaviour.

Our approach to formalizing a logical environment for supporting reasoning about process algebras is based on a purely definitional approach to using higher order logic. The same methodology has been applied in [11], where a proof tool is developed for Milner's π-calculus.

6 Conclusion and Future Work

The goal of the work presented in this paper is to develop a proof system for value-passing CCS in the HOL theorem prover, in order to perform verification of agents which communicate with each other by transmitting data. Value-passing CCS is formalized by embedding the syntax directly in HOL, and then using the translation from value-passing CCS into pure CCS, as defined in [13], and a previous formalization of pure CCS in HOL [14]. This allows us to derive the transition relation over value-passing CCS, its properties, the behavioural equivalences and their laws from the corresponding ones over pure CCS. The notion of polymorphism plays an important role when mechanizing value-passing actions and their translation into pure actions. Furthermore, the formalization in HOL of the indexed summation operator is an example of the use of function types in recursive type definitions.

When reasoning about the value-passing calculus, agents are not encoded into pure agents. The translation into pure CCS, which can lead to infinitely branching expressions, is used only at the meta-level to reason about the properties and the various semantic notions for the value-passing calculus, and to derive them using the HOL formalization for pure CCS.

Current work consists of completing the derivation of the proof tools for value-passing CCS from those for pure CCS. Future work includes the formalization in our HOL-CCS environment of the notation for recursion based on sets of recursive equations. This notation and the rec-notation, adopted in this paper, are equivalent (actually, they are strong equivalent [17]). However, the first one is more convenient, in particular when dealing with parametric recursion. The equations defining a set of recursive agents can be formalized through a function which associates a CCS expression to any agent identifier. This function can then be given as a parameter to the definition of the transition relation using the inductive definition package [12]. In this way, a class of transition systems is defined, one for each set of recursive equations.

Acknowledgements

I am grateful to Mike Gordon for his advice in the formalization of the translation, and to David Shepherd for sharing his definitions with me and being patient with all my subsequent e-mail messages. Thanks are also due to the HUG'93 referees for their helpful comments on this paper.

References

1. Bruns, G., 'A language for value-passing CCS', Technical Report ECS-LFCS-91-175, LFCS, University of Edinburgh, August 1991.
2. Camilleri, J. and T. Melham, 'Reasoning with Inductively Defined Relations in the HOL Theorem Prover', Technical Report No. 265, Computer Laboratory, University of Cambridge, August 1992.
3. Cleaveland, R., J. Parrow, and B. Steffen, 'The Concurrency Workbench', in [18], pp. 24–37.
4. Gordon, M. J. C. and T. F. Melham, 'Introduction to HOL: a theorem proving environment for higher order logic', Cambridge University Press, 1993.
5. Gunter, E. L., 'Why We Can't Have SML Style datatype Declarations in HOL', in Proceedings of the 1992 International Workshop on Higher Order Logic Theorem Proving and Its Applications, L. J. M. Claesen and M. J. C. Gordon (eds.), IFIP Transactions A-20, North-Holland, 1993, pp. 561–568.
6. Hennessy, M., 'A Proof System for Communicating Processes with Value-Passing', in Formal Aspects of Computing, Vol. 3, 1991, pp. 346–366.
7. Hennessy, M. and H. Lin, 'Symbolic bisimulations', Technical Report 1/92, Computer Science, University of Sussex, April 1992.

8. Hennessy, M. and H. Lin, 'Proof Systems for Message-Passing Process Algebras', Technical Report 5/93, Computer Science, University of Sussex, February 1993, also in Proceedings of *CONCUR '93*, E. Best (ed.), Lecture Notes in Computer Science, Springer-Verlag, Vol. 715, 1993.

9. Lin, H., 'A Verification Tool for Value-Passing Processes', in Proceedings of *PSTV XIII*, Liege, Belgium, May 1993.

10. Melham, T. F., 'Automating Recursive Type Definitions in Higher Order Logic', in *Current Trends in Hardware Verification and Automated Theorem Proving*, G. Birtwistle and P. Subrahmanyam (eds.), Springer-Verlag, 1989, pp. 341–386.

11. Melham, T. F., 'A Mechanized Theory of the π-calculus in HOL', Technical Report No. 244, Computer Laboratory, University of Cambridge, January 1992.

12. Melham, T. F., 'A Package for Inductive Relation Definitions in HOL', in Proceedings of the *1991 International Workshop on the HOL Theorem Proving System and its Applications*, P. J. Windley, M. Archer, K. N. Levitt and J. J. Joyce (eds.), IEEE Computer Society Press, 1992, pp. 350–357.

13. Milner, R., *Communication and Concurrency*, Prentice Hall, London, 1989.

14. Nesi, M., 'A Formalization of the Process Algebra CCS in Higher Order Logic', Technical Report No. 278, Computer Laboratory, University of Cambridge, December 1992.

15. Nesi, M., 'Formalizing a Modal Logic for CCS in the HOL Theorem Prover', in Proceedings of the *1992 International Workshop on Higher Order Logic Theorem Proving and Its Applications*, L. J. M. Claesen and M. J. C. Gordon (eds.), IFIP Transactions A-20, North-Holland, 1993, pp. 279–294.

16. Shepherd, D. E., Private communication, March 1993.

17. Taubner, D., 'A note on the notation of recursion in process algebras', *Information Processing Letters*, North-Holland, 1991, Vol. 37, pp. 299–303.

18. Proceedings of the *Workshop on Automatic Verification Methods for Finite State Systems*, Grenoble, 1989, J. Sifakis (ed.), Lecture Notes in Computer Science, Springer-Verlag, Vol. 407, 1990.

19. Proceedings of the *2nd Workshop on Computer Aided Verification*, New Brunswick, New Jersey, 1990, E. M. Clarke and R. P. Kurshan (eds.), Lecture Notes in Computer Science, Springer-Verlag, Vol. 531, 1991.

20. Proceedings of the *3rd Workshop on Computer Aided Verification*, Ålborg University, 1991, K. G. Larsen and A. Skou (eds.), Lecture Notes in Computer Science, Springer-Verlag, Vol. 575, 1992.

TPS: An Interactive and Automatic Tool
for Proving Theorems of Type Theory

Peter B. Andrews[1], Matthew Bishop[1], Sunil Issar[1],
Dan Nesmith[2], Frank Pfenning[1], Hongwei Xi[1]

Contact: Peter B. Andrews
Mathematics Department
Carnegie Mellon University
Pittsburgh, PA 15213, U.S.A.
Andrews@CS.CMU.EDU
412-268-2554

Abstract

This is a demonstration of TPS, a theorem proving system for classical type theory (Church's typed λ-calculus). TPS can be used interactively or automatically, or in a combination of these modes. An important feature of TPS is the ability to translate between expansion proofs and natural deduction proofs.

CATEGORY: Demonstration

1. Introduction

This presentation is a demonstration of TPS, a theorem proving system for classical type theory (Church's typed λ-calculus [14]) which has been under development at Carnegie Mellon University for a number of years.[3] TPS is based on an approach to automated theorem proving called the *mating method* [2], which is essentially the same as the *connection method* developed independently by Bibel [13]. The mating method does not require reduction to clausal form.

TPS handles two sorts of proofs, natural deduction proofs and expansion proofs.

Natural deduction proofs are human-readable formal proofs. An example of such a proof which was produced automatically by TPS is given in Figure 1-1. This is a proof that equality (as defined in the Leibniz sense — see line (2) of the proof) is the intersection of all reflexive relations. To explain the notation briefly, consider line (16):

$$(16) \quad 10 \quad \vdash \quad \sim\!q_{o_\iota}\, x^1_\iota \lor q\, x_\iota \qquad\qquad \text{MP: 15 13}$$

The wff being asserted in this line is $\sim\!q_{o_\iota}\, x^1_\iota \lor q\, x_\iota$; it was inferred by Rule MP (Modus Ponens) from the wffs in lines (15) and (13), and the formula asserted in line (10) is the sole hypothesis for this line. See [1] or [4] for more details about this formulation of natural deduction.

Expansion proofs are concise Herbrand expansions of theorems of type theory. The structure of an expansion proof, which incorporates a mating, is closely and directly related to the structure of the theorem it establishes, and provides a context for search which facilitates concentrating on the essential logical structure of the theorem. Expansion proofs were introduced by Miller [23]. A brief explanation of them may be found in [7], which provides an introduction to many ideas underlying TPS. [23], [24], [25], [27], [28], and [29] contain extensive treatments of expansion proofs, translations between natural deduction proofs and expansion proofs, and ways of improving the natural deduction proofs obtained by these translations. Many of these ideas are implemented in TPS.

---

[1]Carnegie Mellon University, Pittsburgh, PA 15213, U.S.A.

[2]University of the Saarland, D-W-6600 Saarbruecken, Germany

[3]This material is based upon work supported by the National Science Foundation under grants CCR-9002546 and CCR-9201893.

Figure 1-1: A Natural Deduction Proof

(1)	1	⊢	$x^1_i = x_i$	Hyp
(2)	1	⊢	$\forall q_{o\iota} \,.q\, x^1_i \supset q\, x_i$	Equality: 1
(3)	3	⊢	$\forall X_i\, p_{o\iota\iota}\, X\, X$	Hyp
(4)	3	⊢	$p_{o\iota\iota}\, x^1\, x^1$	UI: x^1 3
(5)	1	⊢	$p_{o\iota\iota}\, x^1\, x^1 \supset p\, x^1\, x_i$	UI: $[p\, x^1]$ 2
(6)	1,3	⊢	$p_{o\iota\iota}\, x^1_i\, x_i$	RuleP: 4 5
(7)	1	⊢	$\forall X_i\, p_{o\iota\iota}\, X\, X \supset p\, x^1_i\, x_i$	Deduct: 6
(8)	1	⊢	$\forall p_{o\iota\iota} \,.\forall X_i\, p\, X\, X \supset p\, x^1_i\, x_i$	UGen: p 7
(9)		⊢	$x^1_i = x_i \supset \forall p_{o\iota\iota} \,.\forall X_i\, p\, X\, X \supset p\, x^1\, x$	Deduct: 8
(10)	10	⊢	$\forall p_{o\iota\iota} \,.\forall X_i\, p\, X\, X \supset p\, x^1_i\, x_i$	Hyp
(11)	11	⊢	$q_{o\iota}\, x^1_i$	Hyp
(12)	10	⊢	$\forall X_i\, [\lambda w_i\, \lambda w^1 \,.{\sim}q_{o\iota}\, w \lor q\, w^1]\, X\, X$	
			$\supset [\lambda w\, \lambda w^1 \,.{\sim}q \lor w \lor q\, w^1]\, x^1_i\, x_i$	UI: $[\lambda w\, \lambda w^1 .{\sim}q\, w \lor q\, w^1]$ 10
(13)	10	⊢	$\forall X_i\, [{\sim}q_{o\iota}\, X \lor q\, X] \supset {\sim}q\, x^1_i \lor q\, x_i$	Lambda: 12
(14)		⊢	${\sim}q_{o\iota}\, X_i \lor q\, X$	RuleP
(15)		⊢	$\forall X_i \,.{\sim}q_{o\iota}\, X \lor q\, X$	UGen: X 14
(16)	10	⊢	${\sim}q_{o\iota}\, x^1_i \lor q\, x_i$	MP: 15 13
(17)	10,11	⊢	$q_{o\iota}\, x_i$	RuleP: 11 16
(18)	10	⊢	$q_{o\iota}\, x^1_i \supset q\, x_i$	Deduct: 17
(19)	10	⊢	$\forall q_{o\iota} \,.q\, x^1_i \supset q\, x_i$	UGen: q 18
(20)	10	⊢	$x^1_i = x_i$	Equality: 19
(21)		⊢	$\forall p_{o\iota\iota}\, [\forall X_i\, p\, X\, X \supset p\, x^1_i\, x_i] \supset x^1 = x$	Deduct: 20
(22)		⊢	$[x^1_i = x_i \supset \forall p_{o\iota\iota} \,.\forall X_i\, p\, X\, X \supset p\, x^1\, x]$	
			$\land \,.\forall p\, [\forall X_i\, p\, X\, X \supset p\, x^1\, x] \supset x^1 = x$	RuleP: 9 21
(23)		⊢	$x^1_i = x_i \equiv \forall p_{o\iota\iota} \,.\forall X_i\, p\, X\, X \supset p\, x^1\, x$	ImpEquiv: 22
(24)		⊢	$x^1_i = x_i = \forall p_{o\iota\iota} \,.\forall X_i\, p\, X\, X \supset p\, x^1\, x$	Ext=: 23
(25)		⊢	$\forall x_i \,.x^1_i = x = \forall p_{o\iota\iota} \,.\forall X_i\, p\, X\, X \supset p\, x^1\, x$	UGen: x 24
(26)		⊢	$\forall x_i \,.\; [\lambda v_i \,.x^1_i = v]\, x$	
			$= [\lambda y_i\, \forall p_{o\iota\iota} \,.\forall X_i\, p\, X\, X \supset p\, x^1\, y]\, x$	Lambda: 25
(27)		⊢	$\lambda v_i\, [x^1_i = v] = \lambda y_i\, \forall p_{o\iota\iota} \,.\forall X_i\, p\, X\, X \supset p\, x^1\, y$	Ext=: 26
(28)		⊢	$\forall x^1_i \,.\lambda v_i\, [x^1 = v] = \lambda y_i\, \forall p_{o\iota\iota} \,.\forall X_i\, p\, X\, X \supset p\, x^1\, y$	
				UGen: x^1 27
(29)		⊢	$\forall x_i \,.\lambda v_i\, [x = v] = \lambda y_i\, \forall p_{o\iota\iota} \,.\forall X_i\, p\, X\, X \supset p\, x\, y$	AB: 28
(30)		⊢	$\forall x_i \,.\; [\lambda u_i\, \lambda v_i \,.u = v]\, x$	
			$= [\lambda x\, \lambda y_i\, \forall p_{o\iota\iota} \,.\forall X_i\, p\, X\, X \supset p\, x\, y]\, x$	Lambda: 29
(31)		⊢	$\lambda u_i\, \lambda v_i\, [u = v] = \lambda x_i\, \lambda y_i\, \forall p_{o\iota\iota} \,.\forall X_i\, p\, X\, X \supset p\, x\, y$	
				Ext=: 30
(32)		⊢	$\lambda u_i\, \lambda v_i\, [u = v] = \lambda x_i\, \lambda y_i\, \forall p_{o\iota\iota} \,.\text{REFLEXIVE}\, p \supset p\, x\, y$	EquivWffs: 31
(33)		⊢	$\lambda u_i\, \lambda v_i\, [u = v]$	
			$= \lambda x_i\, \lambda y_i\, \forall p_{o\iota\iota} \,.[\lambda r_{o\iota\iota}\, \text{REFLEXIVE}\, r]\, p \supset p\, x\, y$	Lambda: 32
(34)		⊢	$\lambda u_i\, \lambda v_i\, [u = v]$	
			$= \cap \,.\lambda r_{o\iota\iota}\, \text{REFLEXIVE}\, r$	EquivWffs: 33

TPS is still under development, but it has working facilities for searching for expansion proofs automatically or interactively, translating these into natural deduction proofs, constructing natural deduction proofs interactively, translating natural deduction proofs which are in normal form into expansion proofs, and solving unification problems in higher-order logic, as well as a variety of utilities designed to facilitate research and efficient interaction with the program. The ability to translate between expansion proofs and natural deduction proofs is one of the important and attractive features of TPS. It permits both humans and computers to work in contexts which are appropriate to them.

2. Finding Proofs in TPS

When one uses TPS to prove a theorem interactively, the proof can be displayed in a window called a proofwindow, which is updated automatically whenever a command which changes the proof is executed. Another window displays only the "active lines" of the proof, so that one can concentrate on the essentials of the problem. One can work forwards, backwards, or in a combination of these modes. Online help is available for all commands, as well as for their arguments. One can easily rearrange proofs and delete parts of proofs. One can save proofs in files, and read them in at another time to continue work, or one can save the entire sequence of commands one has executed, and re-execute them later.

The purely interactive facilities of TPS have been used under the name ETPS (Educational Theorem Proving System) by students in logic courses at Carnegie Mellon for a number of years to construct natural deduction proofs. ETPS contains exercises from the textbook [4]. Students quickly learn to use ETPS by reading the manual [30] (which contains several complete examples of how to construct proofs) and doing assigned exercises. ETPS was reviewed in [15].

The basic tools in TPS for automatically applying rules of inference to construct natural deduction proofs are *tactics*, which can be combined using *tacticals* [16]. A tactic applies rules of inference forwards or backwards to derive new proof lines or to justify certain lines of the proof while introducing other lines which may still require justification. The main use of tactics in TPS is to translate an expansion proof into a natural deduction; in this context, the tactics consult the expansion proof for useful information, such as how to instantiate a quantifier. However, one can also use tactics to speed up the process of constructing proofs interactively. The tactics are written in Lisp, and TPS has a command USE-TACTIC for applying tactics. TPS has a command called GO2 which repeatedly calls a number of tactics to apply mundane rules of inference to construct the easy parts of the proof, and quickly bring one to the point where some judgement and insight are needed. The user can choose whether or not to be prompted for approval before each of these tactics is applied.

One can use TPS to prove a theorem completely automatically (if it is not too difficult), or to provide help by automatically proving certain lines of the current proof from other specified lines after the user has made crucial choices interactively. When one asks TPS to find a proof automatically, it starts out by searching for an expansion proof, and then translates this into a natural deduction proof. Search procedures using outermost quantifier duplication [2] and path-focused quantifier duplication [18] [19] are implemented in TPS. TPS uses Huet's higher-order unification algorithm [17], and applies *primitive substitutions* [7] to introduce connectives and quantifiers in substitution terms for set variables. Much must be done to explore these search procedures more thoroughly and to develop better search procedures.

3. Features of TPS

TPS has a number of top levels, each with its own commands, which are listed when one uses the "?" command at that top level. The main top level is for constructing natural deduction proofs, and there is another devoted to matings and expansion proofs.

Another top level is a formula editor which facilitates constructing new formulas from others already known to TPS, and performing various manipulations such as λ-conversion. When one enters the editor, windows display the formula being edited and the particular part of the wff one is focused on. TPS uses a type inference mechanism which enables it to determine the types of many variables when a user is typing a wff into TPS. It also understands various conventions for omitting brackets.

Many aspects of the program's behavior can be controlled by setting flags, and there are over 200 of these flags. TPS has a top level called Review for examining and changing the settings of flags, and for defining and reusing groups of flag settings called *modes*.

Still another top level is a library facility for saving and displaying wffs, definitions, modes, and disagreement pairs for higher-order unification problems. Definitions can be polymorphic (i.e., contain type variables), and can contain other definitions to any level of nesting. When TPS retrieves a definition or theorem, it retrieves all the necessary subsidiary definitions.

Proofs in natural deduction style can be printed in files which are processed by Scribe or Tex, so that familiar notations of logic appear in them. The symbols of logic also appear when wffs are displayed on the screen.

TPS runs in a variety of implementations of Common Lisp, and has been distributed to a number of researchers. It is a large program whose uncompiled source code contains more than 100,000 lines (including comments) and occupies about 3.5 megabytes. The compiled core image for a Decstation 3100 occupies about 14.4 megabytes using Allegro Common Lisp, and about 31.4 megabytes using CMU Common Lisp.

Considerable documentation [11], [12], [20], [21], [26], [30] has been written, though more is needed. The Facilities Guides [11] [12] are produced automatically. An earlier version of TPS, which contributed much to the present version, was described in [22] and [3]. More information about TPS will appear in [10].

4. References

1. Peter B. Andrews, "Transforming Matings into Natural Deduction Proofs," in *5th Conference on Automated Deduction*, edited by W. Bibel and R. Kowalski, Les Arcs, France, Lecture Notes in Computer Science 87, Springer-Verlag, 1980, 281-292.

2. Peter B. Andrews, *Theorem Proving via General Matings*, Journal of the ACM 28 (1981), 193-214.

3. Peter B. Andrews, Dale A. Miller, Eve Longini Cohen, Frank Pfenning, "Automating Higher-Order Logic," in *Automated Theorem Proving: After 25 Years*, edited by W. W. Bledsoe and D. W. Loveland, Contemporary Mathematics series, vol. 29, American Mathematical Society, 1984, 169-192.

4. Peter B. Andrews, *An Introduction to Mathematical Logic and Type Theory: To Truth Through Proof*, Academic Press, 1986.

5. Peter B. Andrews, Frank Pfenning, Sunil Issar, C. P. Klapper, "The TPS Theorem Proving System," in *8th International Conference on Automated Deduction*, edited by Jorg H. Siekmann, Oxford, England, Lecture Notes in Computer Science 230, Springer-Verlag, 1986, 663-664.

6. Peter B. Andrews, Sunil Issar, Daniel Nesmith, Frank Pfenning, "The TPS Theorem Proving System," in *9th International Conference on Automated Deduction*, edited by Ewing Lusk and Ross Overbeek, Argonne, Illinois, Lecture Notes in Computer Science 310, Springer-Verlag, 1988, 760-761.

7. Peter B. Andrews, *On Connections and Higher-Order Logic*, Journal of Automated Reasoning 5 (1989), 257-291.

8. Peter B. Andrews, Sunil Issar, Dan Nesmith, Frank Pfenning, "The TPS Theorem Proving System," in *10th International Conference on Automated Deduction*, edited by M. E. Stickel, Kaiserslautern, FRG, Lecture Notes in Artificial Intelligence 449, Springer-Verlag, 1990, 641-642.

9. Peter B. Andrews, Sunil Issar, Dan Nesmith, and Frank Pfenning, *The TPS Theorem Proving System*, Journal of Symbolic Logic 57 (1992), 353-354. (abstract)

10. Peter B. Andrews, Matthew Bishop, Sunil Issar, Dan Nesmith, Frank Pfenning, Hongwei Xi. TPS: A Theorem Proving System for Classical Type Theory, 1993, unpublished.

11. Peter B. Andrews, Sunil Issar, Dan Nesmith, Frank Pfenning, Hongwei Xi, Matthew Bishop, *TPS3 Facilities Guide for Programmers and Users*, 1993. 160+vii pp.

12. Peter B. Andrews, Sunil Issar, Dan Nesmith, Frank Pfenning, Hongwei Xi, Matthew Bishop, *TPS3 Facilities Guide for Users*, 1993. 94+v pp.

13. Wolfgang Bibel, *Automated Theorem Proving*, Vieweg, Braunschweig, 1987.

14. Alonzo Church, *A Formulation of the Simple Theory of Types*, Journal of Symbolic Logic 5 (1940), 56-68.

15. Doug Goldson and Steve Reeves, *Using Programs to Teach Logic to Computer Scientists*, Notices of the American Mathematical Society 40 (1993), 143-148.

16. Michael J. Gordon, Arthur J. Milner, Christopher P. Wadsworth. *Edinburgh LCF*, Lecture Notes in Computer Science 78, Springer Verlag, 1979.

17. Gerard P. Huet, *A Unification Algorithm for Typed λ-Calculus*, Theoretical Computer Science 1 (1975), 27-57.

18. Sunil Issar, "Path-Focused Duplication: A Search Procedure for General Matings," in *AAAI-90. Proceedings of the Eighth National Conference on Artificial Intelligence*, AAAI Press/The MIT Press, 1990, 221-226.

19. Sunil Issar. *Operational Issues in Automated Theorem Proving Using Matings*, Ph.D. Thesis, Carnegie Mellon University, 1991. 147 pp.

20. Sunil Issar, Peter B. Andrews, Frank Pfenning, Dan Nesmith, *GRADER Manual*, 1991. 23+i pp.

21. Sunil Issar, Dan Nesmith, Peter B. Andrews, Frank Pfenning, *TPS3 Programmer's Guide*, 1992. 99+iii pp.

22. Dale A. Miller, Eve Longini Cohen, Peter B. Andrews, "A Look at TPS," in *6th Conference on Automated Deduction*, edited by Donald W. Loveland, New York, USA, Lecture Notes in Computer Science 138, Springer-Verlag, 1982, 50-69.

23. Dale A. Miller. *Proofs in Higher-Order Logic*, Ph.D. Thesis, Carnegie Mellon University, 1983. 81 pp.

24. Dale A. Miller, "Expansion Tree Proofs and Their Conversion to Natural Deduction Proofs," in *7th International Conference on Automated Deduction*, edited by R. E. Shostak, Napa, California, USA, Lecture Notes in Computer Science 170, Springer-Verlag, 1984, 375-393.

25. Dale A. Miller, *A Compact Representation of Proofs*, Studia Logica 46 (1987), 347-370.

26. Dan Nesmith, Peter B. Andrews, Sunil Issar, Frank Pfenning, *TPS User's Manual*, 1991. 35+ii pp.

27. Frank Pfenning, "Analytic and Non-analytic Proofs," in *7th International Conference on Automated Deduction*, edited by R. E. Shostak, Napa, California, USA, Lecture Notes in Computer Science 170, Springer-Verlag, 1984, 394-413.

28. Frank Pfenning. *Proof Transformations in Higher-Order Logic*, Ph.D. Thesis, Carnegie Mellon University, 1987. 156 pp.

29. Frank Pfenning and Dan Nesmith, "Presenting Intuitive Deductions via Symmetric Simplification," in *10th International Conference on Automated Deduction*, edited by M. E. Stickel, Kaiserslautern, FRG, Lecture Notes in Artificial Intelligence 449, Springer-Verlag, 1990, 336-350.

30. Frank Pfenning, Sunil Issar, Dan Nesmith, Peter B. Andrews, *ETPS User's Manual*, 1992. 48+ii pp.

Modelling Bit Vectors in HOL:
the word Library

Wai Wong

University of Cambridge
Computer Laboratory
New Museums Site
Pembroke Street
Cambridge CB2 3QG

Abstract. The bit vector is one of the fundamental data objects in
hardware specification and verification. The modelling of bit vectors is
a key to the success of a hardware verification project. This paper describes a pragmatic approach to modelling bit vectors in higher-order
logic and an implementation as a system library in the HOL theorem
prover. In this approach, bit vectors are represented using a polymorphic
type. Restricted quantifications are used to simulate dependent types to
model the sizes of the vectors. The library consists of many theorems
asserting the basic properties of the bit vectors and some useful tools for
reasoning about them. Examples showing the effective use of the library
are described.

The bit vector (or word)[1] is one of the fundamental data objects in hardware specification and verification. The modelling of bit vectors is a key to the
success of a hardware verification project. This is one of the main application
areas of the HOL theorem prover, however, there has not yet been a universally
acceptable, general and flexible infrastructure in the form of a library which supports reasoning about words. Prompted by the need of a hardware verification
project[14] carried out by the author, an attempt has been made to develop such
an infrastructure which fulfils the needs of the current project, and in addition,
will benefit others confronting similar tasks. The result is the word library.

A brief survey of different approaches in modelling bit vectors in HOL is presented in the first section. This is followed by a description of the basic abstract
properties of bit vectors and an overview of the approach used in implementing the word library. Then, a more detailed description of the library is given.
Finally, examples of effective use of the library are shown.

1 A Brief Survey

Probably, the earliest application of the HOL theorem prover is in hardware
verification. As early as 1985, Gordon described why higher order logic is a

---

[1] These two terms will be used interchangeably in this paper.

good formalisation for specifying and verifying hardware [7]. Case studies were also carried out to verify simple circuits from n-bit counters [6] [5] to simple processors [9]. These were followed by a large project carried out by Cohn in which the major part of the design of a real microprocessor — VIPER — was formally verified [3] [4].

During this early period, there was a 'lack of a well-developed HOL infrastructure for supporting advanced reasoning about bit-string manipulations' as described by Cohn in [4]. In these projects, bit vectors of size n were represented by types :wordn. These types were simply introduced to the system using new_type without being properly defined. Similarly, constants denoting word operations were introduced to the system using new_constant. Axioms had to be stated in order to reason about simple word manipulations. This was certainly not satisfactory and against the practice of using only definitional extension to the logic which has since been established in the HOL user community.

Since then, much better infrastructure for advance modelling and reasoning has been developed in the HOL system. These include the type definition package by Melham [11] and many system tools and libraries. Various suggestions on how to model bit vectors have been made and tried. These can be grouped into four different approaches described below.

- Use a function of type :num->bool to model a word, i.e., a word w is represented by a function f_w such that for every i less than the size of the word, $f_w(i)$ is the ith bit of w. This was first suggested by Gordon [7] and used by others [2] [10]. This approach requires a parameter to specify the size of the word. The representation function f_w is ambiguous.

- Use a boolean list of length n to model an n-bit word. This was suggested by many researchers. Melham gave a comprehensive account of this approach in [12]. This approach provides a direct and unambiguous concrete representation for words of any finite size. The size of the word is implicit (as the length of the list). This has been used as the underlying representation type in the next approach.

- Use a concrete type for each specific size of words, e.g., an 8-bit word is modelled by a type :word8. This approach abstracts the explicit list into a structured type. Constants denoting functions on words can be defined easily. This has been used in many projects, notably by Graham [8]. An experimental library wordn implementing this approach is available in the HOL contrib directory.[2] One of the drawbacks of this approach is that the types for different sizes of words are disjoint and do not share common properties. In large application, the number of word types and the number of constants needed increase exponentially.

- To overcome the drawback of the above approach, Boulton [1] and Melham [13] suggested dependent types to model bit vectors. In this approach, all words of different sizes are represented by a base type. It is parameterized on

---

[2] The original code of this library was written by Melham. More features have been added by the author.

the word size. The advantage of this approach is that all words share the same common properties inherent in the base type. However, the current version (Version 2.01) of the HOL system does not have the necessary support for implementing dependent types.

2 Abstract Properties

An abstract model of words should encompass all their basic properties. It should be independent of any concrete representation. The basic abstract properties of words are:

- a word is a vector of n elements;
- the size of a given word n is constant;
- all elements are of the same type;
- an individual element is accessed via its index.

Suppose that w is a word of size n, it can be written as

$$w = \|w_{n-1}w_{n-2}\ldots w_1 w_0\|$$

where w_i represents the ith bit of the word w. We adopt the convention that the bits are indexed from the right hand side starting from 0. The index operation $w[i]$ accesses the ith bit of a word for all i less then n. A segment operation extracts a segment from a word. For example,

$$w[m, k] = \|w_{k+m-1}\ldots w_k\| \tag{1}$$

where $(k + m) \leq n$ is a m-bit segment of the word w starting from the kth bit.

A word concatenation operation $\bullet$ can be defined as

$$A_\bullet B = \|a_{n-1}\ldots a_0\| \bullet \|b_{m-1}\ldots b_0\| \tag{2}$$
$$= \|a_{n-1}\ldots a_0 b_{m-1}\ldots b_0\|$$

which builds a word of size $n + m$ from two words of size n and m, respectively.

Since words of all sizes share these basic properties, a base type of some kind would be a starting point for modelling words. This base type should then be parameterized with the size and the type of the elements. This suggests a dependent type of the form

$$: (\alpha, n)\text{word}$$

where α is the type of the elements and n is the size. In the current version of the HOL system, it is possible to define a polymorphic type $: (\alpha)\text{word}$ which takes the element type as a parameter, but it is not possible to parameterize a type with natural numbers. There is also difficulty in defining a real abstract type in the current version of HOL.

3 A Pragmatic Approach

To overcome the difficulties mentioned above, the approach used in implementing the **word** library uses facilities available in the current version of HOL only. First of all, it defines a polymorphic type : (*)**word** to represent generic words. This allows one to use different types to represent the bits according to the requirements of one's applications. For example, : (**bool**)**word** is suitable for many hardware applications using two-value logic.

Dependent types are simulated using restricted universal quantifications. A restricted universal quantification is written in the form

$$\forall x :: P. t[x]$$

where if $x : \alpha$ then P can be any term of type $\alpha \rightarrow bool$; this denotes the quantification of x over those values satisfying P. The semantics of this quantification is defined by the following equation:

$$\vdash_{def} \quad \forall x :: P. t[x] = \forall x. P x \supset t[x] \tag{3}$$

Suppose that P is the predicate PWORDLEN n which returns T when applied to a word w if and only if w is an n-bit word, then the expression

$$\forall w :: \text{PWORDLEN } n. \ldots$$

can be read as '*for all n-bit words* w, ...'. For a specific value of n, say 8, one can define a predicate **word8** by the definition

$$\vdash_{def} \quad \text{word8} = \text{PWORDLEN 8}.$$

This predicate can then be used in expressions, such as $\forall w :: \text{word8}. \ldots$ Since the syntax of restricted quantification resembles the syntax of types closely and the semantics of the quantification is suitably defined, using this to simulate dependent types is very comprehensible.

As we cannot define a real abstract type in HOL, the list type is used as the underlying representation of the polymorphic type : (*)**word**. However, through disciplined use of system functions and properties derived for the new type, direct reference to the underlying representation is minimised. For example, when defining new constants, constant specification is used to specify the abstract properties of the new constant instead of using constant definition which needs access to the representation. In the development of the library, the proofs of some basic theorems about words have to refer to the underlying lists. After a small number of basic theorems are derived, one can proceed to reason about words on a more abstract level without resorting to the underlying representation.

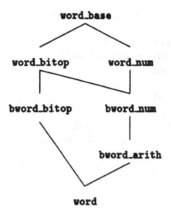

Fig. 1. The ancestry of theories

4   The Library

The **word** library consists of several theories and some ML functions implementing tactics and conversions. The ancestry of the theories is illustrated in Fig. 1. The theories whose names begin with **word_** contains definitions of generic constants and theorems asserting general properties of words. These generic constants are polymorphic and can be applied to words of any types. There are three such theories in the library, namely **word_base**, **word_bitop** and **word_num**. As boolean words are used most often, the theories whose names begin with **bword_** are about this type of words. The subsections below describe individual theories in more detail.

4.1   The Basics: the Theory **word_base**

First of all, the polymorphic type : (*)**word** is defined in this theory. It is defined using **define_type** with the following specification:

    '**word = WORD (*)list**'

The basic constants denoting the functions of indexing, segmenting and concatenation of words described in Sect. 2 are BIT, SEG and WCAT, respectively. The predicate PWORDLEN for discriminating the size of words and a function named WORDLEN returning the size of a word are also defined in this theory. The types and specification of these constants are listed in Table 1. Several constants denoting some simple functions on words are also defined for convenient, such as MSB for most significant bit.

A comprehensive set of theorems stating the basic properties of words and the basic constants have been derived and stored in this theory. Some of the more important ones are discussed below. The theorem SEG_PWORDLEN states

Table 1. Basic constants in the theory word_base

PWORDLEN:num -> ((*)word -> bool)
    PWORDLEN $n\,w$ = T iff w is an n-bit word
WORDLEN:(*)word -> num
    WORDLEN $w = n$
BIT:num -> ((*)word -> *)
    BIT $i\;|a_{n-1}\ldots a_i\ldots a_0| = a_i$
SEG:num -> (num -> ((*)word -> (*)word))
    SEG $m\,k\;|a_{n-1}\ldots a_{k+m-1}\ldots a_k\ldots a_0| = |a_{k+m-1}\ldots a_k|$
WCAT:(*)word # (*)word -> (*)word
    WCAT$(|a_{n-1}\ldots a_0|,|b_{m-1}\ldots b_0|) = |a_{n-1}\ldots a_0 b_{m-1}\ldots b_0|$

that the size of the word resulting from taking an m-bit segment from an n-bit word is m providing that $k + m \leq n$ where k is the starting bit.

HOL Theorem (SEG_PWORDLEN)

$$\vdash \forall n.\,\forall w :: \text{PWORDLEN}\,n.$$
$$\forall m\,k.\,k + m \leq n \supset \text{PWORDLEN}\,m\,(\text{SEG}\,m\,k\,w)$$

A nested SEG expression can be simplified providing that the sizes and starting bits satisfy certain conditions. This is asserted by the theorem SEG_SEG.

HOL Theorem (SEG_SEG)

$$\vdash \forall n.\,\forall w :: \text{PWORDLEN}\,n.$$
$$\forall m_1\,k_1\,m_2\,k_2.\,k_1 + m_1 \leq n \wedge k_2 + m_2 \leq m_1 \supset$$
$$(\text{SEG}\,m_2\,k_2\,(\text{SEG}\,m_1\,k_1\,w) = \text{SEG}\,m_2\,(k_1 + k_2)\,w)$$

The theorem SEG_WCAT_SEG asserts that taking a segment from a word which is built by concatenating two words w_1 and w_2 is equivalent to taking the appropriate segments from each word and then concatenating them provided that the segment spans across the boundary of the two constituent words.

HOL Theorem (SEG_WCAT_SEG)

$$\vdash \forall n_1.\,\forall w_1 :: \text{PWORDLEN}\,n_1.\,\forall n_2.\,\forall w_2 :: \text{PWORDLEN}\,n_2.$$
$$\forall m\,k.$$
$$m + k \leq n_1 + n_2 \wedge k < n_2 \wedge n_2 \leq m + k \supset$$
$$(\text{SEG}\,m\,k\,(\text{WCAT}\,(w_1, w_2)) =$$
$$\text{WCAT}\,(\text{SEG}\,((m + k) - n_2)\,0\,w_1, \text{SEG}\,(n_2 - k)\,k\,w_2))$$

4.2 Generic Bitwise Operators: the Theory word_bitop

This theory defines two predicates PBITOP and PBITBOP. When applied to a suitably typed word function *op*, they will return T if and only if *op* is a bitwise unary or binary operator, respectively. The meaning of *bitwise* is that the operator preserves the size of its operands and the operation on each bit is independent of any other bits. Note that as these predicates are polymorphic the type of the bits can be anything. The exact definitions of these predicates are as follows:

PBITOP:((*)word -> (**)word) -> bool
   PBITOP *op* = T iff *op* is a bitwise unary operator

$$\vdash_{def} \forall op. \text{PBITOP} \, op =$$
$$(\forall n. \forall w :: \text{PWORDLEN} \, n.$$
$$\text{PWORDLEN} \, n \, (op \, w) \land$$
$$(\forall k \, m. \, k + m \leq n \supset (op \, (\text{SEG} \, m \, k \, w) = \text{SEG} \, m \, k \, (op \, w))))$$

PBITBOP:((*)word -> (**)word -> (***)word) -> bool
   PBITBOP *op* = T iff *op* is a bitwise binary operator

$$\vdash_{def} \forall op. \text{PBITBOP} \, op =$$
$$(\forall n. \forall w_1 :: \text{PWORDLEN} \, n. \, \forall w_2 \text{PWORDLEN} \, n.$$
$$\text{PWORDLEN} \, n \, (op \, w_1 \, w_2) \land$$
$$(\forall k \, m. \, k + m \leq n \supset$$
$$(op \, (\text{SEG} \, m \, k \, w_1) \, (\text{SEG} \, m \, k \, w_2) = \text{SEG} \, m \, k \, (op \, w_1 \, w_2)))$$

Also in this theory are the definitions of two generic shift operators: SHL and SHR. Their types and specification is listed in Table 2. Both take three arguments and return a pair. The first argument is a boolean value indicating the kind of operation to be performed. The second and the third arguments to SHR are a single bit and a word, respectively. The order of these two arguments to SHL is reversed. Depending on the value of the boolean and single bit argument, these operators can perform either a logical shift, an arithmetic shift or a rotation operation. If the boolean argument is T, the single bit argument is not used. SHR shifts its operand one bit to the right and the left-most bit is duplicated to fill the vacant position, thus, implementing an arithmetic shift. If the boolean argument is F, SHR fills the vacant position with the single bit argument. If this bit is the right-most bit of the operand, a rotation is performed. If it has value 0, it results in a logical shift. SHL operates in a similar way. The pair returned by these operators consists of a word which is the operation result and a single bit which is the bit shifted out of the operand.

A number of theorems asserting the operational behaviour of these operators and their relationship with the basic constants WCAT and SEG have been derived and stored in this theory.

Table 2. Shift operators

SHR:bool -> * -> (*)word -> ((*)word # *)

$$\text{SHR } f\, b\, |a_{n-1}\dots a_1 a_0| = \begin{cases} (|a_{n-1}a_{n-1}\dots a_1|, a_0) & \text{if } f = \mathsf{T} \\ (|b\, a_{n-1}\dots a_1|, a_0) & \text{if } f = \mathsf{F} \end{cases}$$

SHL:bool -> (*)word -> * -> (* # (*)word)

$$\text{SHL } f\, |a_{n-1}a_{n-2}\dots a_0|\, b = \begin{cases} (a_{n-1}, |a_{n-2}\dots a_0 a_0|) & \text{if } f = \mathsf{T} \\ (a_{n-1}, |a_{n-2}\dots a_0\, b|) & \text{if } f = \mathsf{F} \end{cases}$$

4.3 Boolean Bitwise Operators: the Theory bword_bitop

In this theory, a small set of boolean bitwise operators are defined and theorems asserting that they are bitwise operators are proved. The boolean bitwise operators are:

WNOT :bool word -> bool word   bitwise negation
WAND :bool word -> bool word -> bool word   bitwise AND
WOR   :bool word -> bool word -> bool word   bitwise OR
WXOR :bool word -> bool word -> bool word   bitwise exclusive-OR

The theorems stating that they are bitwise are:

PBITOP_WNOT	⊢ PBITOP WNOT
PBITOP_WAND	⊢ PBITBOP WAND
PBITOP_WOR	⊢ PBITBOP WOR
PBITOP_WXOR	⊢ PBITBOP WXOR

4.4 Natural Numbers and Words: the Theory word_num

Words are often interpreted as natural numbers. In this theory, two constants are defined to map generic words to natural numbers and vice versa:

NVAL:(* -> num) -> num -> (*)word -> num
     NVAL $f\, b\, w$ returns the numeric value of w. f is a function mapping a bit to its numeric value and b is the base or radix of the word.
NWORD:num -> (num -> *) -> num -> num -> (*)word
     NWORD $n\, f'\, b\, m$ returns a n-bit word representing the value of m. f' is a function mapping a number to a bit and b is the base.

The upper bound of the numeric value of a word is stated by the theorem NVAL_MAX.

HOL Theorem (NVAL_MAX)

$$\vdash \forall f\, b.\, (\forall x.\, f\, x < b) \supset$$
$$\forall n.\, \forall w :: \text{PWORDLEN } n.\, \text{NVAL } f\, b\, w < (b \text{ EXP } n)$$

Provided that the bit value function f satisfies $\forall x.\, f\, x < b$, the numeric value of a word w is always less than b^n. The theorem NVAL_WCAT states that the value of a word can be calculated from the values of its segments.

HOL Theorem (NVAL_WCAT)

$$\vdash \forall n.\, \forall w_1 :: \text{PWORDLEN}\, n.$$
$$\forall m.\, \forall w_2 :: \text{PWORDLEN}\, m.$$
$$\forall f\, b.$$
$$\text{NVAL}\, f\, b\, (\text{WCAT}\, (w_1, w_2)) =$$
$$(\text{NVAL}\, f\, b\, w_1 \times (b\, \text{EXP}\, m)) + (\text{NVAL}\, f\, b\, w_2)$$

The theorem stating the size of the result of mapping from natural number to word is NWORD_PWORDLEN.

HOL Theorem (NWORD_PWORDLEN)

$$\vdash \forall n\, f\, b\, m.\, \text{PWORDLEN}\, n\, (\text{NWORD}\, n\, f\, b\, m)$$

4.5 Boolean Words and Numbers: the Theory bword_num

In this theory, two functions mapping between a single bit and number are defined first. Then, the constants denoting the mapping between boolean words and natural numbers are defined in terms of these bit mapping functions and the generic word–num mapping functions described in Sect. 4.4.

```
BV :bool -> num
```
$\quad \vdash \forall b.\, \text{BV}\, b = (b \Rightarrow \text{SUC}\, 0 \mid 0)$
```
VB :num -> bool
```
$\quad \vdash \forall n.\, \text{VB}\, n = \neg((n\, \text{MOD}\, 2) = 0)$
```
BNVAL:bool word -> num
```
$\quad$ BNVAL w returns the numeric value of w. $\vdash_{def}$ BNVAL $w = \text{NVAL BV}\, 2\, w$
```
NBWORD:num -> num -> bool word
```
$\quad$ NBWORD n m returns a n-bit word representing the value of m. $\vdash_{def}$ NBWORD $n\, m = \text{NWORD}\, n\, \text{VB}\, 2\, m$

The functions BNVAL and NBWORD are inverse to each other in the set of numbers less than 2^n where n is the size of the word. The following theorems state the basic properties of these mapping functions.

HOL Theorem (VB_BV)

$$\vdash \forall x.\, \text{VB}\, (\text{BV}\, x) = x$$

HOL Theorem (BV_VB)

$$\vdash \forall x.\, x < 2 \supset (\text{BV}\, (\text{VB}\, x) = x)$$

HOL Theorem (NBWORD_BNVAL)

$$\vdash \forall n. \forall w :: \text{PWORDLEN } n. \text{NBWORD } n \, (\text{BNVAL } w) = w$$

HOL Theorem (BNVAL_NBWORD)

$$\vdash \forall n\, m.$$
$$m < (2 \text{ EXP } n) \supset (\text{BNVAL } (\text{NBWORD } n\, m) = m)$$

HOL Theorem (NBWORD_PWORDLEN)

$$\vdash \forall n\, m. \text{PWORDLEN } n \, (\text{NBWORD } n\, m)$$

HOL Theorem (NBWORD_MOD)

$$\vdash \forall n\, m. \text{NBWORD } n \, (m \text{ MOD } (2 \text{ EXP } n)) = \text{NBWORD } n\, m$$

The theorem NBWORD_SUC asserts the fact that converting a number m to a word can be performed bit by bit recursively.

HOL Theorem (NBWORD_SUC)

$$\vdash \forall n\, m. \text{NBWORD } (\text{SUC } n) \, m =$$
$$\text{WCAT } (\text{NBWORD } n \, (m \text{ DIV } 2), \text{WORD } [\text{VB } (m \text{ MOD } 2)])$$

The theorem SEG_NBWORD states that taking an m-bit segment of an n-bit word mapped to by NBWORD from a number a is equivalent to mapping the quotient of a divided by 2^k to an m-bit word.

HOL Theorem (SEG_NBWORD)

$$\vdash \forall n\, m\, k\, a. \, k + m \leq n \supset$$
$$(\text{SEG } m\, k \, (\text{NBWORD } n\, a) = \text{NBWORD } m \, (a \text{ DIV } (2 \text{ EXP } k)))$$

4.6 Proof Tools

The word library currently has a small set of tools in the form of conversions and tactics for manipulating words. These include the following:

BIT_CONV : conv When applied to a term as the left hand side of the following theorem, this conversion returns the theorem

$$\vdash \text{BIT } k \, (\text{WORD}[w_{n-1}; \ldots; w_k; \ldots; w_0]) = w_k$$

SEG_CONV : conv When applied to a term as the left hand side of the following theorem, this conversion returns the theorem

$$\vdash \text{SEG } m\, k \, (\text{WORD}[w_{n-1}; \ldots; w_k; \ldots; w_0]) = [w_{m+k-1}; \ldots; w_k]$$

SEG_SEG_CONV : `(term -> conv)` When applied to a term as the left hand side of the following theorem, the conversion **SEG_SEG_CONV** `"n"` returns the theorem

$$\text{PWORDLEN}\, n\, w \vdash \text{SEG}\, m_2\, k_2(\text{SEG}\, m_1\, k_1 w) = \text{SEG}\, m_2\, k\, w$$

where $k = k_1 + k_2$ and n, k_1, k_2, m_1 and m_2 are numeric constants and satisfy the following relations: $k_1 + m_1 \leq n$ and $k_2 + m_2 \leq m_1$.

PWORDLEN_CONV : `(term list -> conv)` When applied to a term of the form PWORDLEN $m\, tm$, the conversion **PWORDLEN_CONV** `tms` returns a theorem asserting the size of the word tm. This theorem is in the form

$$A \vdash \text{PWORDLEN}\, m\, tm$$

where the exact form of A, tm and the term list argument **tms** is given in the table below:

tm	tms	theorem
WORD$[b_{n-1};\ldots;b_0]$	[ ]	$\vdash$ PWORDLEN n (WORD$[b_{n-1};\ldots;b_0]$)
SEG $m\, k\, tm'$	["n"]	PWORDLEN $n\, tm'$
		$\vdash$ PWORDLEN m (SEG $m\, k\, tm'$)
WCAT(tm', tm'')	["n1"; "n2"]	PWORDLEN $n_1\, tm'$, PWORDLEN $n_2\, tm''$
		$\vdash$ PWORDLEN n (WCAT(tm', tm''))
		where $n = n1 + n2$
WNOT tm'	[ ]	PWORDLEN $n\, tm'$
		$\vdash$ PWORDLEN n (WNOT tm')
WAND $tm'\, tm''$	[ ]	PWORDLEN $n\, tm'$, PWORDLEN $n\, tm''$
		$\vdash$ PWORDLEN n (WAND $tm'\, tm''$)
WOR $tm'\, tm''$	[ ]	PWORDLEN $n\, tm'$, PWORDLEN $n\, tm''$
		$\vdash$ PWORDLEN n (WOR $tm'\, tm''$)
WXOR $tm'\, tm''$	[ ]	PWORDLEN $n\, tm'$, PWORDLEN $n\, tm''$
		$\vdash$ PWORDLEN n (WXOR $tm'\, tm''$)

PWORDLEN_bitop_CONV : `conv` When applied to a term PWORDLEN $n\, tm$ where tm involves only bitwise operators and simple variables, this conversion returns the theorem

$$\ldots, \text{PWORDLEN}\, n\, w_i, \ldots \vdash \text{PWORDLEN}\, n\, tm$$

where there is one assumption PWORDLEN $n\, w_i$ for each simple variable w_i in tm. This conversion automatically descends into the subterms until it reaches all simple variables.

PWORDLEN_TAC : `(term list -> tactic)` When applied to a goal of the form PWORDLEN `n tm`, the tactic **PWORDLEN_TAC** `tms` solves it if the conversion **PWORDLEN_CONV** `tms` returns a theorem without assumptions. Otherwise, the assumptions of the theorem returned by the conversion become the new subgoals.

5 Working with Words

The basic technique for reasoning about words with the word library is by structural induction on the size of word. Since the structure of words is linear and symmetric, structural induction can be carried out from either ends using the WCAT operation as the basic constructor. In addition, structural analysis can be done at any position of a word. In general, there are three theorems associated with each basic word function: one for each kind of structural analysis. Considering the function NBWORD as an example, the theorem NBWORD_SUC described in Sect. 4.5 is for structural induction from the right hand end. The theorem NBWORD_SUC_LEFT shown below is for structural induction from the left hand end, and the theorem NBWORD_SPLIT is for structural analysis at any position.

HOL Theorem (NBWORD_SUC_LEFT)

$\vdash \forall n\, m.\, \text{NBWORD}\, (\text{SUC}\, n)\, m =$
$\text{WCAT}\, (\text{WORD}\, [\text{VB}\, ((m\ \text{DIV}\ (2)\ \text{EXP}\ (n))\ \text{MOD}\ 2)], \text{NBWORD}\, n\, m)$

HOL Theorem (NBWORD_SPLIT)

$\vdash \forall n_1\, n_2\, m.\, \text{NBWORD}\, (n_1 + n_2)\, m =$
$\text{WCAT}\, (\text{NBWORD}\, n_1\, (m\ \text{DIV}\ (2)\ \text{EXP}\ (n_2)), \text{NBWORD}\, n_2\, m)$

The following example uses structural induction from the right hand end to prove a theorem about taking an n-bit segment of an $(n + 1)$-bit word which is the result of converting a natural number using the function NBWORD. We first set up the goal

$?-\quad \forall n\, m.\, \text{SEG}\, n\, 0 (\text{NBWORD}(\text{SUC}\, n)\, m) = \text{NBWORD}\, n\, m.$

Then, the induction tactic INDUCT_TAC is applied to the size of the word. This generates two subgoals. The first subgoal, corresponding to the base case of the induction, is

$?-\quad \text{SEG}\, 0\, 0 (\text{NBWORD}(\text{SUC}\, 0)\, m) = \text{NBWORD}\, 0\, m.$

This is trivial to solve since a zero-bit segment of a word is WORD[] and converting a number to a zero-bit word always gives the same result. The second subgoal corresponding to the step case of the induction is

$?-\quad \forall m.\, \text{SEG}\, (\text{SUC}\, n)\, 0\, (\text{NBWORD}\, (\text{SUC}\, (\text{SUC}\, n))\, m) = \text{NBWORD}\, (\text{SUC}\, n)\, m$

The right hand end induction theorem for NBWORD, NBWORD_SUC, can now be used to rewrite the goal. Rewriting the resulting goal further with the theorem SEG_WCAT_SEG and simplifying the result reduces it to

$?-\quad \text{SEG}\, n\, 0\, (\text{NBWORD}\, (\text{SUC}\, n)\, (m\ \text{DIV}\ 2)) = \text{NBWORD}\, n\, (m\ \text{DIV}\ 2).$

The induction hypothesis can then be used to solve the goal. However, as the theorem SEG_WCAT_SEG is restricted universally quantified, ordinary rewriting tactics, such as REWRITE_TAC, cannot use it to rewrite the goal. Special tactics are required. The res_quan library provides the basic facilities for manipulating restricted quantifications[15].

6   A large application

The **word** library has been used in the project of formal verification of a bit slice ALU. The formal model of the ALU were developed in HOL in two levels: the specification and the implementation. The specification level views the 32-bit ALU as a black box which is capable of performing 13 logical and arithmetic operations, such as AND, OR, addition and so on. At the implementation level, the ALU is constructed from eight 4-bit slices. Each of these slices can carry out the same set of operations. The goal is to prove these two levels are functionally equivalent.

Taking the bitwise-AND operation as an example, the operation result at the specification level is denoted by the expression

$$\text{WNOT}\,((\text{WNOT}\,r)\,\text{WAND}\,t)$$

where r and t are the 32-bit operands. The corresponding expression at the implementation level is

```
WCAT
(WCAT
(WCAT
 (WNOT
 (WNOT (SEG 4 12 (SEG 16 16 r)) WAND SEG 4 12 (SEG 16 16 t)),
 ...)))
```

In essence, this expression specifies that a bitwise-AND operation can be carried out as a series of bitwise-AND operations on the 4-bit segments of the operands and then concatenating the partial results to form the 32-bit result.

To unify these two expressions, the theorem **SEG_SEG** is used to eliminate the nested SEG operations. Then, using the properties of the bitwise operators WAND and WNOT as specified in the predicates **PBITOP** and **PBITBOP**, the implementation expression can be simplified to the same form of the specification. The details of the verification can be found in [14].

7   Conclusion

A pragmatic approach to modelling bit vectors in HOL has been developed as a system library. This provides an abstract and flexible infrastructure to reason about the fundamental data objects in hardware specification and verification. It has been successfully used in a hardware verification project.

Two enhancements to the library would be very useful. The first is a theory about the mapping between words and integers since words are very often interpreted as integers. The second is to provide more tools in the form of conversions and tactics for manipulating words. These tools should be largely automatic like **PWORDLEN_bitop_CONV**.

References

1. R. J. Boulton. A HOL semantics for a subset of ELLA. Technical Report 254, University of Cambridge Computer Laboratory, 1992.
2. A. J. Camilleri. Executing behavioural definitions in higher order logic. Technical Report 140, University of Cambridge Computer Laboratory, 1988.
3. A. Cohn. A proof of correctness of the VIPER microprocessor: the first level. Technical Report 104, University of Cambridge Computer Laboratory, January 1988.
4. A. Cohn. A proof of correctness of the VIPER microprocessor: the second level. Technical Report 134, University of Cambridge Computer Laboratory, May 1990.
5. A. Cohn and M. J. C. Gordon. A machanized proof of correctness of a simple counter. Technical Report 94, University of Cambridge Computer Laborartory, July 1986.
6. M. J. C. Gordon. Hardware verification by formal proof. Technical Report 74, University of Cambridge Computer Laborartory, 1985.
7. Michael . C. Gordon. Why higher-order logic is a good formalism for specifying and verifying hardware. In G. J. Milne and P. A. Subrahmanyan, editors, *Formal Aspects of VLSI Design*, pages 153–178. Springer-Verlag, 1986.
8. Brian T. Graham. *The SECD Microprocessor—a Verification Case Study*. Kluwer Academic Publishers, 1992.
9. J. Joyce, G. Birtwistle, and M. J. C. Gordon. Proving a computer correct in higher order logic. Technical Report 100, University of Cambridge Computer Laborartory, December 1986.
10. J. J. Joyce. Formal specification and verification of microprocessor systems. In S. Winter and H. Schumny, editors, *EUROMICRO 88:Proceedings of the 14th Symposium on Microprocessing and Microprogramming*. Horth-Holland, 1988.
11. T. F. Melham. Automating recursive type definition in higher-order logic. In Graham Birtwistle and P. A. Subrahmanyam, editors, *Current Trends in Hardware Verification and Automated Theorem Proving*, pages 341–386. Springer-Verlag, 1989.
12. T. F. Melham. Formalizing abstraction mechanisms for hardware verification in higher order logic. Technical Report 201, University of Cambridge Computer Laboratory, 1990.
13. T. F. Melham. The HOL logic extended with quantification over type variables. In L. J. M. Claesen and M. J. C. Gordon, editors, *Higher Order Logic Theorem Proving and its Applications*. Horth-Holland, 1993.
14. W. Wong. Formal verification of VIPER's ALU. Technical Report 300, University of Cambridge Computer Laboratory, University of Cambridge Computer laboratory, New Museums Site, Pembroke Street, Cambridge CB2 3QG, ENGLAND, May 1993.
15. W. Wong. *The HOL res_quan Library*. Computer Laboratory, University of Cambridge, 1993.

Eliminating Higher-Order Quantifiers to Obtain Decision Procedures for Hardware Verification*

Klaus Schneider[1], Ramayya Kumar[2] and Thomas Kropf[1]

[1] Universität Karlsruhe, Institut für Rechnerentwurf und Fehlertoleranz,
(Prof. D. Schmid), P.O. Box 6980, 76128 Karlsruhe, Germany,
e-mail:{schneide|kropf}@ira.uka.de
[2] Forschungszentrum Informatik, Haid-und-Neustraße 10-14, 76131 Karlsruhe, Germany,
e-mail:kumar@fzi.de

Abstract. In this paper, we present methods for eliminating higher-order quantifiers in proof goals arising in the verification of digital circuits. For the description of the circuits, a subset of higher-order logic called hardware formulae is used which is sufficient for describing hardware specifications and implementations at register transfer level. Real circuits can be dealt with as well as abstract (generic) circuits. In case of real circuits, it is formally proved, that the presented transformations result in decidable formulae, such that full automation is achieved for them. Verification goals of abstract circuits can be transformed by the presented methods into goals of logics weaker than higher-order logic, e.g. (temporal) propositional logic. The presented transformations are also capable of dealing with hierarchy and have been implemented in HOL90.

1 Introduction

Higher-order logic is well suited for hardware verification [Gord86, Joyc91], but unfortunately this logic is neither decidable nor recursively enumerable. Hence, no form of automation is available for the entire logic, although automation is an essential prerequisite for getting verification tools usable by VLSI designers. However, in the limited context of hardware verification, first results have been achieved to automate parts of hardware proofs in higher-order logic by using a definite sequence of transformation steps and first-order proving techniques [ScKK91a, KuSK93]. The work presented in this paper establishes a set of efficient algorithms aimed at a further automation of hardware correctness proofs in higher-order logic. To our knowledge, for the first time, decision procedures for a certain class of higher-order formulae are presented. This class is sufficient for describing hardware specifications and implementations at register-transfer level. The basic idea of the presented procedures is the elimination of higher-order quantifiers such that formulae of weaker logics are obtained which can then be proved be appropriate automated theorem proving tools. We always try to achieve the "weakest possible" logic, since the weaker the logic the more efficient the proof techniques are. Indeed, proof goals arising from real circuits

---

* This work has been partly financed by a german national grant, project Automated System Design, SFB No.358.

can always be transformed into equivalent formulae of (temporal) propositional logic and hence the proof can be done automatically and also very efficiently.

Quantifier elimination methods have been an early interest in the research of mathematical logic [Acke35a] and they still belong to recent research in this field [GaOh92]. For example, the SCAN algorithm [GaOh92], is able to eliminate quantifiers in the following ways: $\exists P.[(P \lor Q) \land (\neg P \lor R)]$ is equivalent to $Q \lor R$, and more difficult $\exists P.[(P(a) \lor Q) \land (\neg P(b) \lor R)]$ is equivalent to $(a = b) \to (Q \lor R)$. As another example, HOL's unwind library uses the following elimination theorem for eliminating combinatorial internal lines in circuit descriptions which occur in the formalisation of implementations [HaDa86]:

$$[\exists \ell \forall t. (\ell(t) \leftrightarrow \Delta(t)) \land \Phi[\ell(t)]] \leftrightarrow [\forall t. \Phi[\Delta(t)]]$$

Sequential line *variables*, i.e. outputs of flipflops and registers cannot be eliminated by the above mentioned theorem since they are used to store information. In this paper, further elimination rules are presented for eliminating all internal line *quantifiers* including sequential lines. As a result a decision procedure for real circuits is obtained which is especially useful to verify complex data paths and controllers, since the interactive proofs of these circuits in a higher-order logic environment are quite tedious and time consuming. The main idea of this approach, however, carries over to abstract circuits, too.

In section 2, we define the subset of higher-order logic we use for describing hardware. In section 3, we describe the quantifier elimination methods and in section 4, we prove the decidability of our verification goals for real circuits and indicate a decision procedure. Section 5 demonstrates the presented proof tactics by proving the equivalence of two circuits. In the last section, we summarize the results and discuss our future work.

Before giving the basic definitions, we explain the used notation: $\mathbb{B}$ denotes the set of boolean values and $\mathbb{N}$ is the set of natural numbers which are identified with canonical terms: $0, SUC(0), SUC(SUC(0)), \ldots$. $\mathcal{VAL}^I(\varphi)$ is the truth value of the propositional formula φ under the interpretation I and $\alpha \Rightarrow \beta \mid \gamma$ means: if α then β else γ.

2   Hardware Formulae

In this section we give the definition of a subset of higher-order logic called *hardware formulae* which are used to describe the behaviour of digital circuits. This description is based on the consideration that digital circuits are complex finite state systems, and hence corresponding notions such as output and state transition functions are found. The concept of this description goes back to Hanna and Daeche [HaDa86] and a detailed description of these formulae can also be found in [KuSK93]. In contrast to simple finite state machines, these formulae can contain an additional subformulae $\Re$ for handling hierarchy and complex data types. If we neglect this formula $\Re$, i.e if we assume $\Re = \top$, then the obtained subclass is as powerful as finite state machines. If the formulae $\Re$ are arbitrary quantifier-free formulae, then a subclass called *simple hardware formulae* is obtained, which has some decidable properties (section 4). In an application, these quantifier-free formulae could stem from parts of the implementation which have not been expanded or which have been replaced by their specification according to a previous correctness result. More powerful descriptions can also be obtained by arbitrary formulae $\Re$ as these formulae can themselves be hardware formulae which thus leads to hierarchical hardware formulae. These hierarchical hardware formulae can be used to specify and verify larger circuits, with the advantage that small portions of the entire proof can be done step after step. Last but not least, $\Re$ can be used for the description of generic n-bit circuits which are defined by primitive recursion. In those cases, $\Re$ is the n-bit circuit which is used to implement the $n + 1$-bit circuit in the recursion step.

The main focus of this paper is the set of simple hardware formulae, but the presented quantifier elimination methods are independent of the rest formulae $\Re$.

Definition 2.1 (Hardware Formulae)

Given that $\vec{i} := (i_1, \ldots, i_m), \vec{o} := (o_1, \ldots, o_n)$, $\vec{\ell} := (\ell_1, \ldots, \ell_k)$, $\vec{q} := (q_1, \ldots, q_l)$ are tuples of variables of type $\mathbb{IN} \rightarrow \mathbb{IB}$ and the formulae $\Delta_j(\vec{p}), \Omega_j(\vec{p}), \Phi_j(\vec{p})$ are quantifier-free formulae with the variables $\vec{p} := (p_1, \ldots, p_{m+k+l})$ of type $\mathbb{IB}$, all formulae $\mathcal{H}(\vec{i}, \vec{o})$ of the following form are called hardware formulae[1]:

$$
\begin{aligned}
\mathcal{H}(\vec{i}, \vec{o}) := \exists \vec{\ell}. \exists \vec{q}. \forall t. \\
\left[\vec{\ell}(t) \leftrightarrow \vec{\Delta}(\vec{i}(t), \vec{\ell}(t), \vec{q}(t)) \right] \wedge \\
\left[(\vec{q}(0) \leftrightarrow \vec{\omega}) \wedge \left(\vec{q}(t+1) \leftrightarrow \vec{\Omega}(\vec{i}(t), \vec{\ell}(t), \vec{q}(t)) \right) \right] \wedge \\
\left[\vec{o}(t) \leftrightarrow \vec{\Phi}(\vec{i}(t), \vec{\ell}(t), \vec{q}(t)) \right] \wedge \\
\Re(\vec{i}, \vec{o}, \vec{\ell}, \vec{q})
\end{aligned}
$$

The variables $\text{COMB}(\mathcal{H}(\vec{i}, \vec{o})) := \{\ell_1, \ldots, \ell_k\}$, $\text{SEQ}(\mathcal{H}(\vec{i}, \vec{o})) := \{q_1, \ldots, q_l\}$, $\text{IN}(\mathcal{H}(\vec{i}, \vec{o})) := \{i_1, \ldots, i_m\}$ and $\text{OUT}(\mathcal{H}(\vec{i}, \vec{o})) := \{o_1, \ldots, o_n\}$ are called the combinatorial, sequential, input and output variables, respectively and t denotes the time variable. $\Re$ is an arbitrary rest formula, which can also be a hardware formula itself. Hardware formulae also have to fulfill the following two restrictions:

1. Given the relation R on $\text{COMB}(\mathcal{H}(\vec{i}, \vec{o})) \times \text{COMB}(\mathcal{H}(\vec{i}, \vec{o}))$, such that $\ell_j \mathrel{R} \ell_i$ means ℓ_j occurs in $\Delta_i(\vec{i}(t), \vec{\ell}(t), \vec{q}(t))$, the transitive closure of R (called as the dependency relation) has to be acyclic.
2. For each variable belonging to $\text{COMB}(\mathcal{H}(\vec{i}, \vec{o})) \cup \text{SEQ}(\mathcal{H}(\vec{i}, \vec{o})) \cup \text{OUT}(\mathcal{H}(\vec{i}, \vec{o}))$ there exists exactly one equation[2] in $\mathcal{H}(\vec{i}, \vec{o})$ with the variable occurring on the left hand side applied on the time variable t.

We denote the set of hardware formulae over a given signature Σ by $\mathcal{HW}_\Sigma$.

---

[1] By the equivalence of tuples of formulae $\vec{\varphi} \overset{n}{\leftrightarrow} \vec{\psi}$ we mean the conjunction of the equivalences of the corresponding components $\bigwedge_{j=0}^{n} (\varphi_j \leftrightarrow \psi_j)$.

[2] In case of a sequential line $q_\mu \in SEQ(\mathcal{H}(\vec{i}, \vec{o}))$, we consider the formula $(q_\mu(0) \leftrightarrow \omega_\mu) \wedge (q_\mu(t+1) \leftrightarrow \Omega_\mu(\vec{i}(t), \vec{\ell}(t), \vec{q}(t)))$ as one equation, since it is equivalent to $q_\mu(t) \leftrightarrow (t = 0) \Rightarrow \omega_\mu \mid \Omega_\mu(\vec{i}(t-1), \vec{\ell}(t-1), \vec{q}(t-1))$.

3 Transformations on Hardware Formulae

In general, hardware proof goals look like $\vdash \mathcal{H}(\vec{i}, \vec{o}) \rightarrow \mathcal{P}(\vec{i}, \vec{o})$ where $\mathcal{H}(\vec{i}, \vec{o}) \in \mathcal{HW}_\Sigma$ is the description of a circuit and $\mathcal{P}(\vec{i}, \vec{o})$ is a property or an entire specification of the circuit, e.g. a safety or liveness property [Gupt92]. The only free variables of $\mathcal{P}(\vec{i}, \vec{o})$ are input and output variables $\vec{i}, \vec{o}$ of $\mathcal{H}(\vec{i}, \vec{o})$.

In this section, we will show how the higher-order quantifiers of the hardware formulae can be eliminated in the general proof goals $\vdash \mathcal{H}(\vec{i}, \vec{o}) \rightarrow \mathcal{P}(\vec{i}, \vec{o})$ and also in the special cases of the form $\vdash \mathcal{H}_1(\vec{i}, \vec{o}) \rightarrow \mathcal{H}_2(\vec{i}, \vec{o})$ where $\mathcal{P}(\vec{i}, \vec{o})$ is a hardware formula.

In the next two subsections, we will first outline the principles of quantifier elimination over line variables and in the following subsection, we show how these principles are used in a proof procedure.

3.1 Elimination of Combinatorial Lines

As the dependency relation of the combinatorial line variables has to be acyclic, ℓ_j cannot occur in its definitional equation $\Delta_j(\vec{i}(t), \vec{\ell}(t), \vec{q}(t))$. Thus we can substitute $\ell_j(t)$ in the remaining formula by $\Delta_j(\vec{i}(t), \vec{\ell}(t), \vec{q}(t))$ according to the theorem

$$\left[\exists \ell_j \forall t. \left(\ell_j(t) \leftrightarrow \Delta_j(\vec{i}(t), \vec{\ell}(t), \vec{q}(t))\right) \wedge P(\ell_j(t))\right] \leftrightarrow \left[\forall t. P(\Delta_j(\vec{i}(t), \vec{\ell}(t), \vec{q}(t)))\right]$$

A multiple application of this theorem eliminates all combinatorial lines [KuSK93]. Thus without any loss of generality, we only consider hardware formulae without combinatorial line variables.

3.2 Elimination of Sequential Line Quantifiers

The elimination of sequential line quantifiers is not as easy as in the combinatorial case. Since the sequential line variables are used to store information, they can in general not be eliminated. In some special cases, however, sequential lines can be eliminated [KuSK93], but we will not consider these cases in this paper. Instead, we show how here the sequential line quantifiers can be completely eliminated without eliminating the sequential line variables. For this reason, we define for each hardware formula, the corresponding state transition formula by:

Definition 3.1 (State Transition Formula)
For each hardware formulae $\mathcal{H}(\vec{i}, \vec{o}) \in \mathcal{HW}_\Sigma$ as given in definition 2.1, the corresponding state transition formulae $\mathcal{S}_\mathcal{H}(\vec{i})$ is defined as:

$$\boxed{\begin{aligned} \mathcal{S}_\mathcal{H}(\vec{i}) := \exists_1 \vec{\ell}. \exists_1 \vec{q}. \forall t. \\ \left[\vec{\ell}(t) \leftrightarrow \vec{\Delta}(\vec{i}(t), \vec{\ell}(t), \vec{q}(t))\right] \wedge \\ \left[(\vec{q}(0) \leftrightarrow \vec{\omega}) \wedge \left(\vec{q}(t+1) \leftrightarrow \vec{\Omega}(\vec{i}(t), \vec{\ell}(t), \vec{q}(t))\right)\right] \end{aligned}}$$

where $\exists_1$ denotes unique existence.

In this formula the state transition function of the circuit is described, while the rest of the hardware formula consists of the output function and the rest formula. The procedures of the next section make use of the following theorem:

Theorem 3.1 *For each* $\mathcal{H}(\vec{i},\vec{o}) \in \mathcal{HW}_{\Sigma}$, $\vdash \mathcal{S}_{\mathcal{H}}(\vec{i})$ *is a theorem.*

Proof. State transition theorems are consequents of the primitive recursion theorem of the natural numbers which is given below:

$$\vdash \forall \omega \forall \Omega. \exists_1 f.(f(0) = \omega) \wedge (\forall n.f(n+1) = \Omega (fn)\, n)$$

In this theorem, ω has an arbitrary type α, f has type $\mathbb{N} \rightarrow \alpha$ and Ω has type $\alpha \rightarrow \mathbb{N} \rightarrow \alpha$ (α is a type variable [Andr65]). Given a state transition formula, we first eliminate the combinatorial line quantifiers and obtain a sequential state transition formula which has the following form:

$$\hat{\mathcal{S}}_{\mathcal{H}}(\vec{i}) := \exists \vec{q}.\forall t. \left[(\vec{q}(0) \leftrightarrow \vec{\omega}) \wedge \left(\vec{q}(t+1) \leftrightarrow \vec{\Omega}(\vec{i}(t), \vec{q}(t)) \right) \right]$$

Now we instantiate $\mathbb{B}^n$ for α, $\vec{\omega}$ for ω and $\lambda \vec{b}.\lambda t.\vec{\Omega}(\vec{b}, t)$ for Ω). After applying β-reduction, the following theorem is obtained:

$$\vdash \exists_1 \vec{f}. \left[\vec{f}(0) = \vec{\omega} \right] \wedge \left[\forall t.\vec{f}(t+1) = \vec{\Omega}(\vec{f}(t), t) \right]$$

Now, assume $\Pi_k(B)$ denotes the kth projection of a tuple B. Then we replace $\vec{f}(t)$ by $(\Pi_1(\vec{f}(t)), \ldots, \Pi_n(\vec{f}(t)))$, $\vec{f}(0)$ by $(\Pi_1(\vec{f}(0)), \ldots, \Pi_n(\vec{f}(0)))$ and $\vec{f}(t+1)$ by $(\Pi_1(\vec{f}(t+1)), \ldots, \Pi_n(\vec{f}(t+1)))$. Using the fact that two tuples are equal if and only if the corresponding components are equal, yields finally in:

$$\exists_1 \vec{f}. \left(\bigwedge_{i=1}^{n} \Pi_i(\vec{f}(0)) \leftrightarrow \omega_i \right) \wedge \left(\bigwedge_{i=1}^{n} \Pi_i(\vec{f}(t+1)) \leftrightarrow \Omega_i(\Pi_1(\vec{f}(t)), \ldots, \Pi_n(\vec{f}(t))) \right)$$

In order to prove $\hat{\mathcal{S}}_{\mathcal{H}}(\vec{i})$, we add the theorem above to the assumptions and instantiate a new variable $\vec{f}$ for its quantified variable. Then we instantiate $\lambda t.(\Pi_i(\vec{f}(t)))$ for q_i on the right side of our proof goal. The rest of the proof is easily done by β-reduction and rewriting.

■

Instead of using state transition theorems for eliminating sequential line quantifiers, one might think of instantiating these quantifiers by the solution of the recursion equations of the sequential lines. This solution can be written as a fixpoint as follows: Assume that a given fixpoint operator $\mathcal{F}$ has the following property: $f(\mathcal{F}f) = \mathcal{F}f$, i.e. $\mathcal{F}f$ is a fixpoint of f. Then the solution of $q = \lambda t.(t = 0) \Rightarrow \omega \mid \Omega(q(t-1))$ would be the fixpoint of $H_q := \lambda f.\lambda t.(t = 0) \Rightarrow \omega \mid \Omega(f(t-1))$ and can therefore be written as $\mathcal{F}(\lambda f.\lambda t.(t = 0) \Rightarrow \omega \mid \Omega(f(t-1)))$.

However, the definition of the fixpoint operator $\mathcal{F}$ is not straightforward in HOL, as it has to be shown that a solution of the recursion equation exists. Of course, we can define $\mathcal{F} := \epsilon x.x = fx$, where ϵ is Hilbert's Choice-operator[3]. It is easy to prove with this definition that $\exists x.(x = fx) \vdash f(\mathcal{F}f) = \mathcal{F}f$, so what remains is to show the existence of the fixpoints. In case of the natural numbers, this requires again the proof of the primitive recursion theorem. In books about the theory of functional programming languages like [Jone87, Gord88c], fixpoint operators are often defined as $\mathcal{F} := \lambda h.(\lambda x.h(xx))(\lambda x.h(xx))$, but this does not work in typed lambda calculus as x has no finite type.

In the next subsection we use state transition theorems to eliminate quantification over sequential line variables.

3.3 Elimination of Line Quantifiers in Hardware Proof Goals

General proof goals. The elimination of $\exists$-quantifiers in the antecedent hardware formulae of $\vdash \mathcal{H}(\vec{i}, \vec{o}) \rightarrow \mathcal{P}(\vec{i}, \vec{o})$ is quite trivial and straightforward. Usual sequent calculus rules are used to eliminate the implication and all the $\exists$-quantifiers in the hardware formula. Furthermore, the $\forall$-quantifier is shifted inwards to the conjuncts it binds, and the conjunctions are eliminated according to sequent calculus rules. Thus, we obtain the following result:

$$\left[\begin{array}{l} \forall t. \left[(\vec{q}(0) \leftrightarrow \vec{\omega}) \wedge \left(\vec{q}(t+1) \leftrightarrow \vec{\Omega}(\vec{i}(t), \vec{q}(t)) \right) \right] ; \\ \forall t. \left[\vec{o}(t) \leftrightarrow \vec{\Phi}(\vec{i}(t), \vec{q}(t)) \right] ; \Re(\vec{i}, \vec{o}, \vec{q}) \end{array} \right] \vdash \mathcal{P}(\vec{i}, \vec{o})$$

Now all formulae on the left side of the sequent are formulae of first-order logic provided that $\Re(\vec{i}, \vec{o}, \vec{q})$ is a formula of first-order logic[4] as well. The equivalences on the left side can now be used to rewrite the whole proof goal. Therefore we can eliminate the equations for the output variables and obtain the sequent:

$$\left[\begin{array}{l} \forall t. \left[(\vec{q}(0) \leftrightarrow \vec{\omega}) \wedge \left(\vec{q}(t+1) \leftrightarrow \vec{\Omega}(\vec{i}(t), \vec{q}(t)) \right) \right] ; \\ \Re(\vec{i}, \lambda t.\vec{\Phi}[\vec{i}(t), \vec{q}(t)], \vec{q}) \end{array} \right] \vdash \mathcal{P}(\vec{i}, \lambda t.\vec{\Phi}[\vec{i}(t), \vec{q}(t)])$$

The rest of the proof depends crucially on the structure of $\Re(\vec{i}, \lambda t.\vec{\Phi}[\vec{i}(t), \vec{q}(t)], \vec{q})$ and $\mathcal{P}(\vec{i}, \lambda t.\vec{\Phi}[\vec{i}(t), \vec{q}(t)])$, thus the general proof paradigm ends here but can be extended for special structures as shown in the next subsection.

Proof goals at the same level of abstraction. In this subsection, we consider proof goals of the form $\vdash \mathcal{H}_1(\vec{i}, \vec{o}) \rightarrow \mathcal{H}_2(\vec{i}, \vec{o})$. As these proof goals are specialized forms of the general proof goals, the same operations can be performed to obtain a sequent of the form:

---

[3] Given that P has a type $\sigma \rightarrow \mathbb{B}$, $\epsilon x.Px$ denotes *some* member of the set whose characteristic function is P.

[4] However, it might not be a theorem of first-order logic, since the knowledge of the natural numbers is only implicitly assumed. Note that Peano's axioms are not expressible in first-order logic and that the theory of the natural numbers is not even recursively enumerable.

$$\left[\begin{array}{l} \forall t.\,[(\vec{q_1}(0) \leftrightarrow \vec{\omega}_1) \\ \quad \wedge \left(\vec{q_1}(t+1) \leftrightarrow \vec{\Omega}_1(\vec{i}(t),\vec{q_1}(t))\right)] \;; \\ \Re_1(\vec{i},\lambda t.\vec{\Phi}_1[\vec{i}(t),\vec{q_1}(t)],\vec{q_1}) \end{array}\right] \;\vdash\; \left[\begin{array}{l} \exists \vec{q_2}.\forall t. \\ \quad [(\vec{q_2}(0) \leftrightarrow \vec{\omega}_2) \\ \quad\quad \wedge \left(\vec{q_2}(t+1) \leftrightarrow \vec{\Omega}_2(\vec{i}(t),\vec{q_2}(t))\right)] \wedge \\ \quad \left[\vec{\Phi}_1(\vec{i}(t),\vec{q_1}(t)) \leftrightarrow \vec{\Phi}_2(\vec{i}(t),\vec{q_2}(t))\right] \wedge \\ \quad \Re_2(\vec{i},\lambda t.\vec{\Phi}_1[\vec{i}(t),\vec{q_1}(t)],\vec{q_2}) \end{array}\right]$$

Making use of the fact that all state transition theorems are valid, the lemma $\vdash \exists \vec{q_2}.\forall t.\left[(\vec{q_2}(0) \leftrightarrow \vec{\omega}_2) \wedge \left(\vec{q_2}(t+1) \leftrightarrow \vec{\Omega}_2(\vec{i}(t),\vec{q_2}(t))\right)\right]$ can be introduced as an assumption in the antecedent of the sequent given above. Then usual sequent calculus rules are used to instantiate the higher-order quantifiers on the left side with new variables, i.e. variables which do not occur free in the considered sequent (e.g. choose the variables $\vec{q_2}$ themselves). Now the quantifiers on the right hand side can be instantiated. In contrast to the instantiation on the left side of the $\exists$-quantifiers, the instantiations on the right side can be done with an arbitrary terms. Therefore, we use exactly the same variables for instantiation as on the left side. Additionally, we instantiate the $\forall$-quantifier on the right side with a new variable t. As a result we obtain:

$$\left[\begin{array}{l} \forall t.\,[(\vec{q_1}(0) \leftrightarrow \vec{\omega}_1) \\ \quad \wedge \left(\vec{q_1}(t+1) \leftrightarrow \vec{\Omega}_1(\vec{i}(t),\vec{q_1}(t))\right)] \;; \\ \Re_1(\vec{i},\lambda t.\vec{\Phi}_1[\vec{i}(t),\vec{q_1}(t)],\vec{q_1})); \\ \forall t.\,[(\vec{q_2}(0) \leftrightarrow \vec{\omega}_2) \\ \quad \wedge \left(\vec{q_2}(t+1) \leftrightarrow \vec{\Omega}_2(\vec{i}(t),\vec{q_2}(t))\right)] \end{array}\right] \;\vdash\; \left[\begin{array}{l} [(\vec{q_2}(0) \leftrightarrow \vec{\omega}_2) \\ \quad \wedge \left(\vec{q_2}(t+1) \leftrightarrow \vec{\Omega}_2(\vec{i}(t),\vec{q_2}(t))\right)] \wedge \\ \left[\vec{\Phi}_1(\vec{i}(t),\vec{q_1}(t)) \leftrightarrow \vec{\Phi}_2(\vec{i}(t),\vec{q_2}(t))\right] \wedge \\ \Re_2(\vec{i},\lambda t.\vec{\Phi}[\vec{i}(t),\vec{q_1}(t)],\vec{q_2}) \end{array}\right]$$

Next we use the rule for eliminating conjunctions on the right hand side, thus we obtain the subgoals:

- $\ldots \vdash (\vec{q_2}(0) \leftrightarrow \vec{\omega}_2) \wedge \left(\vec{q_2}(t+1) \leftrightarrow \vec{\Omega}_2(\vec{i}(t),\vec{q_2}(t))\right)$

- $\ldots \vdash \left[\vec{\Phi}_1(\vec{i}(t),\vec{q_1}(t)) \leftrightarrow \vec{\Phi}_2(\vec{i}(t),\vec{q_2}(t))\right] \wedge \Re_2(\vec{i},\lambda t.\vec{\Phi}[\vec{i}(t),\vec{q_1}(t)],\vec{q_2})$

The first goal is immediately solved as the succedent occurs on the left side as well. The remaining goal is therefore:

$$\left[\begin{array}{l} \forall t.\,[(\vec{q_1}(0) \leftrightarrow \vec{\omega}_1) \\ \quad \wedge \left(\vec{q_1}(t+1) \leftrightarrow \vec{\Omega}_1(\vec{i}(t),\vec{q_1}(t))\right)] \;; \\ \Re_1(\vec{i},\lambda t.\vec{\Phi}_1[\vec{i}(t),\vec{q_1}(t)],\vec{q_1}); \\ \forall t.\,[(\vec{q_2}(0) \leftrightarrow \vec{\omega}_2) \\ \quad \wedge \left(\vec{q_2}(t+1) \leftrightarrow \vec{\Omega}_2(\vec{i}(t),\vec{q_2}(t))\right)] \end{array}\right] \;\vdash\; \left[\begin{array}{l} \left[\vec{\Phi}_1(\vec{i}(t),\vec{q_1}(t)) \leftrightarrow \vec{\Phi}_2(\vec{i}(t),\vec{q_2}(t))\right] \wedge \\ \Re_2(\vec{i},\lambda t.\vec{\Phi}_1[\vec{i}(t),\vec{q_1}(t)],\vec{q_2}) \end{array}\right]$$

If we abbreviate the conjunction of the equations for $\vec{q_1}$ and $\vec{q_2}$ by Γ, then the goal $\vdash \mathcal{H}_1(\vec{i},\vec{o}) \leftrightarrow \mathcal{H}_2(\vec{i},\vec{o})$ is first split into the two subgoals $\vdash \mathcal{H}_1(\vec{i},\vec{o}) \rightarrow \mathcal{H}_2(\vec{i},\vec{o})$ and $\vdash \mathcal{H}_2(\vec{i},\vec{o}) \rightarrow \mathcal{H}_1(\vec{i},\vec{o})$ which are then transformed by the above tactics into the following four subgoals:

1. $\Gamma \vdash \Re_1(\vec{i},\lambda t.\vec{\Phi}_1[\vec{i}(t),\vec{q_1}(t)],\vec{q_1}) \rightarrow \Re_2(\vec{i},\lambda t.\vec{\Phi}_1[\vec{i}(t),\vec{q_1}(t)],\vec{q_2})$
2. $\Gamma \vdash \Re_1(\vec{i},\lambda t.\vec{\Phi}_1[\vec{i}(t),\vec{q_1}(t)],\vec{q_1}) \rightarrow \left\{\vec{\Phi}_1(\vec{i}(t),\vec{q_1}(t)) \leftrightarrow \vec{\Phi}_2(\vec{i}(t),\vec{q_2}(t))\right\}$
3. $\Gamma \vdash \Re_2(\vec{i},\lambda t.\vec{\Phi}_2[\vec{i}(t),\vec{q_2}(t)],\vec{q_2}) \rightarrow \Re_1(\vec{i},\lambda t.\vec{\Phi}_2[\vec{i}(t),\vec{q_2}(t)],\vec{q_1})$

4. $\Gamma \vdash \Re_2(\vec{i}, \lambda t.\vec{\Phi}_2[\vec{i}(t), \vec{q}_2(t)], \vec{q}_2) \rightarrow \left\{ \vec{\Phi}_1(\vec{i}(t), \vec{q}_1(t)) \leftrightarrow \vec{\Phi}_2(\vec{i}(t), \vec{q}_2(t)) \right\}$

Note that if the additional formulae $\Re_1$ and $\Re_2$ do not occur in the formulae $\mathcal{H}_1(\vec{i}, \vec{o})$ and $\mathcal{H}_2(\vec{i}, \vec{o})$, all subgoals above are all equivalent to each other. Thus, we have proved the following theorem:

Theorem 3.2 (Equivalence of Simple Hardware Formulae) *Given two hardware-formulae* $\mathcal{H}_1(\vec{i}, \vec{o}), \mathcal{H}_2(\vec{i}, \vec{o})$ *with no additional formulae* $\Re_1(\vec{i}, \vec{o}, \vec{q}_1), \Re_2(\vec{i}, \vec{o}, \vec{q}_2)$ *the following holds:* $\vdash \mathcal{H}_1(\vec{i}, \vec{o}) \rightarrow \mathcal{H}_2(\vec{i}, \vec{o})$ *is a theorem iff* $\vdash \mathcal{H}_2(\vec{i}, \vec{o}) \rightarrow \mathcal{H}_1(\vec{i}, \vec{o})$ *is a theorem.*

According to this theorem, the three theorems $\vdash \mathcal{H}_1(\vec{i}, \vec{o}) \rightarrow \mathcal{H}_2(\vec{i}, \vec{o})$, $\vdash \mathcal{H}_2(\vec{i}, \vec{o}) \rightarrow \mathcal{H}_1(\vec{i}, \vec{o})$ and $\vdash \mathcal{H}_1(\vec{i}, \vec{o}) \leftrightarrow \mathcal{H}_2(\vec{i}, \vec{o})$ are equivalent to each other. The proof of the equivalence of simple hardware formula is therefore halved by this theorem. In general, proof goals of the following form are obtained which we call **Finite State Machine goals** or shorter **FSM-goals**:

$$ \forall t. \left[(\vec{q}(0) \leftrightarrow \vec{\omega}_0) \wedge \left(\vec{q}(t+1) \leftrightarrow \vec{\Omega}(\vec{i}(t), \vec{q}(t)) \right) \right] \vdash \forall t.\Psi(\vec{i}(t), \vec{q}(t)) $$

The left hand side of these goals describes a state transition system which has the state sequence $\vec{q}$ for a given input sequence $\vec{i}$. If this goal has been obtained by the transformations of this section, then the state transition system is the product machine of the two considered hardware formulae. In the next section, we prove the decidability of the validity of these FSM-goals. Moreover, various decision procedures [ScKK93d] have been implemented in HOL90 for proving these goals.

3.4 Completeness of the Transformation

Having proved the correctness of the transformation, the question arises, whether the resulting subgoals are provable for *each* valid input sequent. The elimination of propositional operators is uncritical, since the resulting subgoals are equivalent to the original sequent. However, the deletion of the quantifiers could be critical, as in general there are formulae which can only be proved when several instances of the quantified formulae are made[5]. In the instantiations occurring in the presented procedure, however, exactly one instantiation is always sufficient, as the state transition theorem states that there is *exactly one* tuple of line variables satisfying the state transition formula. As unique existence is defined as $\exists_1 x.A(x) :\leftrightarrow [\exists x.A(x)] \wedge [\forall xy.A(x) \wedge A(y) \rightarrow (x = y)]$, further instances have to be identical to the previous ones to satisfy the state transition formula which is a theorem. Thus further instances add no further information and are therefore useless.

In the domain of hardware verification, often multiple recursion is required. For example, if one wants to describe a periodic signal 110110110..., the following definition for a variable q can be given (which we call 3-recursive for obvious reasons): $(q(0) \leftrightarrow T) \wedge (q(1) \leftrightarrow T) \wedge (q(2) \leftrightarrow F) \wedge \forall t.(q(t+3) \leftrightarrow q(t))$. However, the presented tactics do not work with multiple recursion. But it is well known that any n-recursive definition can be replaced by n 1-recursive definitions.

---

[5] For example, in the proof of the theorem $\vdash (\forall x.P(x) \wedge Q(x)) \rightarrow (P(a) \vee Q(b))$ two instantiations of the quantified formula are necessary.

4 Decidability of Simple Hardware Formulae Equivalence

In this section we present a decision procedure for the goals resulting from the transformations of the previous section provided that $\Re$ is a quantifier-free formula. The procedures might also work for more complicated formulae $\Re$, but then the decidability cannot be guaranteed. Some theoretical background is required for proving the decidability:

Lemma 4.1 (Propositional Abstraction) *If the outermost universal quantifiers of a higher-order formula φ are removed and the atomic and quantified subformulae are replaced by propositional variables, the resulting formula ψ is called the propositional abstraction of φ. If ψ is a tautology, φ is a theorem of higher-order logic as well, otherwise nothing can be said about the validity of φ.*

The next lemma is used to enhance the efficiency of propositional abstraction:

Lemma 4.2 (Rewrite Lemma)
Given a quantifier-free formula $\Upsilon(\alpha_1, \ldots, \alpha_n)$ with atomic formulae $\alpha_1, \ldots, \alpha_n$ and some other formulae $\varphi_1, \ldots, \varphi_n$ which do not depend on the α_j's, the following holds[6] (Υ is allowed to contain more atoms than the α_j's and not all α_j's have to occur in Υ.):

$$\bigwedge_{j=1}^{n}(\alpha_j \leftrightarrow \varphi_j) \vdash \Upsilon(\alpha_1, \ldots, \alpha_n) \text{ is a theorem iff } \vdash \Upsilon(\varphi_1, \ldots, \varphi_n) \text{ is a theorem}$$

As a consequence of this lemma, the efficiency of automated proof procedures can be enhanced, since the antecedent can be totally neglected after rewriting with the assumptions). In order to apply this lemma on FSM-goals where m is the largest constant such that $t + m$ occurs in the right-hand formula $\forall t. \Psi(\vec{q}(t), \vec{i}(t))$, we first make the copies $\vec{q}(t + j + 1) \leftrightarrow \vec{\Omega}(\vec{q}(t + j), \vec{i}(t + j))$ for $j \in \{1, \ldots, m-1\}$ on the left hand side.

As there is no decision procedure for temporal propositional logics in HOL, we convert the FSM-goals into equivalent propositional formulae. For this reason, we prove the following lemma and theorem. The formulae $\vec{\Theta}_j^{0,\vec{i}}$ and $\vec{\Theta}_j^{t,\vec{i}}$ defined in the next lemma represent the states which are reachable in j steps from the initial state and from the state $\vec{q}(t)$, respectively (first item of the lemma) with the inputs $\vec{i}(0), \ldots, \vec{i}(j)$ and $\vec{i}(t), \ldots, \vec{i}(t + j)$, respectively.

Lemma 4.3 (Instance Lemma) *Defining the tuples of formulae $\vec{\Theta}_j^{0,\vec{i}}$ and $\vec{\Theta}_j^{t,\vec{i}}$ recursively by $\vec{\Theta}_0^{0,\vec{i}} := \vec{\omega}_0$, $\vec{\Theta}_{j+1}^{0,\vec{i}} := \vec{\Omega}(\vec{\Theta}_j^{0,\vec{i}}, \vec{i}(j))$, $\vec{\Theta}_0^{t,\vec{i}} := \vec{q}(t)$ and $\vec{\Theta}_{j+1}^{t,\vec{i}} := \vec{\Omega}(\vec{\Theta}_j^{t,\vec{i}}, \vec{i}(t + j))$ and given that $\Gamma := [\vec{q}(0) \leftrightarrow \vec{\omega}_0] \wedge \forall t. \left[\vec{q}(t + 1) \leftrightarrow \vec{\Omega}(\vec{q}(t), \vec{i}(t))\right]$ and t is a variable, the following holds:*

---

[6] It has not been stated that $\vdash \left[\left(\bigwedge_{j=1}^{n}(\alpha_j \leftrightarrow \varphi_j)\right) \to \Upsilon(\alpha_1, \ldots, \alpha_n)\right] \leftrightarrow \Upsilon(\varphi_1, \ldots, \varphi_n)$ is a theorem. This is not the case since the simple example $[(a \leftrightarrow b) \to a] \leftrightarrow b$ is not a tautology (interpret $a := T, b := F$).

1. $\Gamma \vdash \vec{q}(j) \leftrightarrow \vec{\Theta}_j^{0,\vec{\imath}}$ and $\Gamma \vdash \forall t.\vec{q}(t+j) \leftrightarrow \vec{\Theta}_j^{t,\vec{\imath}}$ for all $j \in \mathbb{N}$.

2. $\Gamma \vdash \forall t.\Phi(\vec{q}(t),\vec{\imath}(t))$ iff $\vdash \Phi(\vec{\Theta}_j^{0,\vec{\imath}},\vec{\imath}(j))$ holds for all $j \in \mathbb{N}$.

3. If $\Gamma \not\vdash \forall t.\Phi(\vec{q}(t),\vec{\imath}(t))$, then there is a natural number n such that $\not\vdash \Phi(\vec{\Theta}_n^{0,\vec{\imath}},\vec{\imath}(n))$

Proof. 1 is obtained by induction over j and the definition of $\vec{\Theta}_j^{0,\vec{\imath}}$ and $\vec{\Theta}_j^{t,\vec{\imath}}$. 2 is obtained in two steps, first we have $\Gamma \vdash \forall t.\Phi(\vec{q}(t),\vec{\imath}(t))$ iff $\Gamma \vdash \Phi(\vec{q}(j),\vec{\imath}(j))$ for all $j \in \mathbb{N}$ and second $\Gamma \vdash \Phi(\vec{q}(j),\vec{\imath}(j))$ iff $\vdash \Phi(\vec{\Theta}_j^{0,\vec{\imath}},\vec{\imath}(j))$ for all $j \in \mathbb{N}$ (the first proposition is a conclusion of the fact that the natural numbers have a unique model (up to isomorphisms) and the second is a consequence of the rewrite lemma using 1). Proposition 3 is a consequence of proposition 2. The following theorem can be viewed as a 'pumping lemma' for hardware formulae.

Theorem 4.1 (Decidability) *Given a goal* $\Gamma \vdash \forall t.\Phi(\vec{q}(t),\vec{\imath}(t))$ *and the formulae* $\vec{\Theta}_j^{0,\vec{\imath}}$ *and* $\vec{\Theta}_j^{t,\vec{\imath}}$ *as defined in Lemma 4.3, there is a number* n_0*, such that* $\Gamma \vdash \forall t.\Phi(\vec{q}(t),\vec{\imath}(t))$ *holds iff* $\vdash \Phi(\vec{\Theta}_j^{0,\vec{\imath}},\vec{\imath}(j))$ *holds for all canonical terms* $j \leq n_0$*. Moreover* n_0 *can be effectively computed, thus the validity of the considered goals is decidable. The number* n_0 *is called the sequential depth of the problem.*

Proof. We prove the statement: *there is a number* n_0*, such that* $\vdash \Phi(\vec{\Theta}_j^{0,\vec{\imath}},\vec{\imath}(j))$ *holds for all canonical terms* $j \in \mathbb{N}$ *iff* $\vdash \Phi(\vec{\Theta}_j^{0,\vec{\imath}},\vec{\imath}(j))$ *holds for all canonical terms* $j \leq n_0$ which is equivalent to theorem 4.1 according to item 2 of lemma 4.3.

'$\Rightarrow$:' is trivially true no matter what we choose for n_0.

'$\Leftarrow$:' We show the equivalent formulation: $\exists k \in \mathbb{N}. \not\vdash \Phi(\vec{\Theta}_k^{0,\vec{\imath}},\vec{\imath}(k))$ implies $\exists j \leq n_0. \not\vdash \Phi(\vec{\Theta}_j^{0,\vec{\imath}},\vec{\imath}(j))$. Let $\mathcal{M}_\Theta := \{\vec{\Theta}_k^{0,\vec{\imath}} \mid k \in \mathbb{N}\}$ and $\mathfrak{I}$ be the set of *all* (propositional) interpretations of $\mathcal{M}_\Theta$. Then we define the equivalence relation $\approx$ on $\mathfrak{I} \times \mathcal{M}_\Theta$ by:

$$(I_1,\vec{\Theta}_k^{0,\vec{\imath}}) \approx (I_2,\vec{\Theta}_j^{0,\vec{\imath}}) :\Leftrightarrow \mathcal{VAL}^{I_1}(\vec{\Theta}_k^{0,\vec{\imath}}) = \mathcal{VAL}^{I_2}(\vec{\Theta}_j^{0,\vec{\imath}})$$

As $\mathcal{VAL}^{I_k}(\vec{\Theta}_j^{0,\vec{\imath}}) \in \mathbb{B}^{|\vec{q}|}$, $\approx$ has only a finite number of equivalence classes (at most $2^{|\vec{q}|}$). Let $S := \{s_0,\dots,s_q\}$ be these classes and let $m(s)$ be the least k such that $\exists I \in \mathfrak{I}.(I,\vec{\Theta}_k^{0,\vec{\imath}}) \in s$, then we define $n_0 := max\{m(s) \mid s \in S\}$[7]. Now we can conclude: $\forall (I_1,\vec{\Theta}_k^{0,\vec{\imath}}) \exists (I_2,\vec{\Theta}_j^{0,\vec{\imath}}). \mathcal{VAL}^{I_1}(\vec{\Theta}_k^{0,\vec{\imath}}) = \mathcal{VAL}^{I_2}(\vec{\Theta}_j^{0,\vec{\imath}}) \wedge j \leq n_0$. Thus n_0 has the required property: Assume there is an interpretation I_1 which does not satisfy $\Phi(\vec{\Theta}_k^{0,\vec{\imath}},\vec{\imath}(k))$, then there is also an interpretation I_2 which does not satisfy a formula $\Phi(\vec{\Theta}_j^{0,\vec{\imath}},\vec{\imath}(j))$ with $j \leq n_0$.

$$\frac{\Gamma \vdash \forall t.\Phi(\vec{q}(t),\vec{\imath}(t))}{\vdash \bigwedge_{j=0}^{n_0} \Phi(\vec{\Theta}_j^{0,\vec{\imath}},\vec{\imath}(j))}$$

Fig. 1.: SEQ_TAC

A decision procedure for our goals would thus be the following: Compute n_0 by a state traversal (maybe outside HOL) and prove all $\vdash \Phi(\vec{\Theta}_j^{0,\vec{\imath}},\vec{\imath}(j))$ with $j \leq n_0$. As $\Phi(\vec{\Theta}_j^{0,\vec{\imath}},\vec{\imath}(j))$ has $(j+1) \cdot |\vec{\imath}|$ variables, the proof of $\vdash \Phi(\vec{\Theta}_j^{0,\vec{\imath}},\vec{\imath}(j))$ is of order $O(2^{(j+1) \cdot |\vec{\imath}|})$, thus the complexity of this decision procedure

---

[7] Having a look at finite state machines, these definitions become clearer: S is the set of all possible states and $m(s)$ is the least number of inputs to reach the state s from the initial state. n_0 is the least number of inputs required to reach every reachable state.

is of order $O(2^{(n_0+1)\cdot|\vec{i}|})$. This decision procedure has been implemented as a tactic called SEQ_TAC (figure 1).

5   A Case Study

In this section, we demonstrate the use of the previously described methods by proving the equivalence of two different implementations of a serial adder. Although this example can also be proved with simple finite state machine algorithms, it shows how the method can be used to transform the higher-order proof goal into simple propositional logic.

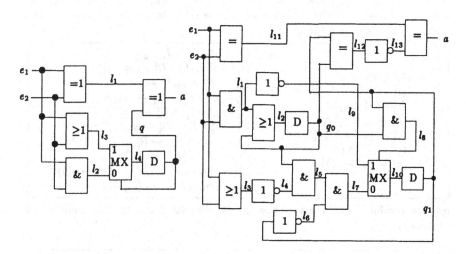

Fig. 2. Equivalent implementations of a serial adder.

The implementations are given in figure 2 and the corresponding state transition diagrams are given in figure 3. The hardware formulae used for the description of the circuits are given in the following table:

$SERIAL1(e_1, e_2, a)$	$SERIAL2(e_1, e_2, a)$
	$\exists l_1 l_2 l_3 l_4 l_5 l_6 l_7 l_8 l_9 l_{10} l_{11} l_{12} l_{13} q_0 q_1.$
$\exists l_1 l_2 l_3 l_4 q.$	$AND(e_1, e_2, l_1) \wedge OR(q_0, l_1, l_2) \wedge$
$XNOR(e_1, e_2, l_1) \wedge$	$DELAY(l_2, q_0) \wedge OR(e_1, e_2, l_3) \wedge$
$AND(e_1, e_2, l_2) \wedge$	$INV(l_3, l_4) \wedge AND(q_0, l_4, l_5) \wedge$
$OR(e_1, e_2, l_3) \wedge$	$INV(q_1, l_6) \wedge AND(l_6, l_5, l_7) \wedge$
$MUX(q, l_3, l_2, l_4) \wedge$	$AND(q_0, q_1, l_8) \wedge INV(l_1, l_9) \wedge$
$DELAY(l_4, q) \wedge$	$MUX(l_8, l_9, l_7, l_{10}) \wedge DELAY(l_{10}, q_1) \wedge$
$XNOR(q, l_1, a)$	$XNOR(e_1, e_2, l_{11}) \wedge XNOR(q_1, q_0, l_{12}) \wedge$
	$INV(l_{12}, l_{13}) \wedge XNOR(l_{13}, l_{11}, a)$

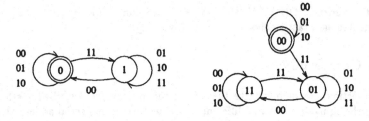

Fig. 3. Transition diagrams of the implementations.

The goal we want to prove is $\forall e_1 e_2 a.SERIAL1(e_1, e_2, a) \leftrightarrow SERIAL2(e_1, e_2, a)$. After the expansion of the definitions of the serial adder predicates and the used basic circuits, we obtain the goal[8]:

$$\exists l_1 l_2 l_3 l_4 q.\forall t.$$
$$(l_1^t \leftrightarrow (e_1^t \leftrightarrow e_2^t)) \wedge$$
$$(l_2^t \leftrightarrow e_1^t \wedge e_2^t) \wedge$$
$$(l_3^t \leftrightarrow e_1^t \vee e_2^t) \wedge$$
$$(l_4^t \leftrightarrow (q^t \Rightarrow l_3^t \mid l_2^t)) \wedge$$
$$(q^0 \leftrightarrow F) \wedge (q^{t+1} \leftrightarrow l_4^t) \wedge$$
$$(a^t \leftrightarrow (q^t \leftrightarrow l_1^t))$$

$$\leftrightarrow$$

$$\exists l_1 l_2 l_3 l_4 l_5 l_6 l_7 l_8 l_9 l_{10} l_{11} l_{12} l_{13} q_0 q_1.\forall t.$$
$$(l_1^t \leftrightarrow e_1^t \wedge e_2^t) \wedge (l_2^t \leftrightarrow q_0^t \vee l_1^t) \wedge$$
$$(q_0^0 \leftrightarrow F) \wedge (q_0^{t+1} \leftrightarrow l_2^t) \wedge$$
$$(l_3^t \leftrightarrow e_1^t \vee e_2^t) \wedge (l_4^t \leftrightarrow \neg l_3^t) \wedge$$
$$(l_5^t \leftrightarrow q_0^t \wedge l_4^t) \wedge (l_6^t \leftrightarrow \neg q_1^t) \wedge$$
$$(l_7^t \leftrightarrow l_6^t \wedge l_5^t) \wedge (l_8^t \leftrightarrow q_0^t \wedge q_1^t) \wedge$$
$$(l_9^t \leftrightarrow \neg l_1^t) \wedge (l_{10}^t \leftrightarrow (l_8^t \Rightarrow l_9^t \mid l_7^t)) \wedge$$
$$(q_1^0 \leftrightarrow F) \wedge (q_1^{t+1} \leftrightarrow l_{10}^t) \wedge$$
$$(l_{11}^t \leftrightarrow (e_1^t \leftrightarrow e_2^t)) \wedge (l_{12}^t \leftrightarrow (q_1^t \leftrightarrow q_0^t)) \wedge$$
$$(l_{13}^t \leftrightarrow \neg l_{12}^t) \wedge (a^t \leftrightarrow (l_{13}^t \leftrightarrow l_{11}^t))$$

Now the combinatorial line variables can be eliminated and the following goal is obtained:

$$\exists q_0 q_1.\forall t. (q_0^0 \leftrightarrow F) \wedge (q_0^{t+1} \leftrightarrow [q_0^t \vee (e_1^t \wedge e_2^t)]) \wedge$$
$$(q_1^0 \leftrightarrow F) \wedge (q_1^{t+1} \leftrightarrow [q_0^t \wedge q_1^t \Rightarrow \neg(e_1^t \wedge e_2^t) \mid \neg q_1^t \wedge q_0^t \wedge \neg(e_1^t \vee e_2^t)]) \wedge$$
$$(a^t \leftrightarrow (\neg [q_1^t \leftrightarrow q_0^t] \leftrightarrow [e_1^t \leftrightarrow e_2^t]))$$
$$\leftrightarrow \exists q.\forall t. (q^0 \leftrightarrow F) \wedge (q^{t+1} \leftrightarrow [q^t \Rightarrow (e_1^t \vee e_2^t) \mid (e_1^t \wedge e_2^t)]) \wedge$$
$$(a^t \leftrightarrow (q^t \leftrightarrow [e_1^t \leftrightarrow e_2^t]))$$

Next we split the above equivalence into two implications and prove the state transitions theorems. The quantifiers are then eliminated by the instantiation of new line variables. Finally, rewriting with the assumptions leads in both cases (cf. theorem 3.2) to the following goal:

$$\forall t.(q_0^0 \leftrightarrow F) \wedge (q_0^{t+1} \leftrightarrow [q_0^t \vee (e_1^t \wedge e_2^t)]) \wedge$$
$$(q_1^0 \leftrightarrow F) \wedge (q_1^{t+1} \leftrightarrow [q_0^t \wedge q_1^t \Rightarrow \neg(e_1^t \wedge e_2^t) \mid \neg q_1^t \wedge q_0^t \wedge \neg(e_1^t \vee e_2^t)]) \wedge$$
$$(q^0 \leftrightarrow F) \wedge (q^{t+1} \leftrightarrow [q^t \Rightarrow (e_1^t \vee e_2^t) \mid (e_1^t \wedge e_2^t)])$$
$$\vdash \forall t. (\neg [q_1^t \leftrightarrow q_0^t] \leftrightarrow [e_1^t \leftrightarrow e_2^t]) \leftrightarrow (q^t \leftrightarrow [e_1^t \leftrightarrow e_2^t])$$

---

[8] We write e_1^t instead of $e_1(t)$.

The sequential depth of this goal is 1, i.e. the equivalence of the two circuits is proved iff the following two subgoals of propositional logic can be proved:

1. $\vdash \underbrace{(\neg[F \leftrightarrow F] \leftrightarrow [e_1^0 \leftrightarrow e_2^0])}_{=e_1^0 \leftrightarrow e_2^0} \leftrightarrow \underbrace{(F \leftrightarrow [e_1^0 \leftrightarrow e_2^0])}_{=e_1^0 \leftrightarrow e_2^0}$

2. $\vdash(\neg\underbrace{[[F \wedge F \Rightarrow \neg(e_1^0 \wedge e_2^0) \mid \neg F \wedge F \wedge \neg(e_1^0 \vee e_2^0)] \leftrightarrow [F \vee (e_1^0 \wedge e_2^0)]]}_{=F} \leftrightarrow [e_1^1 \leftrightarrow e_2^1])$

$$\underbrace{}_{e_1^0 \wedge e_2^0}$$

$$\leftrightarrow (\underbrace{[F \Rightarrow (e_1^0 \vee e_2^0) \mid (e_1^0 \wedge e_2^0)]}_{e_1^0 \wedge e_2^0} \leftrightarrow [e_1^1 \leftrightarrow e_2^1])$$

Both subgoals can be easily proved by a tautology checker.

In section 4, we have proved the decidability for the equivalence problem for simple hardware formulae which allowed an automated proof of the verification of the two serial adder implementations. The decidability result for simple hardware formula does however not carry over to abstract circuits, but the presented tactics are useful in these cases, too. As the primitive recursion theorem has polymorphic types, each $\exists$-quantifier over a function of type $\mathbb{N} \rightarrow \alpha$ can be eliminated this way. For example, if $\mathcal{A}$ is a variable of type $\mathbb{N} \rightarrow \mathbb{N} \rightarrow \mathbb{B}$ which is representing a time dependent bitvector ($\mathcal{A} t n$ denotes the value of the nth bit at time t), it is clear that the primitive recursion theorem can be instantiated such that the following theorem is obtained: $\vdash \forall\omega\forall\Omega.\exists_1\mathcal{A}.(\mathcal{A}\,0 = \omega) \wedge (\forall n.\mathcal{A}\,(n+1) = \Omega\,(\mathcal{A}n)\,n)$. If $\mathcal{A}$ is an internal line in an abstract circuit, then goals of the following form will occur: $\Gamma \vdash \exists\mathcal{A}.\forall t.(\mathcal{A}0 = \omega_0) \wedge (\mathcal{A}(t+1) = \Omega(\mathcal{A}t)t) \wedge \Psi(\mathcal{A}t)$. This goal can be transformed by the quantifier elimination tactic into the subgoal $\Gamma, \forall t.(\mathcal{A}0 = \omega_0) \wedge (\mathcal{A}\,(t+1) = \Omega\,(\mathcal{A}\,t)\,t) \vdash \forall t.\Psi(\mathcal{A}\,t)$ which is easier to prove.

6 Conclusions and Future Work

We have presented methods how hardware correctness proof goals in higher-order logic can be transformed into weaker logics by eliminating the higher-order quantifiers. Proof goals of real combinatorial circuits are transformed into propositional logic and proof goals of real sequential circuits can be transformed into temporal propositional logic, or by a further transformation (SEQ_TAC) into propositional logic, too. The presented tactics have been implemented in HOL90 and as a result, full automation of these proofs has been obtained. All the presented tactics make use of the determinism of the considered circuit. Each internal line of the circuit has at each computation step a definite value which is symbolically computed by the tactics. If the proof can not be done by the instantiation of this value, it can be concluded that the goal which is to be proved is erroneous. The methods can also be carried over to abstract circuits although in general a full automation cannot be achieved for them. However, proofs for special classes can nevertheless be automated.

In our future work, we elaborate proof paradigms for generic circuit classes based on the presented quantifier elimination methods. Problems occur when different classes of circuits are used in an implementation at the next level of design hierarchy.

In these cases, further lemmata have to be proved which are required for the original proof. Our future work will also focus on these lemmata, and we try to automate this lemma generation process by completion methods. Another issue for our future work is to discuss the relations of this approach for real circuits to model checking, as model checking can be viewed as a special case of this approach if no abstract circuits are considered.

References

[Acke35a] W. Ackermann. Untersuchungen über das Eliminationsproblem der mathematischen Logik. *Mathematische Annalen*, 110:390–413, 1935.

[Andr65] P.B. Andrews. *A Transfinite Type Theory with Type Variables*. North-Holland Publishing Company, 1965.

[GaOh92] D. Gabbay and H. J. Ohlbach. Quantifier elimination in second-order predicate logic. Technical Report MPI-I-92-231, Max-Planck-Institut für Informatik, Saarbrücken, July 1992.

[Gord86] M.J.C. Gordon. Why higher-order logic is a good formalism for specifying and verifying hardware. In G. Milne and P.A. Subrahmanyam, editors, *Formal aspects of VLSI Design*. North-Holland, 1986.

[Gord88c] M.J.C. Gordon. *Programming Language Theory and its Implementation*. International Series in Computer Science. Prentice Hall, 1988.

[Gupt92] A. Gupta. Formal hardware verification methods: A survey. *Journal of Formal Methods in System Design*, 1:151–238, 1992.

[HaDa86] F.K. Hanna and N. Daeche. Specification and verification of digital systems using higher-order predicate logic. *IEE Proc. Pt. E*, 133(3):242–254, 1986.

[Jone87] S.L.P. Jones. *The Implementation of Functional Programming Languages*. Prentice Hall, 1987.

[Joyc91] J. Joyce. More reasons why higher-order logic is a good formalism for specifying and verifying hardware. In *International Workshop on Formal Methods in VLSI Design*, Miami,1991.

[KuSK93] R. Kumar, K. Schneider, and Th. Kropf. Structuring and automating hardware proofs in a higher-order theorem-proving environment. *Journal of Formal Methods in System Design*, 2(2):165–223, 1993.

[ScKK91a] K. Schneider, R. Kumar, and Th. Kropf. Structuring hardware proofs: First steps towards automation in a higher-order environment. In P.B. Denyer A. Halaas, editor, *International Conference on Very Large Scale Integration*, pages 81–90, Edinburg, Edingburgh, 1991. North Holland.

[ScKK92c] K. Schneider, R. Kumar, and Th. Kropf. Modelling generic hardware structures by abstract datatypes. In L.J.M. Claesen and M.J.C. Gordon, editors, *Higher Order Logic Theorem Proving and its Applications*, volume A-20 of *IFIP Transactions*, pages 165–176, Leuven, Belgium, 1992. North-Holland.

[ScKK93d] K. Schneider, R. Kumar, and Th. Kropf. Alternative proof procedures for finite-state machines in a higher-order environment. In *International Workshop on Higher-Order Logic Theorem Proving and its Applications*, Vancouver, Canada, 1993.

Toward a Super Duper Hardware Tactic

Mark D. Aagaard[1] Miriam E. Leeser[1] Phillip J. Windley[2]

[1] School of Electrical Engineering Cornell University Ithaca, NY
[2] Laboratory for Applied Logic Department of Computer Science Brigham Young University Provo, UT

Abstract. We present techniques for automating many of the tedious aspects of hardware verification in a higher order logic theorem proving environment. We employ two complementary approaches. The first involves intelligent tactics which incorporate many of the smaller steps currently applied by the user. The second uses hardware combinators to partially automate inductive proofs for iterated hardware structures. We envision a system that captures most of this reasoning in one tactic, SuperDuperHWTac. Ideally, users would use this tactic on a goal for proving that a hardware component meets its specification, and get back a proof documented at a level they would have written by hand. This paper presents preliminary work toward SuperDuperHWTac in both the HOL and Nuprl proof development systems.

1 Introduction

Higher order logic makes specifying hardware designs natural. Unfortunately, it also makes verification tedious. If verification engineers adopt a specific style for doing hardware proofs, they find they do similar steps over and over again. We describe support for a specific style of hardware verification. Our goals are to provide the user with intelligent tactics which automate those steps that are common to proofs done in the given style, and raise the level at which the user interacts with a tactic oriented theorem prover. We also wish to keep the proving process within the context of the theorem prover and to have the system generate a proof which provides documentation of the proof process.

The style of hardware verification we support is one that is very widely used in higher order logic theorem proving. The user describes a hardware component as a relation over inputs and outputs. The component is described by a specification of desired behavior and an implementation in terms of the interconnection of simpler hardware modules. The steps required to verify such a component are 1) set the goal to be verified as implementation $\Longrightarrow$ specification, 2) expand definitions in the goal, and 3) simplify and prove the goal using domain specific and general knowledge.

We view the ideal higher order theorem proving environment as follows. The user defines the goal to be proved and applies one tactic at the top of the proof. We will call this ideal tactic SuperDuperHWTac. SuperDuperHWTac should find a proof in reasonable time, provide a report of the resulting proof which is close to what the user would have written by hand, and only report on those paths taken by the proof system that succeed. This paper reports on preliminary work on writing tactics for hardware proofs that move in the general direction of SuperDuperHWTac. That work has focused on two complementary approaches. The first is implementing smart tactics that incorporate much of the repetitive reasoning found in hardware proofs. The second is to provide hardware combinators to provide automated support for a wide variety of iterated hardware structures.

The most similar work to the work in smart tactics is that done using the Faust and Mephisto packages [4, 7]. They attempt to automate the process of hardware verification by simplifying proofs into subgoals that can be automatically verified using a tool outside the theorem proving environment. Our approach is to use tactics that apply many simpler tactics in order to find a proof of the goal within the theorem proving environment. In addition, they do not attempt to automate aspects of inductive proofs. Hanna [3] has done similar work in hardware combinators. Our work provides iteration schemes and automated support for a wider variety of hardware structures and specifications.

This work was done in both the HOL and Nuprl proof development systems. These systems have a great deal in common. Both are tactic-oriented proof systems using a higher-order logic and are descendents of the LCF project at Edinburgh.Both systems provide large collections of tactics to support goal directed proof and proof management systems that keeps track of the state of an interactive proof session. A metalanguage, ML, for programming and extending the theorem prover allows the user to extend either system. For our purposes, one of the greatest differences between HOL and Nuprl is the type systems. HOL, developed at the University of Cambridge [2], is based on Church's theory of simple types. Nuprl, developed at Cornell University [5], is based on a sequent version of Martin-Löf's constructive type theory. Nuprl uses a rich and expressive type theory with dependent types.

In this paper we discuss the implementation of powerful tactics and hardware combinators in both the HOL and Nuprl systems. While the details of the implementations vary, the same tactics were implemented in both systems. Making use of the tactics and functions provided is similar for a user of either system. This paper is written primarily from a Nuprl perspective. A description of the HOL version of the tacticsand the code that implements the tactics for the HOL system are available.

2 Towards More Intelligent Tactics

Our approach is to provide the user with a small set of tactics that take advantage of the particular structure shared by most combinational and sequential circuits and their implementations. Hardware proofs should largely consist of the repeated application of these tactics. Most of the tactics by themselves are very general, but we have combined them in ways which are tailored for doing hardware verification. These smart tactics are:

RewriteWith convns	Rewrites sequent using the rewrite rules named in convns .
BitCaseSplit	Finds an appropriate term of type bit to do case analysis on, then does the case analysis and calls **RewriteWith** bit_convns to simplify the resulting goals.
RewriteWithHyps	Tries using each of the hypotheses to rewrite the conclusion and other hypotheses.
SpecializeBif	Finds a hypothesis of the form ∀...t1 = if b then e1 else e2 then looks for unambiguous instantiations for the universally quantified variables. If a set is found, then the instantiations are performed, otherwise the tactic fails.
SuperHWTac	A tactic that sets up hardware proofs.

These tactics are discussed in more detail below. In the next section we give examples of how these tactics are used in hardware proofs. Our eventual goal is an even more general tactic, which we have dubbed **SuperDuperHWTac**, comprised of these tactics and others that can be used to make substantial progress in hardware proofs, hopefully completely verifying many simple circuits.

2.1 Standard Tactics and Tacticals

Nuprl has an **Autotactic** which is run after most proof steps. In our smart tactics we use **Autotactic** repeatedly. **Autotactic** automatically solves simple goals, including most membership goals, goals that exactly match a hypothesis, simple contradictions, and simple arithmetic equalities and inequalities. **Autotactic** breaks up complex goals, such as goals of the form $A \implies B$ and $\forall x.P$. It also simplifies hypotheses. For example, the hypothesis $A \land B$ is rewritten as two separate hypotheses. It takes hypotheses of the form $\exists x.A$ and finds a witness for the existential quantifier. **Autotactic** also performs β reduction on hypotheses and the conclusion. We have implemented a tactic in HOL to give essentially the same behavior as **Autotactic** in Nuprl.

With Nuprl and HOL we make extensive use of rewriting. The Nuprl tactic we use most often for rewriting is RewriteWith; it is very similar to REWRITE_TAC in HOL. REWRITE_TAC and RewriteWith take a list of rules and repeatedly apply them until no more progress is made. The most significant difference between the two is that RewriteWith applies the rewrite rules to the conclusion and hypotheses and REWRITE_TAC applies them only to the conclusion.

Our smart tactics are built up from simpler tactics using tacticals. Both HOL and Nuprl have ORELSE, THEN, THENL, Repeat, Try and Progress (named CHANGED_TAC in HOL) tacticals which exhibit essentially the same behavior. In addition, Nuprl provides ORTHEN and Complete which can easily be implemented in HOL. t1 ORTHEN t2 runs tactic t1 then runs tactic t2. Both tactics are always tried whether or not the first one fails. Complete tac succeeds if tac can completely prove the goal it is applied to. Otherwise it fails, and leaves the current goal unaltered.

2.2 Chaining and Equality Reasoning

Much of the work in the hardware proofs requires reasoning with equalities and chaining through implications. We use the tactic RewriteWithHyps to perform these tasks. This tactic is currently implemented using rewriting tactics available in HOL and Nuprl. We commonly rewrite using the following types of formulas:

$\forall \ldots A = B$ — Replace occurrences of A with B

$\forall \ldots P \implies Q$ — If P can be proved trivially then add Q to the hypothesis list.

$\forall \ldots P \implies A = B$ — If P can be proved trivially then replace occurrences of A with B

We have many ideas for using tactics other than RewriteWithHyps to perform similar functions. We are considering using forward chaining, backward chaining and congruence closure rather than relying solely on rewriting. Forward chaining takes a hypothesis of the form $A \implies B$ and tries to automatically prove A. If we can prove A, then B is added to the hypothesis list. Backward chaining takes hypotheses of the form $A \implies B$ and tries to match B against the goal. If the match succeeds, then A is the new goal. In general, this matching process may be complex. We also intend to investigate rewriting expressions to normal forms.

2.3 Reasoning with Bits

Reasoning about bits arises frequently in hardware verification. Our type bit is isomorphic to the Booleans with two values H and L. We provide two primary techniques to deal with reasoning about bits: simplification and induction (case analysis).

We perform bit simplification by repeatedly applying a set of rewrite rules. RewriteWith takes a list of rewrite rules to apply and repeatedly applies these rules to the hypotheses and goal. We have developed a collection of lemmas about bits which we use with RewriteWith. The rules we are currently using include DeMorgan's laws, simplifying bit expressions with constants, and eliminating double negations. This set of rules is guaranteed to terminate. These bit rewrite rules are our first attempt at automated rewriting. This is a well studied area, with known pitfalls, including termination and efficiency.

BitCaseSplitOn is Nuprl's version of case analysis. First, it takes a term t of type bit and brings all hypotheses that contain t into the conclusion as antecedents of an implication. Then it generates two subgoals, one for the goal with all occurrences of t replaced by H, and the other with t replaced by L. This always generates subgoals of bit expressions with constants in them. BitCaseSplitOn is usually followed by RewriteWith bit_convns, which is used to simplify the subgoals.

Picking the term to do induction on is a very difficult problem to solve in general, but we have discovered an effective heuristic for picking terms of type bit to do case analysis on. We do automated case analysis only on terms (b) which occur in if b then t else e. BitCaseSplit looks for terms in this form. If such a term is found, then it performs case analysis on b and then does bit simplification using the bit rewrite rules.

2.4 Specializing Hypotheses

We have defined a tactic that is general purpose, but is especially useful for sequential hardware verification. Hardware modules are almost always specified in the form ∀x. spec(x). We frequently want to eliminate the universal quantifier and replace it with an appropriate instantiation. In general this is a very hard problem, because we cannot always determine a unique instantiation for the variable. We have developed a heuristic which is quick and has worked quite well so far.

Our first step in specializing hypotheses is RewriteWithHyps, which takes care of hypotheses when they are of one the forms listed in Sect 2.2. There will be some times when the hypothesis is not in one of these forms. In hardware, if the specification is not of a form which is amenable for use as a rewrite rule, then it is quite often of the form ∀...out = if b then t else e. In this case, rather than try to match the entire body of the hypothesis with terms in the sequent, we try matching each of **out**, **b**, **t** and **•** in the sequent. If all of these matchings produce the same unique instantiations for the universally quantified variables in the hypothesis, then we perform the specialization. We are working on extending this technique to handle multi-bit case statements as well as other forms of hypotheses that occur frequently in hardware verification.

2.5 Putting It All Together

SuperHWTac is the tactic we use to start our hardware proofs. It is implemented as shown in Def 1. We use Autotactic to set up the proof and then replace instantiations of submodules with the specifications of the submodules (*e.g.* **and(a,b)(c)** is replaced with **c = a ∧ b**). This is done using the tactic **UnfoldToSpecs**. We finish up by calling **RewriteWithHyps**. In addition to **SuperHWTac**, we have begun to experiment with **SuperDuperHWTac** (Def 2) which incorporates all of the techniques that we have discussed in this section.

Def 1 SuperHWTac − initial tactic for hardware proofs
```
let SuperHWTac submodules =
        Autotactic
          THEN UnfoldToSpecs submodules
          THEN RewriteWithHyps
```

Def 2 SuperDuperHWTac − experimental hardware tactic
```
let SuperDuperHWTac submodules =
        Autotactic
          ORTHEN UnfoldToSpecs submodules
          ORTHEN (Repeat (RewriteWithHyps
                          ORTHEN SpecializeBif
                          ORTHEN BitCaseSplit
                          ORTHEN RewriteWith bit_convns
                  )
          )
```

---

3 Verifying Hardware with Smart Tactics

In this section we illustrate the use of our smart tactics with the proof of several hardware components including a selector and an n-bit counter. All proofs are done at the gate level. We use parameterized modules [1] and the standard HOL methodology of describing circuits as relations over inputs and outputs. The top level goal to be proved for each component is of the form:

Vdesign_vars . Vsubmodules .
  ckt_pmod(design_vars, submodules) ∈ ckt_type(design_vars)

Where **ckt_pmod** is the parameterized module implementation of the circuit and **ckt_type** is the type of circuits which meet the specification. In Nuprl the parameterized module design style relies on quantifying over types and dependent types. The HOL implementation achieves similar results using the restricted quantifier library.

3.1 A Bit Selector

This example was chosen to illustrate the use of rewriting with bit expressions. The top level description is **out = if c then a else b**, where a, b and c all have type **bit**. The implementation shown in Fig 1 consists of three **nand** gates and one inverter. The specification for the circuit is shown in Def 3. The only version of this proof was done with smart tactics. The entire proof is: **SuperHWTac ['inv';'nand'] Then BitCaseSplit** BitCaseSplit does case analysis on **a**, then repeatedly applies bit rewrite rules to simplify the goal.

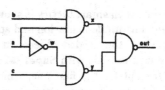

Fig. 1. Schematic of bit selector

Def 3 Bit select specification
   bit_select_spec(a,b,c,out) =
     out = if a then b else c

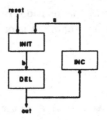

Fig. 2. Implementation of counter

Def 4 Counter specification
```
counter_spec(n)(reset,out) =
    ∀t:ℕ.
        vec2num(2,n)(out(t+1)) =
            if reset(t) then 0 else (vec2num(2,n)(out(t))+1) mod 2^(n+1)
```

3.2 An *n*-bit Counter

The n-bit counter is a more interesting example since it computes an arithmetic function, and uses bit vectors and sequential hardware.

The proof of the counter was originally done before the smart tactics were written. A picture of the proof tree is shown in Fig 3. The proof required fifteen steps. Not only are there a lot of steps where the user must interact with the proof system, but each step requires a fair amount of detail. For example, the user needs to explicitly tell the prover which hardware component to unfold, which hypothesis to rewrite the conclusion with, what rewrite rule to apply, and so on at each of these steps.

The proof of the counter done with smart tactics is shown in Fig 4. The user entered the parameterized module implementation and specification. Utility functions then created and verified the needed objects in the library and started the proof of the parameterized module with **SuperHWTac**. The user entered the lines that begin with **BY** (except for the first one, which is done automatically by the utility functions). The entire proof consists of four tactics applied by the user. The first step done by the user is to specialize the universal quantifier for the time variable (**t**) in hypothesis one. Next she does case analysis on **reset(t)**. **RewriteWithHyps** gets the case where **reset(t)** is false. In the case where **reset(t)** is true, the user must apply **zvec_lemma**. Not only have we reduced the number of rules the user needs to apply, but the level of the rules is more natural and in most cases requires some intelligence.

The implementation of **SuperDuperHWTac** shown in Def 2 is capable of completely proving the counter circuit, except for the application of **zvec_lemma**.

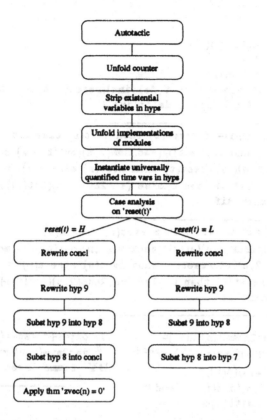

Fig. 3. Proof tree for counter

Also, by using only the tactics used to implement `SuperDuperHWTac` we are able to verify the parity checking circuit in [7]. `SuperDuperHWTac` is not yet capable of doing the complete the proof of the parity circuit by itself. We are continuing to improve `SuperDuperHWTac`. By experimenting with a wider variety of example circuits we believe that we will be able to find a tactic which comes very close to solving many simple hardware circuits.

4 Induction and Hardware Combinators

In doing hardware proofs, we have discovered that the user follows the same reasoning repeatedly. The aim of the smart tactics is to capture this reasoning. The tactics we have developed up to this point work well for combinational circuits and sequential circuits whose proofs do not depend on induction; *i.e.* their structures are not defined in terms of iterators.

$\vdash \forall n:\mathbb{N}.$
    $\forall init:init\_type(n).$
    $\forall del:del\_type(n).$
    $\forall inc:inc\_type(n).$
        $counter\_pmod(n,init,del,inc)(reset,out) \in counter\_type(n)$
BY SuperHWTac ['init';'del';'inc']

---

1. $\forall t:\mathbb{N}.\ out(t+1)=if\ reset(t)\ then\ zvec(n)\ else\ a(t)$
2. $\forall t:\mathbb{N}.\ vec2num(2,n)(a(t))=(vec2num(2,n)(out(t))+1)\ mod\ 2^{n+1}$
$\vdash vec2num(2,n)(if\ reset(t)\ then\ zvec(n)\ else\ a(t)) =$
    $if\ reset(t)\ then\ 0\ else\ (vec2num(2,n)(out(t))+1)\ mod\ 2^{n+1}$
BY SpecializeBif

---

1. $out(t+1)=if\ reset(t)\ then\ zvec(n)\ else\ a(t)$
2. $\forall t:\mathbb{N}.\ vec2num(2,n)(a(t))=(vec2num(2,n)(out(t))+1)\ mod\ 2^{n+1}$
$\vdash vec2num(2,n)(if\ reset(t)\ then\ zvec(n)\ else\ a(t)) =$
    $if\ reset(t)\ then\ 0\ else\ (vec2num(2,n)(out(t))+1)\ mod\ 2^{n+1}$
BY BitCaseSplit

---

1. $\forall t:\mathbb{N}.\ vec2num(2,n)(a(t))=$
  $(vec2num(2,n)(out(t))+1)mod2^{n+1}$
$\vdash vec2num(2,n)(a(t))=$
  $(vec2num(2,n)(out(t))+1)mod2^{n+1}$
BY RewriteWithHyps

---

1. $out(t+1)=zvec(n)$
$\vdash vec2num(2,n)(zvec(n)=0$
BY Lemma 'zvec_lemma'

Fig. 4. Proof of counter using SuperHWTac

We have found that proofs of iterative hardware structures also contain a lot of repetitive reasoning. As with the smart tactics, we would like to capture this reasoning once and for all, and leave users with only those parts of the proof that are unique to the specific structure they are proving. We do this by using *hardware combinators* to capture frequently used iteration schemes. These schemes include one and two dimensional arrays, arrays with a rippled signal, rectangles, triangles, and trees. For each of these structures we provide induction schemes which have been formally derived for each structure. The user can prove properties about an iterated hardware module by making use of the induction schemes provided.

Many circuits are built from regular structures, such as arrays and trees. Figs 5 and 6 show two useful hardware combinators for building arrays of components. The first combinator, array, replicates a module M and connects one copy to each of the signals in two buses. The combinator **arrayRipple** can be used to build circuits using a ripple-carry structure. Each of these structure combinator captures a particular iteration scheme, independent of the internal structure of **cell** and of types of the buses. For example, we can use the function **arrayRipple** and a single bit full adder cell to build an n-bit ripple carry adder.

Fig. 5. array

$$n:\mathbb{N}+ \rightarrow$$
$$((\alpha \times \beta) \rightarrow prop) \rightarrow$$
$$(((n, \alpha)bus \times (n, \beta)bus) \rightarrow prop)$$

Fig. 6. arrayRipple

$$n:\mathbb{N}+ \rightarrow$$
$$((\alpha \times \beta \times \gamma \times \alpha) \rightarrow prop) \rightarrow$$
$$((\alpha \times (n, \beta)bus \times (n, \gamma)bus \times \alpha) \rightarrow prop)$$

Hardware Combinators

We make use of the hardware combinators to organize the process of verifying iterated hardware structures. We define the ripple carry structure recursively, and verify a ripple carry module with induction. The use of the induction lemma structures a proof by breaking down a sequent into two subgoals, one for the base case and one for the inductive case. The induction lemma also captures all of the reasoning required to show that the induction scheme is correct for the hardware combinator. So, for the ripple carry adder, applying the induction theorem would leave the user with two goals, one to show that a single full adder implements a one bit ripple carry adder and one to show that if an n-bit ripple carry adder is correct, then connecting a full adder to the ripple carry adder correctly implements an n+1 bit ripple carry adder.

We prove an induction lemma for each hardware combinator. This induction lemma is proved once and used every time the hardware combinator is used to generate hardware structures. This enables us to capture the reasoning used in regular hardware structures once, and reuse this reasoning many times. We provide more information on hardware combinators elsewhere [6].

Many times, iterated circuits are used to perform arithmetic functions where each cell in the array performs the same operation on a specific digit of the input vector. For a ripple carry array, Hanna [3] has defined the notions of *factorizable* and *proper*. If a specification can be shown to be factorizable and proper, then it can be implemented by iterating a single cell in a ripple carry fashion (called fold in Hanna's work). Applying the factorization theorem in a proof reduces a goal from showing that a ripple carry array of cells satisfies an arithmetic specification to three simpler subgoals. The first two goals are to show that the relation is factorizable and proper. The third goal is to show that a single cell implements the base case of the specification. For the ripple carry adder this means proving that a full adder implements a single bit slice of base two addition.

While our induction lemma works for all properties of a cell iterated using arrayRipple, the factorization theorem only works for a specific class of arithmetic relations. Because it is more restrictive, the factorization theorem is able to encapsulate more reasoning about arrayed circuits. The relevance of the factorization theorem to our work is that it is another example of the power that can be gained by making a few restrictions on the style of hardware design and verification.

5   Discussion

Several issues have arisen in the process of implementing tactics and combinators, including issues in automatic rewriting, documentation provided by the proof system, and what happens when our tactics fail to produce the expected results.

One approach we use is to rewrite with lists of lemmas automatically. Currently, we only use this approach with the bit lemmas but we can envision other applications of the same technique. Our main concerns with automatic rewriting are efficiency and termination. Both HOL and Nuprl use discrimination nets to efficiently match a rule against an expression. We plan to extend the use of discrimination nets to efficiently match sets of rules against sets of expressions. In addition to providing an efficient rewriting package, we must be able to guarantee that it will always terminate. Currently, we guarantee termination through inspection of the rules used, so there is nothing to stop a user from adding rules that will no longer terminate. One way to guarantee termination is to always rewrite toward a normal form. This, however, may cause the expressions to become quite large.

The best approach to automated rewriting within limited domains (such as bits or integer expressions) may be to use decision procedures. Faust [4] is a decision procedure for the restricted sequent calculus in HOL. There are decision procedures for Presburger formulas in both HOL and Nuprl. Nuprl also has a decision procedure for simple arithmetic reasoning.

In our ideal system SuperDuperHWTac proves all hardware proofs completely, or leaves the user with only a few subgoals to prove. One disadvantage of this is that the user no longer gains any deeper understanding of the circuit being verified as a result of doing the proof. Instead the user must place more trust in the correctness of the proof system. In Nuprl we can run a tactic like SuperDuperHWTac and get back a proof tree which explicitly shows all the primitive rules run to prove the goal. Such proofs contain so much detail that users very rarely wade through them. The end result is only marginally better than providing no information at all.

We envision a system where the user gets back a proof at the level she would have written for a hand proof. The proof should show the steps which made progress toward solving the goal and the rewrites that succeeded, preferably by formula rather than by name. We plan to implement such a documentation mode in the future. This mode would provide documentation of a proof that is distinct from the interaction the user has with the proof system at the time of running the proof.

We have some ideas for implementing this. We could mark each tactic with a flag which indicates whether or not that tactic is to be provided as part of the documentation. All of the smart tactics described in this paper would not be part of the documentation. All of the rewrite rules and lemmas would be. We believe that at the time of tactic writing we can tell (in most cases) what level the user wants to see. For example, with the tactic RewriteWith bit_convns, the user wants to see the formulae of the rewrites that succeeded. In addition, we would annotate tactics with the message reported to the user. For BitCaseSplit, the user wants to see "By case analysis on term b", etc.

So far we have assumed that SuperDuperHWTac does exactly what we want; that is, it always makes meaningful progress and only leaves the user with a few subgoals to prove interactively. The tactic will at least simplify the initial goal, so it cannot completely "fail". There are two possible behaviors the user may need to deal with. These roughly correspond to the case where the tactic does not behave as expected, and the case where the goal the user is trying to prove is incorrect. The first case occurs when SuperDuperHWTac does not make as much progress as the user would have expected. In the worst case, the tactic gets no further than Autotactic would have done. The other extreme is that SuperDuperHWTac generates subgoals that the user either cannot prove, or does not understand why they were produced. In either case, the user needs sufficient documentation to see what SuperDuperHWTac actually accomplished. In the second case, seeing the rules that succeeded will inform the user how the proof got to the current state. In the first case, the user may also wish to see which tactics failed in order to get a handle on where things are going wrong. More experience with SuperDuperHWTac will help us give suitable information to the user in both cases. We may also wish to provide a debugging mode where users can single step through the proof at a suitable level of detail.

We have discussed ways of capturing much of the tedious reasoning in hardware proofs. We do this through two techniques. The first is the use of smart tactics to capture low level reasoning that frequently occurs in combinational and sequential hardware verification. The second is to use hardware combinators to capture repetitive reasoning found in iterative hardware structures. We expect that the combination of these two techniques as well as the use of parameterized modules [1] will provide a powerful methodology for hardware verification. The work described so far is in its early stages. We plan to use our approach on a wide variety of hardware modules. In particular we intend to use this approach to develop a library of circuits for floating point arithmetic. As this research develops, we expect some of the details described here to change. The general approach of giving the user more powerful tools to more easily do hardware verification will remain.

Acknowledgements

Mark Aagaard is supported by a fellowship from Digital Equipment Corporation. Miriam Leeser is supported in part by an NSF Young Investigator Award CCR-9257280. Phil Windley is supported in part by the National Science Foundation under Research Initiation Grant MIP-9109618. Windley did most of his work while he was at the University of Idaho.

References

1. M. D. Aagaard and M. E. Leeser. A methodology for reusable hardware proofs. In L. Claesen and M. Gordon, editors, *HOL Theorem Proving System and its Applications*. North-Holland, September 1992.
2. A. Camillieri, M. Gordon, and T. Melham. Hardware verification using higher-order logic. In D. Borrione, editor, *From HDL Descriptions to Guaranteed Correct Circuit Designs*. North Holland, September 1986.
3. F. K. Hanna, N. Daeche, and M. Longley. Specification and verification using dependent types. *IEEE Trans. on Soft. Eng.*, 16(9):949–964, September 1990.
4. R. Kumar, T. Kropf, and K. Schneider. Integrating a first-order automatic prover in the HOL environment. In M. Archer and M. Gordon, editors, *The HOL Theorem Proving System and its Applications*, pages 170–176. IEEE Press, 1991.
5. M. E. Leeser. Using Nuprl for the verification and synthesis of hardware. In C. A. R. Hoare and M. J. C. Gordon, editors, *Mechanized Reasoning and Hardware Design*. Prentice-Hall International Series on Computer Science, 1992.
6. J. O'Leary et al. HML: a hardware description language based on SML. In D. Agnew and L. Claesen, editors, *CHDL*, Apr. 1993.
7. K. Schneider, R. Kumar, and T. Kropf. Structuring hardware proofs: First steps toward automation a higher order environment. In *VLSI*, 1991.

A Mechanisation of Name-carrying Syntax up to Alpha-conversion

Andrew D. Gordon

Programming Methodology Group
Department of Computing Science
Chalmers University of Technology /
University of Gothenburg
412 96 Gothenburg, Sweden
gordon@cs.chalmers.se

Abstract. We present a new strategy for representing syntax in a mechanised logic. We define an underlying type of de Bruijn terms, define an operation of named lambda-abstraction, and hence inductively define a set of conventional name-carrying terms. The result is a mechanisation of the practice of most authors studying formal calculi: to work with conventional name-carrying notation and substitution, but to identify terms up to alpha-conversion. This strategy falls between most previous works, which either treat bound variable names literally or dispense with them altogether. The theory has been implemented in the Cambridge HOL system and used in an experimental application.

There is great interest in using theorem provers to prove properties of lambda-calculi and programming languages [3, 7, 9, 14, 15, 18, 19, 21, 25, 26, 27, 28] One basic issue is the representation in logic of syntax, and in particular, how to represent bound variables. One strategy is to represent bound variables using variable names, as in conventional syntax. Another is to use de Bruijn indices [8]. An advantage of the first is that theorems are expressed in a conventional, human-readable form. An advantage of the second is that substitution is easy to define and the problems of variable name clashes never arise. But neither exactly corresponds to the practice in most written studies[1] of formal calculi where it is convenient to identify syntax up to alpha-conversion for doing proofs, but at the same time retain bound variable names for readability.

This paper presents a novel method for representing syntax, combining these two strategies. The key idea is that conventional lambda-abstraction can be defined on de Bruijn terms, and then one can prove that the conventional inductive definition of name-carrying notation generates exactly the proper de Bruijn terms (that is, terms with no unmatched indices). From this key theorem we can obtain a conventional theory of syntax and substitution [13], in which logical equality corresponds to alpha-conversion. The advantage of this mixed strategy is that theorems can be expressed in a conventional form, without de

---

[1] Church and Rosser [6] and Barendregt [4] are prominent examples.

Bruijn encoding, and although in application proofs renaming of bound variables is sometimes necessary, it is easy to support because logical equality is up to alpha-conversion.

The particular motivation for this paper is to work towards an environment for mechanised proofs about functional programs based on their operational semantics. As a step towards such an environment, we have programmed all the theory given here in the Cambridge HOL system [10], and have used it to support elementary reasoning about Abramsky's lazy lambda-calculus [2]. The point of the paper is not to discuss applications, but simply to explain the syntactic strategy, which may be of use in other theorem provers. Section 1 reviews conventional and de Bruijn notation, and Section 2 gives a type of de Bruijn terms in HOL. Section 3 is the crux of the paper: the inductive definition of name-carrying terms. Section 4 outlines the main components of the derived theory, and Section 5 outlines how it might be applied to represent an example syntax. Section 6 concludes with a discussion of related and future work.

Cambridge HOL supports Gordon's polymorphic formulation of Church's classical higher order logic. Theorems proved in the HOL system are indicated here with a turnstyle $\vdash$ and written in HOL's fairly conventional notation [10]. We use Melham's encoding of sets as logical predicates [16].

1 Review of Name-carrying Versus de Bruijn Notation

As usual, we consider each term of the conventional notation for lambda-calculus—referred to as *name-carrying* by de Bruijn—to be either a constant, a, a variable, x, an application $t_1 t_2$ or a lambda-abstraction $\lambda x. t_0$, where each t_i is a term.

De Bruijn's key idea was to erase the name x from a lambda-abstraction $\lambda x. t$, and replace each free occurrence of x in t by a numeric index, equal to the level of the occurrence. The *level* of an occurrence of a subterm u within a term t, is the number of λ's in t that enclose u. So name-carrying $\lambda x. (\lambda y. x)x$ becomes de Bruijn $\lambda. (\lambda. 1)0$. An occurrence of an index i in a term is *matched* iff i is less than the level of the occurrence. In $\lambda. (01)$ each of the indices occurs at level 1, so index 0 is matched but 1 is not.

A matched occurrence of an index represents a bound variable. How is a free variable represented? For an entirely nameless notation, de Bruijn suggested that the names of free variables could be replaced by unmatched indices interpreted via a fixed enumeration of variable names. Instead, we adopt another convention suggested by de Bruijn, in which "free variables have names but the bound variables are nameless" [8, p392]. We will refer to this simply as de Bruijn notation. For instance, name-carrying $x(\lambda y. axy)$ becomes de Bruijn $x(\lambda. ax0)$.

For each occurrence of an index i at level ℓ define its *degree* to be 0 if it is matched, otherwise the positive number $i - \ell + 1$. The *degree* of a de Bruijn term is the maximum degree of all index occurrences in the term. So $\lambda. (01)$ has degree 1 and $x(\lambda. ax0)$ degree 0. A de Bruijn term is *proper* iff its degree is 0, that is, when no unmatched indices occur in it. The strategy proposed here rests essentially on the fact that the name-carrying terms can be modelled up to alpha-conversion by the proper de Bruijn terms.

2   A HOL Type of de Bruijn Terms

This section shows how de Bruijn terms and their basic operations can be expressed in the HOL logic. The following grammar specifies a HOL type of de Bruijn terms.

$$
\begin{aligned}
db = \ & \text{dCON} * && \text{(constant)} \\
| \ & \text{dVAR string} && \text{(free variable)} \\
| \ & \text{dBOUND num} && \text{(bound variable)} \\
| \ & \text{dABS db} && \text{(abstraction)} \\
| \ & \text{dAPP db db} && \text{(application)}
\end{aligned}
$$

In HOL, Melham's recursive type package [19] proves automatically the existence of a type which is the free algebra given by the grammar, and proves theorems supporting primitive recursive definitions and proofs by structural induction.

The type obtained, * db, is polymorphic. Constants are represented by values in the polymorphic type *. This polymorphic type would be instantiated to a type capable of representing all the symbols used in a particular syntax. Variable names are represented by strings. All we need to know about the type of names is that it is infinite—so that fresh names can always be generated—and this is easy to prove about strings. If desired, one could parameterise db on a type of variable names, as Melham does [18], but then theorems would need to bear the assumption that the type of names was infinite.

We introduce new HOL constants, dDEG and dFV of types * db -> num and * db -> string set, to express respectively the degree and set of free variables of a de Bruijn term.

```
|- (dDEG (dCON c)   = 0) /\
   (dDEG (dVAR x)   = 0) /\
   (dDEG (dBOUND n) = SUC n) /\
   (dDEG (dABS t)   = PRE (dDEG t)) /\
   (dDEG (dAPP t u) = MAX (dDEG t) (dDEG u))

|- (dFV (dCON c)   = {}) /\
   (dFV (dVAR x)   = {x}) /\
   (dFV (dBOUND n) = {}) /\
   (dFV (dABS t)   = dFV t) /\
   (dFV (dAPP t u) = (dFV t) UNION (dFV u))
```

Its not hard to see that that the structural definition of dDEG is equivalent to the one given in Section 2.

To support variable abstraction and substitution we adopt two of de Bruijn's operations in the form given by Paulson [23].[2]

```
|- (Abst i x (dCON c)   = dCON c) /\
   (Abst i x (dVAR y)   = ((x = y) => dBOUND i | dVAR y)) /\
   (Abst i x (dBOUND j) = dBOUND j) /\
   (Abst i x (dABS t)   = dABS (Abst (SUC i) x t)) /\
   (Abst i x (dAPP t u) = dAPP (Abst i x t) (Abst i x u))
```

```
|- (Inst i (dCON c)     u = dCON c) /\
   (Inst i (dVAR x)     u = dVAR x) /\
   (Inst i (dBOUND j)   u = ((i = j) => u | dBOUND j)) /\
   (Inst i (dABS t)     u = dABS (Inst (SUC i) t u)) /\
   (Inst i (dAPP t1 t2) u = dAPP (Inst i t1 u) (Inst i t2 u))
```

The intention is that Abst i x t is the term obtained by replacing each free occurrence of variable x in t with index i (or a greater index within abstractions). Inst i t u is meant to be the term obtained by replacing each bound variable i in t with u (which is assumed to be proper).

3 An Inductive Definition of Name-carrying Terms

This section explains our strategy for mechanising a name-carrying notation. Although the underlying type is one of de Bruijn terms, we wish to recover the conventional operation of free variable abstraction. We can define a new operation dLAMBDA of type string -> * db -> * db from the operations dABS and Abst.

```
|- dLAMBDA x t = dABS (Abst 0 x t)
```

Intuitively dLAMBDA binds free occurrences of variable x in t. With dLAMBDA we have a name-carrying notation that denotes de Bruijn terms. For instance, we can prove the following by rewriting in HOL, that name-carrying $\lambda x.(\lambda y.x)x$ equals de Bruijn $\lambda.(\lambda.1)0$.

```
|- dLAMBDA 'x' (dAPP (dLAMBDA 'y' (dVAR 'x')) (dVAR 'x')) =
   dABS(dAPP(dABS(dBOUND 1))(dBOUND 0))
```

Since the underlying type erases bound variable names, equality on the derived name-carrying notation is up to alpha-conversion.

```
|- dLAMBDA 'x' (dAPP(dLAMBDA 'y' (dVAR 'x'))(dVAR 'x')) =
   dLAMBDA 'y' (dAPP(dLAMBDA 'x' (dVAR 'y'))(dVAR 'y'))
```

---

[2] Abst and Inst correspond to Paulson's abstract and subst. Inst is a slight simplification of subst because we are only interested in instantiating proper terms for bound variables.

We have operations now to express all conventional notation in de Bruijn form, and so we are free to adopt the conventional inductive definition of name-carrying terms.

$$\frac{\qquad}{\text{META (dCON } c)} \qquad \frac{\qquad}{\text{META (dVAR } x)}$$

$$\frac{\text{META } t}{\text{META (dLAMBDA } x \; t)} \qquad \frac{\text{META } t \qquad \text{META } u}{\text{META (dAPP } t \; u)}$$

In HOL, Melham's tool for inductive definitions [17] proves automatically that **META** is the least predicate closed under the given rules. The name of the predicate comes from its intended use as a single logical type able to encode the syntax of a range of object languages. This method is sometimes known as a *meta-theory of syntax* and was promoted by Martin-Löf and others [22], and implemented in systems such as Elf [25] and Isabelle [24]. The point is to define substitution and alpha-conversion once and for all in the meta-theory.

The key result of the implemented theory is that the terms satisfying this conventional inductive definition are precisely the proper de Bruijn terms.

```
|- !t. META t = (dDEG t = 0)
```

The forwards direction is a rule induction on the derivation of the predicate **META**. The backwards direction is a course-of-values induction on the length of the term t (a measure discussed in the next section).

4 A HOL Type of Name-carrying Terms up to Alpha-conversion

The next step is to define a subtype of * db, called * **meta**, whose elements are precisely the proper de Bruijn terms, or equivalently, conventional name-carrying lambda-terms up to alpha-conversion.

Types in HOL can be modelled as non-empty sets. A new type can be introduced in bijection with a given non-empty subset of an existing type. This is sound in the sense that it preserves the property of the logic having a standard model [10]. In the present case, we define the new type * **meta**, whose elements are precisely the set of proper de Bruijn terms in the existing type * db. It is easy to prove that the set of proper de Bruijn terms is not empty. Tools in the HOL system prove automatically some basic properties of the bijection between * **meta** and the elements of type * db satisfying predicate **META**. We omit the details. Operations on type * db that are closed under the predicate **META** can be used to induce equivalent operations on type * **meta**.

The basic proposition of the paper is that applications needing a representation of name-carrying syntax can be built using this new type. In the remainder of this section we outline the basic theory implemented for this type.

Basic Properties of meta-terms

We refer to values of the new type as meta-terms. There are four basic constructors of meta-terms, corresponding to the four rules of the inductive definition of META, that is, constants, variables, applications and lambda-abstractions.

```
mCON    : * -> * meta
mVAR    : string -> * meta
'       : * meta -> * meta -> * meta
mLAMBDA : string -> * meta -> * meta
```

The binary application constructor ' is infix. These four constructors exhaust the meta-terms.

```
|- !m.
    (?c. m = mCON c) \/
    (?x. m = mVAR x) \/
    (?x n. m = mLAMBDA x n) \/
    (?m1 m2. m = m1 ' m2)
```

There are also theorems stating that each of the constructors yields distinct terms, and that the three constructors mCON, mVAR and ' are one-one. The exhaustion theorem follows from the inductive characterisation of meta-terms, META, whereas the others follow from the definition of de Bruijn terms as a free algebra.

To obtain substitution on meta-terms, we first define substitution on de Bruijn terms.

```
|- !t u x. t dSUB (u,x) = Inst 0 (Abst 0 x t) u
```

De Bruijn term t dSUB (u,x) is intended to be term t with each free occurrence of x replaced by u. This operation on the representation type is used to induce a substitution operation on meta-terms that we denote with infix /. The meta-term (m / x) n is meant to be meta-term n with each free occurrence of x replaced by meta-term m.

Given that FV of type * meta -> string set is the operation induced by the free variable function dFV, we can prove alpha-conversion as a logical equation.

```
|- !m x y. ~y IN (FV m) ==>
    (mLAMBDA x m = mLAMBDA y (((mVAR y) / x) m))
```

We can prove the expected distributive laws for substitution,

```
|- !a m x. (m / x)(mCON a) = mCON a
|- !m x. (m / x)(mVAR x) = m
|- !m x y. ~(x = y) ==> ((m / x)(mVAR y) = mVAR y)
|- !m n x. (n / x)(mLAMBDA x m) = mLAMBDA x m
|- !m n x y.
    ~(x = y) /\ ~y IN (FV n) ==>
    ((n / x)(mLAMBDA y m) = mLAMBDA y ((n / x) m))
|- !m n p x. (p / x)(m ' n) = ((p / x) m) ' ((p / x) n)
```

by first proving the equivalent properties about de Bruijn terms.

Structural Induction

The inductive definition of **meta**-terms gives rise to the following induction principle:

```
|- !P: (*)meta -> bool.
   (!x. P(mCON x)) /\
   (!x. P(mVAR x)) /\
   (!m. P m ==> (!x. P(mLAMBDA x m))) /\
   (!m n. P m /\ P n ==> P(m ' n))
   ==>
   (!m. P m)
```

One difficulty found in applications with this induction principle is that the hypothesis for proving P(mLAMBDA x m) is too weak; it allows P m to be assumed, but if the bound variable has to be renamed, one needs to assume P ((mVAR z / x) m) for some fresh variable z.

One solution to this problem rests on the fact that variable renaming does not alter the *length* of a term, the number of syntactic constructors it contains. This is a standard notion [13], that can be induced on **meta**-terms via a primitive recursion on the type of de Bruijn terms.

```
|- (LGH (mCON c)       = 1) /\
   (LGH (mVAR x)       = 1) /\
   (LGH (mLAMBDA x m) = 1 + (LGH m)) /\
   (LGH (m ' n)        = (LGH m) + (LGH n))
```

This measure is invariant under variable renaming [13, Lemma 1.14(d)].

```
|- !m x y. LGH (((mVAR x) / y) m) = LGH m
```

Now we can prove a new induction principle, which strengthens the induction hypothesis for **mLAMBDA**-terms to hold for all **meta**-terms n of the same length as **m**.

```
|- !P: (*)meta -> bool.
   (!x. P(mCON x)) /\
   (!x. P(mVAR x)) /\
   (!m.
     (!n. (LGH n = LGH m) ==> P n)
    ==>
     (!x. P(mLAMBDA x m))) /\
   (!m n. P m /\ P n ==> P(m ' n))
   ==>
   (!m. P m)
```

The proof of this principle is by course-of-values induction on the length of the **meta**-term.

This second induction theorem solves the problem that bound variables may need to be renamed, but does so using the LGH measure. One can avoid explicit mention of the measure by using a third induction theorem, which follows easily from the second.

```
|- !P.
    (!x. P(mVAR x)) /\
    (!x. P(mCON x)) /\
    (?X. FINITE X /\
        (!x m. ~x IN X /\ P m ==> P(mLAMBDA x m))) /\
    (!m n. P m /\ P n ==> P(m ' n))
    ==>
    (!m. P m)
```

This is a convenient formulation of structural induction. To prove the case for mLAMBDA-terms, one first must select a set X of variables, and then prove that predicate P holds for (mLAMBDA x m) from the assumptions that P holds of m, and, crucially, that bound variable x is distinct from everything in X. By specifying a suitable X one gets to assume that x is a fresh variable. The clause for mLAMBDA-terms captures the intuition that one can always assume that bound variables are distinct from everything else of interest. Tactics have been programmed in ML to automate application of this theorem and, for the case of mLAMBDA-terms, to select sets X and prove them finite.

Derived Theorems

The theorems mentioned in the last two subsections about meta-terms are sufficient to derive the following results, without recourse to theorems about the underlying type of de Bruijn terms.

Using the third form of structural induction, we can derive all the properties of substitution used by Hindley and Seldin in their text on lambda-calculus [13]. Here is their Lemma 1.14,

```
|- !x m. ((mVAR x) / x) m = m
|- !x n m. ~x IN (FV m) ==> ((n / x) m = m)
|- !x n m. x IN (FV m) ==>
        (FV((n / x) m) = (FV n) UNION ((FV m) DELETE x))
|- !m x y. LGH(((mVAR x) / y )m) = LGH m
```

and their Lemma 1.15.

```
|- !x y n m. ~y IN (FV m) ==>
        ((n / y)(((mVAR y) / x) m) = (n / x) m)
|- !x y m. ~y IN (FV m) ==>
        (((mVAR x) / y)(((mVAR y) / x) m) = m)
|- !x y p q m.
        ~y IN (FV p) /\ ~(x = y) ==>
        ((p / x)((q / y) m) = (((p / x) q) / y)((p / x) m))
```

```
|- !x y p q m.
      ~y IN (FV p) /\ ~x IN (FV q) /\ ~(x = y) ==>
      ((p / x)((q / y) m) = (q / y)((p / x) m))
|- !x p q m. (p / x)((q / x) m) = (((p / x) q) / x) m
```

The third part of their Lemma 1.14 shows that formalised substitution / never accidentally captures free variables. If some y is free in n it remains free in (n / x) m.

Given alpha-conversion, Hindley and Seldin's lemmas and proof of the infinity of string names, it is possible to derive theorems concerning mLAMBDA analogous to the one-one theorems proved for the other constructors.

```
|- !x y m n.
      (mLAMBDA x m = mLAMBDA y n) ==> (m = ((mVAR x) / y) n)
|- !m n x y.
      (mLAMBDA x m = mLAMBDA y n) =
      (?z.
        ~z IN (FV m) /\
        ~z IN (FV n) /\
        (((mVAR z) / x) m = ((mVAR z) / y) n))
```

5 How the meta Type Can Be Applied

Here is an example grammar to illustrate how meta-terms could play the role of a meta-theory of syntax in HOL.

```
e ::= x
    | NUM n | (e1 PLUS e2)
    | BOOL b | (IF e1 e2 e3)
    | (LAMBDA x e) | (APP e1 e2)
    | LET x = e1 IN e2
```

To encode the syntactic constants and literals we can use the following HOL type

```
con = NumLit num
    | BoolLit bool
    | SynCon string
```

and then represent terms of the grammar as elements of type (con)meta, using the logical constants defined as follows.

```
|- NUM n        = mCON (NumLit n)
|- e1 PLUS e2   = (mCON(SynCon 'PLUS') ' e1) ' e2
|- BOOL b       = mCON (BoolLit b)
|- IF e1 e2 e3  = ((mCON(SynCon 'IF') ' e1) ' e2) ' e3
|- LAMBDA x e   = mCON(SynCon 'LAMBDA') ' (mLAMBDA x e)
|- APP e1 e2    = (mCON(SynCon 'APP') ' e1) ' e2
|- LET x e1 e2  = (mCON(SynCon 'LET') ' e1) ' (mLAMBDA x e2)
```

One can then give the set of terms in the grammar as an inductively-defined predicate on meta-terms. This is not the only way to embed this grammar; it remains future work to investigate alternatives and how syntactic properties can be derived for the new logical constants.

6 Related Work, Status and Future Plans

This paper has presented a strategy for representing syntax within a mechanised logic: construct the type of name-carrying terms—identified up to alpha-conversion—from an inductively defined subset of de Bruijn terms. This method mixes two strategies used in previous work: to represent name-carrying syntax directly [15, 19, 21, 26, 27, 28] and to adopt some variant of de Bruijn notation [3, 14, 27]. Coquand [7] pursues a fourth strategy based on explicit substitutions [1]. More experience with the different strategies is needed before they can be meaningfully compared, but the mixed strategy advocated here is at least a plausible method for directly mechanising the practice of many authors studying formal calculi: to work with name-carrying notation and to identify terms up to alpha-conversion.

Melham's π-calculus work includes another logical formulation of conventional name-carrying notation and substitution, with terms identified up to alpha-conversion. In his approach, a natural alternative to the present one, the underlying type is a free algebra of name-carrying syntax for the π-calculus. On this underlying type he defines substitution and alpha-conversion [18], and then gives a type of syntax up to alpha-conversion as the quotient of the underlying type under alpha-conversion.[3] Of course, this construction would work for lambda-calculus as well.

Hence we have two independently constructed representations for the same abstract type of syntax up to alpha-conversion. Each representation has its merits. Name-carrying syntax up to literal equality would be needed to represent language definitions, such as that of Standard ML [20], for instance, where syntax is not identified up to alpha-conversion. On the other hand, de Bruijn notation is a common implementation technique, and so a logical theory of de Bruijn notation is needed if such implementations are to be verified. An important question left open by the present work is to find an axiomatisation of name-carrying syntax up to alpha-conversion. Melham's quotient construction may be of help here, and in any case a useful extension of the present work would be to add a third type, of name-carrying syntax up to literal equality, and investigate its relation to the other two.

Section 5 showed how a grammar might be embedded in HOL using the meta type. This is reminiscent of how logical frameworks like Isabelle [24] or Elf [25] are used to embed object languages and logics in a fixed meta-logic. One significant difference is that to the best of my knowledge in neither Isabelle nor Elf can theorems about substitution such as the ones from Section 4 be proved in the meta-logic.

---

[3] Private communication with T. F. Melham, June 1993.

De Bruijn notation has been used to implement several theorem provers, such as his own AUTOMATH and Paulson's Isabelle. In Isabelle, for instance, syntax is represented as an ML type using de Bruijn indices. The human interacting with Isabelle sees and types a name-carrying notation, which is mapped to and from the internal de Bruijn notation by ML functions. An important difference from Isabelle is that in our approach the type of de Bruijn terms is represented in the logic rather than the programming metalanguage.

The difficult part of embedding syntax is dealing with bound variables. The approach of this paper is *first-order* in the sense that the variable-binding operation of the embedded syntax (mLAMBDA) is distinct from variable-binding in the logic. One might think of adopting a *higher-order* approach in which logical lambda-abstraction binds variables. A first attempt would be to construct a type with the same four constructors as * meta but with typing (* meta -> * meta) -> * meta for mLAMBDA. Such a type cannot be constructed using Melham's recursive types package, and in fact Gunter [12] has shown by a cardinality argument that this approach is impossible in HOL, if the constructors are injective (which we would want). Roughly speaking, the HOL function space is much bigger than the space of syntactic lambda-abstractions. A better approach is to give mLAMBDA the typing (string -> * meta) -> * meta. Despeyroux, Felty and Hirschowitz are investigating this embedding using Coq [9].

There is another, in some respects simpler, approach to representing languages in logic, pioneered by Gordon [11], in which each phrase of syntax is mapped metalinguistically to a logical constant that represents not the syntax itself but its meaning. When embedding a language of imperative commands, for instance, the meaning might be a relation between two states. The present approach is quite different in that syntax itself is represented in the logic, which is needed for instance to support operational language definitions. The two approaches—meaning versus syntax—are sometimes distinguished respectively with terms 'semantic' versus 'syntactic', or 'shallow' versus 'deep' [5].

The status of this work is that all the theory mentioned has been implemented in the Cambridge HOL system. Using the theory, Abramsky's lazy lambda-calculus has been mechanised in HOL, and elementary facts proved. The meta type can be used to encode other syntaxes using appropriate constants, just as the single ML type of terms in Isabelle can encode a range of syntaxes. The goal of future work is to build a single HOL theory able to encode any name-carrying syntax, together with tools to support definition of new syntaxes, recursive functions and proofs by induction, and hence to offer a general solution to the notorious problems of bound variables and substitution.

Acknowledgements

Comments from Catarina Coquand, Graham Hutton, Karsten Kehler Holst, Konrad Slind and the anonymous referees, and several discussions with Tom Melham were very helpful.

References

1. Martín Abadi, Luca Cardelli, Pierre-Louis Curien, and Jean-Jacques Lévy. Explicit substitutions. *Journal of Functional Programming*, 1(4):375–416, October 1991.
2. Samson Abramsky. The lazy lambda calculus. In David Turner, editor, *Research Topics in Functional Programming*, pages 65–116. Addison-Wesley, 1990.
3. Thorsten Altenkirch. A formalization of the strong normalization proof for System F in LEGO. In *TLCA '93 International Conference on Typed Lambda Calculi and Applications, Utrecht, 16–18 March 1993*, volume 664 of *Lecture Notes in Computer Science*, pages 13–28. Springer-Verlag, 1993.
4. H. P. Barendregt. *The Lambda Calculus: Its Syntax and Semantics*, volume 103 of *Studies in logic and the foundations of mathematics*. North-Holland, revised edition, 1984.
5. Richard Boulton, Andrew Gordon, Mike Gordon, John Harrison, John Herbert, and John Van Tassel. Experience with embedding hardware description languages in HOL. In V. Stavridou, T. F. Melham, and R. T. Boute, editors, *Theorem Provers in Circuit Design: Theory, Practice and Experience: Proceedings of the IFIP TC10/WG 10.2 International Conference, Nijmegen, June 1992*, IFIP Transactions A-10, pages 129–156. North-Holland, 1992.
6. Alonzo Church and J. B. Rosser. Some properties of conversion. *Transactions of the American Mathematical Society*, 36(3):472–482, May 1936.
7. Catarina Coquand. A machine assisted semantical analysis of simply typed lambda calculus. In P. Dybjer, J. Hughes, A. Moran, and B. Nordström, editors, *Proceedings of El Wintermöte*, pages 92–100. Programming Methodology Group, Chalmers University of Technology and University of Gothenburg, June 1993. Available as Report 73.
8. N. G. de Bruijn. Lambda calculus notation with nameless dummies, a tool for automatic formula manipulation, with application to the Church-Rosser theorem. *Indagationes Mathematicae*, 34:381–392, 1972.
9. Joëlle Despeyroux and André Hirschowitz. Higher-order abstract syntax and induction. Transparencies for a talk at the Types BRA Workshop on Proving Properties of Programming Languages, Sophia-Antipolis, September 1993.
10. M. J. C. Gordon and T. F. Melham, editors. *Introduction to HOL: A theorem-proving environment for higher-order logic*. Cambridge University Press, 1993.
11. Michael J. C. Gordon. Mechanizing programming logics in higher order logic. Technical Report 145, University of Cambridge Computer Laboratory, September 1988.
12. Elsa L. Gunter. Why we can't have SML style datatype declarations in HOL. In L. Claesen and M. Gordon, editors, *Higher Order Logic Theorem Proving and its Applications*, pages 365–372, Leuven, 1992. IMEC.
13. J. Roger Hindley and Jonathan P. Seldin. *Introduction to Combinators and λ-Calculus*. Cambridge University Press, 1986.
14. Gérard Huet. Residual theory in λ-calculus: A complete Gallina development. Preprint, 1992.
15. James McKinna and Robert Pollack. Pure Type Systems formalized. In *TLCA '93 International Conference on Typed Lambda Calculi and Applications, Utrecht, 16–18 March 1993*, volume 664 of *Lecture Notes in Computer Science*, pages 289–305. Springer-Verlag, 1993.
16. T. F. Melham. *The HOL pred_sets Library*. University of Cambridge Computer Laboratory, February 1992.

17. Thomas F. Melham. A package for inductive relation definitions in HOL. In *Proceedings of the 1991 International Workshop on the HOL Theorem Proving System and its Applications, Davis, California*, pages 350–357. IEEE Computer Society Press, 1991.

18. Thomas F. Melham. A mechanized theory of the π-calculus in HOL. Technical Report 244, University of Cambridge Computer Laboratory, January 1992.

19. Thomas Frederick Melham. *Formalizing Abstraction Mechanisms for Hardware Verification in Higher Order Logic*. PhD thesis, University of Cambridge Computer Laboratory, August 1990. Available as Technical Report 201.

20. Robin Milner, Mads Tofte, and Robert Harper. *The Definition of Standard ML*. MIT Press, Cambridge, Mass., 1990.

21. Monica Nesi. A formalization of the process algebra CCS in higher order logic. Technical Report 278, University of Cambridge Computer Laboratory, December 1992.

22. Bengt Nordström. Martin-Löf's type theory as a programming logic. Report 27, Programming Methodology Group, Chalmers University of Technology and University of Gothenburg, September 1986.

23. Lawrence C. Paulson. *ML for the Working Programmer*. Cambridge University Press, 1991.

24. Lawrence C. Paulson. The Isabelle reference manual. Internal report, University of Cambridge Computer Laboratory, 1992.

25. Frank Pfenning. A proof of the Church-Rosser theorem and its representation in a logical framework. Technical Report CMU-CS-92-186, Computer Science Dept., Carnegie Mellon University, September 1992.

26. Randy Pollack. Closure under alpha-conversion. Laboratory for Foundations of Computer Science, Department of Computer Science, University of Edinburgh, September 1993.

27. N. Shankar. A mechanical proof of the Church-Rosser theorem. *Journal of the ACM*, 35(3):475–522, July 1988.

28. J. von Wright. Representing higher-order logic proofs in HOL. University of Cambridge Computer Laboratory, July 1993.

A HOL Decision Procedure for Elementary Real Algebra

John Harrison
University of Cambridge Computer Laboratory
New Museums Site
Pembroke Street
Cambridge CB2 3QG
ENGLAND.
+44 223 334760
jrh@cl.cam.ac.uk

November 1, 1993

Abstract

The elementary theory of real algebra, including multiplication, is decidable. More precisely, there is an algorithm to eliminate quantifiers which does not introduce new free variables or new constants other than rational numbers. Therefore if a closed term of elementary real algebra involves no constants other than the rational numbers, its truth or falsity can be determined automatically. Quite a number of interesting algebraic and geometric problems can be expressed in this decidable subset. In this paper we describe a HOL implementation of a quantifier-elimination procedure and give some preliminary results.

1 Introduction

The formulas of elementary (first-order) real arithmetic are those built up, using the usual logical operations and quantification over the real numbers, from atomic formulas involving the usual ordering relations, equality, and the operations of addition, subtraction, negation and multiplication, which are in turn built from real variables and rational constants. A quantifier elimination algorithm is one which takes a term and returns one provably equivalent to it but involving fewer quantifiers. A simple example of such an equivalence (though not usually thought of by schoolchildren as quantifier elimination!) is the test for the existence of a root of a quadratic equation:

$$\vdash (\exists x.\ ax^2 + bx + c = 0) \equiv b^2 - 4ac \geq 0$$

(Strictly speaking this does not fall within our subset but we can make it so by rewriting x^2 to xx.)

It was first proved by Tarski [7] that such a quantifier elimination procedure exists for the elementary theory of reals. His method was rather complicated and inefficient, and better algorithms were developed by Seidenberg [6] and Cohen [2] among others. More recently, researchers in computer algebra have been investigating the use of Collins' method of Cylindrical Algebraic Decomposition [3] to give more efficient quantifier elimination procedures. Even these algorithms tend to be doubly exponential in the number of quantifiers to be eliminated, i.e. of the form:

$$O(2^{2^{kn}})$$

where n is the number of quantifiers and k is some constant. Very recent work [8] has improved this to 'only' being doubly exponential in the number of *alternations* of quantifiers.

In this paper we describe a HOL decision procedure for the subset of higher order terms corresponding to the elementary theory of reals. This is a HOL derived rule, i.e. each step of the reduction is justified by an actual HOL theorem. We broadly follow the algorithm given in Kreisel and Krivine's book [5] (a more leisurely treatment including pictures can be found in [4]). We have made a few simplifications and corrections, which are noted where significant.

2 Preliminary simplification

Here we note how the general problem can be reduced to one particularly simple case by exploiting a few logical identities.

2.1 Arbitrary real closed fields

For the sake of completeness (no pun intended) we should note that none of the following depends on the fact that the underlying field is the real numbers. In fact any so-called *real closed* field will do, although our work in HOL does not make this generalization.

A field is said to be *formally real* if whenever a sum of squares is zero, all the numbers in the sum are zero (equivalently, -1 is not expressible as a sum of squares). A *real closed* field is one which is formally real, but has no formally real proper algebraic extension (an algebraic extension results from adjoining to the field the roots of polynomial equations).

That the reals are real closed is easily seen. In any real closed field one can prove certain key properties (e.g. every polynomial of odd degree has a root) which are enough to make the procedure we describe work [9].

2.2 Adding some eliminable extensions

We can extend the permissible subset to include a few other operations which can be eliminated in terms of those in the strict subset. Terms raised to the power of a numeral can be rewritten in terms of repeated multiplication:

$$x^n = xx \ldots x$$

The abs function can simply be rewritten with its definition:

$$\vdash abs(x) = (0 \leq x) \rightarrow x \mid -x$$

Conditional expressions can be pulled up through terms using the theorems:

$$\vdash f(b \rightarrow x \mid y) = b \rightarrow f(x) \mid f(y)$$

and

$$\vdash (b \rightarrow f \mid g)x = b \rightarrow f(x) \mid g(x)$$

until they reach the 'atomic formula' level (of course HOL really has no syntactic notion of formulas vs. terms, so it would be more accurate to say 'boolean term level'). At this level they can be eliminated using the following theorem:

$$\vdash b \rightarrow x \mid y = (b \wedge x) \vee (\neg b \wedge y)$$

2.3 Eliminating superfluous relations and functions

The ordering relations are heavily redundant, and we can reduce them all to $>$ alone by rewriting with:

$$
\begin{aligned}
\vdash x \geq y &= \neg y > x \\
\vdash x \leq y &= \neg x > y \\
\vdash x < y &= y > x
\end{aligned}
$$

It is convenient to get rid of subtraction and negation, by treating -1 as a constant in the remainder of the elimination procedure:

$$\vdash x - y = x + (-1)y$$
$$\vdash -x = (-1)x$$

Finally we can get rid of boolean equality (shown here as logical equivalence for emphasis) and implication:

$$\vdash x \equiv y = (x \implies y) \land (y \implies x)$$
$$\vdash x \implies y = \neg x \lor y$$

2.4 Normalizing polynomial relations

We can make the right-hand sides of equations and inequalities zero by rewriting with:

$$\vdash x > y = x + (-1)y > 0$$
$$\vdash (x = y) = (x + (-1)y = 0)$$

Now we can distribute multiplication over addition:

$$\vdash x(y + z) = xy + xz$$
$$\vdash (x + y)z = xz + yz$$

So now the 'atomic formulas' are all either equations or inequalities with $>$, in either case with 0 on the right and a polynomial (in general, in several variables) on the left. The only logical operations are conjunction, disjunction and negation.

2.5 Reducing to simple existential elimination, pushing in negations

Clearly it is sufficient to be able to eliminate the innermost quantifier, since we can then apply the procedure repeatedly (we get a logical equivalence at each stage). Furthermore, universal quantifiers can be turned into existential ones via

$$\vdash \forall x. \; P[x] = \neg \exists x. \; \neg P[x]$$

Hence we only need to eliminate a single existential quantifier from a formula transformed according to the above sections. We can push negations down a term using the De Morgan laws:

$$\vdash \quad \neg \neg P = P$$
$$\vdash \quad \neg(P \land Q) = \neg P \lor \neg Q$$
$$\vdash \quad \neg(P \lor Q) = \neg P \land \neg Q$$

Furthermore, when the negations, if any, reach the 'atomic formula' level, they can be eliminated completely, since:

$$\vdash \quad \neg(p(x) = 0) = p(x) > 0 \lor (-1)p(x) > 0$$
$$\vdash \quad \neg p(x) > 0 = (p(x) = 0) \lor (-1)p(x) > 0$$

Now we can distribute conjunction over disjunction:

$$\vdash \quad x \wedge (y \vee z) = (x \wedge y) \vee (x \wedge z)$$
$$\vdash \quad (x \vee y) \wedge z = (x \wedge z) \vee (y \wedge z)$$

and can push the existential quantifier down through the disjuncts:

$$\vdash (\exists x. P[x] \vee Q[x]) = (\exists x. P[x]) \vee (\exists x. Q[x])$$

Therefore our problem is reduced to eliminating a single existential quantifier whose body is a conjunction of atomic formulas of the form $p(x) = 0$ or $p(x) > 0$, where x is the quantified variable and $p(x)$ is a polynomial in x with, in general, coefficients composed of other variables.

2.6 Reduction to bounded existential quantifier

For this step we use the following theorem:

$$\vdash (\exists y.\ P[y]) = (\exists u.\ 0 < u < 1 \wedge (\exists x.\ -1 < x < 1 \wedge P(u^{-1}x)))$$

To see the truth of this, consider left-to-right and right-to-left implications, and pick witnesses for the antecedents. The right-to-left implication is trivial: set $y = u^{-1}x$. For the other direction, choose $u = 1/(|y| + 2)$ and $x = y/(|y| + 2)$.

Now it is sufficient to have a procedure which will eliminate an existential quantifier restricted to a finite open interval, i.e.

$$\exists x.\ a < x < b \wedge P[x]$$

and produce a formula which still corresponds to our rules. We can transform a general existentially quantified goal using the above theorem, then eliminate the inner existential quantifier treating u^{-1} as just another variable or constant. But since the result of this stage is built up from atomic formulas of the form $p(u^{-1}) = 0$ and $p(u^{-1}) > 0$, where the p's are polynomials, and since we have in context that $0 < u$, we can multiply these equations and inequalities through by any necessary power of u giving a logically equivalent form. Then we have another bounded existential quantifier in the chosen subset:

$$\exists u.\ 0 < u < 1 \wedge R[u]$$

to which our procedure can be applied.

3 Degree reduction

By the *degree* of a variable x in a polynomial $p(x)$ we shall mean the highest power of x occurring in some monomial in $p(x)$. For example in $x^3 + 3xy^2 + 8$, the variables x, y and z have degrees 3, 2 and 0 respectively. Note that this might more accurately be called the 'formal degree' since we do not exclude the possibility that the coefficient of the relevant monomial might be zero for some or all values of the other variables.

We need to deal with terms of the following form:

$$p_1(x) = 0 \wedge \ldots \wedge p_k(x) = 0 \wedge q_1(x) > 0 \wedge \ldots \wedge q_l(x) > 0$$

which occur in the body of our existential terms (together with a bound on the quantified variable x, but we can forget about that for the present stage). It is a crucial observation that we can reduce such a term to a logically equivalent disjunction of similar terms, but each disjunct having the following properties:

- There is at most one equation (i.e. $k = 0$ or $k = 1$ in the above scheme)

- The degree of x in the equation, if any, is no more than the lowest degree of x in any of the original equations whose leading coefficient is known to be nonzero[1].

- If there is an equation, then the degree of x in all the inequalities is strictly lower than its degree in the equation.

To see this, observe that first we can repeatedly case-split over whether the leading coefficients of the lowest-degreed equation are zero. This gives a disjunction of similar terms, where either the equations all disappear completely (by this colourful phrase we really mean that x is no longer free in them, and hence we are at liberty to pull them outside the quantifier), in which case we are finished, or we have a situation where the leading coefficient of the lowest-degreed equation is nonzero.

3.1 Degree reduction of other equations

If the lowest-degreed equation, say $p_1(x) = 0$ has degree n, then we have that $p_1(x)$ is $ax^n + r_1(x)$ for some a which by the above remarks we can assume to be nonzero. But now if there is any other equation, say $p_2(x)$, it must have the form $bx^m + r_2(x)$ where, since $p_1(x)$ was chosen to be of least degree, $m \geq n$. However the following, since $a \neq 0$, is easily seen to be a logical equivalence.

$$\vdash (p_1(x) = 0) \wedge (p_2(x) = 0) = (p_1(x) = 0) \wedge (bx^{m-n}p_1(x) - ap_2(x) = 0)$$

But now we have reduced the degree of the latter equation[2]. If $p_1(x)$ still has lower degree, we can repeat the procedure; if the second equation disappears then we can attack the next equation, and finally if the second equation now has lower degree, we can case-split on that one's leading coefficients and start eliminating using it. Eventually we have at most one equation with x free.

3.2 Degree reduction of inequalities

If we have one equation $p_1(x)$ of the form $ax^n + r_1(x)$ left, then we may again suppose that $a \neq 0$. Now if the polynomial on the left of an inequality $q_1(x)$, say, is of the form $bx^m + s_1(x)$ with $m \geq n$, we can reduce its degree using the following:

$$\vdash (p_1(x) = 0) \wedge q_1(x) > 0 = (p_1(x) = 0) \wedge (a^2 q_1(x) + (-1)abp_1(x) > 0)$$

which is again easily seen to be true (observe that $0 < a^2$ because $a \neq 0$). It is convenient to finish off the elimination step by case-splitting over the leading coefficients in the inequalities, so we may thereafter assume them to be nonzero.

4 The main part of the algorithm

We shall find it convenient to consider formulas of the following three kinds:

- I: $\exists x.\ a < x < b \wedge (p(x) = 0) \wedge q_1(x) > 0 \wedge \ldots \wedge q_l(x) > 0$

- II: $\exists x.\ a < x < b \wedge q_1(x) > 0 \wedge \ldots \wedge q_l(x) > 0$

- III: $\forall x.\ a < x < b \implies q_1(x) > 0 \wedge \ldots \wedge q_l(x) > 0$

---

[1] Kreisel and Krivine assert that the degree can be made no greater than the lowest degree in the original equations. This is false if they mean 'formal degree' since an equation all of whose coefficients are zero cannot be used for elimination.

[2] On the other reading of 'degree' the corresponding statement of Kreisel and Krivine that if a is zero then deleting it from $p_1(x)$ reduces its degree is false. So they lose one way or the other; I have not been able to determine for certain which meaning they do attach to 'degree'.

We only want to eliminate quantifiers from the first two kinds (in the former, we allow the case where there are no inequalities, i.e. $l = 0$), but the third form arises as an intermediate result and it is useful to consider it separately. The basic idea of the algorithm is to transform between these forms, making polynomials which occurred in inequalities in one form appear in equations in another. We can then apply the degree-reduction step dealt with above and therefore get a term with a lower overall degree. This step is then repeated. To make the notion of decreasing degree rigorous, we will, following Kreisel and Krivine, make the following formal definition of the degree of a variable in a term:

- The degree of x in $p(x) = 0$ is the degree of x in $p(x)$

- The degree of x in $q(x) > 0$ is *one greater than* the degree of x in $q(x)$

- The degree of x in a non-atomic formula is the highest degree of x in any of its atoms.

It is clear that if the body of a quantifier has zero degree in the quantified variable, the elimination of the quantifier is trivial, the following being the case:

$$\vdash (\exists x.\ A) = A$$

Therefore we will have a terminating algorithm for quantifier elimination if we can show that quantifier elimination from a formula in any of these classes can be reduced to the consideration of other such formulas with lower degree. Let us consider the cases in reverse order.

4.1 Reduction of Type III Terms

Observe that if c lies between a and b, the following is true:

$$\vdash (\forall x.\ a < x < b \Longrightarrow \bigwedge_{i=1}^{l} q_i(x) > 0) = (\bigwedge_{i=1}^{l} q_i(c) > 0 \wedge \neg \exists x.\ a < x < b \wedge \bigvee_{i=1}^{l} q_i(x) = 0)$$

This relies on the Intermediate Value Theorem: if a polynomial is positive at one value of x but not at another, it must attain zero at some intermediate value of x. In particular we may choose $c = 2^{-1}(a + b)$, the midpoint of the interval [3]. Now the (bounded) existential quantifier can be pushed through the disjuncts and we are reduced to several type I terms of lower degree. We have introduced the constant 2^{-1} but this is unproblematic since we can do rational arithmetic when quantifier elimination is finished.

4.2 Reduction of Type II Terms

The key observation here is that, again as a consequence of the continuity of polynomials, the set of points at which a polynomial (and by induction any finite set of polynomials as considered here) is positive is *open* in the topological sense. This means that given any point in the set, there is some nontrivial surrounding region which is also contained entirely in the set:

$$open(S) = \forall x \in S.\ \exists \delta > 0.\ \forall x'.\ |x' - x| < \delta \Longrightarrow x' \in S$$

This means that if a set of polynomials are all positive at a point, as a Type II term asserts, then they are positive in some interval surrounding that point (and conversely, obviously). The idea behind the reduction step is that we can choose this interval to be as large as possible, subject to its being within the bound associated with the quantifier. One possibility is that the polynomials are positive throughout the interval. But then we have a Type III term which, as we have already seen, can be reduced to one of lower degree. Otherwise one end is, or both ends are, properly

---

[3]The first edition of Kreisel and Krivine made the mistake of thinking $\bigwedge_{i=1}^{l} q_i(a) \geq 0$ adequate. The second edition has a rather complicated alternative, roughly, that the first nonzero derivative at a is positive. Apart from being simpler, our solution generalizes to infinite intervals, as noted in the later section on optimizations.

contained within the quantifier bounds, and it is easy to see that some polynomial in the set must be zero at any such points (otherwise, by the open set property, the interval could be made larger). This has allowed us to turn an inequality into an equation as we wanted. Consider the formal statement of the reduction theorem:

$$\vdash \ (\exists x.\ a < x < b \land (p(x) = 0) \land \bigwedge_{i=1}^{l} q_i(x) > 0)$$
$$\equiv \exists a'\ b'.\ a \leq a' < b' \leq b \land$$
$$(\forall x.\ a' < x < b' \implies \bigwedge_{i=1}^{l} q_i(x) > 0) \land$$
$$(a' = a \lor \bigvee_{i=1}^{l} q_i(a') = 0) \land$$
$$(b' = b \lor \bigvee_{i=1}^{l} q_i(b') = 0)$$

We have already seen that Type III terms can be reduced to terms of lower degree, so we can eliminate the inner quantifier over x. We can then distribute over the disjunction and get in general many different clauses. The point is that even if the elimination of x has resulted in higher-degree terms in a' and/or b', they can be reduced to the levels of a simple equality or one of the q_j's using the range-reduction step. (We may assume that each of the q_j's has nonzero leading coefficient because of the previous elimination step.) Thus a' and b' only need be eliminated from lower-degree terms.

4.3 Reduction of Type I terms

The reduction of these terms has the same idea behind it as in the previous section, but the details are slightly more complicated since we need to consider the derivative of the polynomial appearing in the equation. Suppose that we have a point x where $p(x) = 0$ and all the $q_i(x)$ are positive. There are three possibilities for the derivative $p'(x)$. The simplest is that it is zero. In this case we can perform an elimination and get a term of lower degree (note that the leading coefficient of the derivative is known to be nonzero, because it is a positive integer multiple of the leading coefficient of the original polynomial). The more complicated cases arise where the derivative is either strictly positive or strictly negative at x. We shall consider the former; the latter can be treated in essentially the same way.

The derivative is also a polynomial, and is continuous, and is positive at x. We can therefore consider the larger group of inequalities resulting from its addition to the original ones. All these polynomials are strictly positive at x, and so as last time we can find a maximum interval in which this is true. Furthermore, since $p'(x)$ is positive throughout the interval, and $p(x)$ is zero somewhere within it, then $p(x)$ must be strictly negative at the left-hand end, and strictly positive at the right-hand end. Conversely, if this is the case, then there must be a point within the interval at which the polynomial $p(x)$ cuts the x-axis (again using the Intermediate Value Theorem). This allows a similar reduction to the above.

5 The HOL Implementation

When implementing any sort of derived rule in HOL, it is desirable to move as much as possible of the inference from 'run time' to the production of a few proforma theorems which can then be instantiated efficiently. To this end, we have defined encodings of common syntactic patterns in HOL which make it easier to state proforma theorems of sufficient generality.

5.1 Syntactic encodings

Firstly, we define a constant poly which takes a list as an argument and gives a polynomial in one variable with that list of coefficients, where the head of the list corresponds to the constant term, and the last element of the list to the term of highest degree.

```
|- (poly [] x = &0) /\
   (poly (CONS h t) x = h + (x * poly t x))
```

This immediately has the benefit of letting us prove quite general theorems such as 'every polynomial is differentiable' and 'every polynomial is continuous'. Differentiability of polynomials can be defined as a simple recursive function on the list of coefficients. The operations of addition, negation and constant multiplication can likewise be defined in an easy way in terms of the list of coefficients. For example, addition can be executed by rewriting with the following clauses:

```
|- (poly_add [] m = m) /\
   (poly_add l [] = l) /\
   (poly_add (CONS h1 t1) (CONS h2 t2) = (CONS (h1 + h2) (poly_add t1 t2)))
```

and we have the theorems

```
|- !l x. --(poly l x) = poly (poly_neg l) x
```

```
|- !l x c. c * (poly l x) = poly (poly_cmul c l) x
```

```
|- !l m x. poly l x + poly m x = poly (poly_add l m) x
```

The reduction theorems have a recurring theme of 'for all polynomials in a finite list' or 'for some polynomial in a finite list'. Accordingly, we make the following general definitions:

```
|- (FORALL P [] = T) /\
   (FORALL P (CONS (h:*) t) = P h /\ FORALL P t)
```

```
|- (EXISTS P [] = F) /\
   (EXISTS P (CONS (h:*) t) = P h \/ EXISTS P t)
```

Now we need only the following extra definitions:

```
|- POS x l = poly l x > &0
```

```
|- ZERO x l = (poly l x = &0)
```

and we are in a position to state the reduction theorems actually at the HOL object level.

5.2 HOL versions of reduction theorems

The theorem justifying reduction to bounded existential quantifiers is:

```
|- !P. (?x. P x) = (?u. &0 < u /\ u < &1 /\
                        (?x. --(&1) < x /\ x < &1 /\ P(x / u)))
```

The reduction theorem for Type III terms takes the following form:

```
|- !l a b. a < b /\
           (!x. a < x /\ x < b ==> FORALL (POS x) l)
   = a < b /\
     FORALL (POS (inv(&2) * (a + b))) l /\
     (~?x. a < x /\ x < b /\ EXISTS (ZERO x) l)
```

Note that this one, unlike the other two, demands the extra context of $a < b$ to be included, otherwise the left-to-right implication fails. The reduction theorem for Type II terms is:

```
|- !l a b. (?x. a < x /\ x < b /\ FORALL (POS x) l)
           = (?a' b'. a <= a' /\ a' < b' /\ b' <= b /\
                      (!z. a' < z /\ z < b' ==> FORALL (POS z) l) /\
                      ((a' = a) \/ EXISTS (ZERO a') l) /\
                      ((b' = b) \/ EXISTS (ZERO b') l))
```

And the reduction theorem for Type I terms is the following:

```
|- !l r a b.
    (?x. a < x /\ x < b /\ FORALL(POS x) l /\ (poly r x = &0)) =

    (?x. a < x /\ x < b /\ FORALL(POS x)l /\
        (poly r x = & 0) /\ (poly(poly_diff r)x = & 0)) \/

    (?a' b'. a <= a' /\ a' < b' /\ b' <= b /\
        ((a' = a) \/ EXISTS (ZERO a') (CONS (poly_diff r) l)) /\
        ((b' = b) \/ EXISTS (ZERO b') (CONS (poly_diff r) l)) /\
        ((poly (poly_neg r) a' > &0 /\ poly r b' > &0 /\
          !z. a' < z /\ z < b' ==>
            FORALL (POS z) (CONS (poly_diff r) l)) \/
        (poly r a' > &0 /\ poly (poly_neg r) b' > &0 /\
          !z. a' < z /\ z < b' ==>
            FORALL (POS z) (CONS (poly_neg (poly_diff r)) l)))))
```

5.3   Implementation of the algorithm

Although the above theorems have all been proved, and a naive degree-reduction function coded, the system has only been used semi-manually so far. Each step can generate a large number of disjuncts and it is currently intolerably slow for anything beyond one bounded existential quantifier. However we believe there are grounds for optimism that the system can be made more practical by modification of the algorithm and more careful coding (quite apart from the many other interesting algorithms which merit exploring).

Allowing infinite intervals

One unfortunate feature of the algorithm as currently implemented is that to eliminate one unbounded existential quantifier, we generate two bounded ones (as well as the hassle of multiplying out the terms to force the intermediate result back into the subset). However it seems to us that the basic algorithm works in a broadly unchanged manner if we allow $a = -\infty$ and/or $b = +\infty$ in the bounds. Here we are speaking informally of infinity – it might be easy enough to formalize an extended real line in HOL, but we are really anticipating splitting each reduction theorem into four separate theorems according as none, either one or both ends of the existential quantifier are bounded.

Note that another advantage of our use of the midpoint in the reduction of Type I terms is that it can easily be extended to give a simple expression for a nominated point of an interval even if it is semi-infinite or infinite. The various cases are:

$$-\infty < \quad 0 \quad < +\infty$$
$$-\infty < \quad b-1 \quad < b$$
$$a < \quad a+1 \quad < +\infty$$
$$a < \quad \tfrac{a+b}{2} \quad < b$$

The appropriate reduction theorems are still true, mutatis mutandis. For example, to see that the right-to-left implication of the Type I reduction theorem remains true for an interval of the form $(-\infty, +\infty)$, we need to check that a polynomial whose derivative is positive *everywhere* possesses a root *somewhere* (this is untrue for plenty of non-polynomial functions, e.g. e^x).

Actually a stronger result is true: if the derivative of a polynomial is nonzero everywhere then the polynomial has a root. Indeed, every polynomial of odd (actual) degree has a root, so either the antecedent of this conditional statement is trivially false, or the consequent trivially true.

Ordering of disjuncts

Many of the terms split into large numbers of disjuncts. The implementation already exploits the 'laziness' assumption that if one disjunct evaluates to truth, the others need not be considered. This is mostly of benefit when the term contains no free variables and hence the result of quantifier elimination is evaluable. However one possibility which has not been exploited is to use straightforward heuristics to decide the order in which the disjuncts are attacked. Clearly it is generally a good idea to start with the 'simple' ones, e.g. those of lowest degree or containing fewer variables.

Treating simple cases separately

Certain terms can be eliminated in a simple ad-hoc way without proceeding all the way to degree zero. In fact we could stop at degree 1 since after an elimination step only a single equation remains which can be eliminated very easily by observing:

$$\vdash (\exists x.\ ax + b = 0) = \neg a = 0 \lor b = 0$$

More generally, it might be advantageous to delegate some terms, e.g. linear terms lying within the Presburger subset, to more efficient decision procedures. Work is already in progress on extending the HOL Presburger decision procedure [1] to the reals.

Optimizing the implementation

As detailed above, some pains were taken to move the burden of inference onto the production of a few proforma theorems, in order to make the algorithm as far as possible a matter of instantiation and rewriting. However the part of the algorithm doing elimination has been coded rather naively and could probably be speeded up by a moderate factor with careful coding.

6 Conclusion

Though clearly not yet ready for use as a general-purpose tool, we believe this sort of algorithm is a very promising avenue of research. In the future we hope to implement some of the optimizations detailed above in the hope that the procedure can be made much more efficient.

7 Acknowledgements

My thanks go to Konrad Slind, who first pointed out to me the decidability of the elementary theory of reals, and with whom I have had many stimulating discussions on this and other matters. James Davenport explained to me the status of various algorithms and got me started on a literature search. Finally, my supervisor Mike Gordon has never failed to be encouraging and supportive.

References

[1] R. J. Boulton, *The HOL arith library*, in HOL System Libraries reference manual, distributed with HOL version 2.01.

[2] P. J. Cohen, *Decision Procedures for Real and p-adic Fields*, Communications in Pure and Applied Mathematics vol. 22, pp. 131-151, 1969.

[3] G. E. Collins, *Quantifier Elimination for Real Closed Fields by Cylindrical Algebraic Decomposition*, Second GI Conference on Automata Theory and Formal Languages, Springer Lecture Notes in Computer Science vol. 33, pp. 134-183, 1975.

[4] E. Engeler, *The Foundations of Mathematics*, Springer 1993.

[5] G. Kreisel, J. L. Krivine, *Elements of Mathematical Logic*, 2nd revised printing, North-Holland 1971.

[6] A. Seidenberg, *A New Decision Method for Elementary Algebra*, Annals of Mathematics vol. 60, pp. 365-374, 1954.

[7] A. Tarski, *A Decision Method for Elementary Algebra and Geometry*, University of California Press 1951.

[8] N. J. Vorobjov (Jr.), *Deciding Consistency of Systems of Polynomial in Exponent Inequalities in Subexponential Time*, Proceedings of the MEGA-90 Symposium on Effective Methods in Algebraic Geometry, Progress in Mathematics vol. 94, Birkhäuser 1991.

[9] B. L. van der Waerden, *Algebra, vol. I*, 7th edition, Springer-Verlag 1966.

AC Unification in HOL90

Konrad Slind

Institut für Informatik
Technische Universität München
Arcisstrasse 21
80290 München
Germany

Abstract. We report on an implementation of associative-commutative unification in an LCF-style theorem prover. We show how this algorithm can be used to implement rewriting modulo associativity and commutativity. This is an example of the sound incorporation of automatic first order methods into an interactive higher-order logic theorem prover.

1 Introduction

Unification is the study of how to turn equations into identities by applying substitutions. Applications of unification are found in programming languages, computer algebra systems, artifical intelligence, theorem proving, and systems for polymorphic type inference.

A problem with automatic proof procedures is that they have a hard time doing inference with permutative equations, such as those expressing commutativity and associativity. One remedy for this, initiated by Plotkin [Plo72], is to incorporate the permutative axioms into unification and rewriting. This began the study of equational unification, which relaxes the constraint that the unified terms be identical; instead, the unified terms are to be provably equal in equational logic. Equational unification is a very general field, encompassing for instance Hilbert's Tenth problem [JK90], but the fundamental algorithm in this area is undoubtedly associative-commutative (AC) unification, in which the equational axioms express the associative ($\forall xyz.x \ op \ y \ op \ z = (x \ op \ y) \ op \ z$) and commutative ($\forall xy.x \ op \ y = y \ op \ x$) properties of op (written infix here). Some common examples are $\{\cap, \cup\}$ in set theory; $\{\wedge, \vee\}$ in logic; and $\{\times, +\}$ in arithmetic.

The familiar notion of a single most general unifier no longer works in AC unification. Consider $x \underset{\text{AC}}{=} a + b$; it has unifiers $\{a+b/x\}$ and $\{b+a/x\}$, neither of which are instantiations of the other. The best we can do is to find the smallest set of non-overlapping unifiers from which we can produce all unifiers. This *minimal, complete* set of unifiers μCSU has the properties

completeness $unifier(\sigma) \supset \exists \theta \in \mu CSU. \ \exists \gamma. \ \gamma \circ \theta = \sigma$
minimality $\forall \theta_1 \theta_2 \in \mu CSU. \ \exists \gamma. \ \gamma \circ \theta_1 = \theta_2 \supset \theta_1 = \theta_2$

The original AC unification algorithm was discovered by Stickel [Sti81]. The termination of the algorithm was solved by Fages [Fag84]. Fortenbacher [For85] gives an elegant algebraic presentation and some optimizations of Stickel's algorithm. Some recent advances in the study of AC unification are described in [BCD90, KN92]. A test suite and the performance of some of the leading implementations of the time is given by Buerckert *et al.* [BHK+88].

We have implemented Fortenbacher's AC unification algorithm. In this paper, we describe in detail the workings of the algorithm and show how to use it to provide more powerful variants of some popular rules of inference. Among these are identity modulo AC, *modus ponens* modulo AC, AC-rewriting, and a version of HOL-style "resolution" that works modulo AC.

Stickel's algorithm is defined for first order terms. How does this work in HOL's world of higher-order terms? We identify a set of "essentially first order HOL terms" in the following. AC unification is done in HOL by mapping from this set into first order terms, doing AC unification in this first order world, then mapping the resulting substitutions back out into HOL substitutions.

2 First order terms

The set of single-sorted first order terms $\mathcal{F}(\Sigma \cup \mathcal{V})$ is inductively formed from Σ, a set of aritied constants, and $\mathcal{V}$ a countably infinite set of variables:

- if x is in $\mathcal{V}$, then x is in $\mathcal{F}(\Sigma \cup \mathcal{V})$;
- if f of arity $n \geq 0$ is in Σ, and $t_1, \ldots, t_n$ are in $\mathcal{F}(\Sigma \cup \mathcal{V})$, then $f(t_1, \ldots, t_n)$ is in $\mathcal{F}(\Sigma \cup \mathcal{V})$.

Now we describe the HOL analogue of $\mathcal{F}(\Sigma \cup \mathcal{V})$. The key idea in this is to collapse types into arities.

Definition. A *curried type of sort* τ is a function type $\tau \to \tau'$ where τ' is either τ or a curried type of sort τ. The *width* of $\tau_1 \to \tau_2 \to \ldots \to \tau_n$, a curried type of sort τ, is n.

Definition. $\mathcal{F}_\tau$, the set of first order terms of type τ, is the smallest subset of term (HOL terms) formed by the following rules:

variables: if $v : \tau \in$ term, then $v : \tau$ is in $\mathcal{F}_\tau$. There is a countably infinite set of variables at each type.

constants: if $c : \tau \in$ term, then $c : \tau$ is in $\mathcal{F}_\tau$

combinations: if $f : \delta$ is a constant, δ is the curried type of sort τ of width $m+1$, and each member of the list $[t_1, \ldots, t_m]$ is in $\mathcal{F}_\tau$, then the combination $f t_1 \ldots t_m$ is in $\mathcal{F}_\tau$.

abstractions: are not allowed.

By a simple argument [Sli91], this establishes a fragment of term corresponding to single sorted first order terms. In other words, we can confidently map from $\mathcal{F}_\tau$ to first order terms, run standard first order algorithms on those terms, and map the results back into HOL terms. We know that the meaning of the first order operations will be preserved. Note that none of this translation has any impact on provability in HOL; these first order operations are just used to supply *information* to HOL inference rules and it is reassuring to know that the information is accurate.

An interesting aspect of $\mathcal{F}_\tau$ is that the definition of terms is parameterized on an arbitrary τ, which could be polymorphic.

3   The algorithm

Consider an equation built from two terms that are themselves made from applications of an associative and commutative function symbol, e.g.,

$$(x + x) + ((y + 0) + 2) = 1 + (1 + (z + 2))$$

The properties of + allow us to *flatten* the terms and regard each as the application of a variadic + to a multiset:

$$+\{x, x, y, 0, 2\} = +\{1, 1, z, 2\}$$

Stickel's insight was to map the flattened terms to a simpler domain, do unification there, and transform those substitutions back to unifiers in the domain of terms. He introduced the concept of *variable abstraction* whereby all constants and function applications in the flattened term are mapped to new variables.

The result of variable abstraction is then considered as a term in an Abelian semigroup (an algebra with a single associative and commutative operator). Linear Homogeneous Diophantine Equations (LHDE's) over the positive integers are isomorphic to Abelian semigroups, so a solution of the LHDE corresponding to two abstracted terms induces a solution to the unification problem in the Abelian semigroup. To return to the domain of terms requires reconciling the Abelian semigroup unifier with the variable abstraction substitution.

A nice feature of Stickel's algorithm is that it places no restrictions on the first order terms: more than one AC constant is allowed and non-AC constants are also allowed in the terms to be unified. This allows Stickel's algorithm to be embedded into a standard first order unification algorithm [MM82]. In such an embedding, when we wish to unify two applications, say $f(t_1, \ldots, t_n)$ and $f(s_1, \ldots s_m)$, if f is an AC constant, then we must perform the AC unification algorithm, as opposed to the usual decomposition operation. Extra difficulties then arise, since AC unification returns a set of substitutions, instead of the usual, single, substitution.

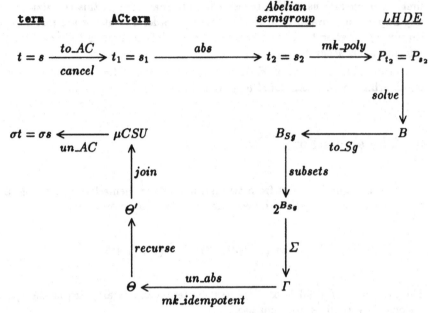

The above diagram summarizes Stickel's approach. Read the right-pointing arrows as problem mappings, and the down-, left-, and up-going arrows as the finding of solutions to those problems. The labels on the arrows denote functions that are used to transform the problem or solutions. In cases where functions are above and below an arrow, the top function is executed first. We shall explain the algorithm by explaining the labels attached to the transitions in the diagram. The algorithm uses four domains: HOL terms, ACterms, an Abelian semigroup, and LHDE. An ACterm is much like an ordinary first order term, except that an application of an AC constant *op* to arguments is flattened into a pairing of *op* with a list of terms, none of which is an application of *op*.

We develop an example along the way, the one given in Stickel's original paper:

$$(x + (y + a)) + b + x + c = b + b + c + z + b.$$

to_AC

Conceptually, we map from a $\mathcal{F}_r$ term to a first order term, and from there to an ACterm. However, this is done directly in the implementation. We obtain

$$+(x, x, y, a, b, c) = +(z, b, b, b, c).$$

cancel

Stickel proved that cancelling common subterms preserves unifiers. Therefore, we obtain

$$+(x, x, y, a) = +(z, b, b).$$

abs

All non variable subterms are replaced by new variables. We remember this operation by forming a substitution that, when applied to the variables, results in the original terms. We are now in an Abelian semigroup. To mirror this, we also map the AC operator to $\star$, the semigroup operator. The new variables are w and u; the substitution θ is $\{a/w, b/u\}$. The unification problem is now

$$x \star x \star y \star w = u \star u \star z.$$

mk_poly

Sum the multiplicities of each variable in each term. This gives us a linear homogeneous diophantine equation (LHDE) (a diophantine equation with no exponents and constant coefficients). Abelian semigroups are isomorphic to LHDE's [Fag84]. Therefore, the problem is now

$$2x + y + w = 2u + z.$$

solve

We can compute a finite basis set of solutions to the equation. A basis B has the property that every solution to the equation can be expressed as a linear combination of basis solutions. The basis set is computed by a backtracking algorithm that simply enumerates the solutions up to a bound. We use bounds on solutions that are derived in [Hue78]. The 7 basis solutions for our example form the rows in the following table.

	$2x$	y	w	$2u$	z
	(a)			(b)	
a	0	0	1	0	1
b	0	1	0	0	1
c	0	0	2	1	0
d	0	1	1	1	0
e	0	2	0	1	0
f	1	0	0	0	2
g	1	0	0	1	0

We have tagged each column with the fact of whether it corresponds to a variable or non-variable in the ACterm. This information is used in later optimizations.

to_Sg

Now we must transfer LHDE unifiers back to semigroup unifiers. Generate new variables: one for each basis solution. Consider a basis solution b_i and v_i its newly generated variable: for each box in b_i regard the number there as a multiplicity for v_i. If the box has a 1, put v_i; if the box has a 2, put a $v_i \star v_i$, etc. If the box has 0, don't put anything. So we have

	new var.	2x	y	(a) w	(b) 2u	z
a	v_1	-	-	v_1	-	v_1
b	v_2	-	v_2	-	-	v_2
c	v_3	-	-	$v_3 \star v_3$	v_3	-
d	v_4	-	v_4	v_4	v_4	-
e	v_5	-	$v_5 \star v_5$	-	v_5	-
f	v_6	v_6	-	-	-	$v_6 \star v_6$
g	v_7	v_7	-	-	v_7	-

subsets

It is our aim to transform the basis solutions into unifiers in the semigroup, but there is one catch—a 0 in a basis solution does not correspond to any element of a semigroup; semigroups do not have an identity element. The next best thing is to consider all the subsets of B that have non-zero column sums. There are of course exponentially many subsets, so we impose some constraints based on optimizations noticed by Fortenbacher:

- For a subset with all column sums non-zero, it is not allowed that a column is headed by a constant or a function application and the sum of that column is > 1. This implies that a row with an entry ≥ 2 in a non-variable column can be eliminated from the process of subset construction. (This allows us to delete row c.)
- Within a row of the basis, all boxes containing a 1 must have their corresponding terms AC-unifiable. (This allows us to delete row d.)

Using these criteria, we are left with 4 subsets, out of a possible $2^7 = 128$:

$$\{ \{a, e, f\}, \quad \{a, b, e, f\}, \quad \{a, b, g\}, \quad \{a, b, f, g\} \}$$

Σ

For each subset, make a substitution by binding the column variable with the sum of the columns. The result Γ is the set of semigroup unifiers.

$$\Gamma = \begin{cases} \{v_6/x; \; v_5 \star v_5/y; \; v_1/w; \; v_5/u; \; v_1 \star v_6 \star v_6/z\}, \\ \{v_6/x; \; v_2 \star v_5 \star v_5/y; \; v_1/w; \; v_5/u; \; v_1 \star v_2 \star v_6 \star v_6/z\}, \\ \{v_7/x; \; v_2/y; \; v_1/w; \; v_7/u; \; v_1 \star v_2/z\}, \\ \{v_6 \star v_7/x; \; v_2/y; \; v_1/w; \; v_7/u; \; v_1 \star v_2 \star v_6 \star v_6/z\} \} \end{cases}$$

un_abs

We recall the substitution θ from **abs**: $\{a/w; \; b/u\}$. We must reconcile the bindings of θ with each member of Γ. Every binding in θ is of the form tm/v, where tm is a constant or an application and v is a variable created to be new. If there is a binding s/v in a $\gamma \in \Gamma$, then s is a variable, since tm can not unify with a $\star$-term. Hence we replace the binding s/v in γ with tm/s.

mk_idempotent

Idempotence is essential for minimality. Recall that an idempotent unifier σ is one that has the property $\sigma \circ \sigma = \sigma$. A unifier is made idempotent by composing it with itself until a fixed point is reached. In our example, we now have Θ as

$$\{v_6/x;\ b + b/y;\ a/v_1;\ b/v_5;\ a + v_6 + v_6/z\}$$
$$\{v_6/x;\ v_2 + b + b/y;\ a/v_1;\ b/v_5;\ a + v_2 + v_6 + v_6/z\}$$
$$\{b/x;\ v_2/y;\ a/v_1;\ b/v_7;\ a + v_2/z\}$$
$$\{v_6 + b/x;\ v_2/y;\ a/v_1;\ b/v_7;\ a + v_2 + v_6 + v_6/z\}$$

recurse

Let us take stock of the situation. We have a set of "unifiers" Θ that are not quite unifiers since they have not taken account of the results of any subproblems. What are the subproblems? They arise from those application terms not headed by $+$. Fortunately for the reader, no such subterms exist in this example: a and b are the only non-variable subterms. A unification problem like $x + x + a + a = g(y) + g(a) + z + z$ does have subproblems: one is $g(y) = g(a)$.

join

The recursive calls comprise one of the most expensive parts of the algorithm since, in general, a subproblem may have a μCSU Θ'. Each σ in Θ' must get composed with each member of Θ to get the final set. The combinatorial aspects of this are exacerbated by the fact that a σ_1 in Θ' may have a binding tm_1/v and σ_2 in Θ' may have a binding tm_2/v. This requires solving the new subproblem $tm_1 = tm_2$ with its own μCSU, and folding those solutions in with the others.

un_AC

This is the trivial mapping of AC terms in the unifiers to HOL terms.

3.1 Other approaches

Recent approaches to AC unification solve sets of equations rather than the single-equation-at-a-time approach found here. Also, the complexity of merging recursive solutions is largely avoided. However, the power set operation must still be done, and the complexity of solving systems of equations is also exponential, so the algorithm has double exponential complexity[KN92].

4 Using AC unification in hol90

The main use of AC unification in hol90 is for AC rewriting, the implementation of which requires an algorithm for deleting common terms from both sides of an equality theorem. If we have two terms, t_1 and t_2, in $\mathcal{F}_\tau$ (for some τ), that we unify with the AC unification algorithm, we obtain a minimal, complete set of unifiers Θ. If the implementation of the algorithm is correct, then each instance: $\theta t_1 \underset{AC}{=} \theta t_2$ can be proved automatically with the following simple algorithm (expressed as a tactic).

4.1 AC cancellation and equivalence

We begin by associating everything to the right, by (ordinary) rewriting with the associativity theorem. Then we have $X + t_1' = Y + t_2'$, for an AC operator $+$. The algorithm repetitively deals with three cases:

$lhs = rhs$ Then we are done.

$X \underset{AC}{=} Y$ If X is identical to Y, we cancel X and Y off their respective sides with **AP_TERM_TAC**, and continue. Otherwise, we make a recursive call to the algorithm to prove $\vdash X \underset{AC}{=} Y$, use that theorem to rewrite X to Y, cancel Y from both sides, and continue.

$X \underset{AC}{\neq} Y$ In this case, we look for an AC-equivalent to X in t_2'.
  - If it is found, we rearrange the rhs so that cancellation is possible. This is performed by using the commutativity property on the rhs to get $X + t_1' = t_2' + Y$. Then we reassociate the rhs. This has the effect of "rotating" the rhs by one element. We rotate by one element at a time, until a term AC-equivalent to X appears in the left-most position of the rhs, whereupon we continue.
  - If it is not found, we "rotate" the left hand side and continue.

The algorithm terminates when all the arguments to $+$ on the lhs have been seen. The actual algorithm used in the implementation is more complicated, since it can handle cases where full cancellation is not possible, in which case it will cancel as much as possible.

Example. Consider

$$(x + 0) \times (((y + (y + (y + (0 + 0)))) + 0) \times (2 \times 2))$$
$$= ((y + 0) + ((y + 0) + (y + 0))) \times (2 \times (2 \times (x + 0)))$$

We wish to prove this is a theorem. Use the association laws to associate everything to the right:

$$(x + 0) \times (y + (y + (y + (0 + (0 + 0))))) \times (2 \times 2))$$
$$= (y + (0 + (y + (0 + (y + 0))))) \times (2 \times (2 \times (x + 0))))$$

The multiplicands are not AC-equivalent, so the rhs must be rotated until they are. When the rotation is done, we find that they are identical:

$$(x + 0) \times ((y + (y + (y + (0 + (0 + 0))))) \times (2 \times 2))$$
$$= (x + 0) \times (y + (0 + (y + (0 + (y + 0))))) \times (2 \times 2))$$

Now we cancel with **AP_TERM_TAC**:

$$(y + (y + (y + (0 + (0 + 0))))) \times (2 \times 2)$$
$$= (y + (0 + (y + (0 + (y + 0))))) \times (2 \times 2))$$

The multiplicands are AC-equivalent, but not identical; we recursively prove them equivalent and replace the multiplicand on the *rhs* by that of the left.

$$(y + (y + (y + (0 + (0 + 0))))) \times (2 \times 2)$$
$$= (y + (y + (y + (0 + (0 + 0))))) \times (2 \times 2))$$

Now we have an identity and have proved the theorem. In the system, the cancellation tactic is called AC_CANCEL_TAC[1]and it is parameterized by a set of AC theorems.

```
type ac_eqns = {A:thm,C:thm} list
(* val AC_CANCEL_TAC : ac_eqns -> tactic *)
local
val get = theorem "arithmetic"
val AC_thms = [{A=get"ADD_ASSOC", C=get"ADD_SYM"},
               {A=get"MULT_ASSOC",C=get"MULT_SYM"}]
in
fun ARITH_AC_REFL tm = prove(tm,AC_CANCEL_TAC AC_thms)
end;

val ARITH_AC_REFL = fn : term -> thm

- ARITH_AC_REFL(--'(x + 0) * ((y + y + y + 0 + 0) + 0) * (2 * 2) =
                    ((y + 0) + (y + 0) + y + 0) * 2 * 2 * (x + 0)'--);
val it =
  |- (x + 0) * ((y + y + y + 0 + 0) + 0) * 2 * 2 =
     ((y + 0) + (y + 0) + y + 0) * 2 * 2 * (x + 0) : thm
```

4.2 AC rewriting

Recall that standard HOL rewriting takes a list L of equality theorems and a term tm and if an instantiation θl of a left hand side of a theorem of L is equal to tm, then a theorem $\vdash tm = \theta r$ is returned. This is the basis of HOL's congruence-based rewriting package[Pau83]. The substitution θ is found is by means of a matching algorithm. Matching is "one-sided" unification; when solving the matching problem $lhs = rhs$, only lhs is allowed to be substituted into. (This can be ensured by replacing all variables in the rhs by constants and then using the ordinary unification algorithm.) We perform *AC rewriting* in the following manner:

- Let θ be an AC match between $\vdash l = r$ of a theorem in the rewrite list and tm, i.e., $\theta l \underset{AC}{=} tm$. At this point, it is not valid to replace tm by θr; we only know that we *should* be able to prove $\vdash \theta l = tm$ using only AC theorems.

---

[1] AC_CANCEL_TAC is an example of a *tactic generator*: it first computes a tactic by analyzing the structure of the goal, then applies that tactic to the goal.

- Prove that $\vdash \theta l = tm$, by using AC_CANCEL_TAC.
- Use a rule of inference that performs transitivity of equality to replace tm by θr.

Example The basic AC rewriting function is **AC_REWR_CONV**. In the following assume that $th = \vdash x + SUC\ x = 2 * x + 1$

```
(*     val AC_REWR_CONV : ac_eqns -> th -> conv     *)

AC_REWR_CONV AC_thms th (--'SUC (4*3) + 3*4'--);
val it = |- SUC (4 * 3) + 3 * 4 = 2 * 4 * 3 + 1 : thm
```

Using **AC_REWR_CONV**, we have constructed a family of AC rewriting tools in the same style as the standard HOL rewriting tools. There are some differences worth discussing:

- In standard HOL rewriting, discrimination nets implement a form of hashing for rules: they are used for efficient rule lookup. When at a subterm, we search for a simplified form of it in the net. The result of a successful lookup will be a set, usually singleton, of rules. The first rule in the set that matches will be applied. For AC rewriting, an implementation of nets that accomodates flattened terms is desirable. In the meantime, we just use a simple lookup along a list of rewrite rules.
- Consider using the theorem $\vdash x + x = 2 * x$ as a rewrite rule on the term "$0 + 3 + 0$". This entails finding a substitution for x such that $x + x$ can be made equal to "$0 + 3 + 0$". This is impossible though, since the "3" is not catered for; we need a *placeholder* in the rewrite rule to shunt all the unmatched parts of the AC term into. When rewriting with a theorem $\vdash x\ op \ldots op\ y = rhs$, where op is an AC constant with type $\tau \to \tau \to \tau$, a new rule is automatically built by generating a new variable v of type τ and inferring a new theorem $\vdash v\ op\ x\ op \ldots op\ y = v\ op\ rhs$. Both the original rule and the new rule are used for rewriting. The new rule is the *variable extension* discovered [PS81] in the context of extending the Knuth-Bendix completion procedure to work modulo an equational theory.
- AC rewriting is easy to fold in with the standard HOL rewriting strategies, expressed using *conversionals* [Pau83]. However, this can be too naive in some cases: the standard strategy performs a match at every node in the term. A single AC match operation, on the other hand, effectively searches for a match through all nodes "down to the next node headed by a different or non-AC constant". To avoid useless repetition of matching, a more sophisticated traversal strategy is required: a match failure at an AC node requires search for the next useful node to try.

We have also used AC matching to provide a stronger form of *modus ponens*: if P in a theorem $\forall x_1 \ldots x_n . P \supset Q$ has an AC match with a theorem P', the instantiated consequent Q' is derived. This form of *modus ponens* has been used to provide AC versions of the HOL "resolution" facilities[GM93].

So far, applications of unification and AC unification in hol90 have been restricted to building tools based on matching; however, one application of unification is in solving existential goals. Given an existential goal $\exists x.P$, one would find all the constraints on x in P, and turn them into a set S of equations. If a unifier θ could be computed for S, then instantiating x by θ in P would not in general guarantee a solution to the goal, but would often ease the problem of finding existential witnesses[2]. This unification-based approach is generally supported in *Isabelle* [Pau90] via the use of Prolog-style *logic variables*; to get the same benefits in HOL would require a smart goal-directed proof management system.

Extensions There is further work to be done in the area of expanding the set of HOL terms that the algorithm (or its descendants) can work over. The "Miller" subcase of lambda calculus terms[Mil91, Nip93], augmented with types, seems like an interesting starting point for this. Also, efficiency optimizations are very important, in view of the high complexity of the operations. For example, AC rewriting uses only one substitution from the whole μCSU: the lazy generation of subsets from the basis (and perhaps the lazy computation of the basis itself) should be investigated. Finally, there are simple modifications to the AC unification algorithm to accomodate theories with an identity law $\vdash x \ op \ 1 = x$ (such as various brands of arithmetic) or an idempotence law $\vdash x \ op \ x = x$ (such as instances of boolean algebra, like set theory and logic). These extensions would require corresponding slight modifications to the equivalence checking algorithm.

5 Conclusion

Using the *Isabelle* theorem prover [Nip89a, Nip89b], Nipkow has done something similar to our work. He presents matching, unification, and rewriting for various equational theories by means of presenting the algorithms as proof rules. The machinery of *Isabelle* does the rest. The advantage of this is that it is very simple to get an algorithm going. As he notes, the disadvantage of this is that it is slow and the AC unification algorithm need not terminate (although the AC matching algorithm does). In contrast, the implementation presented here should be comparable in speed to others, terminates, and is therefore more suitable for production use.

The main contribution of this paper is the implementation of the AC unification algorithm and its incorporation into the hol90 system. Useful tools springing from this are an AC cancellation tactic, AC rewriting, and AC "resolution". In view of the complexity of the underlying algorithms, an AC extension of a proof

---

[2] Quantifier elimination methods are decision procedures that can be seen as theory-specific generalizations of this approach.

procedure should not be considered as a replacement for the standard version: it is a heavyweight tool to be used in special cases.

The work done in this paper is an example of soundly linking a theorem prover up with an *extralogical expert* or *oracle*. Using ML computation to supplement or guide logical operations has long been part of the LCF "ethos", but it has only recently been expounded in print [GM93, chapter 6]. (There is no reason for the extralogical expert to be written in ML: [HT93] documents a linkage to a computer algebra system.) Sound linkage with an oracle consists in not trusting it: the oracle is merely a source of raw material for the proof procedures of the theorem prover. In particular, "facts" coming from the oracle should *not* blithely be accepted as theorems; they must be *proved* or at the very least tagged using Gordon's scheme[Gor93].

The usefulness of sound linkage is that one can, without effort and without compromising the integrity of the theorem prover, take advantage of progress in the extralogical expert. For example, the AC unification algorithm described in this paper is almost ten years old; subsequent research has produced faster and cleaner methods—our insistence on checking AC substitutions by performing AC equivalence proofs means that we could switch to a new implementation of AC unification without having to convince ourselves of the correctness of the new implementation, *i.e.*, without loss of confidence in our theorem prover.

6 Acknowledgements

Thanks to Tobias Nipkow for pointing out that what I thought was an *ad hoc* trick was in fact variable extension. Part of this work was done at the University of Calgary in 1988; thanks to Graham Birtwistle and the Canadian government for financial support then. Part of this work has been done at the Technische Universität München in 1993 while being supported by the Esprit TYPES project. The diagram was laid out with Paul Taylor's diagram package.

References

[BCD90] A. Boudet, E. Contejean, and H. Devie. A new AC unification algorithm with an algorithm for solving systems of Diophantine equations. In *Proceedings of the Fifth Annual Symposium on Logic in Computer Science*, pages 289–299, Philadelphia, Pennsylvania, June 1990. Computer Society Press.

[BHK+88] Hans-Juergen Buerckert, Alexander Herold, Deepak Kapur, Jorg Siekmann, Mark Stickel, Michael Tepp, and Hantao Zhang. Opening the AC-unification race. Technical Report SR-88-11, SEKI, July, 1988.

[Fag84] Francois Fages. Associative-commutative unification. In Robert Shostak, editor, *Proceedings of the Seventh International Conference on Automated Deduction (LNCS 170)*, pages 194–208, Napa, California, USA, May 1984. Springer-Verlag.

[For85] Albrecht Fortenbacher. An algebraic approach to unification under associativity and commutativity. In Jean-Pierre Jouannaud, editor, *Proceeding* of the First International Conference on Rewriting Techniques and Applications (LNCS 202), pages 381–397, Dijon, France, May 1985. Springer-Verlag.

[GM93] Mike Gordon and Tom Melham. *Introduction to HOL, a theorem proving environment for higher order logic.* Cambridge University Press, 1993.

[Gor93] Michael Gordon. Note to info-hol mailing list, archive number 0914. Tagging theorems for proof acceleration, January 1993.

[HT93] John Harrison and Laurent Thery. Extending the hol theorem prover with a computer algebra system to reason about the reals. In *LPAR 93 (LNCS ??)*, St. Petersburg, Russia, 1993.

[Hue78] Gerard Huet. An algorithm to generate the basis of solutions to homogeneous linear Diophantine equations. *Information Processing Letters*, 7(3):144–147, April 1978.

[JK90] Jean-Pierre Jouannaud and Claude Kirchner. Solving equations in abstract algebras: A rule-based survey of unification. Technical Report 561, Unite' Associee' au CNRS UA 410, March 1990.

[KN92] Deepak Kapur and Paliath Narendran. Double-exponential complexity of computing a complete set of AC-unifiers. In *Proceeding of the Seventh Annual IEEE Symposium on Logic in Computer Science*, Santa Cruz, California, June 1992. Computer Society Press.

[Mil91] Dale Miller. A logic programming language with lambda-abstraction, function variables, and simple unification. In Peter Schroeder-Heister, editor, *Exensions of Logic Programming (LNCS 475)*, pages 253–281. Springer-Verlag, 1991.

[MM82] Alberto Martelli and Ugo Montanari. An efficient unification algorithm. *Transactions on Programming Languages and Systems*, 4(2):258–282, April 1982.

[Nip89a] Tobias Nipkow. Equational reasoning in Isabelle. *Science of Computer Programming*, 12:123–149, 1989.

[Nip89b] Tobias Nipkow. Term rewriting and beyond—theorem proving in Isabelle. *Formal Aspects of Computing*, 1:320–338, 1989.

[Nip93] Tobias Nipkow. Functional unification of higher order patterns. In *Proceedings of the Eighth Annual IEEE Symposium on Logic in Computer Science*, pages 64–74, Montreal, Canada, June 1993. IEEE Computer Society Press.

[Pau83] Lawrence Paulson. A higher order implementation of rewriting. *Science of Computer Programming*, 3:119–149, 1983.

[Pau90] Lawrence Paulson. Isabelle: The next 700 theorem provers. In P.G. Oddifreddi, editor, *Logic and Computer Science*, pages 361–386. Academic Press, 1990.

[Plo72] Gordon Plotkin. Building-in equational theories. *Machine Intelligence*, (7):73–90, 1972.

[PS81] Gary Peterson and Mark Stickel. Complete sets of reductions for some equational theories. *Journal of the Association for Computing Machinery*, 28:233–264, 1981.

[Sli91] Konrad Slind. An implementation of higher order logic. Technical Report 91-419-03, University of Calgary Computer Science Department, 1991.

[Sti81] Mark Stickel. A unification algorithm for associative-commutative functions. *Journal of the Association for Computing Machinery*, 28(3):423–434, July 1981.

Server-Process Restrictiveness in HOL

Stephen H. Brackin[1] * and Shiu-Kai Chin[2] ** ***

[1] Odyssey Research Associates, 301 Dates Drive, Ithaca NY 14850
[2] Department of Electrical and Computer Engineering,
Syracuse University, Syracuse NY 13244

Abstract. Restrictiveness is a security property of multilevel systems, guaranteeing that a system's behavior visible at one security level does not reveal the existence of inputs or outputs at other levels not dominated by this level. This paper gives a convenient language for specifying processes, an operational semantics for this language, and a collection of easy to understand, inductively defined security properties that can be used to substantially automate proving restrictiveness for buffered server processes specified in this language.

1 Introduction

Designing and evaluating the most trusted computer systems requires formally specifying and proving that these systems have specified security properties [7]. *Restrictiveness*, a security property developed by McCullough [10, 12, 11] is of particular interest for such as analysis. Restrictiveness is *composable*, meaning that a system composed of properly connected restrictive parts is itself restrictive.

The original version of restrictiveness, which we will call *trace restrictiveness*, was behavioral [10]; it defined system security in terms of possible sequences of input, output, and internal events. Unfortunately, proving theorems with this form of restrictiveness was extremely difficult, as it required complicated inductions over possible extensions to sequences [11, 3]. Different researchers chose two different approaches to this problem, both of which involve strengthing trace restrictiveness to provide more useful induction hypotheses:

- Alves-Foss and Levitt developed *incremental restrictiveness*, a condition on single-event extensions to event sequences; it implies trace restrictiveness and is itself composable [1, 2, 3, 4].
- McCullough [11, 13] developed what we will call *state restrictiveness*, which is usually simply called restrictiveness. State restrictiveness defines conditions on machine states that produce all of the system behavior observable at a security level, but allow no deductions about other behavior.

* Supported by Rome Laboratory Contract F30602-90-C-0092
** Supported by Rome Laboratory Contract F30602-92-C-0120
*** The authors wish to thank Daryl McCullough and David Rosenthal for identifying subtle errors in earlier versions of this work.

Rosenthal [18, 19] identified conditions sufficient to guarantee state restrictiveness in the broad case of buffered server processes. A buffered process consists of a FIFO queue and a process being buffered; it saves its inputs on the queue until the process being buffered is ready to receive them. The process being buffered is a server process if it waits for input in a parameterized state and processes each input by producing zero or more outputs and then calling itself to again wait for input in a possibly different parameterized state.

Sutherland, McCullough, Rosenthal, and others [16] developed process specification languages (similar to subsets of CSP [9]) for conveniently defining state machines, and they identified syntactic analyses that could be performed on such specifications to generate conditions sufficient for establishing state restrictiveness of buffered server processes. They and others at Odyssey Research Associates developed the Romulus (nee Ulysses) design analysis tool, which can take a specification of a buffered server process and compute from it a list of conditions sufficient for establishing Rosenthal's conditions for guaranteeing state restrictiveness [16].

For largely historical reasons, the HOL version of the Romulus specification language was defined only partially, with axioms, giving extensibility but making the language and its semantics hard to understand [17]. The Romulus implementation also depended on a "meta"-level approach, generating Rosenthal-style verification conditions with an impure tactic that was itself hard to understand [17]. Further, the language's semantics required that all state-machine parameters and all messages received as inputs or sent as outputs be coerced to a single type :datatype, losing the advantages of HOL's strong typing and significantly complicating proofs of the verification conditions [17].

We developed the results presented here, using Slind's HOL90 Release 5, to address these limitations in Romulus. We modeled our process specification language, PSL, on the Romulus specification language [17], but defined it as a concrete recursive type using Melham's automated type definition facility [14] and gave it an operational semantics using Camilleri and Melham's inductive definitions package [6].

We defined security properties analogous to Rosenthal-style verification conditions using the inductive definitions package, and we proved simple theorems about these properties that can be used as rewrite rules in proofs and taken as alternative definitions of the security properties, definitions requiring no extensive knowledge of security theory or HOL. Finally, we used polymorphic concrete recursive types to impose strong typing on state-machine parameters and message contents, in the process developing a simple technique for effectively defining processes in terms of process-valued functions.

All of our definitions are conservative extensions of the HOL logic. In addition, having the actual security definitions, abstract syntax, and operational descriptions available at the object level clarifies the intended meaning of the security properties and simplifies the proofs.

We do not intend to argue here that one version of restrictiveness is preferable to another, or that our version is equivalent to another. For the sake of simplifying the implementation, our version of restrictiveness was made intermediate in strength between trace and state restrictiveness; we believe that it could be made equivalent to Rosenthal's conditions for state restrictiveness of buffered server processes with a modest additional implementation effort.

The results presented here were part of an experiment to test the practicality of reasoning about state restrictiveness at the object level, that is, reasoning about PSL descriptions when PSL was formally embedded in HOL. These results indicate that the definitional approach is both practical and preferable.

We give an intuitive explanation of conditions giving state restrictiveness for buffered server processes in Section 2, we formally define the process specification language PSL and give its operational semantics in Section 3, we give security definitions in HOL using PSL in Section 4, we specify and prove secure a simple message sorter in Section 5, and we conclude in Section 6.

2   Server-Process Restrictiveness

This section gives an informal description of conditions similar to those identified by Rosenthal [19] for showing state restrictiveness of buffered server processes.[3] Section 4 contains the formal definitions. The theories of trace, incremental trace, and state restrictiveness presented in [11, 3] are more general, and incremental trace restrictiveness has HOL proofs showing that it guarantees trace restrictiveness [1, 2], HOL proofs of the sort not yet produced for our conditions. We believe, though, that our conditions are much easier to establish in the broad class of cases where they apply.

2.1   Views and Projection Functions

The processes we analyze deal with multiple security levels (e.g., unclassified, confidential, secret, and top secret). We model these processes as rated state machines, which assign security levels to each input and output event. The raw information that can be obtained, or *viewed*, by an observer depends on that observer's security level.

What an observer at a given security level can know about a system is described by three functions: functions giving the security level associated with input and output events; and a *projection function* hiding state information.

A projection function induces an equivalence partition on process states, for the process being buffered, based on security level. Two states in the same equivalence partition with respect to one level must also be in the same equivalence partition with respect to any other level dominated by that level. The projection of a state to a level determines all the information, but only the information, necessary to determine the system's behavior that can be viewed at that level.

---

[3] Restrictiveness for server processes should not be confused with "server restrictiveness," a different concept introduced by Rosenthal in [21].

2.2 Restrictiveness Conditions

Informally, these conditions are sufficient for guaranteeing, for buffered server processes, that an observer at a particular security level can never deduce anything about higher- or incomparable-level events:

1. Any output produced in response to an input is at a level that dominates the level of that input.
2. If the level of an input is not dominated by the observer's level, then the projection function induces a state partition such that the process being buffered appears not to have changed state (i.e., the process' state after responding to this input is in the same partition block as it was before receiving this input).
3. If the security level of an input is dominated by that of the observer, then any two states of the process being buffered that seemed equivalent to the observer are not distinguished by the response to this input:
   - All outputs possible for one state are also possible for the other.
   - Any possible next state for one state is in the same partition block as any possible next state for the other state.
   - The previous two conditions hold for any *combination* of visible outputs and seemingly equivalent next states.

The buffering guarantees that the full process is always ready to accept input, so there is never any information conveyed about the process' state by its ability or inability to accept input.

3 Process Specification Language

This section defines PSL, our process specification language. By "process" we mean either a state machine or a program-like description of a machine state. PSL is a simple language with the basic processes Skip, Send, Receive, and Call. Skip is the finished process that does nothing. Send transmits an output event. Receive takes a predicate on input events determining which input events it receives and a function determining how it responds to each event it receives. Call invokes the process associated with a function name and possible arguments.

Processes are combined using the PSL operators ;;, Orselect, If, and Buffered. Infix operator ;; is like the sequence operator in CSP[9]. Orselect is like the non-deterministic choice operator + in CCS [15]. If is the if-then-else operator. Buffered takes a predicate on input events, a buffer, and a process to be buffered. It returns the process that puts the input events satisfying the predicate onto the buffer, then passes them on to the process being buffered when that process is ready to receive them.

3.1 Syntax

The abstract syntax of PSL is given below. It is embedded in HOL in the usual way using **define_type**. Recall that `:'name` in type signatures in HOL90 is the same as `:*name` in HOL88.

```
process =
  Skip | ;; of process # process | Orselect of process # process |
  If of bool # process # process | Send of 'outev |
  Receive of ('inev -> bool) # ('inev -> 'invoc) | Call of 'invoc |
  Buffered of ('inev -> bool) # ('inev)list # process'
```

The type variables `'inev`, `'outev`, and `'invoc` are meant to stand for possibly polymorphic concrete recursive types of input events, output events, and function invocations. The type constructors for input and output events are meant to correspond to ports where messages enter or leave the process, and the types of the arguments to these constructors are meant to be the types of these messages; this imposes strong typing constraints on message contents. Function invocations are meant to correspond to names for calls to process-valued functions; as described in the following section on PSL semantics, they effectively allow processes to be defined in terms of process-valued functions.

3.2 Semantics

There are three kinds of events: input events; output events; and silent or internal events called **Tau**. We introduce event types into HOL using **define_type**.

```
event = Out of 'outev | In of 'inev | Tau
```

In defining the semantics of PSL, the function *invocval* is a parameter to the function returning the transition relation that shows how PSL processes are transformed by events. The meaning of a PSL process thus varies with the mapping of invocations to PSL processes.

There are 14 rules that define the behavior of PSL. We use SML variable **proc** to abbreviate the polymorphic type `:('outev,'inev,'invoc)process`. The rules are defined inductively using **new_inductive_definition**. The function **transition** takes as a parameter a function *invocval* of type `:'invoc -> ^proc` assigning interpretations to invocations, and returns a transition relation of type `:^proc->('outev,'inev)event->^proc->bool` that shows how PSL processes of type `^proc` are transformed by events. We denote the transition from state p to state q via event e by $p \xrightarrow{e} q$. **SNOC** *element list* puts *element* onto the end of *list*.

$$\frac{}{(\text{Skip} ;; p2) \xrightarrow{\text{Tau}} p2} \qquad \frac{p1 \xrightarrow{e} px}{(p1 ;; p2) \xrightarrow{e} (px ;; p2)}$$

$$\frac{}{(\text{Orselect } p1\ p2) \xrightarrow{\text{Tau}} p1} \qquad \frac{}{(\text{Orselect } p1\ p2) \xrightarrow{\text{Tau}} p2}$$

$$\frac{b}{(\text{If } b\ p1\ p2) \xrightarrow{\text{Tau}} p1} \qquad \frac{\neg b}{(\text{If } b\ p1\ p2) \xrightarrow{\text{Tau}} p2}$$

$$\frac{}{(\text{Send } outev) \xrightarrow{(\text{Out } outev)} \text{Skip}}$$

$$\frac{(received\ inev)}{(\text{Receive } received\ response) \xrightarrow{(\text{In } inev)} (invocval\ (response\ inev))}$$

$$\frac{}{(\text{Call } invocation) \xrightarrow{\text{Tau}} (invocval\ invocation)}$$

$$\frac{(buffering\ inev)}{(\text{Buffered } buffering\ buf\ p) \xrightarrow{(\text{In } inev)} (\text{Buffered } buffering\ (\text{SNOC } buf\ inev)\ p)}$$

$$\frac{p \xrightarrow{(\text{In } inev)} px}{(\text{Buffered } buffering\ (\text{CONS } inev\ buf)\ p) \xrightarrow{\text{Tau}} (\text{Buffered } buffering\ buf\ px)}$$

$$\frac{p \xrightarrow{(\text{In } inev)} px,\ \neg(buffering\ inev)}{(\text{Buffered } buffering\ buf\ p) \xrightarrow{(\text{In } inev)} (\text{Buffered } buffering\ buf\ px)}$$

$$\frac{p \xrightarrow{(\text{Out } outev)} px}{(\text{Buffered } buffering\ buf\ p) \xrightarrow{(\text{Out } outev)} (\text{Buffered } buffering\ buf\ px)}$$

$$\frac{p \xrightarrow{\text{Tau}} px}{(\text{Buffered } buffering\ buf\ p) \xrightarrow{\text{Tau}} (\text{Buffered } buffering\ buf\ px)}$$

We have also defined the semantics for the composition of processes (i.e., when the output of one process is the input of another), but for brevity omit the description. The details can be found in [5].

4 Security Definitions

This section defines security properties sufficient for showing security of non-parameterized PSL server processes when these processes are buffered; these security properties do not involve the projection function. Space limitations prevent giving the similarly defined properties for showing security with parameterized processes (see [5]), but the predicates Loopsback and AllWritesUp given here apply to both parameterized and non-parameterized processes.

We establish security for buffered non-parameterized server processes by examining the process *being* buffered; we do not consider the case in which the process being buffered is itself buffered. The buffered process Buffered $(\lambda inev.\ \text{T})$ [ ] P is secure if P is a non-parameterized server process that satisfies "all outputs are up." A projection function need not be considered, since a non-parameterized process always starts in the same state each time it takes a new input off the buffer.

The section includes collections of theorems in if-and-only-if form that effectively define the security properties by induction on the structural complexity of PSL processes. Our tactics for showing security properties of PSL processes use these theorems as rewrite rules. These theorems can also be taken as the definitions of these properties by those unfamiliar with HOL or security theory. We derived the theorems from their implicative forms after defining the security properties with **new_inductive_definition**. We chose the definitions so that security properties hold only when they hold independently of any nondeterministic choice made and the occurrence of any input event.

A process is a server process if it receives any possible input, finishes its processing of this input, and then loops back by calling itself to wait for the next input. The predicates **Terminates**, **Loopsback**, and **BNPSP_rightform** confirm that a process is a server processes. The predicate **AllWritesUp** confirms that this process only produces "up" outputs.

4.1 Terminates

Terminates tells whether a PSL process finishes its processing. Informally, the rules for termination are the following:

1. **Skip** always terminates.
2. $(p1 \;;\; p2)$ terminates if both $p1$ and $p2$ terminate.
3. (**Orselect** $p1$ $p2$) terminates if both $p1$ and $p2$ terminate.
4. (**If** b $p1$ $p2$) terminates if $p1$ terminates when b is true and $p2$ terminates when b is false.
5. **Send** always terminates.
6. **Receive** never terminates.
7. (**Call** $invocation$) terminates if the process invoked by ($invocval\ invocation$) terminates.
8. **Buffered** never terminates.

The formal rules for termination follow:

$(\forall invocval.\ \textbf{Terminates}\ invocval\ \textbf{Skip})$
$(\forall invocval\ p1\ p2.$
 $\textbf{Terminates}\ invocval\ (p1 \;;\; p2)\ =$
 $(\textbf{Terminates}\ invocval\ p1) \wedge (\textbf{Terminates}\ invocval\ p2))$
$(\forall invocval\ p1\ p2.$
 $\textbf{Terminates}\ invocval\ (\textbf{Orselect}\ p1\ p2)\ =$
 $(\textbf{Terminates}\ invocval\ p1) \wedge (\textbf{Terminates}\ invocval\ p2))$
$(\forall invocval\ b\ p1\ p2.\ \textbf{Terminates}\ invocval\ (\textbf{If}\ b\ p1\ p2)\ =$
 $b \Rightarrow (\textbf{Terminates}\ invocval\ p1)\ |\ (\textbf{Terminates}\ invocval\ p2))$
$(\forall invocval\ outev.\ Terminates\ invocval\ (\textbf{Send}\ outev))$
$(\forall invocval\ received\ response.$
 $\neg(\textbf{Terminates}\ invocval\ (\textbf{Receive}\ received\ response)))$
$(\forall invocval\ invocation.$
 $\textbf{Terminates}\ invocval\ (\textbf{Call}\ invocation)\ =$
 $\textbf{Terminates}\ invocval\ (invocval\ invocation))$
$(\forall invocval\ buffering\ buf\ p.$
 $\neg(\textbf{Terminates}\ invocval\ (\textbf{Buffered}\ buffering\ buf)))$

4.2 Loopsback

Loopsback tells whether a process eventually calls a named process, where the "name" is an invocation for a non-parameterized process and is an invocation type constructor for a parameterized process. Informally, the rules for looping back are the following:

1. **Skip** never loops back.
2. (p1 ;; p2) loops back if p1 terminates and p2 loops back.
3. (**Orselect** p1 p2) loops back if both p1 and p2 loop back.
4. (**If** b p1 p2) loops back if p1 loops back when b is true and p2 loops back when b is false.
5. **Send** never loops back.
6. **Receive** never loops back.
7. Whether (**Call** invocation) loops back depends on a case analysis; see the comments following this list.
8. **Buffered** never loops back.

To handle both parameterized and non-parameterized processes consistently with HOL's typing requirements, we take the (polymorphic) type of the name of a process, pname, to be : 'invoc+('par->'invoc), where 'par can be instantiated with an arbitrary process-parameter type. If pname names a non-parameterized process, (ISL pname) is true, and if pname names a parameterized process, (ISRpname) is true. (**Call** invocation) loops back, when pname is a non-parameterized process, if invocation = (OUTL pname) or (invocval invocation) loops back. A similar set of conditions apply for parameterized processes. The formal rules for looping back follow:

$(\forall invocval\ pname.\neg($**Loopsback** $invocval\ pname\ $**Skip**$))$
$(\forall invocval\ pname\ p1\ p2.$
  Loopsback $invocval\ pname\ (p1\ ;;\ p2) =$
  (**Terminates** $invocval\ p1) \wedge ($**Loopsback** $invocval\ pname\ p2))$
$(\forall invocval\ pname\ p1\ p2.$
  Loopsback $invocval\ pname\ ($**Orselect** $p1\ p2) =$
  (**Loopsback** $invocval\ pname\ p1) \wedge ($**Loopsback** $invocval\ pname\ p2))$
$(\forall invocval\ pname\ b\ p1\ p2.$
  Loopsback $invocval\ pname\ ($**If** $b\ p1\ p2) =$
  $b \Rightarrow ($**Loopsback** $invocval\ pname\ p1)\ |\ ($**Loopsback** $invocval\ pname\ p2))$
$(\forall invocval\ pname\ outev.\ \neg($**Loopsback** $invocval\ pname\ ($**Send** $outev)))$
$(\forall invocval\ pname\ received\ response.$
  $\neg($**Loopsback** $invocval\ pname\ ($**Receive** $received\ response)))$
$(\forall invocval\ pname\ invocation.$
  (**Loopsback** $invocval\ pname\ ($**Call** $invocation)) =$
  $((ISL\ pname) \Rightarrow (invocation = (OUTL\ pname)) \vee$
  (**Loopsback** $invocval\ pname\ (invocval\ invocation)))\ |$
  $((\exists param.\ invocation = ((OUTR\ pname)\ param)) \vee$
  (**Loopsback** $invocval\ pname\ (invocval\ invocation))))$

$(\forall invocval\ pname\ buffering\ buf\ p.$
   $\neg(\textbf{Loopsback}\ invocval\ pname\ (\textbf{Buffered}\ buffering\ buf\ p)))$

4.3 BNPSP_rightform

BNPSP_rightform, where **BNPSP** stands for "buffered, non-parameterized server process," is true of a function mapping invocations to processes and an invocation naming a non-parameterized process if this non-parameterized process is a server process. For instantiating the predicate **Loopsback** in the definition of **BNPSP_rightform**, non-parameterized processes are treated as having the parameter $(--{}`one'--)$. For brevity, we do not give **receivesall** and **reaction** here, but they are straightforward. The predicate **receivesall** is true if it is applied to a PSL **Receive** process which receives arbitrary input events. The conversion rule **reaction** is applied to an input event, a function mapping invocations to PSL processes, and a PSL process. If this process is a **Receive**, **reaction** returns the process that the **Receive** process changes into in response to the input event.

$(\forall invocval\ nppname.$
  $\textbf{BNPSP\_rightform}\ invocval\ nppname\ =$
  $\textbf{receivesall}\ (invocval\ nppname)\ \wedge$
  $(\forall inev.$
  $\textbf{Loopsback}$
   $invocval\ (\textbf{INL}\ nppname)\ (\textbf{reaction}\ inev\ invocval\ (invocval\ nppname))))$

4.4 AllWritesUp

AllWritesUp is defined for a dominance relation, a level-assignment function for outputs, a security level assumed to be the level of some input, a function mapping invocations to processes, a process name, and a PSL process. It holds if, before the process ends with a call to the named process, the level of every output produced dominates the input level. Informally, the rules for always producing "up" outputs are the following:

1. **Skip** satisfies **AllWritesUp**.
2. $(p1\ ;;\ p2)$ satisfies **AllWritesUp** if both $p1$ and $p2$ satisfy it.
3. $(\textbf{Orselect}\ p1\ p2)$ satisfies **AllWritesUp** if both $p1$ and $p2$ satisfy it.
4. $(\textbf{If}\ b\ p1\ p2)$ satisfies **AllWritesUp** if $p1$ satisfies **AllWritesUp** when b is true and $p2$ satisfies **AllWritesUp** when b is false.
5. $(\textbf{Send}\ outev)$ satisfies **AllWritesUp** if the level of $outev$ dominates the level of the input.
6. **Receive** never satisfies **AllWritesUp**.
7. $(\textbf{Call}\ invocation)$ satisfies **AllWritesUp** if the invocation is a call to $pname$ or if $(invocval\ invocation)$ satisfies **AllWritesUp**. The cases are similar to those for **Loopsback**.
8. **Buffered** processes never satisfy **AllWritesUp**.

The formal rules defining **AllWritesUp** follow:

($\forall dom$ *outlev level invocval pname.*
 AllWritesUp *dom outlev level invocval pname* **Skip**)
($\forall dom$ *outlev level invocval pname p1 p2.*
 AllWritesUp *dom outlev level invocval pname* ($p1$;; $p2$) =
  (**AllWritesUp** *dom outlev level invocval pname p1*) $\wedge$
  (**AllWritesUp** *dom outlev level invocval pname p2*))
($\forall dom$ *outlev level invocval pname p1 p2.*
 AllWritesUp *dom outlev level invocval pname* (**Orselect** $p1$ $p2$) =
  (**AllWritesUp** *dom outlev level invocval pname p1*) $\wedge$
  (**AllWritesUp** *dom outlev level invocval pname p2*))
($\forall dom$ *outlev level invocval pname b p1 p2.*
 AllWritesUp *dom outlev level invocval pname* (**If** b $p1$ $p2$) =
  $b \Rightarrow$ (**AllWritesUp** *dom outlev level invocval pname p1*) |
   (**AllWritesUp** *dom outlev level invocval pname p2*))
($\forall dom$ *outlev level invocval pname outev.*
 AllWritesUp *dom outlev level invocval pname* (**Send** *outev*) =
 (*dom* (*outlev outev*) *level*))
($\forall dom$ *outlev level invocval pname received response.*
 ¬(**AllWritesUp**
  dom outlev level invocval pname (**Receive** *received response*)))
($\forall dom$ *outlev level invocval pname invocation.*
 AllWritesUp *dom outlev level invocval pname* (**Call** *invocation*) =
  ((ISL *pname*) $\Rightarrow$
  ((*invocation* = OUTL *pname*) $\vee$
  (**AllWritesUp** *dom outlev level invocval pname* (*invocval invocation*))) |
  ((∃*param*. *invocation* = ((OUTR *pname*) *param*)) $\vee$
  (**AllWritesUp** *dom outlev level invocval pname* (*invocval invocation*))))))
($\forall dom$ *outlev level invocval pname buffering buf p.*
 ¬(**AllWritesUp**
  dom outlev level invocval pname (**Buffered** *buffering buf p*)))

4.5 BNPSP_restrictive

BNPSP_restrictive is a relation between a dominance relation on security levels, a function mapping input events to security levels, a function mapping output events to security levels, a function mapping invocations to processes, and an atomic invocation.

($\forall dom$ *inlev outlev invocval nppname.*
 BNPSP_restrictive *dom inlev outlev invocval nppname* =
  (**BNPSP_rightform** *invocval nppname*) $\wedge$
  (**AllWritesUp** *dom inlev outlev invocval nppname*))

5 Example

This section presents an example using our techniques to specify a process and prove it secure. The example is a sorter for a token-ring station. It sends out signals showing the receipt of the token or a message from the host station and sends on messages from other stations.

We first define input events, output events, sorter invocations, and sorter PSL processes. For generality, we let the types of token-ring stations and message contents be given by type variables. There are two invocations, one for the sorter itself and the other for giving the sorter's response to input events.

```
val SortInEv_Def =
 define_type{name = "SortInEv_Def",
   type_spec =
    'SortInEv = All of bool # 'station # 'station # 'data',
   fixities = [Prefix]};
val SortInEv = ty_antiq(=='::('station,'data)SortInEv'==);
val SortOutEv_Def =
 define_type{name = "SortOutEv_Def",
   type_spec =
    'SortOutEv = Others of bool # 'station # 'station # 'data |
                 Host of one | Tokens of one',
   fixities = [Prefix,Prefix,Prefix]};
val SortOutEv = ty_antiq(=='::('station,'data)SortOutEv'==);
val SortInvoc_Def =
 define_type{name = "SortInvoc_Def",
   type_spec = 'SortInvoc = Sorter | SortInput of ^SortInEv',
   fixities = [Prefix,Prefix]};
val SortInvoc = ty_antiq(=='::('station,'data)SortInvoc'==);
val SortProc =
 ty_antiq(=='::(^SortOutEv,^SortInEv,^SortInvoc) process'==);
```

Now we introduce a constant for an unspecified function assigning security levels to token-ring stations, a constant for the host station, and a constant for the lowest security level. Again for generality, we let the type of security levels be given by a type variable. With these, we define the functions assigning security levels to input and output events. For input events, the token is assigned systemlow and other messages are assigned station_level of the station that sent them. For output events, messages passed on from another station are assigned the level of the message's sender, signals showing receipt of a message from the current station are assigned the level of the current station, and signals showing receipt of the token are assigned systemlow.

```
new_constant{Name="station_level",Ty= =='::'station->'level'==};
new_constant{Name="this_station",Ty = =='::'station'==};
new_constant{Name="systemlow",Ty= =='::'level'==};
new_recursive_definition{name="SortInLevel",
```

```
fixity=Prefix, rec_axiom=SortInEv_Def,
def= let val SortInLevel= --'SortInLevel:^SortInEv->'level'-- in
  --'(^SortInLevel (All tokenflag sender receiver data) =
      (tokenflag => systemlow | (station_level sender)))'-- end};
new_recursive_definition{name="SortOutLevel",
 fixity=Prefix, rec_axiom=SortOutEv_Def,
 def= let val SortOutLevel= --'SortOutLevel:^SortOutEv->'level'--;
      val station_level= --'station_level:'station->'level'--; in
  --'(^SortOutLevel (Others tokenflag sender receiver data) =
      (^station_level sender)) /\
    (^SortOutLevel (Host x) = (^station_level this_station)) /\
    (^SortOutLevel (Tokens x) = systemlow)'-- end};
```

We define the dominance relation on security levels as the reflexive-transitive closure of an unspecified basic order on them, define the process-valued functions interpreting the invocations, and define the function mapping the invocations to their interpretations. The process **sorter**, interpreting **Sorter**, waits for an arbitrary input event and invokes **SortInput** on that event. The function **sortInput**, interpreting **SortInput**, sends event (Tokens (--'one'--)) after receiving the token, sends event (Host (--'one'--)) after receiving a message from the host station, sends an event passing on the received message otherwise, and then calls the sorter process to wait for the next input.

```
new_constant{Name="basic_order",Ty= ==':'level->'level->bool'==};
new_definition("dom",
  let val dom = --'dom:'level -> 'level -> bool'-- in
  --'(^dom x y) = RTC basic_order x y'-- end);
new_definition("sorter",
 --'sorter:^SortProc = (Receive (\ev:^SortInEv. T) SortInput)'--);
new_recursive_definition{name="sortInput",
 fixity=Prefix, rec_axiom=SortInEv_Def,
 def =
  let val sortInput = --'sortInput:^SortInEv -> ^SortProc'-- in
  --'(^sortInput (All tokenflag sender receiver data) =
      (If tokenflag
        (Send (Tokens one))
        (If (sender = this_station)
          (Send (Host one))
          (Send (Others tokenflag sender receiver data)))) ;;
      (Call Sorter))'-- end};
new_recursive_definition{name="SortInvocVal",
 fixity=Prefix, rec_axiom=SortInvoc_Def,
 def =
  let val SortInvocVal= --'SortInvocVal:^SortInvoc->^SortProc'--in
  --'(^SortInvocVal Sorter = sorter) /\
    (^SortInvocVal (SortInput inev) = (sortInput inev))'-- end};
```

Now we give the proof that the sorter is secure, omitting the trivial proof that the dominance relation on levels is transitive as it was defined to be. For brevity, we do not give our specialized tactics here (see [5]), but they are straightforward; they expand out definitions, do case splits on possible input events, and apply structural-complexity rewrite rules until *all PSL constructs and all security predicates defined in terms of them disappear*. The only subgoals not proved automatically are "all outputs are up" conditions that all follow from the reflexivity of the dom relation.

```
val BNPSP_restrictive =
 --'BNPSP_restrictive:
     ('level ->'level->bool) -> (^SortInEv->'level) ->
     (^SortOutEv->'level) -> (^SortInvoc->^SortProc) ->
     ^SortInvoc -> bool'--;
g('^BNPSP_restrictive
     dom SortInLevel SortOutLevel SortInvocVal Sorter');
use "/home/projects/romulus/chin/new/romtactics.sml";
add_definitions_to_sml "-";
e(BNPSP_restrictive_TAC THEN
  ASM_REWRITE_TAC [SortOutLevel, SortInLevel] THEN
  MATCH_ACCEPT_TAC dom_reflexive);
```

The three-line proof here is as simple as the intuitive reason why the sorter is secure: Every output has the same level as the input that causes it. In a parameterized-process example comparing our techniques to the earlier approach of using type coercions and impure tactics, a "high water mark" file system, our techniques produced a clearer, more securely founded proof that was shorter by roughly a factor of four [5].

6   Conclusions

We have shown that making HOL object-language definitions of a process specification language and security properties sufficient for guaranteeing security of buffered server processes is both possible and practical, and seems preferable to an earlier, meta-level approach. In doing so, we have developed techniques for imposing strong typing restrictions on security specifications and effectively defining processes in terms of process-valued functions.

Possible future work includes strengthening our security conditions to make them equivalent to the conditions identified by Rosenthal [19] for establishing state restrictiveness, and then constructing a HOL proof that these conditions guarantee state restrictiveness for buffered server processes.

References

1. J. Alves-Foss and K. Levitt. A Model of Event Systems in Higher Order Logic: Sequence and Event System Theories. Technical Report CSE-90-45, Division of Computer Science, University of California, Davis, November 1990.

2. J. Alves-Foss and K. Levitt. A Security Property in Higher Order Logic: Restrictiveness and Hook-Up Theories. Technical Report CSE-90-46, Division of Computer Science, University of California, Davis, December 1990.

3. J. Alves-Foss and K. Levitt. Mechanical Verification of Secure Distributed Systems in Higher Order Logic. In *Proceedings of the 1991 International Meeting on Higher Order Logic Theorem Proving and its Applications*, pages 263–278, 1991.

4. J. Alves-Foss and K. Levitt. Verification of Secure Distributed Systems in Higher Order Logic: A Modular Approach Using Generic Components. In *Proceedings of the Symposium on Security and Privacy*, pages 122–135, Oakland, CA, 1991. IEEE.

5. S. H. Brackin and S-K Chin. Technical Report (in progress).

6. J. Camilleri and T. Melham. Reasoning with Inductively Defined Relations in the HOL Theorem Prover. Technical Report 265, University of Cambridge Computer Laboratory, August 1992.

7. Department of Defense. *Trusted Computer System Evaluation Criteria*, December 1985. DoD-5200.28-STD.

8. M. J. C. Gordon. The HOL System Description. Cambridge Research Centre, SRI International.

9. C. A. R. Hoare. *Communicating Sequential Processes*. Series in Computer Science. Prentice-Hall International, Englewood Cliffs, NJ, 1985.

10. D. McCullough. Specifications for multilevel security and a hook-up property. In *Proceedings of the Symposium on Security and Privacy*, pages 161–166, Oakland, CA, April 1987. IEEE.

11. D. McCullough. Foundations of Ulysses: The theory of security. Technical Report RADC-TR-87-222, Rome Air Development Center, May 1988.

12. D. McCullough. Noninterference and the composability of security properties. In *Proceedings of the Symposium on Security and Privacy*, pages 177–186, Oakland, CA, April 1988. IEEE.

13. D. McCullough. A hookup theorem for multilevel security. *IEEE Transactions on Software Engineering*, 16(6):563–568, June 1990.

14. T. F. Melham. Automating Recursive Type Definitions in Higher Order Logic. Technical Report 146, University of Cambridge Computer Laboratory, January 1989.

15. R. Milner. Communication and Concurrency. Prentice-Hall, New York, 1990.

16. ORA. Romulus: A computer security properties modeling environment, final report - volume 1: Overview. Technical report, ORA, June 1990.

17. ORA. Romulus, Release 2.0. Odyssey Research Associates, December 1992.

18. D. Rosenthal. An approach to increasing the automation of the verification of security. In *Proceedings of Computer Security Foundations Workshop*, pages 90–97, Franconia, NH, June 1988. The MITRE Corporation, M88-37.

19. D. Rosenthal. Implementing a verification methodology for McCullough security. In *Proceedings of Computer Security Foundations Workshop II*, pages 133–140, Franconia, NH, June 1989. IEEE Computer Society Press.

20. D. Rosenthal. Security models for priority buffering and interrupt handling. In *Proceedings of Computer Security Foundations Workshop III*, pages 91–97, Franconia NH, June 1990. IEEE Computer Society Press.

21. D. Rosenthal. Modeling Restrictive Processes That Involve Blocking Requests. In *Proceedings of Computer Security Foundations Workshop VI*, pages 27–38, Franconia NH, June 1993. IEEE Computer Society Press.

Safety in Railway Signalling Data: A Behavioural Analysis

M. J. Morley (mjm@lfcs.ed.ac.uk)

Laboratory for Foundations of Computer Science, Department of Computer Science, University of Edinburgh, Edinburgh EH9 3JZ

Abstract. Higher Order Logic is used to analyse safety properties of a computer based railway signal control system. British Rail's Solid State Interlocking is a data driven controller whose behaviour is governed by rules stored in a geographic database. The correctness of these data are highly critical to the system's safe operation. Taking as our starting point Gordon's implementation of program logics in Higher Order Logic, we formalise the data correctness problem in a natural way, designing tactics to automate much of the verification task. Our approach requires quadratic time and linear space, and is thought to be adequate for the current generation of Solid State Interlockings. Properly formalised, the compositional proof strategy we suggest will allow us to scale up our analyses arbitrarily.

Keywords: Safety-critical systems, application specific languages, railway signalling, higher-order logic

1 Introduction

A Solid State Interlocking [Cri87] is a generic signal control system developed by British Rail for use on mainline railways to replace the more traditional, more expensive and less flexible electromechanical systems currently in use. Each installation consists of one or more central control processors (SSIs) connected to a number of trackside modules via a reliable, high speed communications link. Messages from the trackside modules to the central processor keep the SSI informed as to the current state of the railway; messages from the SSI are commands that effect changes in signal aspects, points switches, and so on. For reliability and fault tolerance, the major hardware components are duplicated; to ensure early detection of communications failures a cyclic polling strategy is implemented so that trackside hardware can "fail safe" in the absence of a timely message from the central computer. This system is generic in that the main control loop is implemented in a software interpreter, while the SSI's "map" of the network under its control is compiled into a geographic database containing all the interlocking logic necessary to operate the network. While the system maintains safety, the operator's task is primarily to input requests for the setting and cancellation of routes for trains to proceed in the efficient running of the network.

From the perspective of engineering safety-critical systems there are several interesting aspects of the system sketched above that merit consideration. In his overview, Cribbens [Cri87] discusses many of these issues, but one remains a serious problem: the correctness of the geographic data [Mit90]. These data are ultimately responsible for the the outgoing commands from the SSI to the signals and points modules – inconsistencies here could lead to unsafe states of the railway in the otherwise correct functioning of the SSI. These data, and the Geographic Data Language in which they are expressed, are therefore the focus our safety analysis. In the next section we describe our underlying model of the system, arguing that many important safety properties of the system are independent of the particular polling strategy implemented, and of the control interpreter itself. In the following section we describe how HOL is used to formalise the data correctness problem and how tactics automate our analysis of geographic data invariants. We believe our proof method is practical for the current generation of Solid State Interlockings which cover a relatively small geographic area (up to about 30 km, depending on the complexity of the network) but it is unlikely to scale well beyond this. We therefore conclude this paper with an outline formulation of how one might decompose the verification task in general, but leave open the issue of how to formalise this in HOL.

2 Geographic Data

During the execution of its main control loop the SSI will address each of the trackside modules to which it is attached with a command telegram and will expect a immediate reply from each module in turn. This command/reply cycle, known as a minor cycle, also involves a number of other activities: in particular an input request from the signalman's control panel may be processed, and a small block of data will be read in order to calculate the next state of the SSI's image of the railway. This image, or the internal state of the SSI, is recorded in a collection of state variables, some of which are updated on receipt of reply telegrams from the trackside hardware, and others of which are simply control flags adjusted by the SSI. Crucial to our understanding here is the idea that command telegrams are prepared by directly referring to the current image of the railway.

Given the above operational understanding of the SSI it is natural to model the system as a finite state automaton. In an earlier article [Mor91] we explored the next state behaviour of the SSI by modelling it as a finite transition system, while in a recent paper from British Rail Research [IM92] the model presented is that of an input/output automaton. Analyses of a more axiomatic character, of similar problems, may be found in [AC91] and [SS90]. The approach we take here is rather different from these, although the inductive proof method developed in Section 3 originated in [Mor91]. We seek to avoid both the enormous spatial complexities of the automata, and the difficulty of scaling methods which are essentially based on tautology checking. In particular, we abstract away entirely from the input/output behaviour of the SSI and focus instead only on the transitions of the internal state that are induced by the geographic data.

It would not be appropriate for us here to embark on a detailed explanation of the principles of railway signalling, but in order for the reader to gain a feel for the data we are so concerned about it is worth trying to understand a simple example. Note that in presenting this example we have here, as in the rest of this paper, taken a few liberties with the precise notation, and the terminology, used by signalling engineers. Figure 1 illustrates a typical, though fairly simple loop. Each named, physical entity in the network, be it a signal or a point switch, etc.,

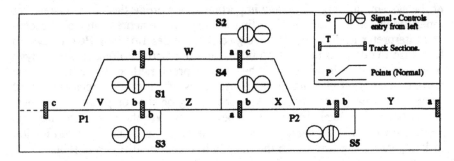

Fig. 1. Example network layout: route R_{53} is from signal S_5 to signal S_3.

is associated with a state variable of appropriate type in the SSI. For example, the variable P_2 records the current state of that point switch (normal for east-west traffic, reverse for east-northwest traffic) while the variable Z records the state of that track section (clear if no train is standing there, occupied otherwise.) Control variables also configure the internal state of the system, there being a control variable for each logical entity defined. For example *routes* are logical entities, R_{53} being the route from the entry signal (S_5) to the entry signal of the next route (S_3). *Sub-routes* are also logical entities, these being route components that traverse a single track section. R_{53} comprises the two sub-routes X_{ab} and Z_{ab}; sub-routes such as Y_{ba} are components of more than one route. Routes and their component sub-routes will be set in response to requests from the signal control panel, while sub-routes will be cleared automatically in the order in which they are traversed. These actions are controlled by rules in the geographic data which directly encode the interlocking logic.

2.1 Geographic Data Language

Rules in the geographic data are conceptually organised according to their function. We have rules to control signal aspects; rules for preparing command telegrams; rules for setting routes; rules responsible for clearing control flags once a route has been cancelled, and so on. Motivated by the safety properties we wish to examine in Section 3, we focus here on the latter two classes of data which we can conveniently interpret as guarded commands. While some rules in the data do involve short sequences of commands, we note that the language is too weak to express iteration or looping of any kind.

Route Request Data: These rules are executed whenever a request from the signal control panel is received to set up a route – in effect, to set the points switches appropriately and to set the signal to green. These data do not change the signal aspect; they merely set up some of the correct conditions for that to occur. For example:

Q_{53}	*if* P_2 *cnf,* Z_{ba} 0, X_{ba} 0
	then P_2 *cn,* R_{53} 1, X_{ab} 1, Z_{ab} 1

The 'guards' here are to test whether it is safe to set up this route: the two sub-routes mentioned must be clear and the points should either be normal, or free to go normal. (The 'free to go normal' condition is a test on the reverse direction sub-routes, but we do not dwell on these details here.) Only if the test succeeds are the variables in the 'command' part updated. As the execution of these rules is uninterruptable in the minor execution cycle of the SSI we model the updating action as a multi-assignment. We wish to verify that the tests in such rules are sufficient to preserve safety.

Sub-Route Release Data: Whereas the route request rules are executed in random order, these are executed in a fixed, though arbitrary order during the polling cycle. We make no assumptions here about the execution order as the safety properties we wish to demonstrate should hold irrespectively. Sub-route release rules are drawn from a more general class of flag operations:

X_{ba} 0	*if* X 0, R_4 0
Y_{ba} 0	*if* Y 0, X_{ca} 0, X_{ba} 0

These too are guarded commands, and they always follow the same pattern: the first sub-route on a route may be cleared once the route has been cancelled and the train has left the first track section; subsequent sub-routes may be cleared once all immediately preceding sub-routes are cleared. For safety we wish to ensure, for example, that no sub-route may be cleared unless the route has already been cancelled.

The execution model we now adopt is as follows: in each iteration of the infinite control loop of the SSI we suppose that one of these guarded commands will be chosen for execution. If the guard is true then the internal state will be updated accordingly; otherwise no action is taken. Since, at present, we are only concerned with safety we do not impose any fairness constraints on the model and, since our commands are always finite, we need not concern ourselves with termination. We emphasise that this is a model of the system's behaviour. The SSI's control loop is entirely deterministic and is of course subject to many more constraints than supposed here. Nevertheless, if we can demonstrate some interesting properties of this system, they are guaranteed to hold when we restrict the non-determinism through imposing a particular polling strategy on the SSI and execution order in the geographic data. Our approach to demonstrating properties of this system is through an analysis of invariants as described in the next section. There we use Floyd-Hoare logic to reason about a subset of the GDL embedded in Higher Order Logic and provide tactics to automate the verification task for the above classes of data. First, we briefly describe the manner in which this model is encoded in HOL.

2.2 Geographic Data in Higher Order Logic

In formalising the data correctness problem in HOL we have not been entirely faithful in our treatment of the Geographic Data Language. We have, for example, chosen to represent all state variables as natural numbers rather than providing specific datatypes for signals, points, etc. In any case, these datatypes are not complex record structures in the GDL; they are fixed length bit-patterns and bit-wise logical operations are all that one has with which to manipulate them. The sub-language of GDL we consider is:

$$C ::= \tilde{\mathcal{X}} := \tilde{\mathcal{E}} \mid if\ B\ then\ C\ \{else\ C\} \mid C\ ;\ C$$
$$B ::= \mathcal{X} = \mathcal{E} \mid \mathcal{X} < \mathcal{E} \mid B \vee B \mid B \wedge B$$

This language is slightly more general than is needed for the examples considered in this paper, but slightly less so than the GDL. Here expressions ($\mathcal{E}$) are functions involving program variables and natural number constants – normally just constants. We follow Gordon's example [Gor89] very closely, giving this language a simple relational semantics, interpreting object language variables as strings, and states as functions : string -> num. We then derive the usual rules for Hoare logic with partial correctness specifications from our semantics, denoting the pre- and post-conditions by functions : state -> bool. The main deviation from [Gor89] is the assignment axiom which now deals with (parallel) multi-assignments in general:

$$\vdash \{\mathcal{P}[\tilde{\mathcal{E}}/\tilde{\mathcal{X}}]\}\ \tilde{\mathcal{X}} := \tilde{\mathcal{E}}\ \{\mathcal{P}\}\ .$$

In turning the ideas of this paper into a mature tool for checking geographic data we should certainly reinforce the intuition of signalling engineers with a more direct embedding of GDL constructs and datatypes in Higher Order Logic. We also need a parser for the routine task of translating GDL into HOL's internal syntax, not to mention a prettyprinter to hide that syntax. For the time being these niceties can be ignored.

3 Verifying Safety Properties

Trains may come to grief in many ways, even when drivers obey signals and speed limits. They may collide head-on or one into the rear of another, or points may derail a train if incorrectly aligned. A train moving in direction X_{ba} (Figure 1) may be derailed if the points are set to reverse, or a train on R_{53}, moving through the same points, may collide with a train waiting at S_2. Signalling engineers test their designs very thoroughly and have an excellent record in building interlockings that avoid these kinds of errors. Now, with the advent of the

GDL, we can offer formal, automated support in the behavioural analysis of their designs. Our innovation is to conduct this analysis in a framework independent of any model of the physical railway, concentrating solely on the interlocking logic. Let us now be specific about the properties of the data we wish to demonstrate:

SP1 No more than one of the sub-routes over a track section is set at any one time;

SP2 If a route is set, then all its component sub-routes are set;

SP3 If a sub-route over a track section containing points is set, then the points are correctly aligned with that sub-route.

These properties are formally characterised below in the predicate I which should hold invariantly. In fact I does not hold in *all* SSI states for in certain initialisation modes all sub-routes are *deliberately* set, with all routes clear and signals at red – this being the most restrictive state of the system. Our analysis therefore only demonstrates that *if* the SSI is in a state satisfying the invariant, its future evolution, governed by the rules in the geographic data, maintains that invariant. Our behavioural analysis concentrates on safety: special arguments, and a modal logic, are needed to demonstrate that a safe state may be reached from an arbitrary (unsafe) initial state.

3.1 Geographic Data Invariants

The first property of interest, **SP1**, is a mutual exclusion property designed to ensure that no two routes inadvertently permit simultaneous access to the same section of track. In a new HOL theory we make some definitions to help write down the invariant, and prove a few simple lemmas:

$$\vdash \,!a\,b.\ \mathsf{MX}\,(a, b) = (a = 0) \ \lor \ (b = 0)$$
$$\vdash \,!a.\ \mathsf{MX}\,(a, 0) = \mathsf{T}$$
$$\vdash \,!a\,b.\ \mathsf{MX}\,(a, b) \implies \mathsf{MX}\,(a, 0)$$

and so on. The idea is that a term like $\mathsf{MX}\,(Zab, Zba)$ is only false if both variables are non-zero – i.e. if both sub-routes are set. We can now characterise the first of the properties above in a conjunctive term like:

$$I_1 \equiv \mathsf{MX}\,(Yab, Yba) \ \land \ \mathsf{MX}\,(Zab, Zba) \ \land \ \cdots$$
$$\mathsf{MX4}\,(Xab, Xba, Xac, Xca) \ \land \ \cdots$$

there being one conjunct for each track section defined in the interlocking. MX4 is defined analogously to MX above.

SP3 above perhaps needs a little interpretation, but a glance at Figure 1 should convince the reader that: if either X_{ac} or X_{ca} are set then P_2 should be reverse; if either X_{ab} or X_{ba} are set then P_2 should be normal. With this in mind we construct two further conjunctive terms, here expanding definitions like MX:

$$I_2 \equiv R53 = 1 \implies (Xab = 1 \ \land \ Zba = 1) \ \land$$

$$R51 = 1 \implies (Xac = 1 \wedge Wba = 1) \wedge \cdots$$
$$I_3 \equiv (Xab = 1 \vee Xba = 1) \implies P2 = cn \wedge$$
$$(Xac = 1 \vee Xca = 1) \implies P2 = cr \wedge \cdots$$

Thus the invariant for the loop in Figure 1 is expressed in the term: $I \equiv I_1 \wedge I_2 \wedge I_3$. In general the invariants with which we expect to deal are rather large conjunctive terms such as these, each conjunct expressing some *local* property of the data.

3.2 Tactics to Demonstrate Invariance

The invariant having been fixed, we set about the proof in Hoare logic that a command C satisfies this by setting a goal in HOL like: $? \vdash \{I\}\, C\, \{I\}$. A similar goal is set for each of the guarded commands specified in the route request and sub-route release data for the system under consideration. Notice that when C is *if B then $\widetilde{X} := \widetilde{\mathcal{E}}$* the top goal simplifies immediately to a verification condition like: $I \wedge B \implies I[\widetilde{\mathcal{E}}/\widetilde{X}]$.

Sub-route Release Data: In order to see how to automate these proof obligations we consider, as an example, the second of the sub-route release rules quoted in Section 2.1, viz: $Y_{ba}\, 0$ *if* $Y\, 0$, $X_{ca}\, 0$, $X_{ba}\, 0$. Although we have not defined all the terms in I explicitly it is clear that only a few terms are affected by this rule. Thus most of the subgoals of the conjunctive goal $I[0/Yba]$ are solved by recognising the appropriate term among the hypotheses. The remainder, on rewriting, are:

$$? H \vdash MX\,(Yab, 0) \tag{1}$$
$$? H \vdash R2 = 1 \implies (Xca = 1 \wedge 0 = 1) \tag{2}$$
$$? H \vdash R4 = 1 \implies (Xba = 1 \wedge 0 = 1) \tag{3}$$

where H is $I \wedge Y = 0 \wedge Xca = 0 \wedge Xba = 0$. All these goals are essentially solved in the same manner: by looking up in the hypothesis list the term (in I) from which they are derived and applying a suitable tactic to the goal generated by undischarging that term. In the case of (1) this is a matter of matching a pre-proven lemma (e.g. the third listed in Section 3.1) while a little rewriting using the hypotheses is all that is required in the other two goals. In particular, from:

$$? H' \vdash (R2 = 1 \implies (Xca = 1 \wedge Yba = 1)) \implies R2 = 1 \implies F$$

we obtain $(R2 = 1 \implies F) \implies R2 = 1 \implies F$ by virtue of the hypothesis $Xca = 0$ which is derived from the guard in the command. If this hypothesis were absent the goal (2) would remain unproven and our sub-route data tactic would fail. Such an omission would most likely be due to an error in the data for Y_{ba} – say, instead of the correct rule quoted above we had: $Y_{ba}\, 0$ *if* $Y\, 0$, $X_{ac}\, 0$, $X_{ba}\, 0$.

The failure of a tactic does not necessarily imply that the data are erroneous. The failure encourages us only to look harder at the rules rejected, and perhaps to design better tactics. We note that the error in Y_{ba} 0 *if* U 0, X_{ca} 0, X_{ba} 0 would not be identified by the sub-route data tactic since the invariant makes no reference to track state variables like U and Y. However, this particular kind of error is one that can be identified through a purely syntactic analysis. In general, we suppose that this computationally inexpensive syntactic analysis precedes the more difficult behavioural analysis discussed here.

Route Request Data: Route request rules are much less uniform than the sub-route release rules considered above and rewriting from the hypotheses in I leaves more sub-goals to solve in general. By considering the verification condition for the rule Q_{53} above:

$$I \wedge Zba = 0 \wedge Xba = 0 \wedge (P2 = cn \vee (Xac = 0 \wedge Xca = 0)) \implies I'$$

where I' abbreviates $I[cn/P1, 1/R53, 1/Xab, 1/Zab]$, we see that the proof tree bifurcates due to the disjunctive term in the antecedent – a term that arises from the 'points free to move' condition alluded to in Section 2.1. Furthermore, it is clear from the sub-goal MX4 $(1, Xba, Xac, Xca)$ that some resolution among the hypotheses will be necessary in order to infer the values of the uninstantiated variables. In this case, on the left hand branch of the proof tree, we use $(Xac = 1 \vee Xca = 1) \implies P2 = cr$ and $P2 = cn$ to add $Xac = 0 \wedge Xca = 0$ to the assumptions, so proving the goal. For this tactic we therefore find it necessary to exploit dependencies between terms of the invariant.

3.3 Experimental Results

Before drawing any firm conclusions we need a sanity check: when we move on to consider real data do the rules not become substantially less regular than those of the example considered here, so causing our tactics to fail more often than we would wish? For the sub-route release data, the geographic data manual [Bri90] answers this question in the negative: these data *always* follow the pattern described in Section 2.1. However, routes can become more complex in two ways: they can become longer, or they can traverse more points. Although we do not have the space to elaborate, we note that complexity of the first kind does lead to difficulty in verifying the mutual exclusion property (**SP1**) in general, given the form of invariant described above. Our solution is to strengthen the invariant. Complexity in the route structure that arises in crossing points leads to greater branching in the proof tree. This is not a problem in principle, merely more computationally intensive. However, it appears from the wealth of examples in [Bri90] that signalling engineers generally avoid disjunctive tests in geographic data: or-branching is apparently too expensive, in real-time, in the SSI's minor execution cycle.

For a particular SSI database, the total number (r) of route and sub-route rules will be fixed. It turns out that the number of terms in the invariant sketched in Section 3.1 is proportional to $r + p$, where p is the number of points, suggesting that our verification method has quadratic time complexity. This is most easily seen by considering the sub-route data tactic in Section 3.2 which forms $O(r)$ sub-goals for each of the r rules to check. Space requirements are linear in the size of the invariant. On an SS/10, with 160 Mb of memory, we found that the loop in Figure 1 ($r = 22$) could be checked in a matter of three to four minutes; a system a little more than three times larger ($r = 72$) succumbed in about 40 minutes. Given the data, we would expect the Leamington Spa signalling scheme [Cri87], where $r \approx 200$, to require about seven hours. This latter system is typical, occupying 2/3 of the maximum capacity of a single SSI.

Experienced HOL users may complain that our tactics are rather slow. Indeed, this is the case since they are implemented using the very general tactics and tacticals of the HOL system. We would expect to implement specific tactics which could manipulate the large number of assumptions in the invariant efficiently, providing a faster data structure for assumption 'lists', thereby gaining a substantial speed-up. Nevertheless, the quadratic time complexity, which is the price we pay for a simple-minded approach to proving local invariance properties globally, remains the limiting factor of our technique. We now briefly consider how we could improve this situation.

4 Problem Decomposition

We noted above that the global invariant describes a collection of local safety properties. The rules too, refer only to local data – e.g. in setting a route like R_{53} one need not inquire about remote areas of the network, to the left of Figure 1. Denoting the global invariant by $\mathcal{I}$, and the data by $\mathcal{G}$, what we now seek are some heuristics for decomposing the global correctness problem $\forall d \in \mathcal{G}, \vdash \{\mathcal{I}\} d \{\mathcal{I}\}$. Here we consider two possible heuristics.

Firstly, note that from $\vdash \{\mathcal{P}\} C \{\mathcal{P}\}$ and $\vdash \{\mathcal{Q}\} C \{\mathcal{Q}\}$ we can derive the theorem $\vdash \{\mathcal{P} \land \mathcal{Q}\} C \{\mathcal{P} \land \mathcal{Q}\}$. Now, given a decomposition like $\mathcal{I} \equiv \mathcal{I}_1 \land \mathcal{I}_2$, and datum $d \in \mathcal{G}$ which we suppose is of the form *if* $\mathcal{B}_d$ *then* $\widetilde{\mathcal{X}_d} := \widetilde{\mathcal{E}_d}$, we can proceed with the two separate goals: $? \vdash \{\mathcal{I}_1\} d \{\mathcal{I}_1\}$ and $? \vdash \{\mathcal{I}_2\} d \{\mathcal{I}_2\}$. For a good choice of heuristic we would hope to find, reasonably often, that the free variables in d and $\mathcal{I}_2$ (say) would be disjoint – i.e. $fv(d) \cap fv(\mathcal{I}_2) = \emptyset$. In this case, the second of the above two goals is trivial since the verification condition $\mathcal{I}_2 \land \mathcal{B}_d \implies \mathcal{I}_2$ matches a simple theorem of propositional logic. Note that this test is a syntactic constraint which is sufficient to prove the top goal. Also, whenever this constraint is met, Gordon's VC_TAC generates the above verification condition so we need only match the appropriate theorem. However, when the syntactic constraint fails to hold we revert to solving: $? \vdash \{\mathcal{I}\} d \{\mathcal{I}\}$.

For definiteness, the troublesome data for the system represented in Figure 2 are those associated with the cross-boundary routes (those from region 2 to S_1, and conversely) and the sub-route release rules for Y_{ab} and U_{ba}. The difficult question then is how to separate $\mathcal{I}$ into its two (or more) components. In this example we place the notional boundary at a point corresponding to a natural interlocking boundary as identified in the guidelines set out in [Bri90]. Thus, we

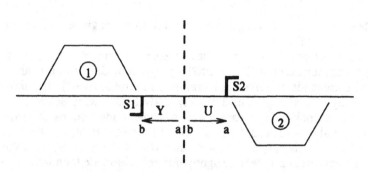

Fig. 2. Routes over interlocking boundaries: From region 1 they terminate at S_2.

use intuitions from signalling engineering to decompose the invariant according to this geographic separation.

The other decomposition heuristic we consider here is more dynamic in character. We can proceed from the top goal as follows:

$$? \vdash \{\mathcal{I}\} \, d \, \{\mathcal{I}\}$$
$$\Longleftrightarrow ? \vdash \mathcal{I}_\alpha \wedge \mathcal{I}_\beta \wedge \mathcal{B}_d \Longrightarrow (\mathcal{I}_\alpha \wedge \mathcal{I}_\beta)[\widetilde{\mathcal{E}_d}/\widetilde{\mathcal{X}_d}]$$
$$\Longleftrightarrow ? \vdash \mathcal{I}_\alpha \wedge \mathcal{I}_\beta \wedge \mathcal{B}_d \Longrightarrow \mathcal{I}_\alpha[\widetilde{\mathcal{E}_d}/\widetilde{\mathcal{X}_d}]$$

once again, assuming $\mathsf{fv}(d) \cap \mathsf{fv}(\mathcal{I}_\beta) = \emptyset$. Here we do not impose the separation into $\mathcal{I}_\alpha$ and $\mathcal{I}_\beta$; we allow the tactic implementing these steps to perform the task *for each* $d \in \mathcal{G}$. It follows from the fact that $(a \Longrightarrow c) \Longrightarrow (a \wedge b \Longrightarrow c)$ that *if* we can show $\mathcal{I}_\alpha \wedge \mathcal{B}_d \Longrightarrow \mathcal{I}_\alpha[\widetilde{\mathcal{E}_d}/\widetilde{\mathcal{X}_d}]$ the top goal is solved. There may be some doubt as to whether $\mathcal{I}_\alpha \wedge \mathcal{B}_d$ is strong enough for us to do this in general, but if not we can fall back to the goal in the third line above. Note that the heuristics we envisage only take sub-sets of the assumptions in the goal so the monotonicity of the logic guarantees their correctness.

5 Conclusions

Although we have not been able to experiment with full-sized examples, the methods described work well for those we have tried. The decomposition heuristics mentioned should be easy to implement, so we believe our overall approach to be generally useful – subject to the provision of some software to support translation between GDL and the HOL syntax. It is also highly flexible, which is important as there are other classes of geographic data, along with their respective safety properties, that we ultimately wish to consider. Moreover, the method can be packaged into a system that novice proof engineers, but expert signalling engineers, will find convenient. This entails more work than just providing powerful tactics since HOL's sub-goal package needs tailoring to handle the large numbers of assumptions dealt with in the proof tree. One could also provide support in generating the invariant from the engineers' scheme plans – drawings such as Figure 1.

There is nothing particular about our execution model of the SSI: its roots can be found, for example, in the UNITY program model, or in the language of guarded commands discussed, for example, in [Tre93, Gri81]. The model, for its simplicity, seems natural for the safety analysis of geographic data. However, if we want to look beyond safety, and consider some notions of progress or weak eventuality – such as demonstrating that a safe state may eventually be reached from an arbitrary initial state – then we may have to give our solution to the data correctness problem a slightly different semantic foundation, perhaps along the lines of [Tre93] where a thorough account of program logics for non-deterministic, sequential programs is given.

Acknowledgements: Thanks mainly to George Cleland for carefully reading the draft of this paper. Ian Mitchell at British Rail Research provided much helpful insight into railway signalling and the SSI. No formal collaboration with British Rail has been entered into in conducting this research.

References

[AC91] Will Atkinson and Jim Cunningham. Proving properties of a safety critical system. *Software Engineering Journal*, pages 41–50, March 1991.

[Bri90] British Railways Board. SSI data preparation guide. Issue SSI8003, February 1990.

[Cri87] A. H. Cribbens. Solid State Interlocking (SSI): an integrated electronic signalling system for mainline railways. *IEE Proc.*, 134(3):148–158, May 1987.

[Gor89] M. J. C. Gordon. Mechanizing Programming Logics in Higher Order Logic. In P. A. Subrahmanyam and Graham Birtwistle, editors, *Current Trends in Hardware Verification and Automated Theorem Proving*. Springer-Verlag, 1989.

[Gri81] David Gries. *The Science of Programming*. Spriner-Verlag, 1981.

[IM92] Michael Ingleby and Ian Mitchell. Proving Safety of a Railway Signalling System Incorporating Geographic Data. In *Proceedings of IFAC SAFECOMP '92*, October 1992.

[Mit90] I. H. Mitchell. Proposal for an SSI data checking tool. Internal publication by B.R. Research, Safety Systems Unit, June 1990.

[Mor91] M. J. Morley. Modelling British Rail's Interlocking Logic – Geographic Data Correctness. Technical Report ECS-LFCS-91-186, University of Edinburgh, November 1991.

[SS90] G. Stålmark and M. Säflund. Modelling and verifying systems and software in propositional logic. In *Proceedings of SAFECOMP '90*, pages 31–36, 1990.

[Tre93] Gavan Tredoux. Mechanizing nondeterministic programming logics in higher order logic. Monograph, March 1993.

On the Style of Mechanical Proving

ISWB Prasetya

Rijksuniversiteit Utrecht, Vakgroep Informatica
Postbus 80.089, 3508 TB Utrecht, Nederland
Email: wishnu@cs.ruu.nl

Abstract. It is quite surprising that —after 5 years of experience— there has been still very little written about how to present and construct mechanical proofs. Such guidelines would definitely help a novice. Since mechanical proving is very different from proving on paper, it may take a long time before one gets accustomed to it and develops an efficient style for constructing and presenting such proofs. Traditional styles, like e.g. decomposing a problem into lemmas and incorporating hints to make the logic of a proof visible, are not straightforwardly taken over to the HOL world. In this paper we present two extensions to HOL, the DERIVATION and LEMMA packages, by which proofs can be written in very much the same format as one would on paper. They are not intended as a replacement of existing HOL mechanisms, but rather as an extension for enabling work in a higher level proof environment.

Category: Research Report

1 Combining the Man and the Machine

To many users of the HOL system, even after gaining extensive experience, the convenience of a pencil-and-paper proof would still be very tempting. Of course such proofs do not have the *consistency preserving* property which HOL provides, but on the other hand one is not bothered by complaints if one deliberately has left out the type of some function, or if one deliberately abuse some notation. Even if one is willing to accept the latter as a fair price to pay for the rigorousness offered by the machine, there is still another wonderful aspect of pencil-and-paper proofs missing: the ability to adhere to a specific proof style, resulting in well documented and easily readable proofs.

Machines are very good in doing brute resolution and rewriting, and this makes them unsurpassed for proving properties of finite domains, with tautology checking for propositional logic being the prime example. However when one wants to use machines to assist reasoning about abstract domains, things will very soon explode; especially as abstract domains are usually —and for a good reason too— infinite. One has to manually direct the machine in selecting what and how to rewrite, and which theorems exactly to use for resolution. When the proof is finally completed, a HOL veteran will not be surprised when overlooking the long and obscure code developed during the process. Showing this result to more conventional mathematicians may cause the remark that they have written PASCAL code better than that, and one will not be able to construct even a clue about the essential proof steps taken.

Without the help of machines people only use resolution and rewriting when it becomes a triviality, or common knowledge, or something which can be convincingly motivated despite the informality. People profit most from their intelligence by systematically breaking up the proof into smaller lemmas. A proof proceeds by proving one lemma after another until one finally comes to the desired one. As the inference of a lemma may depend on any —or even all— previously proven lemmas, a proof is most naturally set structured.

On the other hand, HOL with its Subgoal Package seems to enforce that a proof has a tree-like structure, and if not it has to be molded into such a structure by duplicating parts of the proof. But people (and computer scientists especially) do not like duplication, and so it becomes a sport as how to construct tree-like proofs without duplicating. We do not consider this as a preferable route. A proof should be structured as it seems most natural to one's mind. To influence this naturality with the machine's restrictions is to force man to think machine-like.

Since the lemmas reflect how one's mind views and decomposes the problem and since this is something a machine cannot do, it is best to keep it this way. The machine should however provide an excellent environment for the human to decompose the problem in any desirable way. The machine can contribute at any time whenever a lemma is sufficiently simple to be proven more or less automatically. Section 3 discusses the LEMMA package, an extension to HOL to support set structured proofs.

A special class of proofs is formed by the linear proofs. In a linear proof the derivation of the next expression depends solely on the last derived expression. Linear proofs read nice and are preferred by many. Quite a large number of useful theorems can be proven with linear proofs. Figure 1 displays an example of this style. It shows how $\nu.y \ll \nu.x$ can be derived from $x \ll y$ assuming the *Unit Design Equation*.

Fig. 1. Example of Derivation: ν Inversion

$$
\begin{aligned}
& x \ll y \\
={} & \{\ (1)\ \nu \text{ is an involution, that is: } \forall x \,.\ \nu.(\nu.x) = x\ \} \\
& \nu.(\nu.x) \ll y \\
={} & \{\ (2)\ \text{Unit Design Equation: } \forall x, y\,.\ d \ll x + y\ =\ \nu.x \ll y\ \} \\
& d \ll \nu.x + y \\
={} & \{\ (3)\ \text{Unit Design Equation}\ \} \\
& \nu.y \ll \nu.x
\end{aligned}
$$

The hints tell us how to derive each expression from the previous one although it may not be very precise by only mentioning the most important trick of the particular derivation step. The step is usually small enough for the reader to reconstruct the missing information —usually they are either trivialities or well known facts. For example hint (1) is very precise while hint (3) assumes the reader is aware that + is commutative. Indeed, such a triviality is best verified by a machine.

This style has been successfully applied, especially in the calculational derivation of algorithms from specifications [3]. We ourselves have practiced it with a great pleasure over the years and we think it will be a great enhancement if mechanical proving environments would also support such proofs. This is not to suggest to only use the machine as a piece of paper. We readily admit that paper and machine are two basically different proof mediums. Still, when dealing with abstract domains where finiteness is not a necessary property —or not even desired— mechanical proving becomes an extensive interaction between man and machine. That is why we advocate that a successful, mechanically supported proof style should be a combination of the typical machine proof styles and paper-and-pencil proof styles.

HOL is definitely powerful enough to derive the ν inversion. A typical (non-interactive) proof is given in Figure 2.

Fig. 2. Typical HOL Proof of ν Inversion

```
#let thm = TAC_PROOF
    (([UNIT_DE; PLUS_COM; NU_INVO],
    "!x (y:*dom). x<<y = (NU y) << (NU x)"),
    REWRITE_TAC [SYM(SPEC_ALL (ASSUME UNIT_DE))]
    THEN ONCE_REWRITE_TAC[ASSUME PLUS_COM]
    THEN REWRITE_TAC (map ASSUME [NU_INVO; UNIT_DE])) ;;

thm = ... |- !x y. x << y = (NU y) << (NU x)
```

The proof in Figure 2 is definitely short, but unfortunately it does not tell you very much about *how* the theorem is proved. It does tell you about what facts are used in the proof —for example the commutativity of + is used (PLUS_COM) and the Unit Design Equation is used twice (UNIT_DE)— but how the derivation is constructed, one has to decipher it from the code, or to find out by running the code in interactive mode. Either way, this is not a convenient way to follow the logic of the proof and the linear proof in Figure 1 —despite its informality at some points— will be significantly more appreciated by most human readers.

There is another significant difference. It may be the case that deriving $\nu.y \ll \nu.x$ from $x \ll y$ is more natural than the other way around. Yet in HOL it can be very tempting to follow a different order —as in the proof in Figure 2— or to introduce intermediate expressions which are less natural because, for example, some rewritings can be compacter written than others or because some resolution will create fewer garbage assumptions. A good style should not make explicit such mechanical considerations.

One may indeed ask why we bother about the logic of the proof at all. What counts is that the theorem is proven. The *'how it is proven'* is only interesting to display to let other people verify its correctness. Since HOL guarantees correctness, such *display for verification* is no longer necessary. However, verification is not the only purpose of displaying proofs. The proof may give more insight into the nature of a problem than the theorem itself. It may give a good example of an elegant use of the calculus involved. It may also involve some clever trick which may inspire other proofs. Also, one may wish to change or extend the representation of an abstract domain. This implies re-doing the proofs and modifying them accordingly. Unfortunately if the proofs were done quite some time ago —or worse, they were written by someone else— then this philosophy of *'a theorem, once proven, is all one needs to know'* will definitely cause a lot of annoying re-thinking.

There is of course a neater way of giving proofs using HOL. One can use the WINDOW library by Jim Grundy to mimic the proof style in Figure 1. One can also use the SUBGOAL_THEN tactical, a style which is displayed below.

```
#let thm = TAC_PROOF
  (([UNIT_DE; PLUS_COM; NU_INVO],
   "!x (y:*dom). x<<y = (NU y) << (NU x)"),
   REPEAT GEN_TAC
   THEN SUBGOAL_THEN
     "(x:*dom) << y = NU (NU x) << y"
     (\thm. REWRITE_TAC[thm])
   THENL [ REWRITE_TAC [ASSUME NU_INVO] ; ALL_TAC]
   THEN SUBGOAL_THEN
     "NU (NU (x:*dom)) << y  = d << (NU x) PLUS y"
     (\thm. REWRITE_TAC[thm])
   THENL [ REWRITE_TAC [ASSUME UNIT_DE] ; ALL_TAC]
   THEN SUBGOAL_THEN
     "(d:*dom) << ((NU x) PLUS y)  =  (NU y) << (NU x)"
     (\thm. REWRITE_TAC[thm])
   THEN  ONCE_REWRITE_TAC [ASSUME PLUS_COM]
     THEN REWRITE_TAC[ASSUME UNIT_DE]) ;;
```

Indeed, the proof is considerably larger than the proof in Figure 2, but do not be discouraged by its size. For a sufficiently complicated theorem this is a more than fair price to pay as the logic and flow of the proof is now made explicit.

Still, this SUBGOAL_THEN mechanism can be enhanced further in order to facilitate and encourage even better proof styles, especially by removing the necessity to provide excessive information and by improving the visual appearance. Section 2 discusses the DERIVATION package, an extension to HOL to support the linear proof style as in Figure 1.

2 The DERIVATION Package

The idea is quite similar to the WINDOW library by Jim Grundy. The actual package however is simpler to use and we expect that one will find it easier to learn.

A derivation, viewed as an object, is a list

$$E_1, \ldots, E_i, \ldots$$

such that $E_i = E_{i+1}$. In particular, by transitivity of $=$, $E_1 = E_{last}$ holds, where E_{last} is the last expression in the derivation. *Extending the derivation* with a new expression E basically corresponds to proving $E_{last} = E$.

Such a sequence construction can be generalized to a derivation with respect to some transitive relation R, which is not necessarily an equality. In this case, either $E_i \ R \ E_{i+1}$ or $E_i = E_{i+1}$ should hold. The transitive relation in the derivation in of ν inversion however is just the boolean equality.

The basic idea is to extend HOL with derivation as a new class of objects. The primitive operations are: initialization, extension, undoing, pretty printing, and extracting a theorem. These are the services provided by the package, and are supported both in interactive as well as non-interactive modes.

The mechanism is built on top of the existing proof mechanisms. This means one can use HOL as one is used to, but with the additional possibility to prove a theorem in a derivation-style. The mechanism employs global variables, which prevents the construction of nested derivations. We argue that people should keep each derivation step sufficiently small such that a one or two line hint is enough for others to see why and how the step was taken. If one should need a sub-derivation to justify a derivation step, this is a good indication that the step taken might be to big.

The global variables represent the current derivation and the transitive relation upon which the derivation is based, and furthermore the assumptions of the current derivation.

A good habit is to leave a hint for every derivation step. If the step is trivial, then say so in the hint such that those who later read it will be at least aware of the triviality and can feel safe to skip it. We thus included such hints in the HOL internal representation of a derivation, and consequently the user is required to explicitly supply such hints when extending a derivation.

This is all we say here about the internal mechanism involved, since most users are not required to be aware of it. For our purpose here, *'how to use'* the package is more important than *'how it works'*. People eager enough for the latter will probably find no difficulty in reading the code.

2.1 How a Proof with the DERIVATION Package Looks Like?

Suppose we wish to prove $E \ll G$, where $\ll$ is some transitive relation (examples of such a relation: $=, \Rightarrow, \Leftarrow$, and $<$). Let us assume that we are allowed to assume $A1$ and $A2$. Then a proof with the DERIVATION package looks like:

```
Set_Assumptions [ ('asm1', A1); ('asm2', A2) ] ;;
BD "<<" E ;;
DERIVE ("<<", E1, 'HINT: + is monotonic', Tac1) ;;
DERIVE ("<<", E2, 'HINT: asm2 and e is unit of +', Tac2) ;;
...
DERIVE ("<<", G,'HINT: follows from asm1', TacLast) ;;
```

Do not bother with the precise details yet. The first line sets up the assumptions of the derivation. The second line indicates the base relation of the derivation and the starting expression. As we are interested in proving $E \ll G$ then $\ll$ is the obvious choice of the base relation. The starting expression is of course E.

Each **DERIVE** statement is a derivation step. For example the second **DERIVE** derives $E2$ from $E1$ and **Tac2** is the tactic which does the work —more precisely the tactic should prove ?- E1 << E2. The third argument of **DERIVE** is just the hint associated with the derivation step.

Use **DERIVE** *as if you prove your theorem on paper. The hint part of* **DERIVE** *contains a human-readable information that motivates the justification of the derivation step. The tactic part is the machine codes detailing the precise rewriting and resolution that justify the step.*

One can also at any time pretty-print the derivation so far. The format is very much like in Figure 1.

In further subsections we will discuss each function in further detail. How to use the DERIVATION package is best explained by an example. We will again prove the ν inversion theorem. We consider a general triple $(d, +, \nu)$. Under the assumptions that:

1. ν is an involution, that is, $\forall x \cdot \nu.(\nu.x) = x$
2. $+$ is commutative
3. Unit Design Equation is satisfied, that is, $\forall x, y. \; d \ll x + y \; = \; \nu.x \ll y$

we will prove the ν inversion theorem, that is: $x \ll y \; = \; \nu.y \ll \nu.x$.

2.2 Setting up the Derivation

The first thing to do is to *set up your assumptions*. This is done with function **Set_Assumptions**.

```
2.1 : Set_Assumptions : (string # term) list -> void
```

The argument is a list of pair (L_i, A_i) where A_i is a boolean term representing i-th assumption and L_i is a string representing the label of assumption A_i. The label is used to refer to the assumption. The following sets up the assumptions of our example derivation:

```
#Set_Assumptions
   [('UNIT_DE', "!(x:*dom) y. d << (x PLUS y) = (NU x) << y") ;
    ('PLUS_COM',"!(x:*dom) y. x PLUS y = y PLUS x") ;
    ('NU_INVO', "!x:*dom. NU(NU x) = x") ] ;;
```

An assumption (in theorem form) can be accessed using the function RECALL.

```
2.2 : RECALL : string -> thm
```

The argument is a string which is the label of the assumption we wish to access. This should be the same label set up by function Set_Assumptions. For example RECALL 'NU_INVO' will return . |- !x. NU(NU x) = x.

Having set the assumptions, to start a derivation one has to tell HOL both the base transitive relation and the starting expression. This is done with the function BD ('Begin Derivation').

```
2.3 : BD : term -> term -> void
```

The function has two arguments. The first is the preorder relation upon which the derivation should be based. The second is the starting expression. This function initializes various global variables, and as a side effect it redisplays the list of assumptions, the base relation, and the begin expression. *The base relation has to be transitive* for the derivation to make sense. The following starts the derivation of ν inversion:

```
BD "=:bool->bool->bool" "(x:*dom) << y" ;;
```

2.3 Extending the Derivation

To extend the derivation with a new expression one has to provide (a) the new expression, (b) the relation between the new expression and the last expression derived so far —this is either the base relation or equality—, (c) the hint, and (d) the tactic that justifies the derivation step. This is done with the function DERIVE.

```
2.4 : DERIVE : (term # term # string # tactic) -> void
```

Let say that X is the last derived expression and that << is the base preorder of the derivation. Then the call DERIVE (R,Y,Hint,Tac), if it succeeds —that is, if Tac proves ?- R X Y—, will add Y to the derivation. For the derivation to make sense R should be either << or =. Hint is a string representing the hint associated with the derivation of Y from X. In case of a successful DERIVE, the theorem |- R X Y is also printed.

The complete HOL proof of ν inversion using DERIVE is displayed below. It definitely looks neater then the proof with SUBGOAL_THEN as displayed in the earlier section.

```
DERIVE(":=:bool->bool->bool",
       "(NU (NU x)) << (y:*dom)",
       'HINT: NU is an involution',
       REWRITE_TAC[RECALL 'NU_INVO']) ;;

DERIVE(":=:bool->bool->bool",
       "d << (NU x) PLUS (y:*dom)",
       'HINT: by Unit Design Equation',
       REWRITE_TAC [RECALL 'UNIT_DE']) ;;

DERIVE(":=:bool->bool->bool",
       "NU y << (NU (x:*dom))",
       'HINT: PLUS is commutative and Unit Design Equation',
       ONCE_REWRITE_TAC [RECALL 'PLUS_COM']
       THEN REWRITE_TAC[RECALL 'UNIT_DE']) ;;
```

Let X be the starting expression, Y the last derived expression, and $\ll$ the base relation. Thus Y is derivable from X. Assuming that each derivation step consistently preserves the $\ll$ relation, it follows from the transitivity of $\ll$ that $X \ll Y$ holds. To extract this theorem from the derivation use the function ETD (Extract Theorem from Derivation).

```
2.5 : ETD : void -> thm
```

For example, after the complete derivation of ν inversion is fed to HOL, ETD will return the ν inversion theorem:

```
#ETD();;
... |- x << y = (NU y) << (NU x)
```

At any moment one can request a pretty print of the derivation constructed so far with the function DERIVATION.

```
2.6 : DERIVATION : void -> void
```

For example after the complete proof of ν inversion is fed to HOL, DERIVATION will print as displayed below. Notice how much it looks like the derivation in Figure 1.

```
#DERIVATION();;

"Derivation" (* *)
    "x << y"
"$=" (*HINT: NU is an involution*)
    "(NU(NU x)) << y"
"$=" (*HINT: by Unit Design Equation*)
    "d << ((NU x) PLUS y)"
"$=" (*HINT: PLUS is commutative and Unit Design Equation*)
    "(NU y) << (NU x)"

() : void
```

2.4 Interactive Derivation

The last argument in the DERIVE statement is a tactic that justifies the derivation
step. The question is of course how one can construct this tactic. The answer is:
in the usual way, by working interactively. I assume the reader is familiar with
this practice; otherwise he might consult the HOL Tutorial book [1].

There are only three operations for interactive derivation : Dg (Derivation
step set Goal) to set the target expression of a derivation step, De (Derivation
Extend) to replace the standard goal expander e, and DU (Derivation Undo) to
undo a derivation step.

```
2.7 : Dg : (term # term # string) -> void
```

Let X be the last derived expression so far, << be the base relation, and Asml
be the list of assumptions of the derivation. Dg (R,Y,Hint) invokes the subgoal
package and makes ?- R X Y as the goal and Asml as the assumptions. For the
derivation to make sense R must be either << or =. Hint is a string representing
the hint associated with the derivation step.

One is free to manipulate the goal thus created by using b,e and r as usual.
One can however opt to use De instead of e to automatically extend the derivation.

```
2.8 : De : tactic -> void
```

De works in the same as e, except that when the goal is proven, the derivation
is automatically extended with the expression which was mentioned in the call
to Dg. Also the resulting derivation is then printed.

The figure below shows how to use Dg to setup the derivation of $\nu.(\nu.x) \ll y$
from $x \ll y$. Notice that this actually means proving the equality $x \ll y \; =$

$\nu.(\nu.x) \ll y$. A simple rewriting will prove this. The resulting derivation is shown automatically. Notice that a hint can —as in this case— provide more information than the tactic that proves a derivation step.

```
#Dg(":bool->bool->bool",
    "(NU (NU x)) << (y:*dom)",
    'HINT: NU is an involution') ;;

"x << y = (NU(NU x)) << y"
[ "!x. NU(NU x) = x" ]
[ "!x y. x PLUS y = y PLUS x" ]
[ "!x y. d << (x PLUS y) = (NU x) << y" ]
() : void

#De(ASM_REWRITE_TAC□) ;;
OK..
goal proved
. |- x << y = (NU(NU x)) << y

  "Derivation" (* *)
    "x << y"
  "$=" (*HINT: by reflexivity of NU *)
    "(NU(NU x)) << y"
() : void
```

To undo the last derived expression, use DU. *It is however not a replacement of* b, *the backup function, of the subgoal package.*

```
2.9 : DU : void -> void
```

3 The LEMMA Package

As said before in Section 1, a proof is most naturally a set structure and as such we want to to represent proofs in HOL. A *Proof Space* is essentially a set of the lemmas proven so far. To access the lemmas it is necessary that a unique name is associated with each lemma. It is up to the user to choose the names. When proving a new lemma, one is free to use any lemmas existing in the Proof Space. Assumptions can be considered as initialization of the Proof Space.

The package uses global variables too, so nesting is again not possible. This has been done so accessing lemmas becomes simpler. Indeed, we trade flexibility for simplicity and hope that in practice people will benefit more from the latter.

Let us assume we want to prove |- E, given A1 and A2 as assumptions. A proof using the LEMMA Package looks essentially as follows:

```
Set_Assumptions [ ('Asm1',A1) ; ('Asm2',A2) ] ;;
LEMMA ('lem10',E1,'HINT: Asm1, Asm2, PSP Law', Tac1) ;;
LEMMA ('lem20',E2,'HINT: Asm1, Monotonicity of |-->', Tac2) ;;
...
LEMMA ('lem100',E,'HINT: lem10, lem90, Cancel Law', Tac10) ;;
```

Do not bother with details yet. The first line setup the assumptions of the proof. Each **LEMMA** statement proves a lemma and adds it to the Proof Space. For example the second **LEMMA** derives E2 and Tac2 is the tactic to prove this. The second argument of **LEMMA** is a string that serves as a hint as how the lemma is inferred.

The only primitive operations are: initialization of, addition of a lemma to, deleting a lemma from, and pretty printing the Proof Space. These will be detailed in the next subsections. The package again supports an interactive mode.

The best way to explain how to use the package is again by an example. We will consider an example from programming logic [2]. Consider a non terminating program. For predicates p and q, p unless q means that once p holds it will hold at least until q becomes true, and $p \mapsto q$ means that once p holds then eventually q will hold. The inference rules below hold for these operators. What they operationally mean is not important for our purpose. Our interest is only in the calculation.

Monotonicity of $\mapsto$: $\dfrac{p \mapsto q \, , \, [q \text{ IMP } r]}{p \mapsto r}$

Transitivity of $\mapsto$: $\dfrac{p \mapsto q \, , \, q \mapsto r}{p \mapsto r}$

PSP Law : $\dfrac{p \mapsto q \, , \, r \text{ unless } b}{p \text{ AND } r \mapsto (q \text{ AND } r) \text{ OR } b}$

Cancellation Law : $\dfrac{p \mapsto q \text{ OR } r \, , \, q \mapsto b}{p \mapsto b \text{ OR } r}$

We are interested in proving the following property, which for convenience we will refer as the *Jade Property*.

Jade Property:

$\forall p,q,a,b,c. \ (p \mapsto q) \wedge (b \mapsto c) \wedge (c \mapsto q) \wedge (a \text{ unless } b) \ \Rightarrow \ (p \text{ AND } a \mapsto q)$

A paper proof of the Jade Property goes as follows.
Assume:

asm0	asm1	asm2
$p \mapsto q$	$(b \mapsto c) \wedge (c \mapsto q)$	a unless b

Prove p AND $a \mapsto q$. To do so we derive:

lem10: { asm1, Transitivity of $\mapsto$ } $b \mapsto q$
lem20: { asm0, asm2, PSP Law } p AND $a \mapsto (q$ AND $a)$ OR b
lem30: { lem10, lem20, Cancellation Law } p AND $a \mapsto (q$ AND $a)$ OR q
lem40: { lem30, Monotonicity of $\mapsto$ } p AND $a \mapsto q$
$\square$

Notice the non-linearity in the proof above: lem30 is inferred from lem10 *and* lem20.

3.1 Setting Up Assumptions

The first thing to do before actually starting the proof is to tell HOL the assumptions the proof may use. This is accomplished by function Set_Assumptions, as has already been explained in Section 2. We only want to add that the assumptions are the first lemmas you get for free, and so they are included in the Proof Space. The following will set up the assumptions to prove the Jade Property:

```
#Set_Assumptions
 [('asm0',"(p:^Pred)|-->q") ;
  ('asm1',"(b:^Pred)|-->c /\ c|-->q") ;
  ('asm2',"(a:^Pred) unless b") ] ;;
```

3.2 Extending Proof Space

To extend the Proof Space with a new lemma we have to tell HOL, using the function LEMMA, (a) the new lemma, (b) how to prove it, (c) its name for future references, and (d) a hint as how the lemma was inferred.

```
3.1 : LEMMA : (string # term # string # tactic) -> void
```

LEMMA (name,lem,hint,tac) if successful —that is, if tac can prove ?- lem— then it will add |- lem to the Proof Space and associate the name name to it. hint is a string that hints how |- lem is inferred.

To access the lemmas already in the Proof Space, RECALL is used, which has already been explained in Section 2. Figure 3 shows part of the proof of the Jade Property (lem30 and lem40) in the LEMMA-style. Compare the format with the paper proof given earlier.

One can at any time request a pretty print of the Proof Space, complete with the hints with the function PROOF_SPACE.

```
3.2 : PROOF_SPACE : void -> void
```

Fig. 3. Example of LEMMA style

```
LEMMA ('lem30',"((p:^Pred) AND a) |--> (q AND a) OR q",
       'HINT: lem10, lem20, Cancellation Law',
    ONCE_REWRITE_TAC [pOR_SYM]
    THEN ASSUME_TAC (RECALL 'lem10')
    THEN ASSUME_TAC (ONCE_REWRITE_RULE [pOR_SYM] (RECALL 'lem20'))
    THEN IMP_RES_TAC CANCELATION_LAW) ;;

LEMMA ('lem40',"((p:^Pred) AND a) |--> q",
       'HINT: Lem30, |--> is monotonic',
    SUBGOAL_THEN "|== (((q AND a) OR q) IMP (q:^Pred))"
       (\thm. ASSUME_TAC thm)
    THENL [pred_JADE_TAC 3; ALL_TAC]
    THEN ASSUME_TAC (RECALL 'lem30')
    THEN IMP_RES_TAC LT_MONOTONIC) ;;
```

For example if after feeding HOL with the complete proof of the Jade Property, **PROOF_SPACE** will show as in Figure 4. Compare the pretty print with the paper proof given earlier.

3.3 Interactive Mode

The last argument in the **LEMMA** statement is a tactic that actually proves the lemma. The question is of course how one can construct this tactic. The answer again is: the usual way, by working hard in interactive mode. There are only three interactive Proof Space operations: **Lg** (Lemma set Goal) to set a new lemma to be proven, **Le** (Lemma Extend) to replace the standard goal expander **e**, and **LU** (Lemma Undo) to delete a lemma from the Proof Space. Their use is analogous to that of **Dg**, **De**, and **DU** of the DERIVATION package.

4 Conclusion

What the DERIVATION and LEMMA packages offer is, first of all, a convenient way to write your proof as you would do it on paper. This will spare you from having to fuss about machine restrictions at a too early stage. Secondly, it improves proof readability as the proof is now formatted very much in the same way as paper-and-pencil proofs. Thirdly, it keeps record on the expressions or lemmas you have proven so far. You can at any moment request a pretty print of your proof —complete with hints— which looks as close as typewriter font can be to a nicely printed proof on paper.

Finally, I wish to say that it has to be admitted that in mechanical theorem proving —as in any conjunction between man and machine— man should learn to

Fig. 4. Pretty Printing Proof Space

```
#PROOF_SPACE();;

asm0 : (* Assumption *)
  . |- p |--> q
asm1 : (* Assumption *)
  . |- b |--> c /\ c |--> q
asm2 : (* Assumption *)
  . |- a unless b
lem10 : (* HINT: asm1, |--> is transitive *)
  ... |- b |--> q
lem20 : (* HINT: asm0, asm2, PSP Law *)
  .. |- (p AND a) |--> ((q AND a) OR b)
lem30 : (* HINT: lem10, lem20, Cancellation Law *)
  ... |- (p AND a) |--> ((q AND a) OR q)
lem40 : (* HINT: Lem30, |--> is monotonic *)
  ... |- (p AND a) |--> q
() : void
```

think machine-like to increase his productivity. Likewise, to increase convenience the machine should be equipped with facilities to support human-level reasoning. Then one will at least have the choice whether to be productive or to work conveniently. Moreover this will make mechanical proving more accessible for the uninitiated. A good style of mechanical proving is, in my opinion, to be found in the border between productivity and convenience.

5 Acknowledgement

I wish to thank Doaitse Swierstra for his patience and resourcefulness in improving this paper. I also thank Carl Seger for encouraging me to start writing this paper.

References

1. Cambridge University. *The HOL System Tutorial*, 1991. source included in the standard HOL package.
2. K.M. Chandy and J. Misra. *Parallel Program Design – A Foundation*. Addison-Wesley Publishing Company, Inc., 1988.
3. J.T. Jeuring. *Theories for Algorithm Calculation*. PhD thesis, Utrecht University, 1993.

From Abstract Data Types to Shift Registers:

A Case Study in Formal Specification and Verification at Differing Levels of Abstraction using Theorem Proving and Symbolic Simulation

Sreeranga Rajan, Jeffrey Joyce and Carl-Johan Seger

Department of Computer Science
University of British Columbia
6356 Agricultural Road
Vancouver, B.C V6T 1Z2, Canada
sree@cs.ubc.ca OR sree@cs.stanford.edu

Abstract. The stack is an ubiquitous component in both software and hardware. It is used as an ADT in data structures, while it serves as a component from floating point units to instruction execution units. In this paper, we explore the specification and verification of a simple bounded stack at differing levels of abstraction. We use HOL theorem proving at higher abstraction levels, while using VOSS symbolic trajectory evaluation at the switch-level. This case study provided a simple context to study various issues involved in combining theorem proving and symbolic simulation. Symbolic trajectory evaluation gives us access to accurate circuit models as well as the advantage of automating most of the proof effort. Theorem-proving gives us the expressive power for specification, than would generally be possible in a model checker approach. Further, we add executability to this combination to evaluate HOL theorem prover specifications. Executing HOL specifications helps detect errors before embarking on tedious proofs.

1 Introduction

Symbolic trajectory evaluation gives us access to accurate circuit models as well as the advantage of automating most of the proof effort. Theorem-proving gives us the expressive power for specification than, would generally be possible in a model checker approach. In this paper, we explore the specification and verification of a simple bounded LIFO (Last In First Out stack) at differing levels of abstraction. The stack is an ubiquitous component in both sofwtare and hardware. It is used as an Abstract Data Type (ADT) in data structures, while it serves as a component from floating point units to instruction execution units. We use Higher Order Logic (HOL)[5] theorem proving at higher abstraction levels, while using VOSS symbolic simulation at switch-level[10]. The simplicity of the stack provides an illustrative context to study different issues that arise in combining two different formal methods: theorem proving and symbolic simulation[6]. However, our work is in contrast to starting from an intermediate HCL level specification suggested in[7]. In an HCL specification, the nodes of the switch-level implementation are referenced. Whereas, in our approach, the higher level HOL specification describes the abstract behavior without reference to the structure of the circuit. This abstraction leads to additional transformational proofs besides VOSS_TAC[7], the main tactic used at the HCL level. To explain this further, we can think of VOSS specification as a quadruple $< M, A, C, G >$, where M, A, C, G are respectively switch level Model, Antecedent, Consequent and Guard expression. When expressed in HCL, it is in the form $M \models (G \implies (A \implies C))$. If A, C, G satisfy certain restrictions[1], we can infer that $(M \models (A \land C)) \implies G$. G is a HOL boolean expression that does not contain any references to the structure of M. Thus G can be directly mapped to more abstact HOL specifications. This allows us to connect a switch level model M with its higher level HOL specification.

Specification and verification of stacks and other ADTs has been a typical target of formal methods. At higher abstraction levels, algebraic methods and theorem proving methods[4] have been common place. At the hardware implementation level, symbolic simulation[11] has been used successfully. However, the problem of tying higher-level ADT style specifications to switch-level hardware implementation level has not been addressed.

We start out with a simple, but non-trivial specification at the most abstract level. We then add details to reach an implementation level specification by stepwise refinement in HOL. Finally, we verify that a structural (netlist) shift register implementation satisfies the implementation level specification using VOSS. The VOSS symbolic simulator is used from inside HOL with VOSS specifications written in HCL[6], a language embedded in HOL. The overall verification plan is shown in Fig.1.

2 Switch Level Design

We implemented a 4 bit wide and 32 bit deep stack at the switch-level. We followed a standard implementation of LIFO as a simple serial shifter as shown in Fig.2. Its associated description in the netlist language is shown in Appendix 1. It consists of a series of registers connected as in Fig.2. Each register consists of a D-Latch to hold data for a period of time, and a multiplexer (MUX) to select one signal out of 4 signal inputs. The implementation uses Complementary Metal Oxide Semiconductor (CMOS) logic. Signals SHR and SHL, and a clock signal, control the movement of data among the registers as shown in table below:

The netlist implementation is then compiled into a simulatable model by binding the design to an internal CMOS circuit model. The VOSS symbolic simulator is then used to build special state sequences called trajectories from compiled structural CMOS netlist description. This trajectory model of the structure serves to represent the behavior of the circuit at the switch-level.

---

[1] An explanation of the restrictions is beyond the scope of this paper

Table 1. Shift Register Truth Table

SHR	SHL	CLOCK	ACTION
Low	Low	High	Noop
High	Low	High	Shift Right
Low	High	High	Shift Left
High	High	High	-
-	-	Low	Noop

3 Abstract Specification

We specify a stack at the highest level as an abstract machine operating on a stack ADT, with a PUSH and a POP operation. The specification simply consists of saying a PUSH operation followed by a POP operation leaves the stack unchanged, except for the last element. Since we are dealing with a bounded stack, the last element gets pushed out on a PUSH operation. The finiteness of the stack is thus specified without introducing a concrete depth for the stack. This, not only simplifies the specification, but also provides for a stronger verification that involves a stack of arbitrary maximum depth. Also, we use the polymorphic types to indicate that contents of the stack could be later instantiated to be either a single bit or a vector of bits.

We first introduce a polymorphic recursive data type "stack" by an axiomatic definition. Polymorphism allows us to specify a generic stack. of arbitrary bitwidth and then anchor it to a boolean type at the implementation level.

Definition 3.1 $stack = EMPTY \mid STACK * stack$

where EMPTY and STACK are type constructors and the asterisk indicates polymorphism. The existence of such a type is automatically proved[8]. Then we define the equality of two stack structures upto the last element by an infix predicate "equl": i.e two stacks are identical except when their last elements are excluded from comparison.

```
⊢    (equl EMPTY (bv:* stack) = empty bv) ∧
     (equl (STACK (h:*) t) bv =
          (empty t => ((¬empty bv) ∧ (empty (pop bv))) |
                      ((¬empty bv) ∧ (h = top bv) ∧
                                     (equl t (pop bv)))))


⊢    PUSH (t:time) PUSHb POPb = PUSHb t ∧ ¬POPb t


⊢    POP (t:time) PUSHb POPb = ¬PUSHb t ∧ POPb t


⊢    ABS_STACK (PUSHb,POPb,abs_stack) =
          ∀t. (PUSH t PUSHb POPb) ∧
              (POP (t+1) PUSHb POPb) ⟹
                   (abs_stack (t+2) equl (abs_stack t))
```

4 Verification

We carry out the verification that the switch-level implementation satisfies the abstract specification in several stages. We first introduce an intermediate specification that adds more details to the stack operations. At this level we still maintain an abstract data type and a relational-style specification. We verify that this specification satisfies the abstract specification. This specification is then refined to a shift-register specification. The specification at this level is in functional-style with a concrete list data type representing the stack. This specification is then proved to implement the intermediate level specification. We finally prove that the trajectory representation of the switch-level netlist implementation satisfies the shift-register specification. The correctness results are then combined to prove that the switch-level implementation satisfies the abstract specification by a simple transitivity of implication.

4.1 Intermediate Specification

In the next level, we introduce more details of the machine by actually specifying what happens for each operation on the stack: A PUSH moves every stack element down while a POP moves every element up. The specification here uses "push" and "pop" operations defined on a stack type. The "push" operation adds in a new element and thus extends by one element, an object of stack type. The "pop" operation removes the top element and contracts the stack by one element. Since we are dealing with a finite stack with a constant maximum depth,

- The stack after a PUSH is equal to the suffix of the stack derived from the "push" operation.
- The suffix of the stack after a POP is equal to the prefix of the stack derived from the "pop" operation.

```
⊢    suffix1_eq (STACK h t) av =
          empty t ⇒ empty av | (¬empty av ∧ (h = (top av)) ∧
                                            (suffix1_eq t (pop av)))

⊢    prefix1_eq av bv =  bv suffix1_eq av

⊢    STACK_IMP (PUSHb,POPb,IN,abs_stack) =
          (∀t. (PUSHb t ∧ ¬POPb t ⟹
                          (abs_stack (t+1)) prefix1_eq
                                   (push (IN t, abs_stack t))) ∧
               (¬PUSHb t ∧ POPb t ⟹
                          (abs_stack (t+1)) suffix1_eq
                                   (pop (abs_stack t))))
```

4.2 Verification

We characterize the correctness condition as is usually done by verifying that the intermediate specification implies the abstract specification. To carry out the verification, we first prove the following theorems on prefix and suffix equalities:

- "equal upto the last element" is an equivalence relation.
- Two stacks which are prefixes of the same stack which is non-empty are equal.
- If a single stack is a prefix of two different non-empty stacks, then the two stacks are equal except for the last element.
- If a stack is equal to the prefix of a non-empty stack derived by "push", and another non-empty stack whose suffix if equal to the stack derived by "pop", then the stacks are equal upto the last element.

We conduct the verification for the stacks quantified universally. This in effect provides the correctness result for stacks of arbitrary depth and arbitrary maximum depth, since we can keep instantiating the result to concrete stacks of varied depths. The conditions that the stacks whose prefix is considered be non-empty propagates into the final correctness result:
Intermediate specification implies abstract specification provided that the stack is never empty.
This side condition, which can also be called an abstraction constraint has an interesting interpretation that the stack has a non-zero finite maximum depth. This is true in hardware - we cannot have a stack without atleast one storage structure.

Before we actually carry out the proof, we translate the HOL specifications at the two levels into ML[9] and execute with specific values for the stack structures. The conjectures we are required to prove are checked for different values by execution.

```
⊢    ((∀t. ¬empty (abs_stack t) ∧
      STACK_IMP (PUSHb,POPb,IN,abs_stack)) ⟹
      ABS_STACK (PUSHb,POPb,abs_stack)
```

4.3 Shift register specification and Verification

At this level, we "implement" the stack as a shift register by using a list data type with a shift-left corresponding to "push", shift-right corresponding to "pop" and a NOOP operation leaves the stack unchanged.

The verification that the shift register specification implies the intermediate specification proceeds as follows:

- Introducing an abstraction function from the list type to the stack abstract type.
- Correspondence between Shift-left, shift-right operations and "push", "pop" operations by showing that isomorphism of list stack and abstract stack: the abstraction function is thus a one-to-one mapping.

Thus, we can prove that

```
[stack = abstraction(list) ⟹]
Shift Register Specification ⟹ STACK_IMP
```

4.4 Switch Level Verification

This verification is conducted in the VOSS symbolic simulation domain. A restricted form of temporal logic with conjunction, next time modal operator and domain constraints can be used to specify Hoare-style antecedents and consequents. The antecedents assert values on nodes and consequents are used to check values on nodes after the simulator is run. We verify each of the 4 bit stack separately and conjunct the results. This makes the BDD sizes smaller and thus verification more efficient. The antecedents take the form of assertions on the input node as a symbolic boolean variable, a symbolic vector for the 32-bits of the stack nodes, symbolic boolean variable for SHR and SHL control signals, and a clock signal. The consequent consists of checking the 32-bit stack node vector is a symbolic vector shifted after a definite number of clock cycles corresponding to the vector asserted in the antecedent according to the different values of SHR and SHL control signals. The specification is expressed in HCL[6], a language defined in HOL. The VOSS checker is directly called from HOL and takes in the HCL specification, and returns a corresponding theorem on success. We give a succinct form of the verification result:

```
Switch Level Model ⊨
     (Shift Register Specification ⟹ Antecedent ⟹ Consequent)
```

4.5 Combining Verification Results

We combine the verification results at the higher abstraction levels by combining
the side-conditions and using transitivity of implication:

```
[(stack = abstraction(list)) ∧ ¬ empty(stack) ⟹]
       Shift Register Specification ⟹ ABS_STACK.
```

Combining the switch-level result to the Shift Register in HOL is more involved:
The symbolic variables are replaced by functions of abstract time. Given the
deterministic nature of the circuit and thus the uniqueness of excitation function
on the nodes, we prove that two symbolic variables assigned to a node during
the same time interval must be equal. Using this fact, we arrive at a theorem
that if the consequent holds given that the antecedent holds for legal trajectories
of the circuit, then the symbolic variables have the relationship exhibited by the
Shift Register specification in HOL. In other words, we show that

```
(Switch Level Model ⊨
    (Shift Register Specification ⟹ Antecedent ⟹ Consequent))
⟹
(Switch Level Model ⊨
    (Antecedent ∧ Consequent) ⟹ Shift Register Specification)
```

Thus we arrive at the following result:

```
Switch Level Model ⊨ (Antecedent ∧ Consequent)
    ⟹ Shift Register Specification
```

Note that the **Shift Register Specification** serves as the lowest level HOL
specification. Since we have[2]

```
    Shift Register Specification ⟹ ABS_STACK
```

We establish that

```
    (Switch Level Model ⊨ (Antecedent ∧ Consequent)) ⟹ ABS_STACK
```

This paves the way for the final verification result. By transitivity of implication, for all legal trajectories of the circuit, if the consequent holds (given the
antecedent holds), then the stack specification at the ADT level in HOL is true.
The only side-condition is the existence of the abstraction mapping from lists
to stack data type. The non-emptyness condition is subsumed by the imposition of a specific non-zero finite length on the symbolic boolean vectors at the
switch level. Further, we also prove that there is atleast one legal trajectory of
the circuit in which the antecedent holds. This also provides a solution to the
"false implication" problem, because there is atleast one model that satisfies the
switch-level specification and thus the higher-level specification.

---

[2] please note that we haven't shown side conditions for clarity

5 Effect on Specification Style and Connections to Executability

We had to modify the HOL and VOSS-HCL specifications several times in order to combine the verification results in the form of an implication. Some of the notable modifications that we had to adopt are:

- We had to remove explicit references to the depth of the stack in the specification because the arithmetic operations involving depth were not translatable into VOSS-style specifications.
- We had to abandon the idea of using functional abstractions for lists for the shift register specification in HOL.
- We had to remove the quantification on time and make time a parameter of all the specifications in HOL at one stage.
- We had to transform a relational style specification into a functional style.

Executability of HOL specifications precludes them having universal quantifications in them. This forces the specifications to be modified so that the quantified variables such as time be made a parameter in the equational definitions. This exact same restriction comes about while establishing the verification between switch-level HCL specification and the higher-level shift-register specification in HOL. Further, the imposition of a functional style specification is common to both executability in general and in combining the verification results of the switch-level to the abstract level in HOL.

6 Conclusions and Future Work

We have shown the specification and verification of a stack at differing levels of abstraction: from an ADT level to switch-level using theorem proving and symbolic simulation. The form of the HOL specification and VOSS specification had to modified several times to combine the verification results. In particular, we had to parametrize the HOL specification by time instead of the typical style of hiding the time by a universal quantification. The VOSS specification had to be modified so that the relations among the symbolic variables are expressed in a functional style. Then, the functional style specification had to be verified with a relational style specification.

Hardware specification and verification has increasingly demanded automation without sacrificing high level abstraction mechanisms. Automation is desired for speed and ease of verification process, while abstraction mechanisms are needed for studying the properties of the specification itself. Also, abstraction mechanisms are desired for making a specification general, so as to map the specification on to multiple implementations. This ability to map a specification on to several implementations also enables specifications to be extended to larger systems. We wish to study the issues in using hybrid methods to verify larger systems in our future work.

References

1. Albert Camilleri, "Executing behavioural definitions in higher order logic", Technical Report 140, University of Cambridge, Computer Laboratory, February 1988.
2. Juanito Camilleri, "Symbolic compilation and execution of programs by proof: a case study in HOL", Technical Report 240, University of Cambridge, Computer Laboratory, December 1991.
3. Paul Curzon, "Of what use is a verified compiler specification?", Technical Report 274, University of Cambridge, Computer Laboratory, November 1992.
4. Goguen, Joseph A., "OBJ as a Theorem Prover with Applications to Hardware Verification", in: Birtwistle, G. and Subrahmanyam, P., eds., Current Trends in Hardware Verification and Automated Theorem Proving, Springer Verlag, 1989, pp 218-267.
5. Gordon, M.J.C, "Why Higher Order Logic is a Good Formalism for Specifying and Verifying Hardware", in: Milne, G. and Subrahmanyam, P., eds., Formal Aspects o f VLSI Design, Proceedings of the 1985 Edinburgh Conference on VLSI, North Holland, 198 6, pp 153-177.
6. Joyce, Jeffrey and Seger, C-J, "Linking BDD-Based Symbolic Evaluation to Interactive Theorem-Proving", *To Appear in* Proceedings of the 30th ACM-IEEE Design Automation Conference, IEEE, 1993
7. Joyce, Jeffrey and Seger, C-J, "The HOL Voss System: Model-Checking in a General Purpose Theorem Prover", *To Appear in* Proceedings of the International Workshop on Higher Order Logic Theorem Proving and its Applications, Vancouver BC, Canada, 1993, *Springer Verlag*.
8. Melham, Tom, "Automating Recursive Type Definitions in HOL", Current Trends in Hardware Verification and Automated Theorem Proving, Springer Verlag, 1989
9. Rajan, Sreeranga, "Executing HOL Specifications: Towards an Evaluation Semantic s for Classical Higher Order Logic", In L. Claesen and M. Gordon, editors, *Higher Order Logic Theorem Proving and its Applications.* North-Holland, 1993.
10. Seger, C-J and Bryant, Randall, "Formal Verification of Digital Circuits by Symbolic Evaluation of Partially Ordered Trajectories", UBC Computer Science Department Technical Report 93-8, April 1993.
11. Bryant, Randall, "Formal Verification of Digital Circuits by Logic Simulation" *Hardware Specification Verification and Synthesis: Mathematical Aspects:* Proceedings / Mathematical Sciences Institute workshop, Cornell University Ithaca, New York, 1989 ; M.Leeser, G. Brown (eds.).

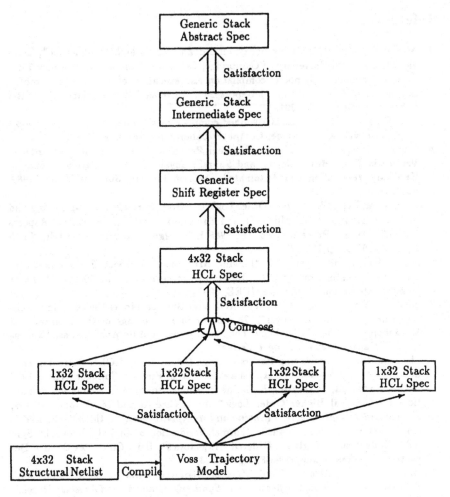

Fig.1. Verification Plan

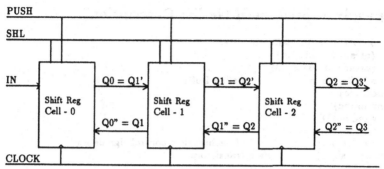

Fig.2a. Serial Shifter

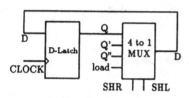

Fig.2b. Inside Shift Register Cell

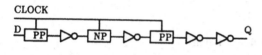

Fig.2c. Inside D-Latch

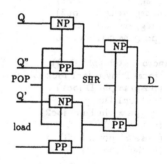

Fig.2d. Inside 4 to 1 MUX

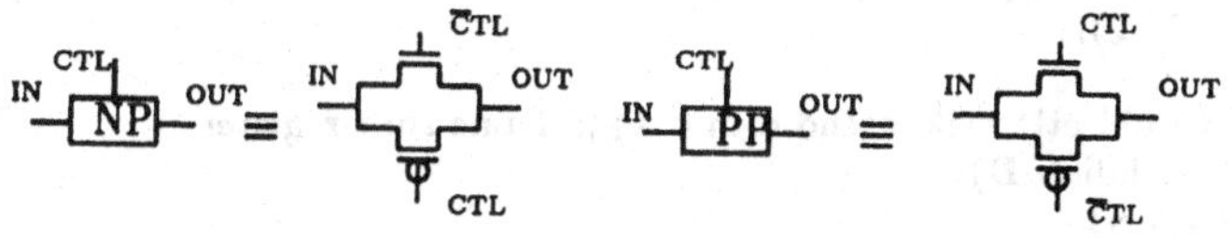

Fig.2e.

Appendix 1: Structural Netlist Description

```
(macro inv (in out) ;; define Inverter
     (ntrans in out gnd)
     (ptrans in vdd out))
(macro buffer (in out)          ;; define Buffer
   (local loc)
   (inv in loc)
   (inv loc out))

(macro mkctl (sig)              ;; Make high/low control signals for
   (inv sig sig.L)        ;; pass transistors
   (connect sig sig.H))

(macro pospass (ctl in out)     ;; define positive pass transistor
     (ntrans ctl.H in out)
     (ptrans ctl.L in out))
(macro negpass (ctl in out)     ;; define negative pass transistor
   (ntrans ctl.L in out)
   (ptrans ctl.H in out))

(macro mux4 (ctl1 ctl2 a b c d out) ;; define 4-1 Multiplexor
   (local loc1 loc2)
   (negpass ctl2 a loc1)
   (pospass ctl2 b loc1)
   (negpass ctl2 c loc2)
   (pospass ctl2 d loc2)
   (negpass ctl1 loc1 out)
   (pospass ctl1 loc2 out))

(macro dlatch (phi D Q)              ;; define D-Latch
   (local Dinv q loc1 loc2 loc3)
   (pospass phi D loc1)
   (inv loc1 Dinv)
   (negpass phi Dinv loc2)
   (inv loc2 q)
   (pospass phi q loc3)
   (buffer loc3 Q))

(macro stack`cell (phi ctl1 ctl2 D Q`minus`1 Q`plus`1 Q) ;; define 1-bit Register
   (local load)
   (mux4 ctl1 ctl2 Q Q`plus`1 Q`minus`1 load D)
   (dlatch phi D Q))

(macro stack (n phi ctl1 ctl2 latchdin in out) ;; define shift-register
   (local end latchdinb D)
   (repeat i 0 (1- n)
        (buffer latchdin.i latchdinb.i))
   (stack`cell phi ctl1 ctl2 D.0 in latchdinb.1 latchdin.0)
   (repeat i 1 (- n 2)
        (stack`cell phi ctl1 ctl2 D.i latchdinb.(1- i)
                      latchdinb.(1+ i) latchdin.i))
   (stack`cell phi ctl1 ctl2 D.(1- n) latchdinb.(- n 2)
                      end latchdin.(1- n))
   (connect out latchdin.0))
```

Verification in Higher Order Logic of Mutual Exclusion Algorithm

Victor A. Carreño

NASA Langley Research Center, MS 130, Hampton, VA 23681-0001 U.S.A.
vac@air16.larc.nasa.gov

Abstract. In this work a mutual exclusion algorithm is modeled using Transition Assertions. The main feature of a mutual exclusion algorithm is to prevent simultaneous access of a shared resource by two or more systems. The specification of the algorithm is a collection of transition assertions with each transition assertion containing a precondition and postcondition. The Transition Assertions model is formalized in higher order logic and the HOL mechanized theorem prover is used to show that the Transition Assertions model complies with the mutual exclusion requirement.

1 Introduction

The problem of mutual exclusion deals with preventing two or more systems or processes from simultaneously accessing a shared resource. For example, if two processes read and write data to a magnetic storage, it will be desirable, for proper operation, to limit access to one process at a time. Several algorithms have been suggested to implement this constraint. The objective of this work is not to evaluate mutual exclusion algorithms, but rather to use one as an example of a procedure in which the quantitative characteristics of time are critical for proper operation.

The mutual exclusion algorithm verified in this work was first presented by Lamport in [4] and is attributed to Fischer. Since Lamport's first paper, the algorithm has been revisited in other works by Schneider, Bloom and Marzulo in [6], Abadi and Lamport in [1] and Shankar in [7]. Although the later papers make use of, and refer to the algorithm in [4], the mutual exclusion algorithm in [6], [1], and [7] has been defined differently from the original. Specifically, deadlock is possible in the later versions.

The algorithm selected is modeled using the Transition Assertions specification method [2]. The Transition Assertions method is based on the assumptions that all events in a system take some non-zero amount of time to complete. Transition Assertions models are defined in higher order logic and a graphical representation is also used.

In this paper, a Transition Assertions model of Fischer's mutual exclusion algorithm is verified using the HOL mechanized theorem prover by showing that the model implies the mutual exclusion requirement. The general algorithm is represented by 12 assertions and 1 assumption. The verification is accomplished by representing two systems each running an instance of the algorithm and showing that both systems can not be in the critical section simultaneously. The total representation consists of 22 assertions and 1 assumption.

2 The Algorithm

Each process which has access to the shared resource must follow the protocol implemented by the following algorithm. Each process is given a distinct identification number which is used in place of i below. The *critical section* is the sequence of instructions which links the process to the shared resource. A memory register called x is read and written by each process using atomic instructions. It is assumed that the critical section as well as instructions outside of the mutual exclusion algorithm do not modify register x or any variable used by the mutual exclusion algorithm. Register x should have value zero when all processes are first started. The general algorithm is:

> **repeat**   **await** $\langle x = 0 \rangle$;
> $\langle x := i \rangle$;
> $\langle delay \rangle$
> **until** $\langle x = i \rangle$;
> *critical section*;
> $x := 0$;

The instruction **await** b blocks progress until condition b holds. If the register x has value zero, the shared resource is idle and can be used by a process. If register x does not have a value zero, a process is in the critical section or entering the critical section. After a process reads x and determines that the value is zero, it changes the value of x to its identification number to block any other process. However, the time needed to write to x is finite and another process can read x to be zero in this interval and also proceed to its critical section. To prevent this condition from happening, the algorithm waits (using *delay*) for a time longer than the time needed to write to x before checking the value of x a second time and entering the critical section. Therefore, the basic assumption for the algorithm to guarantee mutual exclusion is that the atomic action *delay* must always take longer than the assignment $x := i$.

3 Transition Assertions Model

Transition Assertions (TA) is an experimental modeling method to represent or specify dynamic systems. The method is intended to be used where timing is a critical element. Time is discrete and is modeled by the set of natural numbers. Timing constraints are represented explicitly and verified using mathematical logic. Variables are represented as functions from time to values in some appropriate domain. System conditions or predicates are represented as functions of time and variables to the boolean domain. The state of a TA model is characterized by the set of values of system variables.

In this section, the generic transition, Single Transition (ST) is defined and explained. ST is one of nine types of transitions defined in [2]; It is the only type of transition used in the specification of the mutual exclusion algorithm.

The generic transition ST is represented graphically in Fig. 1.

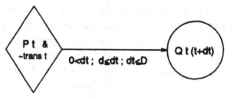

Figure 1. Single Transition Graphical Representation

The formal definition of the transition is given by:

$$ST \ d \ D \ P \ Q \ trans =$$
$$(\forall t.$$
$$\quad (P \ t) \wedge (\neg trans \ t) \Rightarrow$$
$$\quad \exists dt.$$
$$\qquad (0 < dt) \wedge$$
$$\qquad (d \leq dt) \wedge$$
$$\qquad (dt \leq D) \wedge$$
$$\qquad (Q \ t(t + dt)) \wedge$$
$$\qquad (\forall k.(0 < k) \wedge (k < dt) \Rightarrow trans(t + k)) \wedge$$
$$\qquad (\neg trans(t + dt))) \wedge$$
$$(\forall t.(\neg P \ t) \wedge (\neg trans \ t) \Rightarrow (\neg trans(t + 1)))$$

In both representations,

dt is the transition duration,
d is the minimum duration of the transition,
D is the maximum duration of the transition,
P is a predicate which must be true for the transition to start,
Q is a predicate which must be true if the transition occurs, and
trans is a variable which is true when the transition is taking place and false otherwise.

The ST model is a single transition model. Once a transition is taking place, a new transition, specified by the *same* assertion, cannot start, even if the predicate P is satisfied. Other types of Transition Assertions, not used in this work, can represent multiple transitions with the same assertion. An example of a multiple transition system is the postal service; Every time a package is sent, it arrives at its destination in five days [1]. Multiple packages can be sent the same day and packages can be sent every day without waiting for the previous packages to arrive. If the postal service was a single transition system, then when a package

---

[1] Assuming an ideal postal service.

is sent, it would be necessary to wait five days before another package could be sent. A transition effectively 'ties-up' a single transition system until finished.

If two or more single transition assertions are used in the specification of a system, the assertions (and therefore the transitions) are completely independent of each other, unless there are common variables amongst assertions. Thus, Transition Assertion models are inherently parallel (concurrent). Although with the ST model the same transition cannot occur simultaneously, two different transitions, specified by different assertions can happen at the same time.

To illustrate the behaviour of the model, an example is presented. A control is to be specified to operate a soft-drink fountain which fills cups at a fast food restaurant. A cup is placed under the soft-drink source and a button is pushed. When the button is pushed, the fountain will dispense the drink for at least the next 6 seconds and no more than the next 6.3 seconds. If the button is pushed a second time while the fountain is dispensing, this will not cause an extended dispensing time and an overflow. (That is, a new transition will not start before the previous one is completed)

The soft-drink fountain control is now represented by an assertion. The units of time are selected to be 1 time-unit equals 0.1 seconds. *button_pushed* and *dispensing* are functions from time to {*true,false*} and T is the constant *true*.

$$ST\ 60\ 63\ button\_pushed\ (\lambda\ t\ t'.T)\ dispensing$$

Using the definition of ST above, the assertion is:

$$(\forall t.$$
$$(button\_pushed\ t) \land (\neg dispensing\ t) \Rightarrow$$
$$\exists dt.$$
$$(0 < dt) \land$$
$$(60 \leq dt) \land$$
$$(dt \leq 63) \land$$
$$((\lambda\ t\ t'.T)\ t\ (t + dt)) \land$$
$$(\forall k.(0 < k) \land (k < dt) \Rightarrow (dispensing(t + k))) \land$$
$$(\neg dispensing(t + dt))) \land$$
$$(\forall t.(\neg button\_pushed\ t) \land (\neg dispensing\ t) \Rightarrow (\neg dispensing(t + 1)))$$

We now add a new feature to the system. A sound is made when dispensing is finished.

$$ST\ 60\ 63\ button\_pushed\ (\lambda\ t\ t'.beep\ t')\ dispensing$$

In the next section the mutual exclusion algorithm is specified using the single transition model.

4 Transition Assertions Version of Mutual Exclusion Algorithm

The Transition Assertions model is created by translating the algorithm instructions into a set of transitions. Each instruction in the algorithm denotes an action. Actions must start, perform an implicit or explicit task [2], and come to an end or loop indefinitely. Thus, transition models can be used to represent the instructions.

The algorithm is represented graphically in Fig. 2. There is roughly one assertion per instruction. The assertions are annotated with the instructions on the right side.

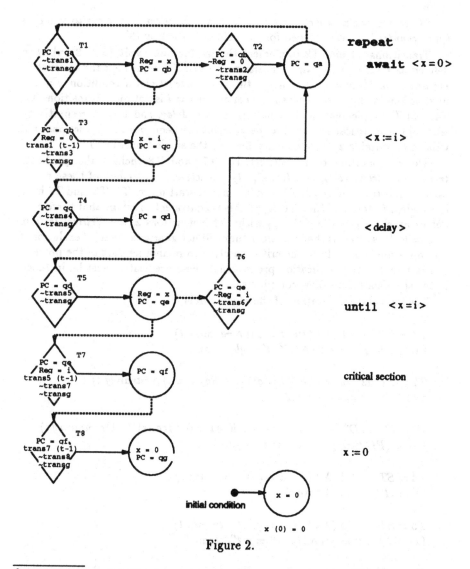

Figure 2.

---

[2] Even instructions that are regarded as *no operation* instructions perform an implicit task since they alter the internal state of the device. In the case of a microprocessor, the change could be increasing the program counter.

All variables in the graph are functions from time to value. Their time arguments have been left out for simplicity. The statement $Reg = x$, for example, should read $Reg\ t' = x\ t'$.

The dashed lines connecting one transition postcondition to another transition precondition indicate that both conditions could be satisfied simultaneously. If both conditions are satisfied then one transition will "flow" onto the next transition.

PC represents a program counter and is a function from time to value. A type *:counter* has been defined for the value returned by PC.

Transitions T1 and T2 implement the instruction **await** $\langle x = 0 \rangle$ which is defined as **while** $\neg b$ **do** *skip*. Transitions T5 and T6 implement the instruction **repeat ... until** $x = i$ by going to the next instructions if condition $x = i$ is met, or looping back to T1 when the condition $x = i$ is not met. Transitions T3, T4, and T8 implement the instructions $x := i$, *delay*, and $x := 0$, respectively. Although the critical section may be an arbitrary sequence of instructions (not writing to variable x), it is modeled here by the single transition T7.

The precondition of transitions T3, T7, and T8 include the predicate *trans1(t-1)*, *trans5(t-1)*, and *trans7(t-1)* respectively. The effect of these predicates is that transitions T3, T7, and T8 cannot start unless T1, T5, and T7 have just ended. In terms of the algorithm, this is equivalent to saying that there cannot be an entry point into the algorithm at instructions $x := i$, *critical section*, or $x := 0$. It is obvious that the algorithm will not guarantee mutual exclusion if a jump is made directly to the critical section from another part of the process. It was found in the verification process that these predicates must be included in the specification to show compliance.

The formal representation of the algorithm is:

$T1 = ST\ d1\ D1\ (\lambda\ t.(PC\ t = qa) \wedge \neg transg\ t)$
$(\lambda\ t\ t'.(Reg\ t' = x\ t') \wedge (PC\ t' = qb))\ trans1$

$T2 = ST\ d2\ D2\ (\lambda\ t.(PC\ t = qb) \wedge \neg (Reg\ t = 0) \wedge \neg transg\ t)$
$(\lambda\ t\ t'.PC\ t' = qa)\ trans2$

$T3 = ST\ d3\ D3\ (\lambda\ t.(PC\ t = qb) \wedge (Reg\ t = 0) \wedge trans1(t-1) \wedge \neg transg\ t)$
$(\lambda\ t\ t'.(PC\ t' = qc) \wedge (x\ t' = i))\ trans3$

$T4 = ST\ d4\ D4\ (\lambda\ t.(PC\ t = qc) \wedge \neg transg\ t)$
$(\lambda\ t\ t'.PC\ t' = qd)\ trans4$

$T5 = ST\ d5\ D5\ (\lambda\ t.(PC\ t = qd) \wedge \neg transg\ t)$
$(\lambda\ t\ t'.(PC\ t' = qe) \wedge (Reg\ t' = x\ t'))\ trans5$

$T6 = ST\ d6\ D6\ (\lambda\ t.(PC\ t = qe) \wedge \neg (Reg\ t = i) \wedge \neg transg\ t)$
$(\lambda\ t\ t'.PC\ t' = qa)\ trans6$

$T7 = ST \ d7 \ D7 \ (\lambda \ t.(PC \ t = qe) \wedge (Reg \ t = i) \wedge trans5(t-1) \wedge \neg transg \ t)$
$(\lambda \ t \ t'.PC \ t' = qf) \ trans7$

$T8 = ST \ d8 \ D8 \ (\lambda \ t.(PC \ t = qf) \wedge trans7(t - 1) \wedge \neg transg \ t)$
$(\lambda \ t \ t'.(PC \ t' = qg) \wedge (x \ t' = 0)) \ trans8$

$ic = x \ 0 = 0$

$sequential = \forall \ t.trans1 \ t \vee trans2 \ t \vee trans3 \ t \vee trans4 \ t \vee trans5 \ t \vee$
$trans6 \ t \vee trans7 \ t \vee trans8 \ t = transgt$

$Reg\_inert = \forall \ t. \neg trans1 \ t \wedge \neg trans5 \ t \Rightarrow (Reg \ (t + 1) = (Reg \ t))$

$x\_inert = \forall \ t. \neg END \ trans3 \ t \wedge \neg END \ trans8 \ t \Rightarrow (x(t + 1) = x \ t)$

$assumption = (D3 < d4)$

There are four assertions in the formal representation which are not explicitly shown in the graphical model. The first additional assertion, *sequential*, links all transitions to force the transitions to be sequential. The algorithm being modeled is a purely sequential one. Therefore, transitions within the same algorithm should not happen simultaneously. For example, transition T1 reads the value of x and puts it in register *Reg*. If the precondition for transition T2 is satisfied before T1 is over, transition T2 should not start. (This would be a form of microprocessor pipe line which will nullify the algorithm.) To make the model strictly sequential, the variable *transg* is used. *transg* is true if any transition is taking place. *transg* must be false for a transition to start.

The next two assertions implement inertia. They will maintain constant the values of the registers unless a transition is changing their value. Without the inertia assertions, the values of the registers will be undefined at times where a transition is not explicitly giving the registers a value.

The last assertion is the assumption. Assumption $D3 < d4$ is the basic assumption given in all previous works. It states that transition T4 (*delay*) always takes longer than T3 ($x:=i$). It is important to note that the assumption that the critical section as well as instructions outside of the mutual exclusion algorithm do not modify register x is explicitly contained in the Transition Assertions formal specification: the assertion x_inert allows x to be changed only by transitions T3a, T3b, T8a, and T8b.

5 Mutual Exclusion with Two Processes

The Mutual exclusion model is now instantiated for a system with two processes sharing a resource. The processes are called a and b, with identification numbers 1 and 2 respectively. The following syntax is the input for the HOL theorem prover. The symbols !, $^-$, and \ are equivalent to the mathematical symbols $\forall$, $\neg$, and λ.

```
let T1a = "ST d1 D1 (\t.(PCa t = qa)/\~transga t)
(\t t'.((Rega:num -> num) t' = x t')/\(PCa t' = qb)) trans1a";;

let T2a = "ST d2 D2 (\t.(PCa t = qb)/\~(Rega t = 0)/\~transga t)
(\t t'.PCa t' = qa) trans2a";;

let T3a = "ST d3 D3
(\t.(PCa t = qb)/\(Rega t = 0)/\(trans1a (t-1))/\~transga t)
(\t t'.(PCa t' = qc)/\(x t' = 1)) trans3a";;

let T4a = "ST d4 D4 (\t.(PCa t = qc)/\~transga t)
(\t t'.PCa t' = qd) trans4a";;

let T5a = "ST d5 D5 (\t.(PCa t = qd)/\~transga t)
(\t t'.(PCa t' = qe)/\((Rega:num -> num) t' = x t')) trans5a";;

let T6a = "ST d6 D6 (\t.(PCa t = qe)/\~(Rega t = 1)/\~transga t)
(\t t'.PCa t' = qa) trans6a";;

let T7a = "ST d7 D7
(\t.(PCa t = qe)/\(Rega t = 1)/\(trans5a (t-1))/\~transga t)
(\t t'.PCa t' = qf) trans7a";;

let T8a = "ST d8 D8
(\t.(PCa t = qf)/\trans7a (t-1)/\~transga t)
(\t t'.(PCa t' = qg)/\(x t' = 0)) trans8a";;

let T1b = "ST d1 D1 (\t.(PCb t = qa)/\~transgb t)
(\t t'.((Regb:num -> num) t' = x t')/\(PCb t' = qb)) trans1b";;

let T2b = "ST d2 D2 (\t.(PCb t = qb)/\~(Regb t = 0)/\~transgb t)
(\t t'.PCb t' = qa) trans2b";;

let T3b = "ST d3 D3
(\t.(PCb t = qb)/\(Regb t = 0)/\(trans1b (t-1))/\~transgb t)
(\t t'.(PCb t' = qc)/\(x t' = 2)) trans3b";;

let T4b = "ST d4 D4 (\t.(PCb t = qc)/\~transgb t)
(\t t'.PCb t' = qd) trans4b";;

let T5b = "ST d5 D5 (\t.(PCb t = qd)/\~transgb t)
(\t t'.(PCb t' = qe)/\((Regb:num -> num) t' = x t')) trans5b";;

let T6b = "ST d6 D6 (\t.(PCb t = qe)/\~(Regb t = 2)/\~transgb t)
(\t t'.PCb t' = qa) trans6b";;
```

```
let T7b = "ST d7 D7
(\t.(PCb t = qe)/\(Regb t = 2)/\(trans5b (t-1))/\~transgb t)
(\t t'.PCb t' = qf) trans7b";;

let T8b = "ST d8 D8
(\t.(PCb t = qf)/\trans7b (t-1)/\~transgb t)
(\t t'.(PCb t' = qg)/\(x t' = 0)) trans8b";;

let sequentiala =
"!t:num.trans1a t\/trans2a t\/trans3a t\/trans4a t\/trans5a t\/
trans6a t\/trans7a t \/ trans8a t = transga t";;

let sequentialb =
"!t:num.trans1b t\/trans2b t\/trans3b t\/trans4b t\/trans5b t\/
trans6b t\/trans7b t \/ trans8b t = transgb t";;

let ic = "x 0 = 0";;

let Rega_inert = "!t.~trans1a t /\~trans5a t ==>
((Rega:num -> num) (t+1) = (Rega t))";;

let Regb_inert = "!t.~trans1b t /\~trans5b t ==>
((Regb:num -> num) (t+1) = (Regb t))";;

let x_inert = "!t.~(END trans3a t) /\ ~(END trans3b t) /\
~(END trans8a t) /\ ~(END trans8b t) ==>
((x:num -> num) t = x (t-1))";;

let assumption = "D3 < d4";;

let fischer_spec = "~T1a /\ ~T2a /\ ~T3a /\ ~T4a /\ ~T5a /\
~T6a /\ ~T7a /\ ~T8a /\ ~T1b /\ ~T2b /\ ~T3b /\ ~T4b /\ ~T5b /\
~T6b /\ ~T7b /\ ~T8b /\ ~sequentiala /\ ~sequentialb /\ ~ic /\
~registera_inert /\ ~registerb_inert /\ ~x_inert /\ ~assumption";;
```

Transitions T1a-T8a correspond to process 'a' and T1b-T8b to process 'b'. Assertion *ic* is the initial condition for register 'x' which must have value zero at time zero.

The assertions sequentiala and sequentialb make each process strictly sequential. Transitions T1a and T4a, for example, cannot occur simultaneously. However, transition T1a and any transition of process b, T1b, T2b, T3b ..., can happen at the same time.

The next three statements implement inertia. The values of variables *Rega*, *Regb*, and *x* will not change unless the given transitions in the assertions are active.

It is worth noting that processes *a* and *b* are only linked by register *x* and the values D3 and d4. They are completely independent otherwise.

The statement *assumption* contain the timing assumption for the algorithm. This assumption is essential to show mutual exclusion.

The last statement, *fischer_spec*, is the system model. It is the conjunction of all the individual assertions[3].

6   Verification

Verification is accomplished by showing that a system behaves according to its requirements. A model of the system written in formal logic, a formal requirement, and a deduction calculus are needed for the verification. A discussion of what constitutes a formal verification and the limitations of formal verification can be found in [5] and [3].

The Transition Assertions version of the mutual exclusion algorithm, instantiated for two processes in section 5, is the formulation of the system model in formal logic. The mutual exclusion requirement is now formalized in the Transition Assertions context. Since the critical sections are represented by transitions T7a and T7b, these transitions cannot be active at the same time. The requirement is then:

$$\forall t. \neg(trans7a\ t \land trans7b\ t) = mutual\ exclusion$$

The system model is verified with respect to the formal requirement in the higher order logic language. The verification consists in proving that the system model *fischer_spec* implies the requirement.

$$fischer\_spec \Rightarrow mutual\ exclusion$$

The HOL theorem prover is used to develop the proof of implication. The proof strategy is based on the condition that if a process is in its critical section (T7a or T7b), then the value of *x* must not change. For example if process *a* enters its critical section, *x* will have value 1. As long as *x* has value 1, process *b* is locked out of its critical section. Since *x* can only be changed by transitions T3a, T3b, T8a and T8b then we must show that these transitions are inactive while a process is in its critical section.

$$\forall t.trans7a\ t \Rightarrow \neg trans3a\ t \land \neg trans3b\ t \land \neg trans8a\ t \land \neg trans8b\ t$$

$$\forall t.trans7b\ t \Rightarrow \neg trans3b\ t \land \neg trans3a\ t \land \neg trans8b\ t \land \neg trans8a\ t$$

---

[3] The symbol ^ is called an antiquotation in the HOL system. It extract the constructs of a term to be used within another term. For example, let term1 = "x + 1";; and let term2 = "(^term1)+5";; is equivalent to let term2 = "x + 1 + 5";;

The first and third conjuncts follow from the *sequentiala* and *sequentialb* components of the specification. For the second conjunct, 3 lemmas are necessary. The first lemma states that if the condition for transition T7a (T7b) is satisfied, then $x = 1$ ($x = 2$) for at least a time $d4$. Recall that the conditions for T7a and T7b to start are $(\lambda\ t.(PCa\ t = qe) \wedge (Rega\ t = 1) \wedge trans5a(t - 1)$ and $(\lambda\ t.(PCb\ t = qe) \wedge (Regb\ t = 2) \wedge trans5b(t - 1)$ which are abbreviated by P7a and P7b below.

$$\forall\ t.P7a\ t \Rightarrow ATLEAST\ (\lambda\ t.x\ t = 1)\ d4\ t$$

$$\forall\ t.P7b\ t \Rightarrow ATLEAST\ (\lambda\ t.x\ t = 2)\ d4\ t$$

The function $ATLEAST$ is defined by:
$$ATLEAST\ A\ dt\ t = dt \le t \wedge (\forall\ i.(t - dt) < i \wedge i \le t \Rightarrow A\ i)$$
The second lemma uses the assumption $D3 < d4$ to show that if $x = 1$ or $x = 2$ for at least $d4$, transitions T3a and T3b are inactive.

$$ATLEAST\ (\lambda\ t.x\ t = 1)\ d4\ t \Rightarrow \neg trans3a\ t \wedge \neg trans3b\ t$$

$$ATLEAST\ (\lambda\ t.x\ t = 2)\ d4\ t \Rightarrow \neg trans3a\ t \wedge \neg trans3b\ t$$

The third lemma, TRANS_START_OR_CONT, states that if a transition is active at time t then it must have started at time $t\text{-}1$ or it must have been active at time $t\text{-}1$. The three lemmas are used with mathematical induction to show:

$$\forall\ t.trans7a\ t \Rightarrow \neg trans3b\ t$$

$$\forall\ t.trans7b\ t \Rightarrow \neg trans3a\ t$$

Several attempts to directly prove the fourth conjunct, showing that T8a and T8b are inactive while a process is in its critical section, failed.[4]

The difficulty in the proof arises from the fact that no constraints are placed on the upper bound duration of transitions T8a and T8b and the precondition of transition T8a and T8b does not depend explicitly on the value of register x.

In order to overcome this obstacle without the need for another assumption, transitions T7 and T8 are combined into a single transition using Transition Assertions composition rules. The new single transition is active if T7 is active, T8 is starting or T8 is active. P8 is again a predicate abbreviation for the condition to start T8.

$$\forall\ t.trans\_conca\ t = (\lambda\ t.trans7a\ t \vee P8a\ t \vee trans8a\ t)$$

---

[4] The algorithms used in [1, 6, 7] did not include the assignment instruction $x := 0$ at the end of the critical section and therefore did not have transitions T8a and T8b.

$$\forall t.trans\_concb \; t = (\lambda \; t.trans7b \; t \lor P8b \; t \lor trans8b \; t)$$

With the single transition defined from the conjunction of T7 and T8, it is possible to prove the lemma EXCLUSION3 which shows:

$$\forall t.trans\_conca \; t \Rightarrow \neg trans\_concb \; t \land x \; t = 1 \land \neg trans3a \; t \land \neg trans3b \; t$$

and

$$\forall t.trans\_concb \; t \Rightarrow \neg trans_c onca \; t \land x \; t = 1 \land \neg trans3a \; t \land \neg trans3b \; t$$

The lemma EXCLUSION3 with 9 principal lemmas and several sublemmas is used to prove the theorem

$$fischer\_spec \Rightarrow mutual \; exclusion$$

as desired. Appendix A contains the lemma EXCLUSION3. The proof of lemma EXCLUSION3 is approximately 6 pages long and could not be included in the paper. The main proof as well as the proof of all principal lemmas and sublemmas are available by request from the author.

7 Conclusion

A mutual exclusion algorithm was modeled using Transition Assertions. The verification process was lengthy and involved. Many of the sublemmas used in the mutual exclusion proof deal with temporal properties of events and are not specific to the mutual exclusion problem. A great part of the effort was devoted to these basic lemmas.

The Transition Assertions method forces a high level of rigor. Assumptions that were made about the algorithm, but not included in the algorithm itself, had to be explicitly stated in the Transition Assertions model. These were the restricted access to register x, and the preclusion of entry points at some instructions.

For systems that are not of the highest criticality, the method imposes too high a penalty in terms of the effort needed to show compliance with requirements.

On the positive side, transition models can be used to model different levels of details. Specification and implementation can be modeled by the same technique which makes the development process more coherent.

8 Acknowledgements

The idea for Transition Assertions are based on Mike Gordon's work on State Transition Assertions. Thanks to James Caldwell for reviewing the paper and making valuable suggestions.

References

1. Martin Abadi and Leslie Lamport, *An Old-Fashioned Recipe for Real Time*, in proceedings 1991 Rex Workshop, Real-time : Theory in Practice, J.W. de Bakker et al., editors, Springer-Verlag, 1992.

2. Victor Carreño, *The Transition Assertions Specification Method*, University of Cambridge Computer Laboratory, Technical Report No. 279, January 1993.

3. Avra Cohen, *Correctness Properties of the Viper Microprocessor: The Second Level*, in: Current Trends in Hardware Verification and Automated Theorem Proving, edited by G. Birtwistle and P.A. Subrahmanyam, Springer-Verlag, 1989, Pages 1-91.

4. Leslie Lamport, *A Fast Mutual Exclusion Algorithm* ACM transactions on Computer Systems, Vol. 5 no. 1, February 1987, Pages 1-11.

5. Thomas Melham, *Formalizing Abstraction Mechanisms for Hardware Verfication in Higher Oreder Logic*, University of Cambridge Computer Laboratory, Technical Report No. 201, August 1990.

6. Fred Schneider, Bard Bloom, and Keith Marzullo, *Putting Time Into Proof Outlines* Cornell University, Department of Computer Science Technical Report TR 91-1238, September 1991.

7. N. Shankar, *Mechanized Verification of Real-Time Systems Using PVS* SRI International Computer Science Laaboratory Technical Report SRI-CSL-92-12, November 1992.

Appendix A

```
EXCLUSION3 =
|- (!t."(END trans3a t) /\ "(END trans3b t) /\
"END trans_conca t /\ "END trans_concb t ==> (x (t-1) = x t))/\
(!t.trans3a t ==> trans3a (t-1)\/("trans3a(t-1)/\ (x(t-1) = 0)))/\
(!t.trans3b t ==> trans3b (t-1)\/("trans3b(t-1)/\ (x(t-1) = 0)))/\
(!t."trans_conca t /\ trans_conca(t+1) ==>
"trans3a t /\ "trans3b t)/\
(!t."trans_concb t /\ trans_concb(t+1) ==>
"trans3a t /\ "trans3b t)/\
"trans_conca 0 /\"trans_concb 0 /\
(!t.trans_conca t ==>
trans_conca(t-1)\/("trans_conca(t-1)/\ (x(t-1) = 1)))/\
(!t.trans_concb t ==>
trans_concb(t-1)\/("trans_concb(t-1)/\ (x(t-1) = 2))) ==>
(!t.
(trans_conca t ==>
"trans_concb t /\ (x t = 1) /\ "trans3a t /\ "trans3b t) /\
(trans_concb t ==>
"trans_conca t /\ (x t = 2) /\ "trans3a t /\ "trans3b t))
```

Using Isabelle to prove simple theorems

Sara Kalvala (sk@cl.cam.ac.uk)

University of Cambridge, UK

1 Introduction

In a recent paper, Parnas presents a few mathematical conjectures and queries the theorem-proving community as to how well current theorem provers deal with such problems [1]. This question can be partially answered by examining how well the Isabelle prover [2] performs on his example conjectures.

This note describes my attempt at formalisation and proof of the given conjectures using Isabelle. I present first a very brief introduction to Isabelle, followed by the proof scripts themselves, and a short discussion of the results.

2 Isabelle

Isabelle is a *generic* proof system: it provides a framework in which different logics can be specified and subsequently used for proof development. Isabelle uses ML as an implementation language for coding data-types in the logic and functions to manipulate these data-types. Isabelle also provides a logic meta-level, which allows users to code different formal logics and use them as the basis for proof development.

Isabelle has often been seen as a tool for implementing different logics, and examining exotic proof systems. However, it also lends itself to applications as a practical proof system. More specifically, higher-order logic and Zermelo-Fraenkel set theory have been implemented and sufficient tools have been added to these implementations so that they can be easily used for proof efforts. The proofs in this paper make use of both these logics (which will be denoted in this paper by $\mathcal{I}_{HOL}$ and $\mathcal{I}_{ZF}$, respectively).

It is beyond the scope of this communication to describe Isabelle in detail, this information can be found elsewhere [2, 3]. The notation used in this paper should be evident for users of other systems such as HOL. One point to remember is the distinction between *meta* level and *object* level operators—for example, implication in the meta-logic (represented as '$\Rightarrow$' and appearing in *rules* by which other theorems can be derived) and in any object logic (often represented as '$\rightarrow$', and representing the logical connective).

Isabelle is a tactic-based proof system. While in HOL theorems and inference rules are implemented as objects of different types, in Isabelle inference rules are simply theorems containing a meta-level implication. There is a fairly small repertoire of *tactics* to be used, the proof control arising mostly from a range of *theorems* to be combined with the tactic. There are three main proof methods supported in Isabelle:

- Resolution: a given inference rule is matched against either the conclusion or an assumption of the goal (or both) resulting in a change to the goal.
- Simplification, or rewriting: consists in using equalities in replacing equals for equals in the goal.
- Classical reasoning: natural deduction-style rules are used through a simple search algorithm to find tautologies in the goal.

The use of these proof methods can be best examined in the proofs below.

3 The theorems

The conjectures provided by Parnas are divided into several groups, based on the application from which they are generated. In this section, their proofs in Isabelle are given. There are of course several different ways in which the informal problem descriptions could be formalised; my emphasis was on simplicity and speed of development.

3.1 Introductory example

Arithmetic The first 2 theorems to be proved are straightforward arithmetic examples. They were proved within $\mathcal{I}_{HOL}$. The formulation has been simplified somewhat by the use of natural numbers instead of reals. Here are the proofs:

```
prove_goal Arith.thy " x < 0 | x = 0 | 0 < x"
(fn _ => [resolve_tac [less_linear] 1]);

prove_goal Arith.thy
  "~ ((x < 0 & x = 0) | (x < 0 & 0 < x) | (0 < x & x = 0))"
(fn _ => [fast_tac (HOL_cs addEs [less_anti_refl, less_anti_sym]) 1]);
```

The first lemma is proved directly using resolution with an existing theorem; the second is also an easy proof, using the tactic for classical reasoning and including two pre-proved rules in the set of rules supplied to the classical prover.

Domain and range The next conjecture is concerned with reasoning about the domain and range of the square root function. As this involves the notion of partial functions, I have represented types as sets within $\mathcal{I}_{ZF}$. The first step was to formalise neg and non_neg as sets, and sqrt and minus as functions from and to the appropriate sets:

```
val sqrt_theory = extend_theory Arith.thy "sqrt"
([],[],[],[],[(["neg","non_neg"],"i"),
            (["sqrt","minus"], "i=>i")], None)
[("sqrt_dom", "x :non_neg ==> sqrt(x):non_neg"),
 ("minus_r1", "x :neg ==> minus(x):non_neg"),
 ("minus_r2", "x :non_neg ==> minus(x):neg")];
```

With this specification of the operators, it becomes easy to prove the conjectures:

```
prove_goal sqrt_theory "x :neg --> sqrt(minus(x)) : non_neg"
(fn _ => [fast_tac (ZF_cs addIs [sqrt_dom,minus_r1,minus_r2]) 1]);

prove_goal sqrt_theory "x :non_neg --> sqrt(x) : non_neg"
(fn _ => [fast_tac (ZF_cs addIs [sqrt_dom,minus_r1,minus_r2]) 1]);
```

3.2 Array search

The simple conjectures described are proved easily, using arithmetic built within $\mathcal{I}_{HOL}$:

```
prove_goal Arith.thy
"(? i . B(i) = x) | (! i . ((0 < i) & (i < Suc(N))) --> ~ (B(i) = x))"
(fn _ => [fast_tac HOL_cs 1]);

prove_goal Arith.thy
" ~ ((? i . ((0 < i) & (i < Suc(N))) & (B(i) = x)) &      \
\    (! i . ((0 < i) & (i < Suc(N))) --> ~ (B(i) = x)))"
(fn _ => [fast_tac HOL_cs 1]);
```

3.3 Palindrome

As this example is concerned with the property of inclusion in a set, I thought I would use $\mathcal{I}_{ZF}$ rather than $\mathcal{I}_{HOL}$.

The first theorem is trivial. I simplified the goal a bit because I used natural numbers rather than integers.

```
prove_goal Arith.thy
"(EX l:nat. (ALL i. (i : n #/ succ(succ(0))) \
                    --> (A(l #+ i) = A(l #+ n #- 1 #- i)))) --> \
\ ~ ({l:nat. (ALL i. (i : n #/ succ(succ(0))) \
                    --> (A(l #+ i) = A(l #+ n #- 1 #- i)))} <= 0)"
(fn _ => [fast_tac ZF_cs 1]);
```

For the second one I needed two lemmas, one of them because the informal characterisation of the lemma uses a particular way of specifying non-empty sets, and the other one to access a simple arithmetic property. Both lemmas were quite easy to prove:

```
val non_empty =
(prove_goal Arith.thy "(EX x:y. P(x)) --> ~({x:y. P(x)} <= 0)"
(fn _ => [fast_tac ZF_cs 1])) RS mp;

val div_1_2 =
prove_goal Arith.thy "succ(0) #/ succ(succ(0)) = 0"
(fn _ => [resolve_tac [quo_less] 1,
         ALLGOALS (fast_tac (ZF_cs addIs [nat_0I,nat_succI]))]);
```

Apart from these lemmas, the proof also depends on an explicit instantiation of a variable, so as to state that there always exists a palindrome of size 1. The proof was finished easily once this instantiation was coded:

```
prove_goal Arith.thy
" ~ ({n :nat . (EX l. (ALL i. (i : n #/ succ(succ(0))) --> \
\     (A(l #+ i) = A(l #+ n #- 1 #- i))))} <= 0)"
(fn _ => [resolve_tac [non_empty] 1,
            res_inst_tac [("x","succ(0)")] bexI 1,
            SIMP_TAC (ZF_ss addrews [div_1_2]) 1,
            fast_tac (ZF_cs addIs [nat_0I,nat_succI]) 1]);
```

4 Discussion

As can be seen above, most of the given goals were proved very easily by the application of a single tactic, and others involved a small amount of proof exploration. None of the proof steps involved a vast amount of expertise; at most, all that was needed was a knowledge of less than a dozen tactics and perusal of the database of pre-proved theorems. Both $\mathcal{I}_{HOL}$ and $\mathcal{I}_{ZF}$ were used, to show that one can even choose the underlying logic which fits the problem at hand best. In the given examples any of the two could have been used with equal ease.

My formulation of some of the problems could be considering over-simplified: I used natural numbers rather than reals (in Sect. 3.1), and I represented the square root function only by the types it involves rather than by a complete interpretation of the function (in Sect. 3.1). The complete formulation of these problems would involve much more work in developing real numbers formally within one of the logics, which has not been completed to date. However, for the conjectures actually presented such formulation is not really needed, so instead the simplified version was used.

The interest expressed by Parnas is on finding quick, "automatic" proofs for a variety of simple theorems. I believe this has been achieved in the proof efforts described in this note. Insofar as the set of conjectures supplied was quite small, this experiment allows Isabelle to be seen as a versatile and powerful enough system to be employed by non-experts in verification efforts, comparable in complexity and efficiency to other systems in widespread use.

References

1. David Parnas. Some theorems we should prove. In *Higher Order Logic and its Applications*, 1993.
2. Lawrence Paulson. Introduction to Isabelle. Technical Report 280, University of Cambridge, Computer Lab, 1993.
3. Lawrence Paulson. The Isabelle reference manual. Technical Report 283, University of Cambridge, Computer Lab, 1993.

Author Index

Springer-Verlag
and the Environment

Lecture Notes in Computer Science

For information about Vols. 1–714
please contact your bookseller or Springer-Verlag

Vol. 751: B. Jähne, Spatio-Temporal Image Processing. XII, 208 pages. 1993.

Vol. 752: T. W. Finin, C. K. Nicholas, Y. Yesha (Eds.), Information and Knowledge Management. Proceedings, 1992. VII, 142 pages. 1993.

Vol. 753: L. J. Bass, J. Gornostaev, C. Unger (Eds.), Human-Computer Interaction. Proceedings, 1993. X, 388 pages. 1993.

Vol. 754: H. D. Pfeiffer, T. E. Nagle (Eds.), Conceptual Structures: Theory and Implementation. Proceedings, 1992. IX, 327 pages. 1993. (Subseries LNAI).

Vol. 755: B. Möller, H. Partsch, S. Schuman (Eds.), Formal Program Development. Proceedings. VII, 371 pages. 1993.

Vol. 756: J. Pieprzyk, B. Sadeghiyan, Design of Hashing Algorithms. XV, 194 pages. 1993.

Vol. 757: U. Banerjee, D. Gelernter, A. Nicolau, D. Padua (Eds.), Languages and Compilers for Parallel Computing. Proceedings, 1992. X, 576 pages. 1993.

Vol. 758: M. Teillaud, Towards Dynamic Randomized Algorithms in Computational Geometry. IX, 157 pages. 1993.

Vol. 759: N. R. Adam, B. K. Bhargava (Eds.), Advanced Database Systems. XV, 451 pages. 1993.

Vol. 760: S. Ceri, K. Tanaka, S. Tsur (Eds.), Deductive and Object-Oriented Databases. Proceedings, 1993. XII, 488 pages. 1993.

Vol. 761: R. K. Shyamasundar (Ed.), Foundations of Software Technology and Theoretical Computer Science. Proceedings, 1993. XIV, 456 pages. 1993.

Vol. 762: K. W. Ng, P. Raghavan, N. V. Balasubramanian, F. Y. L. Chin (Eds.), Algorithms and Computation. Proceedings, 1993. XIII, 542 pages. 1993.

Vol. 763: F. Pichler, R. Moreno Díaz (Eds.), Computer Aided Systems Theory – EUROCAST '93. Proceedings, 1993. IX, 451 pages. 1994.

Vol. 764: G. Wagner, Vivid Logic. XII, 148 pages. 1994. (Subseries LNAI).

Vol. 765: T. Helleseth (Ed.), Advances in Cryptology – EUROCRYPT '93. Proceedings, 1993. X, 467 pages. 1994.

Vol. 766: P. R. Van Loocke, The Dynamics of Concepts. XI, 340 pages. 1994. (Subseries LNAI).

Vol. 767: M. Gogolla, An Extended Entity-Relationship Model. X, 136 pages. 1994.

Vol. 768: U. Banerjee, D. Gelernter, A. Nicolau, D. Padua (Eds.), Languages and Compilers for Parallel Computing. Proceedings, 1993. XI, 655 pages. 1994.

Vol. 769: J. L. Nazareth, The Newton-Cauchy Framework. XII, 101 pages. 1994.

Vol. 770: P. Haddawy (Representing Plans Under Uncertainty. X, 129 pages. 1994. (Subseries LNAI).

Vol. 771: G. Tomas, C. W. Ueberhuber, Visualization of Scientific Parallel Programs. XI, 310 pages. 1994.

Vol. 772: B. C. Warboys (Ed.),Software Process Technology. Proceedings, 1994. IX, 275 pages. 1994.

Vol. 773: D. R. Stinson (Ed.), Advances in Cryptology – CRYPTO '93. Proceedings, 1993. X, 492 pages. 1994.

Vol. 774: M. Banâtre, P. A. Lee (Eds.), Hardware and Software Architectures for Fault Tolerance. XIII, 311 pages. 1994.

Vol. 775: P. Enjalbert, E. W. Mayr, K. W. Wagner (Eds.), STACS 94. Proceedings, 1994. XIV, 782 pages. 1994.

Vol. 776: H. J. Schneider, H. Ehrig (Eds.), Graph Transformations in Computer Science. Proceedings, 1993. VIII, 395 pages. 1994.

Vol. 777: K. von Luck, H. Marburger (Eds.), Management and Processing of Complex Data Structures. Proceedings, 1994. VII, 220 pages. 1994.

Vol. 778: M. Bonuccelli, P. Crescenzi, R. Petreschi (Eds.), Algorithms and Complexity. Proceedings, 1994. VIII, 222 pages. 1994.

Vol. 779: M. Jarke, J. Bubenko, K. Jeffery (Eds.), Advances in Database Technology — EDBT '94. Proceedings, 1994. XII, 406 pages. 1994.

Vol. 780: J. J. Joyce, C.-J. H. Seger (Eds.), Higher Order Logic Theorem Proving and Its Applications. Proceedings, 1993. X, 518 pages. 1994.

Vol. 781: G. Cohen, S. Litsyn, A. Lobstein, G. Zémor (Eds.), Algebraic Coding. Proceedings, 1993. XII, 326 pages. 1994.

Vol. 782: J. Gutknecht (Ed.), Programming Languages and System Architectures. Proceedings, 1994. X, 344 pages. 1994.

Vol. 783: C. G. Günther (Ed.), Mobile Communications. Proceedings, 1994. XVI, 564 pages. 1994.

Vol. 784: F. Bergadano, L. De Raedt (Eds.), Machine Learning: ECML-94. Proceedings, 1994. XI, 439 pages. 1994. (Subseries LNAI).

Vol. 785: H. Ehrig, F. Orejas (Eds.), Recent Trends in Data Type Specification. Proceedings, 1992. VIII, 350 pages. 1994.

Vol. 786: P. A. Fritzson (Ed.), Compiler Construction. Proceedings, 1994. XI, 451 pages. 1994.

Vol. 787: S. Tison (Ed.), Trees in Algebra and Programming – CAAP '94. Proceedings, 1994. X, 351 pages. 1994.

Vol. 788: D. Sannella (Ed.), Programming Languages and Systems – ESOP '94. Proceedings, 1994. VIII, 516 pages. 1994.

Vol. 789: M. Hagiya, J. C. Mitchell (Eds.), Theoretical Aspects of Computer Software. Proceedings, 1994. XI, 887 pages. 1994.

Vol. 790: J. van Leeuwen (Ed.), Graph-Theoretic Concepts in Computer Science. Proceedings, 1993. IX, 431 pages. 1994.